Classics in Mathematics

Kiyosi Itô   Henry P. McKean, Jr.   Diffusion Processes and their Sample Paths

Springer
Berlin
Heidelberg
New York
Barcelona
Budapest
Hong Kong
London
Milan
Paris
Santa Clara
Singapore
Tokyo

Kiyosi Itô was born on September 7, 1915, in Kuwana, Japan. After his undergraduate and doctoral studies at Tokyo University, he was associate professor at Nagoya University before joining the faculty of Kyoto University in 1952. He has remained there ever since and is now Professor Emeritus, but has also spent several years at each of Stanford, Aarhus and Cornell Universities and the University of Minnesota.

Itô's fundamental contributions to probability theory, especially the creation of stochastic differential and integral calculus and of excursion theory, form a cornerstone of this field. They have led to a profound understanding of the infinitesimal development of Markovian sample paths, and also of applied problems and phenomena associated with the planning, control and optimization of engineering and other random systems.

Professor Itô has been the inspirer and teacher of an entire generation of Japanese probabilists.

Henry McKean was born on December 14, 1930, in Wenham, Massachusetts. He studied mathematics at Dartmouth College, Cambridge University, and Princeton University; he received his degree from the last in 1955. He has held professional positions at Kyoto University, MIT, Rockefeller University, Weizmann Institute, Balliol College, Oxford, and the Courant Institute of Mathematical Sciences (1969 to present). His main interests are probability, Hamiltonian mechanics, complex function theory, and nonlinear partial differential equations.

Kiyosi Itô Henry P. McKean, Jr.

Diffusion Processes and their Sample Paths

Reprint of the 1974 Edition

Springer

Kiyosi Itô
Kyoto University, RIMS
Sakyo-Ku
606 Kyoto
Japan

Henry P. McKean, Jr.
New York University
Courant Institute of Mathematical Sciences
New York, NY 10012
USA

Originally published as Vol. 125 of the
Grundlehren der mathematischen Wissenschaften

Cataloging-in-Publication Data applied for

Die Deutsche Bibliothek - CIP-Einheitsaufnahme

Itô, Kiosi:
Diffusion processes and their sample paths / Kiosi Itô ; Henry P. McKean, Jr. - Reprint of the 1974 ed. - Berlin ; Heidelberg ; New York ; Barcelona ; Budapest ; Hong Kong ; London ; Milan ; Paris ; Santa Clara ; Singapore ; Tokyo : Springer, 1996
(Grundlehren der mathematischen Wissenschaften ; Vol. 125) (Classics in mathematics)
ISBN 3-540-60629-7
NE: MacKean, Henry P.:; 1. GT

Mathematics Subject Classification (1991):
28A65, 60-02, 60G17, 60J60, 60J65, 60J70

ISBN 3-540-60629-7 Springer-Verlag Berlin Heidelberg New York

Printed in Germany

SPIN 10518453 41/3144- 5 4 3 2 1 0 – Printed on acid-free paper

K. Itô H. P. McKean, Jr.

Diffusion Processes and their Sample Paths

Second Printing, Corrected

Springer-Verlag
Berlin Heidelberg New York 1974

Kiyosi Itô
Cornell University, Ithaca, N. Y. 14850, USA

Henry P. McKean, Jr.
Courant Institute of Mathematical Sciences,
New York University, New York, USA

AMS Subject Classifications (1970):
28 A 65, 60-02, 60 G 17, 60 J 60, 60 J 65, 60 J 70

ISBN 3-540-03302-5 Springer-Verlag Berlin Heidelberg New York
ISBN 0-387-03302-5 Springer-Verlag New York Heidelberg Berlin

DEDICATED TO

P. LÉVY

WHOSE WORK HAS BEEN

OUR SPUR AND ADMIRATION

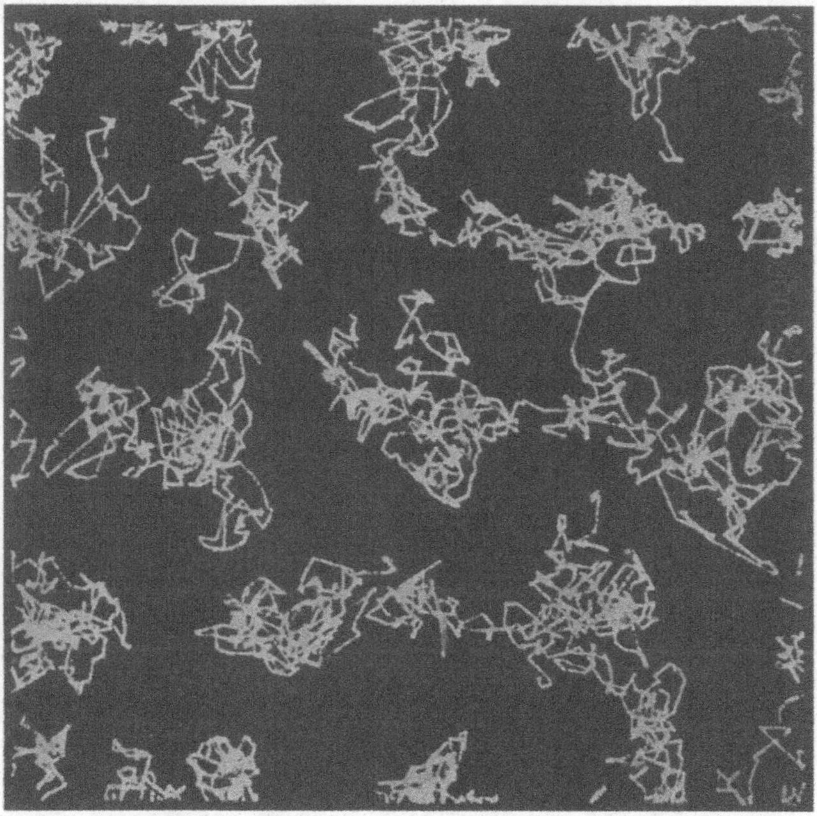

Computer-simulated molecular motions, reminiscent of 2-dimensional BROWNian motion. [From ALDER, B. J., and T. E. WAINWRIGHT: Molecular motions. Scientific American **201**, no. 4, 113–126 (1959)].

Preface

ROBERT BROWN, an English botanist, observed (1828) that pollen grains suspended in water perform a continual swarming motion (see, for example, D'ARCY THOMPSON [*1*: 73—77]).

L. BACHELIER (1900) derived the law govering the position of a single grain performing a 1-dimensional BROWNian motion starting at $a \in R^1$ at time $t = 0$:

$$1)\quad P_a[x(t) \in db] = g(t, a, b)\, db \qquad (t, a, b) \in (0, +\infty) \times R^2,$$

where g is the source (GREEN) function

$$2)\quad g(t, a, b) = \frac{e^{-(b-a)^2/2t}}{\sqrt{2\pi t}}$$

of the problem of heat flow:

$$3)\quad \frac{\partial u}{\partial t} = \frac{1}{2}\frac{\partial^2 u}{\partial a^2} \qquad (t > 0).$$

BACHELIER also pointed out the MARKOVian nature of the BROWNian path expressed in

$$4)\quad P_a[a_1 \leqq x(t_1) < b_1, a_2 \leqq x(t_2) < b_2, \ldots, a_n \leqq x(t_n) < b_n]$$

$$= \int_{a_1}^{b_1} \int_{a_2}^{b_2} \cdots \int_{a_n}^{b_n} g(t_1, a, \xi_1)\, g(t_2 - t_1, \xi_1, \xi_2) \ldots$$

$$g(t_n - t_{n-1}, \xi_{n-1}, \xi_n)\, d\xi_1\, d\xi_2 \ldots d\xi_n \qquad 0 < t_1 < t_2 < \cdots < t_n$$

and used it to establish the law of maximum displacement

$$5)\quad P_0\left[\max_{s \leqq t} x(s) \leqq b\right] = 2\int_0^b \frac{e^{-a^2/2t}}{\sqrt{2\pi t}}\, da \qquad t > 0, b \geqq 0$$

(see BACHELIER [*1*]).

A. EINSTEIN (1905) also derived 1) from statistical mechanical considerations and applied it to the determination of molecular diameters (see, for example, A. EINSTEIN [*1*]).

BACHELIER was unable to obtain a clear picture of the BROWNian motion and his ideas were unappreciated at the time; nor is this surprising because the precise definition of the BROWNian motion involves a measure on the path space, and it was not until 1909 that E. BOREL published his classical memoir [1] on BERNOULLI trials. But as soon as the ideas of BOREL, LEBESGUE, and DANIELL appeared, it was possible to put the BROWNian motion on a firm mathematical foundation; this was achieved in 1923 by N. WIENER [1].

Consider the space of *continuous* paths $w: t \in [0, +\infty) \to R^1$ with coordinates $x(t) = w(t)$ and let B be the smallest BOREL algebra of subsets B of this path space which includes all the simple events $B = (w: a \leqq x(t) < b)$ $(t \geqq 0, a < b)$. WIENER established the existence of non-negative BOREL measures $P_a(B)$ $(a \in R^1, B \in \mathsf{B})$ for which 4) holds; among other things, this result attaches a precise meaning to BACHELIER's statement that *the Brownian path is continuous.*

P. LÉVY [2] found another construction of the BROWNian motion and, in his 1948 monograph [3], gave a profound description of the fine structure of the individual BROWNian path. LÉVY's results, with several complements due to D. B. RAY [4] and ourselves, will be explained in chapters 1 and 2, with special attention to the standard BROWNian *local time* (the *mesure du voisinage* of P. LÉVY):

6) $$\mathfrak{t}(t, a) = \lim_{b \downarrow a} \frac{\text{measure}\,(s: a \leqq x(s) < b, s \leqq t)}{2(b-a)}.$$

Given a STURM-LIOUVILLE operator $\mathfrak{G} = (c_2/2)\,D^2 + c_1\,D$ $(c_2 > 0)$ on the line, the source (GREEN) function $p = p(t, a, b)$ of the problem

7) $$\frac{\partial u}{\partial t} = \mathfrak{G}\,u \qquad (t > 0)$$

shares with the GAUSS kernel g of 2) the properties

8a) $$0 \leqq p$$

8b) $$\int_{R^1} p(t, a, b)\,db = 1$$

8c) $$p(t, a, b) = \int_{R^1} p(t-s, a, c)\,p(s, c, b)\,dc \qquad t > s > 0.$$

Soon after the publication of WIENER's monograph [3] in 1930, the associated stochastic motions (diffusions) analogous to the BROWNian motion ($\mathfrak{G} = D^2/2$) made their debut; the names of W. FELLER and A. N. KOLMOGOROV stand out in this connection. At a later date (1946), K. ITÔ [2] proved that if

9) $$|c_1(b) - c_1(a)| + |\sqrt{c_2(b)} - \sqrt{c_2(a)}| < \text{constant} \times |b-a|,$$

then the motion associated with $\mathfrak{G} = (c_2/2) D^2 + c_1 D$ is identical in law to the *continuous* solution of

$$10) \qquad a(t) = a(0) + \int_0^t c_1(a)\, ds + \int_0^t \sqrt{c_2(a)}\, db$$

where b is a standard BROWNian motion.

W. FELLER took the lead in the next development.

Given a MARKOVian motion with sample paths $w: t \to x(t)$ and probabilities $P_a(B)$ on a linear interval Q, the operators

$$11) \qquad H_t: f \to \int P_a[x(t) \in db]\, f(b)$$

constitute a *semi-group*:

$$12) \qquad H_t = H_{t-s} H_s \qquad (t \geqq s),$$

and as E. HILLE [*1*] and K. YOSIDA [*1*] proved,

$$13) \qquad H_t = e^{t\mathfrak{G}} \qquad (t > 0)$$

with a suitable interpretation of the exponential, where $\mathfrak{G}$ is the so-called *generator*.

D. RAY [*2*] proved that if the motion is *strict Markov* (*i.e.*, if it *starts afresh* at certain stochastic (MARKOV) times including the passage times $\mathfrak{m}_a = \min(t : x(t) = a)$, *etc.*), then $\mathfrak{G}$ is *local* if and only if the motion has *continuous* sample paths, substantiating a conjecture of W. FELLER; this combined with FELLER's papers [*4, 5, 7, 9*] implies that the generator of a strict MARKOVian motion with continuous paths (diffusion) can be expressed as a *differential operator*

$$14) \qquad (\mathfrak{G}\, u)(a) = \lim_{b \downarrow a} \frac{u^+(b) - u^+(a)}{m(a, b]},$$

where m is a non-negative BOREL measure positive on open intervals and, with a change of scale, $u^+(a) = \lim_{b \downarrow a} (b - a)^{-1}[u(b) - u(a)]$, except at certain singular points where $\mathfrak{G}$ degenerates to a differential operator of degree $\leqq 1$.

E. B. DYNKIN [*1*] also arrived at the idea of a strict MARKOV process, derived an elegant formula for $\mathfrak{G}$, and used it to make a simple (probabilistic) proof of FELLER's expression for $\mathfrak{G}$; the names of R. BLUMENTHAL [*1*] and G. HUNT [*1*] and the monographs of E. B. DYNKIN [*6, 8*] should also be mentioned in this connection.

Our plan is the following.

BROWNian motion is discussed in chapters 1 and 2 and then, in chapter 3, the general linear *diffusion* is introduced as a strict MARKOVian motion with continuous paths on a linear interval subject to possible annihilation of mass. $\mathfrak{G}$ is computed in great detail in chapter 4 using probabilistic methods similar to those of E. B. DYNKIN [5]; in the so-called non-singular case it is a differential operator

$$15)\qquad (\mathfrak{G} u)(a) = \lim_{b \downarrow a} \frac{u^+(b) - u^+(a) - \int_{(a,b]} u\, d k}{m(a, b]}$$

with u^+ and m as in 14), where now k is the (non-negative) BOREL measure that governs the annihilation of mass (such generators occur somewhat disguised in W. FELLER [9]).

Given $\mathfrak{G}$ as in 14) and a standard BROWNian motion with sample paths $w: t \to x(t)$, if $\mathfrak{t}$ is *P.* LÉVY's *Brownian local time* and if $\mathfrak{f}^{-1}$ is the inverse function of the local time integral

$$16)\qquad \mathfrak{f}(t) = \int \mathfrak{t}(t, \xi)\, m(d\xi),$$

then the motion $x(\mathfrak{f}^{-1})$ is identical in law to the diffusion attached to $\mathfrak{G}$, as will be proved in chapter 5, substantiating a suggestion of H. TROTTER; B. VOLKONSKIĬ [1] has obtained the same *time substitution* in a less explicit form.

Given $\mathfrak{G}$ as in 15), the associated motion can be obtained by ***killing*** the paths $x(\mathfrak{f}^{-1})$ described above at a (stochastic) time $\mathfrak{m}_\infty$ with conditional law

$$17)\qquad P_\bullet[\mathfrak{m}_\infty > t \mid x(\mathfrak{f}^{-1})] = e^{-\int \mathfrak{t}(\mathfrak{f}^{-1}(t), \xi)\, k(d\xi)}$$

as is also proved in chapter 5; in the special case of the *elastic Brownian motion* on $[0, +\infty)$ with generator $\mathfrak{G} = D^2/2$ subject to the condition $\gamma u(0) = (1-\gamma)\, u^+(0)$ $(0 < \gamma < 1)$,

$$\mathfrak{f} = \int_0^{+\infty} \mathfrak{t}(t, \xi)\, 2 d\xi = \text{measure}\,(s: x(s) \geqq 0,\, s \leqq t),$$

$x(\mathfrak{f}^{-1})$ is identical in law to the classical *reflecting Brownian motion* $x^+ = |x|$, and 17) takes the simple form

$$18)\qquad P_\bullet[\mathfrak{m}_\infty > t \mid x^+] = e^{-\frac{\gamma}{1-\gamma} \mathfrak{t}^+(t, 0)} \qquad (\mathfrak{t}^+ = 2\mathfrak{t}),$$

substantiating a conjecture of W. FELLER: that the elastic BROWNian motion ought to be the same as the reflecting BROWNian motion killed at the instant a certain increasing functional $e(\mathfrak{Z}_+ \cap [0, t))$ of the time t and the visiting set $\mathfrak{Z}^+ = (t: x^+(t) = 0)$ hits a certain level.

Details about the fine structure of the sample paths of the general linear diffusion with emphasis on local times, will be found in chapter 6. BROWNian motion in several dimensions is treated in chapter 7, and in chapter 8, the reader will find some glimpses of the general higher-dimensional diffusion.

Acknowledgements. W. FELLER has our best thanks, his ideas run through the whole book, and we shall think it a success if it pleases him.

We have also to thank R. BLUMENTHAL and G. HUNT who placed at our disposal their then unpublished results on MARKOV times, and H. TROTTER with whom we had many helpful conversations.

This book was begun at Princeton and the Institute for Advanced Study (1954/56) with the partial support of the Office of Ordnance Research, and continued at Kyôto (1957/58) with the aid of a Fulbright grant, at Hanover, N. H. and Cambridge, Mass. (1960) with the support of the Office of Naval Research, and at Stanford, Calif. with the support of the National Science Foundation (1962/63); to those institutions and agencies, also, we extend our warmest thanks. Finally, we must thank the staff of Springer-Verlag for their meticulous labors and cordial attitude to what must have been a difficult job.

Kyôto, Japan, and Cambridge, Mass., November 1964

K. Itô H. P. McKean, Jr.

The present edition is the same as the first except for the correction of numerous errors. Among those who helped us in this task, we would particularly like to thank F. B. KNIGHT.

September 1973 K. I., H. P. McK.

Contents

Numbering: 1.2 means section 2 of chapter 1; 1.2.3) or problem 1.2.3 or diagram 1.2.3 means formula or problem or diagram 3 of 1.2; 3) or problem 3 or diagram 3 means formula or problem or diagram 3 of the *current* section. R. BROWN [*1*: 2] means page 2 of the item R. BROWN [*1*] of the *bibliography*.

Problems accompanied by some indication of their solutions are placed at the end of each section; these often contain additional information needed below and are an essential part of the exposition.

Diagrams do not pretend to photographic faithfulness; for example, the BROWNian path is often pictured as if it had isolated zeros, which is not at all the acual case.

A list of *notations* is placed at the end of the book.

Warning: positive means >0, while non-negative means $\geqq 0$; it is the same for negative (<0) and non-positive ($\leqq 0$). $a \wedge b$ means the smaller of a and b, $a \vee b$ the larger of the two.

Prerequisites

The reader is expected to have about the same mathematical background as is needed to read COURANT-HILBERT [*1, 2*]. Besides this, he should have mastered most of W. FELLER's book on probability [*3*] plus the topics listed below (see A. N. KOLMOGOROV [*2*] for a helpful outline and some of the proofs).

Algebras. Given a space W, a class A of its subsets is said to be an *algebra* if

1) $$W \in \mathsf{A}$$

2) $$A - B, A \cup B, A \cap B \in \mathsf{A} \quad \textit{in case} \quad A, B \in \mathsf{A}.$$

A is said to be a *Borel algebra* if, in addition,

3) $$\bigcup_{n \geq 1} B_n, \ \bigcap_{n \geq 1} B_n \in \mathsf{A} \quad \textit{in case} \quad B_n \in \mathsf{A} \ (n \geq 1).$$

Borel extension. A class A of subsets of W is contained in a least *Borel algebra* B. B is the so-called *Borel extension* of A.

Probability measures. Consider a non-negative set function $P(C)$ defined on an algebra A. P is said to be a *probability measure* if

1) $$P(W) = 1$$

and

2) $$P(A \cup B) = P(A) + P(B) \qquad A, B \in \mathsf{A}, \quad A \cap B = \varnothing.$$

P is said to be a *Borel probability measure* if, in addition,

3a) $$P(B_n) \downarrow 0 \quad (n \uparrow +\infty) \quad B_n \downarrow \varnothing, B_n \in \mathsf{A} \ (n \geq 1)$$

or, what is the same,

3b) $$P\Big(\bigcup_{n \geq 1} B_n\Big) = \sum_{n \geq 1} P(B_n) \qquad B_n \cap B_m = \varnothing \ (n < m), \quad B_n \in \mathsf{A} (n \geq 1), \ \bigcup_{n \geq 1} B_n \in \mathsf{A}.$$

Kolmogorov extension. Given a BOREL probability measure P on an algebra A, there is a unique BOREL probability measure Q on the BOREL extension B of A that coincides with P on A: $Q(B) = \inf \sum_{n \geq 1} P(A_n)$, where the infemum is taken over all coverings $\bigcup_{n \geq 1} A_n$ of $B \in \mathsf{B}$ with $A_n \in \mathsf{A} \ (n \geq 1)$; the triple (W, B, Q) is said to be a *probability space.*

Measurable functions. Given a space W and a BOREL algebra B of its subsets, a function $f: W \to [-\infty, +\infty]$ is said to be *measurable* B (or BOREL) if the ordinate set $f^{-1}[a, b) \in \mathsf{B}$ for each choice of $a < b$.

Integrals. Given a probability space (W, B, Q), the *integral* or *expectation* of a non-negative B measurable function f is defined to be

$$\begin{aligned} E(f) &= \int_W f\, dQ \\ &= \lim_{n\uparrow+\infty} \sum_{l\geq 1} l\, 2^{-n} Q[f^{-1}[(l-1)\,2^{-n}, l\, 2^{-n})] \qquad Q[f^{-1}(+\infty)] = 0 \\ &= +\infty \qquad\qquad\qquad\qquad\qquad\qquad\qquad\qquad Q[f^{-1}(+\infty)] > 0. \end{aligned}$$

E, applied to such non-negative functions, satisfies

1) $E(f) \geq 0$
2) $E(1) = 1$
3) $E(f_1 + f_2) = E(f_1) + E(f_2)$
4a) $E(f_n) \uparrow E(f) \qquad f_n \uparrow f$
4b) $E(f_n) \downarrow E(f) \qquad f_n \downarrow f,\ E(f_1) < +\infty,$
5) $E(\underline{\lim}\, f_n) \leq \underline{\lim}\, E(f_n).$

$E(f, B) = E(B, f)$ is short for $\int_B f\, dQ$.

Products. Given probability spaces (W_1, B_1, Q_1) and (W_2, B_2, Q_2), the class A of finite sums A of disjoint rectangles $B_1 \times B_2$ ($B_1 \in \mathsf{B}_1$, $B_2 \in \mathsf{B}_2$) is an algebra and

$$Q(B_1 \times B_2) \equiv Q_1(B_1) \times Q_2(B_2)$$

can be extended to a BOREL probability measure on A; the *product* $Q_1 \times Q_2$ is the KOLMOGOROV extension of this Q to the BOREL extension $\mathsf{B} \equiv \mathsf{B}_1 \times \mathsf{B}_2$ of A.

Fubini's theorem. Given a $\mathsf{B}_1 \times \mathsf{B}_2$ measurable function f from $W_1 \times W_2$ to $[0, +\infty)$,

$$\int_{W_1 \times W_2} f\, dQ_1 \times Q_2 = \int_{W_1} dQ_1 \int_{W_2} f\, dQ_2 = \int_{W_2} dQ_2 \int_{W_1} f\, dQ_1.$$

Infinite products. Given probability spaces (W_n, B_n, Q_n) $(n \geq 1)$,

$$\mathsf{A}_n: A = B \times W_{n+1} \times W_{n+2} \times \textit{etc.}, \quad B \in \mathsf{B}_1 \times \mathsf{B}_2 \times \cdots \times \mathsf{B}_n$$

is a BOREL algebra and

$$Q(A) \equiv Q_1 \times Q_2 \times \cdots \times Q_n(B)$$

is a BOREL probability measure on the algebra $\mathsf{A} = \bigcup_{n\geq 1} \mathsf{A}_n$; the *infinite product* $\times_{n\geq 1} Q_n$ is defined as the KOLMOGOROV extension of Q to the BOREL extension $\times_{n\geq 1} \mathsf{B}_n$ of A.

Independence. Given a probability space (W, B, P), the BOREL algebras $\mathsf{B}_1, \mathsf{B}_2 \subset \mathsf{B}$ are said to be independent if

$$P(B_1 \cap B_2) = P(B_1)\, P(B_2) \qquad B_1 \in \mathsf{B}_1, B_2 \in \mathsf{B}_2;$$

the BOREL algebras B_n $(n \geqq 1)$ are independent, if, for each $n \geqq 1$, B_n is independent of the BOREL extension of $\bigcup_{l \neq n} \mathsf{B}_l$; the sets $B_n \in \mathsf{B}$ $(n \geqq 1)$ are independent if the algebras $\mathsf{B}_n = [\varnothing, B_n, W - B_n, W]$ $(n \geqq 1)$ are such; the B measurable functions f_n $(n \geqq 1)$ are independent if the BOREL extensions F_n of their ordinate sets $f_n^{-1}[a, b)$ $(a < b)$ are such; the B measurable function f is independent of the BOREL algebra $\mathsf{A} \subset \mathsf{B}$ if the BOREL extension of its ordinates sets is independent of A; *etc.* $E(f_1 f_2) = E(f_1)\, E(f_2)$ if f_1 is independent of f_2.

Kolmogorov 01 law. If A_n $(n \geqq 1)$ are independent subalgebras of B, if B_n is the BOREL extension of $\bigcup_{l \geqq n} \mathsf{A}_l$, and if $B \in \bigcap_{n \geqq 1} \mathsf{B}_n$, then $P(B) = 0$ or 1.

Strong law of large numbers. Given independent, non-negative, B measurable functions $f_n (n \geqq 1)$ with the same distribution,

$$P\left[\lim_{n \uparrow +\infty} n^{-1} \sum_{k \leqq n} f_k = E(f_1)\right] = 1.$$

Borel-Cantelli lemmas. Given events $B_n \in \mathsf{B}$ $(n \geqq 1)$, if

$$\sum_{l \geqq 1} P(B_l) < +\infty, \quad \text{then} \quad P\left[\bigcap_{n \geqq 1} \bigcup_{l \geqq n} B_l\right] = 0; \quad \text{if} \quad \sum_{l \geqq 1} P(B_l) = +\infty$$

and if the events are independent, then $P\left[\bigcap_{n \geqq 1} \bigcup_{l \geqq n} B_l\right] = 1$.

Conditional expectations. Given a BOREL subalgebra A of B and a non-negative B measurable function f, *the conditional expectation* $E(f \mid \mathsf{A})$ *of f relative to* A is defined to be the class of A measurable functions g such that $E(g, A) = E(f, A)$ $(A \in \mathsf{A})$; two such functions differ on a null set $\in \mathsf{A}$. $E(f \mid \mathsf{A})$ is the RADON-NIKODYM derivative of $Q(A) = E(f, A)$ with respect to the restriction of P to A. $E(f \mid \mathsf{A})$ (applied to non-negative f) satisfies

1) $$E(f \mid \mathsf{A}) \geqq 0$$

2) $$E(1 \mid \mathsf{A}) = 1$$

3) $$E(f_1 + f_2 \mid \mathsf{A}) = E(f_1 \mid \mathsf{A}) + E(f_2 \mid \mathsf{A})$$

4) $$E(f_n \mid \mathsf{A}) \uparrow E(f \mid \mathsf{A}) \qquad f_n \uparrow f$$

5) $$E(E(f \mid \mathsf{A}_2) \mid \mathsf{A}_1) = E(f \mid \mathsf{A}_1) \qquad \mathsf{A}_2 \supset \mathsf{A}_1$$

and

6) $$E(E(f \mid \mathsf{A})) = E(f);$$

in addition,

7) $$E(f \mid \mathsf{A}) = E(f)$$

if f and A are independent, and

8) $$E(ef \mid \mathsf{A}) = e\,E(f \mid \mathsf{A})$$

if e is measurable A. In all this, it is understood that $E(f \mid \mathsf{A}) = (\geqq)\, f_1$ is short for the statement that $f_2 = (\geqq)\, f_1$ for some f_2 from the class $E(f \mid \mathsf{A})$.

Conditional probabilities. Given $B \in \mathsf{B}$, the conditional expectation $E(e \mid \mathsf{A})$ of its indicator function e is the so-called *conditional probability* $P(B \mid \mathsf{A})$ *of* B *relative to* A; the rules for its manipulation are evident from the preceding article.

Gaussian distributions. A class of measurable functions f is said to be *Gaussian* and *centered* if each choice of $x = (f_1, f_2, \ldots, f_d)$ and $(\gamma_1, \ldots, \gamma_d) \in R^d$, the inner product $\gamma \cdot x$ is GAUSS distributed with mean $E(\gamma \cdot x) = 0$, *i.e.*,

1) $$P(a \leqq \gamma \cdot x < b) = \int_a^b \frac{e^{-c^2/2Q}}{\sqrt{2\pi Q}}\, dc,$$

where Q is the quadratic form $Q = Q(\gamma) = E[(\gamma \cdot x)^2] = \sum_{ij} E(f_i f_j)\, \gamma_i \gamma_j$ and the integral is interpreted as the unit mass at $c = 0$ if the determinant $|Q| = 0$. Suppose that the determinant $|Q| \neq 0$; then one can invert the FOURIER transform $E(e^{i\gamma \cdot x})$ $(= e^{-Q/2})$ to obtain the probability density of x:

2) $$p(x) = (2\pi)^{-d} \int_{R^d} e^{-i\gamma \cdot x} e^{-Q/2}\, d\gamma = \frac{e^{-Q^{-1}(x)/2}}{(2\pi)^{d/2}\sqrt{|Q|}}.$$

where Q^{-1} is the inverse quadratic form. Under the norm $\|f\|_2 = \sqrt{E(f^2)}$, the functions f span out a HILBERT space, and in this geometrical picture, statistical independence and perpendicularity are the same.

1. The standard Brownian motion

1.1. The standard random walk

Consider a person walking in the integers according to the following rule: start at time $n = 0$ at $s(0) = 0, \pm 1, \pm 2, \ldots$ and flip a true coin; if tails comes up, step at time $n = 1$ to $s(1) = s(0) - 1$; if heads comes up, step to $s(1) = s(0) + 1$. Coming, thus, at time $n - 1$ to $s(n-1)$, flip the coin; if tails comes up, step at time n to $s(n) = s(n-1) - 1$, if heads comes up, step to $s(n) = s(n-1) + 1$; *etc.*

We need a few definitions.

W is the space of sample paths

1) $$w : n = 0, 1, 2, \ldots \to s_n = s(n) = 0, +1, +2, \ldots;$$

$s_n(w) = s(n, w)$ is used in place of $s_n = s(n)$ if the path w needs to be emphasized.

B is the BOREL algebra generated by the sets

2) $$(w : s(n) = l) \quad n \geqq 0, \quad l = 0, \pm 1, \pm 2, \ldots$$

$P_l(B)$ is the probability that the sample path $s(n) : n \geqq 0$ lies in $B \in \mathsf{B}$ as a function of its starting point $s(0) = l = 0, \pm 1, \pm 2, \ldots$; for example,

3) $$P_l[s(n) = k] = \binom{n}{n_+} 2^{-n}.$$

where

$$n_+ = \frac{n}{2} + \frac{1}{2}(k - l), \quad \binom{n}{n_+} = n!/(n - n_+)!\, n_+! \ \text{ for } n_+ = 0, 1, \ldots, n,$$

and $\binom{n}{n_+} = 0$ otherwise.

$\mathsf{D} = [W, \mathsf{B}, P_l : l = 0, \pm 1, \pm 2, \ldots]$ is the *standard random walk.*

Given $m \geqq 0$,

4) $$P_\bullet[s(m+1) = l \mid s(n) : n \leqq m] = P_{s(m)}[s(1) = l] \qquad l = 0, \pm 1, \pm 2, \ldots;$$

i.e., if $s(m)$ is known, then the next position $s(m+1)$ is independent of the past $s(n) : n < m$; *in other words,* D *is a Markov process.*

D is in fact a strict MARKOV process as will now be explained.

Define $\mathsf{B}_m = \mathsf{B}[s(n) : n \leqq m]$ to be the smallest subfield of B including all the events $(a \leqq s(n) < b)$ $(n \leqq m, a < b)$, let $\mathfrak{m} = \mathfrak{m}(w) = 0, 1, 2, \ldots, +\infty$ be a *Markov time* in the sense of

$$5) \qquad (\mathfrak{m} \leqq m) \in \mathsf{B}_m \qquad (m \geqq 0),$$

let $\mathsf{B}_\mathfrak{m}$ be the field of events $B \in \mathsf{B}$ such that $B \cap (\mathfrak{m} \leqq m) \in \mathsf{B}_m$ for each $m \geqq 0$, and let $w_\mathfrak{m}^+$ be the *shifted path* $s_n(w_\mathfrak{m}^+) = s_{n+\mathfrak{m}}(w)$. $\mathsf{B}_\mathfrak{m}$ should be thought of as measuring the *past* $s(n) : n \leqq \mathfrak{m}$, and the statement is that if $\mathfrak{m} < +\infty$, then, *conditional on the present* $s(\mathfrak{m})$, *the past* $s(n) : n \leqq \mathfrak{m}$ *and the future* $s(n + \mathfrak{m}) : n \geqq 0$ *are independent, the future being a standard random walk starting at* $s(\mathfrak{m})$:

$$6a) \qquad P_\bullet[w_\mathfrak{m}^+ \in B \mid \mathsf{B}_\mathfrak{m}] = P_{s(\mathfrak{m})}(B) \qquad \mathfrak{m} < +\infty, B \in \mathsf{B},$$

i. e.,

$$6b) \quad P_\bullet[A, w_\mathfrak{m}^+ \in B, \mathfrak{m} < +\infty] = E_\bullet[A \cap (\mathfrak{m} < +\infty), P_{s(\mathfrak{m})}(B)] \qquad A \in \mathsf{B}_\mathfrak{m}, B \in \mathsf{B};$$

in other words, *if* $\mathfrak{m} < +\infty$, *then the walk starts from scratch at time* $n = \mathfrak{m}$.

Here is a simple proof of this.

Given $m = 0, 1, 2, \ldots$, the constant $\mathfrak{m} = m$ is a MARKOV time, $\mathsf{B}_\mathfrak{m} = \mathsf{B}_m = \mathsf{B}[s(n) : n \leqq m]$, and 6a) is just

$$7) \qquad P_\bullet[w_m^+ \in B \mid s(n) : n \leqq m] = P_{s(m)}(B) \qquad B \in \mathsf{B},$$

which is evident on iterating 4); this is the so-called *simple* MARKOV property.

Consider the general MARKOV time $\mathfrak{m}$ and an event $A \in \mathsf{B}_\mathfrak{m}$.

Because the indicator of A is a BOREL function

$$e[s(1), s(2), \ldots, s(\mathfrak{m}), s(\mathfrak{m}), s(\mathfrak{m}), \ldots] \text{ of } s(n \wedge \mathfrak{m}) \ (n \geqq 0),$$

$$e[s(1), \textit{etc.}] = e(s(1), s(2), \ldots, s(m), s(m), s(m), \ldots], \text{ on } (\mathfrak{m} = m),$$

and since $(\mathfrak{m} = m) \in \mathsf{B}_m$, $A \cap (\mathfrak{m} = m) \in \mathsf{B}_m$, and therefore

$$
\begin{aligned}
8) \qquad & P_\bullet[A, w_\mathfrak{m}^+ \in B, \mathfrak{m} < +\infty] \qquad\qquad B \in \mathsf{B} \\
&= \sum_{m \geqq 0} P_\bullet[A \cap (\mathfrak{m} = m), w_m^+ \in B] \\
&= \sum_{m \geqq 0} E_\bullet[A \cap (\mathfrak{m} = m), P_\bullet(w_m^+ \in B \mid \mathsf{B}_m)] \\
&= \sum_{m \geqq 0} E_\bullet[A \cap (\mathfrak{m} = m), P_{s(m)}(B)] \\
&= E_\bullet[A \cap (\mathfrak{m} < +\infty), P_{s(\mathfrak{m})}(B)],
\end{aligned}
$$

as desired.

Consider the spatial shift $w \to w + l$:

$$s(n, w + l) = s(n, w) + l \qquad (n \geqq 0)$$

and the reflection $w \to -w$:

$$s(n, -w) = -s(n, w) \qquad (n \geqq 0);$$

using 3), it is easily seen that

9) $$P_0(w + l \in B) = P_l(B), P_l(-w \in B) = P_{-l}(B) \qquad B \in \mathsf{B}.$$

Problem 1. Sketch the RADEMACHER functions

$$e_1(t) = -1 \quad (0 \leqq t < \tfrac{1}{2}) \quad = +1 \quad (\tfrac{1}{2} \leqq t < 1)$$
$$e_{n+1}(t) = e_1((2^n t)) \qquad (0 \leqq t < 1, n \geqq 1),$$

$(2^n t)$ being the fractional part of $2^n t$, and check that, relative to the classical BOREL measure on $[0, 1)$,

$$s(n) = 0 \quad (n = 0) \quad = e_1 + e_2 + \cdots + e_n \quad (n \geqq 1)$$

is a standard random walk starting at 0.

1.2. Passage times for the standard random walk

MARKOV times of special importance are the *passage times*

1) $$\mathfrak{m}_k = \min(n : s_n = k) \qquad k = 0, \pm 1, \pm 2, \ldots$$

We will use the strict MARKOVian character of the standard random walk to compute the generating functions

2) $$E_l(\alpha^{\mathfrak{m}_k}) = \sum_m \alpha^m P_l(\mathfrak{m}_k = m) \quad l, k = 0, \pm 1, \pm 2, \ldots, \quad 0 < \alpha < 1.$$

Because $\mathfrak{m}_k(w + l) = \mathfrak{m}_{k-l}(w)$ and $\mathfrak{m}_k(-w) = \mathfrak{m}_{-k}(w)$, it is clear from 1.1.9) that

3) $$E_l[\alpha^{\mathfrak{m}_k(w)}] = E_0[\alpha^{\mathfrak{m}_k(w+l)}] = E_0[\alpha^{\mathfrak{m}_{k-l}(w)}]$$
$$E_0[\alpha^{\mathfrak{m}_{-k}(w)}] = E_0[\alpha^{\mathfrak{m}_k(-w)}] = E_0[\alpha^{\mathfrak{m}_k(w)}].$$

Using the strict MARKOVian character, it is also clear that, for $k > 0$, $\mathfrak{m}_k = \mathfrak{m} + \mathfrak{m}_k(w_{\mathfrak{m}}^+)$ with $\mathfrak{m} = \mathfrak{m}_1$ so that

4) $$\begin{aligned} E_0(\alpha^{\mathfrak{m}_k}) &= E_0\left[\alpha^{\mathfrak{m}_1 + \mathfrak{m}_k(w_{\mathfrak{m}}^+)}\right] \\ &= E_0\left[\alpha^{\mathfrak{m}_1} E_0\left(\alpha^{\mathfrak{m}_k(w_{\mathfrak{m}}^+)} \,\middle|\, \mathsf{B}_{\mathfrak{m}_1}\right)\right] \\ &= E_0(\alpha^{\mathfrak{m}_1}) E_1(\alpha^{\mathfrak{m}_k}) \\ &= E_0(\alpha^{\mathfrak{m}_1}) E_0(\alpha^{\mathfrak{m}_{k-1}}) \\ &= E_0(\alpha^{\mathfrak{m}_1})^k; \end{aligned}$$

in addition, $\mathfrak{m}_k = 1 + \mathfrak{m}_k(w_1^+)$, and so

$$5) \quad \begin{aligned} E_0(\alpha^{\mathfrak{m}_k}) &= E_0\left[\alpha^{1+\mathfrak{m}_k(w_1^+)}\right] \\ &= \alpha\, E_0\left[E_0\left(\alpha^{\mathfrak{m}_k(w_1^+)} \,\middle|\, \mathsf{B}_1\right)\right] \\ &= \alpha\, E_0(E_{s_1}(\alpha^{\mathfrak{m}_k})) \\ &= \frac{\alpha}{2} E_{-1}(\alpha^{\mathfrak{m}_k}) + \frac{\alpha}{2} E_{+1}(\alpha^{\mathfrak{m}_k}) \\ &= \frac{\alpha}{2} E_0(\alpha^{\mathfrak{m}_{k+1}}) + \frac{\alpha}{2} E_0(\alpha^{\mathfrak{m}_{k-1}}). \end{aligned}$$

Putting $k = 1$ in 5) and using 4),

$$6) \quad \gamma^2 - 2\alpha^{-1}\gamma + 1 = 0 \qquad \gamma = E_0(\alpha^{\mathfrak{m}_1}), \qquad 0 < \alpha < 1.$$

Solving 6) gives

$$7) \quad \gamma = E_0(\alpha^{\mathfrak{m}_1}) = \alpha^{-1}(1 \pm \sqrt{1-\alpha^2}),$$

and here the $+$ sign is excluded owing to $E_0(\alpha^{\mathfrak{m}_1}) < 1$. Expanding 7) in a MACLAURIN series,

$$8) \quad E_0(\alpha^{\mathfrak{m}_1}) = \sum_{n \geqq 1} \alpha^{2n-1} \frac{(2n-2)!}{n!(n-1)!} 2^{-2n+1},$$

and thus

$$9) \quad \begin{aligned} &P_0(m_1 = 2n) = 0 \\ &P_0(\mathfrak{m}_1 = 2n-1) = \frac{(2n-2)!}{n!(n-1)!} 2^{-2n+1} \qquad n = 1, 2, 3, \ldots \end{aligned}$$

Using 3) and 4), it follows that

$$10) \quad E_l(\alpha^{\mathfrak{m}_k}) = \gamma^{|l-k|}.$$

Because $\gamma \uparrow 1$ as $\alpha \uparrow 1$,

$$11) \quad P_l(\mathfrak{m}_k < +\infty) = 1,$$

and, using the fact that $s_n : n \geqq 0$ starts from scratch at a passage time, it is seen that *the walk visits* $0, \pm 1, \pm 2, \ldots$ *each an infinite number of times.*

Problem 1 after K. L. CHUNG and W. FELLER [*1*]. Consider the number t_{2n} of integers $k = 1, 2, \ldots, 2n$ such that the greater of s_{k-1} and s_k is >0; prove that

$$\sum_{n \geqq 0} \beta^{2n} E_0[\alpha^{t_{2n}}, s_{2n} = 0] = \frac{2}{\sqrt{1-\alpha^2\beta^2} + \sqrt{1-\beta^2}} \qquad 0 < \alpha, \beta < 1$$

and deduce the so-called arcsine law:

$$P_0[t_{2n} = 2k \mid s_{2n} = 0] = (n+1)^{-1}, \quad k = 0, 1, 2, \ldots, n, \quad n \geqq 1.$$

[Given $0 < \alpha < 1$, if $e_{2n} = E_0(\alpha^{t_{2n}}, s_{2n} = 0)$ and if $\mathfrak{m}$ is the least integer $n = 2, 4, \ldots$ such that $s_n = 0$, then

$$\begin{aligned} e_{2n} &= \sum_{k \leq n} E_0[\alpha^{t_{2n}}, \mathfrak{m} = 2k, s_{2n} = 0] \\ &= \sum_{k \leq n} E_0(\alpha^{t_{2k}}, \mathfrak{m} = 2k)\, E_0(\alpha^{t_{2n-2k}}, s_{2n-2k} = 0) \\ &= \sum_{k \leq n} \tfrac{1}{2}(1 + \alpha^{2k})\, P_0(\mathfrak{m} = 2k)\, e_{2n-2k} \quad \text{for} \quad n \geq 1; \end{aligned}$$

thus, if $0 < \beta < 1$, then

$$\begin{aligned} f = \sum_{n \geq 0} \beta^{2n} e_{2n} \\ = 1 + \sum_{n \geq 1} \tfrac{1}{2}(1 + \alpha^{2n})\, \beta^{2n}\, P_0(\mathfrak{m} = 2n)\, f, \end{aligned}$$

and using the formula

$$\sum_{n \geq 1} \alpha^{2n} P_0(\mathfrak{m} = 2n) = 1 - \sqrt{1 - \alpha^2},$$

it is seen that

$$f = \frac{2}{\sqrt{1 - \alpha^2 \beta^2} + \sqrt{1 - \beta^2}} = 2\beta^{-2}(1 - \alpha^2)^{-1}\left(\sqrt{1 - \alpha^2 \beta^2} - \sqrt{1 - \beta^2}\right).$$

$P_0(t_{2n} = 2k, s_{2n} = 0)$ is the coefficient of $\alpha^{2k}\beta^{2n}$ in f; thus, for $k \leq n$, $P_0(t_{2n} = 2k, s_{2n} = 0)$ is the coefficient of $\alpha^{2k}\beta^{2n+2}$ in $-\dfrac{2\sqrt{1 - \beta^2}}{1 - \alpha^2}$, and since that coefficient is independent of k, the proof is complete.]

Problem 2. Use 10) and the successive passage times:

$$\begin{aligned} \mathfrak{m}^1 &= \min(n : s_n = k) \\ \mathfrak{m}^2 &= \min(n : n > \mathfrak{m}^1, s_n = k) \\ \mathfrak{m}^3 &= \min(n : n > \mathfrak{m}^2, s_n = k) \\ &\textit{etc.} \end{aligned}$$

to evaluate the generating function $\sum_n \alpha^{n+1} P_l(s_n = k)$.

$$\begin{aligned} \Big[\sum_{n \geq 0} \alpha^{n+1} P_l(s_n = k) \\ &= \sum_{n \geq 0} \alpha^{n+1} \sum_{m \geq 1} P_l(\mathfrak{m}^m = n) \\ &= \alpha \sum_{m \geq 1} E_l(\alpha^{\mathfrak{m}^m}) \\ &= \alpha \sum_{m \geq 1} E_l(\alpha^{\mathfrak{m}_k})\, E_0(\alpha^{\mathfrak{m}})^{m-1} \\ &= \alpha\, \gamma^{|l-k|} [1 - E_0(\alpha^{\mathfrak{m}})]^{-1} \end{aligned}$$

with $\mathfrak{m}$ as in the solution of problem 1; now use the evaluation

$$\alpha^{-1}[1 - E_0(\alpha^{\mathfrak{m}})] = \alpha^{-1}(1 - \alpha\gamma) = \alpha^{-1}\sqrt{1 - \alpha^2} = \tfrac{1}{2}(\gamma^{-1} - \gamma).]$$

1.3. Hinčin's proof of the de Moivre-Laplace limit theorem

Our next topic is the ingeneous proof of the DE MOIVRE-LAPLACE limit theorem:

1) $$\lim_{n\uparrow+\infty} P_0\left[a \leqq \frac{s(n)}{\sqrt{n}} < b\right] = \int_a^b \frac{e^{-c^2/2}\,dc}{\sqrt{2\pi}}$$

due to A. YA. HINČIN [*1*: 3—5] (see W. FELLER [*3*: 133—137], H. TROTTER [*2*], and problem 2 below for different proofs, and 1.10 for a bigger and better limit theorem).

Define

2) $$u_n(t, a) = E_{[\sqrt{n}\,a]}\left[f(n^{-\frac{1}{2}} s_{[nt]})\right] \qquad t \geqq 0,\ a \in R^1$$

for $f \in C(R^1)$ and $n \geqq 1$*. Because $u_n(t, a) = u_n(m/n, l/\sqrt{n})$ for $m = [n\,t]$ and $l = [\sqrt{n}\,a]$,

3) $$\begin{aligned} u_n\left(t + \frac{1}{n}, a\right) &= E_l[f(n^{-\frac{1}{2}} s_{m+1})] \\ &= E_l[E_{s_1}[f(n^{-\frac{1}{2}} s_m)]] \\ &= \frac{1}{2} u_n\left(t, \frac{l-1}{\sqrt{n}}\right) + \frac{1}{2} u_n\left(t, \frac{l+1}{\sqrt{n}}\right) \\ &= \frac{1}{2} u_n\left(t, a - \frac{1}{\sqrt{n}}\right) + \frac{1}{2} u_n\left(t, a + \frac{1}{\sqrt{n}}\right), \end{aligned}$$

and subtracting $u_n(t, a)$ from both sides of 3), it is found that

4a) $$\begin{aligned} &n \times \left[u_n\left(t + \frac{1}{n}, a\right) - u_n(t, a)\right] \\ &\quad = \frac{1}{2} (\sqrt{n})^2 \times \left[u_n\left(t, a + \frac{1}{\sqrt{n}}\right) - 2u_n(t, a) + u_n\left(t, a - \frac{1}{\sqrt{n}}\right)\right] \end{aligned}$$

4b) $$u_n(0, b) = f([\sqrt{n}\,b]/\sqrt{n}).$$

HINČIN's idea is to compare u_n with the solution

5) $$u = u(t, a) = \int_{R^1} \frac{e^{-(b-a)^2/2t}}{\sqrt{2\pi t}} f(b)\,db$$

of the continuous analogue of 4):

6a) $$\frac{\partial u}{\partial t} = \frac{1}{2} \frac{\partial^2 u}{\partial a^2}$$

6b) $$u(0+, \cdot) = f$$

and to conclude that

7) $$\lim_{n\uparrow+\infty} u_n = u.$$

* $[n\,t]$, $[\sqrt{n}\,a]$, $[\sqrt{n}\,b]$ indicate integral parts; otherwise [ ] has its usual meaning.

Consider with HINČIN the function $u^\bullet = u + n^{-1/3} t$ assuming $f \in C^4(R^1)$, note the bounds

8a) $$\frac{\partial^3 u^\bullet}{\partial a^3} = \frac{\partial^3 u}{\partial a^3} = \int \frac{e^{-(b-a)^2/2t}}{\sqrt{2\pi t}} d^3 f/db^3 \, db \leqq c_1 < +\infty$$

8b) $$\frac{\partial^2 u^\bullet}{\partial t^2} = \frac{1}{4} \frac{\partial^4 u}{\partial a^4} = \frac{1}{4} \int \frac{e^{-(b-a)^2/2t}}{\sqrt{2\pi t}} d^4 f/db^4 \, db \geqq -c_2 > -\infty$$

and choosing $n > (c_1 + c_2)^6$, let us check the estimates

$$Q_m : u_n\left(\frac{m}{n}, \frac{l}{\sqrt{n}}\right) \leqq u^\bullet\left(\frac{m}{n}, \frac{l}{\sqrt{n}}\right)$$

for $m \geqq 1$.

Q_0 is automatic, and supposing that Q_m holds and using 4a), 6a), and 8), one finds

9) $$\begin{aligned} u_n\left(\frac{m+1}{n}, \frac{l}{\sqrt{n}}\right) &= \frac{1}{2} u_n\left(\frac{m}{n}, \frac{l-1}{\sqrt{n}}\right) + \frac{1}{2} u_n\left(\frac{m}{n}, \frac{l+1}{\sqrt{n}}\right) \\ &\leqq \frac{1}{2} u^\bullet\left(\frac{m}{n}, \frac{l-1}{\sqrt{n}}\right) + \frac{1}{2} u^\bullet\left(\frac{m}{n}, \frac{l+1}{\sqrt{n}}\right) \\ &\leqq u^\bullet\left(\frac{m}{n}, \frac{l}{\sqrt{n}}\right) + \frac{1}{2n} \frac{\partial^2 u^\bullet}{\partial a^2}\left(\frac{m}{n}, \frac{l}{\sqrt{n}}\right) + n^{-3/2} c_1 \\ &= u^\bullet\left(\frac{m}{n}, \frac{l}{\sqrt{n}}\right) + \frac{1}{n} \frac{\partial u^\bullet}{\partial t}\left(\frac{m}{n}, \frac{l}{\sqrt{n}}\right) - n^{-4/3} + n^{-3/2} c_1 \\ &\leqq u^\bullet\left(\frac{m+1}{n}, \frac{l}{\sqrt{n}}\right) + \frac{1}{2} n^{-2} c_2 - n^{-4/3} + n^{-3/2} c_1 \\ &< u^\bullet\left(\frac{m+1}{n}, \frac{l}{\sqrt{n}}\right) + n^{-3/2} \left[c_1 + c_2 - n^{1/6}\right] \\ &< u^\bullet\left(\frac{m+1}{n}, \frac{l}{\sqrt{n}}\right); \end{aligned}$$

in other words, Q_m implies Q_{m+1}, and this completes the induction. Now

10) $$u_n(t, a) \leqq u^\bullet\left(\frac{m}{n}, \frac{l}{\sqrt{n}}\right) \quad m = [n t], \, l = [\sqrt{n}\, a],$$

and letting $n \uparrow +\infty$, it appears that

11) $$\overline{\lim_{n \uparrow +\infty}} \, u_n \leqq u;$$

the proof of $\underline{\lim}_{n \uparrow +\infty} u_n \geqq u$ is left to the reader as well as the proof of 1) using 7) for $f \in C^4(R^1)$.

On the basis of 1), it is natural to conjecture that $n^{-1/2} s[nt]$ $(t \geqq 0)$ approximates some stochastic (BROWNian) motion for which 6a) takes the place of 4a). 1.4 and 1.5 are devoted to the construction of this BROWNian motion, 1.10 to the discussion of the approximation.

Problem 1. Check that

$$\lim_{n\uparrow+\infty} P_0\Big[\max_{\theta\leqq t} n^{-1/2} s_{[n\theta]} \geqq a\Big] = \int_0^t \frac{a}{\sqrt{2\pi\theta^3}} e^{-a^2/2\theta}\, d\theta \qquad t, a \geqq 0$$

(see problem 1.10.1 for another proof).

[Because $\min(\theta : n^{-1/2} s_{[n\theta]} \geqq a) = n^{-1}\mathfrak{m}_{[\sqrt{n}a]+1}$, it is obvious that

$$P_0\Big[\max_{\theta\leqq t} n^{-1/2} s_{[n\theta]} \geqq a\Big] = P_0\big[n^{-1}\mathfrak{m}_{[\sqrt{n}a]+1} \leqq t\big]; \quad \text{also, using 1.2.7)},$$

$$\int_0^{+\infty} e^{-\alpha t} P_0\big[n^{-1}\mathfrak{m}_{[\sqrt{n}a]+1} \in dt\big] = E_0\Big(e^{-\alpha n^{-1}\mathfrak{m}_{[\sqrt{n}a]+1}}\Big)$$

$$= \big(e^{\alpha/n} - \sqrt{e^{2\alpha/n} - 1}\big)^{[\sqrt{n}a]+1} \to e^{-\sqrt{2\alpha}\,a} \qquad \text{as } n\uparrow+\infty;$$

now use the formula

$$e^{-\sqrt{2\alpha}\,a} = \int_0^{+\infty} e^{-\alpha t} \frac{a}{\sqrt{2\pi t^3}} e^{-a^2/2t}\, dt$$

(see A. ERDÉLYI [*1* (1): 245]).]

Problem 2. Prove the DE MOIVRE-LAPLACE limit theorem using STIRLING's approximation:

$$n! = \sqrt{2\pi}\, n^{n+1/2} e^{-n+\theta/12n} \qquad 0 < \theta = \theta(n) < 1, n > 1.$$

[Bound $P_0\Big[a \leqq \frac{s(n)}{\sqrt{n}} < b\Big]$ for $0 < a < b < +\infty$ first and then estimate the tail $P_0[s(n)/\sqrt{n} > b]$ using ČEBYŠEV.]

1.4. The standard Brownian motion

Consider the space W of continuous functions $w: t \to x_t = x(t)$ from $[0, +\infty)$ to R^1* and introduce the class C of subsets

1) $$C = x_{\mathfrak{t}}^{-1}(B) = x_{t_1 t_2 \ldots t_n}^{-1}(B) \quad \mathfrak{t} = (t_1, t_2, \ldots, t_n),$$
$$0 < t_1 < t_2 < \cdots < t_n,\ B \in \mathsf{B}(R^n)^{**},\ n \geqq 1$$

of W, where $x_{\mathfrak{t}}^{-1}$ is the map inverse to

2) $$x_{\mathfrak{t}} : w \to (x_{t_1}(w), x_{t_2}(w), \ldots, x_{t_n}(w)) \in R^n.$$

---

* $x_t(w) = x(t, w)$ is used in place of $x_t = x(t)$ if the path w needs to be emphasized.

** $\mathsf{B}(R^n)$ is the algebra of BOREL subsets of R^n.

C is an algebra; in fact,

3) $$W = x_t^{-1}(R^n),$$

4) $$W - x_t^{-1}(B) = x_t^{-1}(R^n - B)$$

and

5) $$x_t^{-1}(B_1) \cup x_t^{-1}(B_2) = x_t^{-1}(B_1 \cup B_2),$$

i.e., $x_t^{-1}\,\mathsf{B}(R^n)$ is an algebra, and to complete the proof, it is sufficient to note that two such algebras $x_t^{-1}\,\mathsf{B}(R^n)$ are contained in a third.

Consider, next, the GAUSS kernel

6) $$g(t, a, b) = \frac{e^{-(b-a)^2/2t}}{\sqrt{2\pi t}} \qquad t > 0,\ a, b \in R^1$$

and the set functions

7) $$P_t(C) = \int_B \cdots \int g(t_1, 0, b_1)\, db_1\, g(t_2 - t_1, b_1, b_2)\, db_2 \ldots g(t_n - t_{n-1}, b_{n-1}, b_n)\, db_n$$
$$C = x_t^{-1}(B), \quad B \in \mathsf{B}(R^n), \qquad n \geqq 1.$$

P_t is a probability measure on $x_t^{-1}\,\mathsf{B}(R^n)$; indeed, since $x_t^{-1}(B_1) = x_t^{-1}(B_2)$ implies $B_1 = B_2$, P_t is well-defined; since $x_t^{-1}(B_1) \cap x_t^{-1}(B_2) = \emptyset$ implies $B_1 \cap B_2 = \emptyset$, it is clear from 5) and 1) that P_t is additive; and, since

8) $$\int_{-\infty}^{+\infty} g(t, a, b)\, db = 2 \int_0^{+\infty} \frac{e^{-b^2/2}}{\sqrt{2\pi}}\, db$$
$$= \left(\frac{2}{\pi} \int_0^{+\infty} da \int_0^{+\infty} db\, e^{-a^2/2} e^{-b^2/2}\right)^{\frac{1}{2}}$$
$$= \left(\frac{2}{\pi} \int_0^{\pi/2} d\theta \int_0^{+\infty} e^{-r^2/2}\, r\, dr\right)^{\frac{1}{2}}$$
$$= 1,$$

$P_t(W) = 1$, which completes the proof.

P_t coincides with $P_{\mathfrak{s}}$ on $x_{\mathfrak{s}}^{-1}\,\mathsf{B}(R^m)$* if $\mathfrak{s} \subset \mathfrak{t}$; indeed,

9) $$\int_{-\infty}^{+\infty} g(t - s, a, c)\, g(s, c, b)\, dc$$
$$= \int_{-\infty}^{+\infty} \frac{e^{-(a-c)^2/2(t-s)}}{\sqrt{2\pi(t-s)}} \frac{e^{-(c-b)^2/2s}}{\sqrt{2\pi s}}\, dc$$
$$= \frac{e^{-(b-a)^2/2t}}{\sqrt{2\pi t}} = g(t, a, b) \qquad t > s > 0,\ a, b \in R^1,$$

* m is the number of points in $\mathfrak{s}$.

and taking, for example,

$$C = x^{-1}_{t_1 t_2 t_3 t_4 \ldots t_n}(B_1) = x^{-1}_{t_1 t_2 t_4 \ldots t_n}(B_2), \qquad B_1 \in \mathsf{B}(R^n), \quad B_2 \in \mathsf{B}(R^{n-1}),$$

one finds

10) $$P_{t_1 t_2 t_3 t_4 \ldots t_n}(C) = \int_{(b_1, b_2, b_3, b_4, \ldots, b_n) \in B_1} g(t_1, 0, b_1)\, g(t_2 - t_1, b_1, b_2) \times$$
$$\times g(t_3 - t_2, b_2, b_3)\, g(t_4 - t_3, b_3\, b_4) \ldots g(t_n - t_{n-1}, b_{n-1}, b_n) \times$$
$$\times db_1\, db_2\, db_3\, db_4 \ldots db_n$$
$$= \int_{(b_1, b_2, b_4 \ldots, b_n) \in B_2} g(t_1, 0, b_1)\, db_1\, g(t_2 - t_1, b_1, b_2)\, db_2 \times$$
$$\times \int_{b_3 \in R^1} g(t_3 - t_2, b_2, b_3)\, db_3\, g(t_4 - t_3, b_3, b_4)\, db_4 \ldots g(t_n - t_{n-1}, b_{n-1}, b_n)\, db_n$$
$$= \int_{B_2} g(t_1, 0, b_1)\, db_1\, g(t_2 - t_1, b_1, b_2)\, db_2\, g(t_4 - t_2, b_2, b_4)\, db_4 \ldots$$
$$\ldots g(t_n - t_{n-1}, b_{n-1}, b_n)\, db_n$$
$$= P_{t_1 t_2 t_4 \ldots t_n}(C).$$

Define $P(C) = P_{\mathfrak{t}}(C)$ for $C \in x_{\mathfrak{t}}^{-1}\, \mathsf{B}(R^n)$. P is a probability measure on C, and, as we will now prove, *it is extensible to a Borel probability measure on the Borel extension* B *of* C.

Consider, for the proof, $C_n \in \mathsf{C}$ such that $C_1 \supset C_2 \supset \ldots$ and $P(C_n) \geqq c_1 > 0$*; it is to be shown that $\bigcap_{n \geqq 1} C_n$ is non-void.

Suppose, as we can, that

11) $$C_n = x_{\mathfrak{t}_n}^{-1}(B_n),$$

where

$$\mathfrak{t}_n = (t_1, t_2, \ldots, t_n, 2^{-n}, 2 \cdot 2^{-n}, \ldots, n \cdot 2^n \cdot 2^{-n}),$$
$$B_n \in \mathsf{B}(R^m),\ m = \text{the number of points of } \mathfrak{t}_n,\ n \geqq 1,$$

and, noting the estimate

12) $$P\Big[\max_{\substack{0 < s_2 - s_1 \leqq 2^{-n} \\ s_1, s_2 \in \mathfrak{t}_n}} |x_{s_2} - x_{s_1}| > c_2\, 2^{-n/3}\Big]$$
$$\leqq \sum_{\substack{0 < s_2 - s_1 \leqq 2^{-n} \\ s_1, s_2 \in \mathfrak{t}_n}} P[|x_{s_2} - x_{s_1}| > c_2\, 2^{-n/3}]$$
$$= \sum_{\substack{0 < s_2 - s_1 \leqq 2^{-n} \\ s_1, s_2 \in \mathfrak{t}_n}} 2 \int_{c_2 2^{-n/3}}^{\infty} \frac{e^{-b^2/2(s_2 - s_1)}}{\sqrt{2\pi(s_2 - s_1)}}\, db$$
$$\leqq \frac{3n}{c_2}\, 2^{5n/6}\, e^{-c_2^2 2^{n/3 - 1}},\ **$$

---

* c_1, c_2, *etc.* are positive constants.

** $\int_a^{+\infty} e^{-b^2/2t}\, db \leqq \frac{t}{a} \int_a^{+\infty} e^{-b^2/2t} \frac{b}{t}\, db = \frac{t}{a} e^{-a^2/2t}$; see problem 1.4.1. for a finer estimate.

choose $c_2 > 0$ so great that

13) $$\sum_{n\geqq 1} \frac{3n}{c_2} 2^{5n/6} e^{-c_2^2 2^{n/3-1}} \leqq \frac{c_1}{3}.$$

Choose, also, compact $B_n'' \subset B_n' \subset B_n$ such that

14) $$P(x_{t_n}^{-1}(B_n - B_n')) < c_1 3^{-n}$$

and

15) $$x_{t_n}^{-1}(B_n') \cap (w: \max_{\substack{0 < s_2 - s_1 \leqq 2^{-n} \\ s_1, s_2 \in t_n}} |x_{s_2} - x_{s_1}| \leqq c_2 2^{-n/3}) = x_{t_n}^{-1}(B_n''),$$

and note that

16) $$P(\bigcap_{m\leqq n} x_{t_m}^{-1}(B_m''))$$
$$\geqq P(C_n) - \sum_{m\leqq n} P(x_{t_m}^{-1}(B_m - B_m')) - \sum_{m\leqq n} P(x_{t_m}^{-1}(B_m' - B_m''))$$
$$\geqq c_1 - \frac{c_1}{2} - \frac{c_1}{3} = \frac{c_1}{6} > 0.$$

$\bigcap_{m\leqq n} x_{t_m}^{-1}(B_m'')$ is therefore never void, and, since B_n'' was compact, we can select a point $(x(s): s \in \mathrm{t}) \in R^\infty$, $\mathrm{t} = \bigcup_n \mathrm{t}_n$ such that

17) $$(x(s): s \in \mathrm{t}_n) \in B_n'' \qquad n \geqq 1.$$

But then

18) $$|x(s_2) - x(s_1)| \leqq c_2 2^{-n/3} \quad 0 < s_2 - s_1 \leqq 2^{-n}, s_1, s_2 \in \mathrm{t}_n, n \geqq 1,$$

and it follows that $x(s)$ is uniformly continuous on each bounded subset of t; in fact, if

19) $$s_1, s_2 \in \mathrm{t}, \quad 0 < s_2 - s_1 < 2^{-m}, \quad 0 < s_1 < s_2 < m,$$

then

20) $$|x(s_2) - x(s_1)| < c_3 2^{-m/3} \qquad c_3 = 4c_2 \sum_{n\geqq 1} 2^{-n/3}.$$

As to the proof of this, take n such that $s_1, s_2 \in \mathrm{t}_{n+m}$ and consider $s_1' = k_1 2^{-n-m}$ $s_2' = k_2 2^{-n-m}$ such that

21) $$s_1 < s_1' < s_2' < s_2, \quad s_1' - s_1 < 2^{-n-m}, \quad s_2 - s_2' < 2^{-n-m};$$

and, using 18) and the terminating expansions

22) $$s_1' = k 2^{-l} - 2^{-p_1} - 2^{-p_2} - etc. \quad m \leqq l < p_1 < p_2 < etc.$$
$$s_2' = k 2^{-l} + 2^{-q_1} + 2^{-q_2} + etc. \quad m \leqq l < q_1 < q_2 < etc.,$$

estimate as follows:

23) $$|x(s_2) - x(s_1)|$$
$$\leqq |x(s_1') - x(s_1)| + |x(s_1') - x(s_1')| + |x(s_2) - x(s_2')|$$
$$< c_2 2^{-(m+n)/3} + 2 \sum_{l\geqq 1} c_2 2^{-(m+l)/3} + c_2 2^{-(m+n)/3}$$
$$< c_3 2^{-m/3}.$$

$x(s)$ $(s \in \mathfrak{t})$ can now be extended to a continuous function on $[0, +\infty)$, and since this function is a member of $\bigcap_{n \geq 1} x_{\mathfrak{t}_n}^{-1}(B_n'') \subset \bigcap_{n \geq 1} C_n$, the proof is complete. Now P can be extended to a BOREL probability measure on the BOREL extension B of C; with this extension, *the triple* $[W, \mathsf{B}, P]$ *is called standard Brownian motion starting at* 0. *P is the so-called Wiener measure.*

Given $a \in R^1$, the same proof justifies the extension of the set function

24) $$P_a(C) = \int_B g(t_1, a, b_1)\, db_1\, g(t_2 - t_1, b_1, b_2)\, db_2 \ldots g(t_n - t_{n-1}, b_{n-1}, b_n)\, db_n$$

$$C = x_{\mathfrak{t}}^{-1}(B), \quad B \in \mathsf{B}(R^n), \quad n \geq 1$$

to B.

Because $g(t, a, b) = g(t, 0, |b - a|)$,

25) $$P_a(B) = P_0(w + a \in B), \; P_a(-w \in B) = P_{-a}(B) \qquad B \in \mathsf{B},$$

where $w + a$ is the translated path $x(t, w + a) = x(t) + a$ and $-w$ is the reflected path $x(t, -w) = -x(t)$.

Because

26) $$P_a[x(0) = a] = P_0[x(0) = 0] = \lim_{n\uparrow\infty} \lim_{t\downarrow 0} P_0[|x(t)| < n^{-1}] = \lim_{n\uparrow\infty} \lim_{t\downarrow 0} 2 \int_0^{n^{-1}t^{-1/2}} \frac{e^{-b^2/2}}{\sqrt{2\pi}}\, db = 1,$$

$P_a(B)$ is to be thought of as *the chance that the event B occurs for the Brownian path starting at $a \in R^1$.*

$\mathsf{D} = [W, \mathsf{B}, P_a : a \in R^1]$ is the *standard Brownian motion*; it is a collection of individual stochastic processes, one to each point $a \in R^1$, knit together in a certain manner as we will now explain.

With the help of 24), it is clear that

27) $$P_a[x(t) \in db \mid x(t_1), x(t_2), \ldots, x(t_n)] = P_{x(t_n)}[x(t - t_n) \in db] = g(t - t_n, x(t_n), b)\, db \qquad t > t_n \geq \cdots \geq t_2 \geq t_1, n \geq 1, a, b \in R^1,$$

or, what is the same,

28) $$P_a[x(t_2) \in db \mid \mathsf{B}_{t_1}] = P_c[x(t_2 - t_1) \in db] \qquad c = x(t_1), \; t_2 \geq t_1, \; a, b \in R^1,$$

where $\mathsf{B}_{t_1} = \mathsf{B}[x(s) : s \leq t_1]$ is the smallest BOREL subalgebra of B including all the events $(w : a \leq x(s) < b)$ $(s \leq t_1)$; this is the so-called *simple* MARKOV property.

28) exhibits the *knitting together* of the individual BROWNian motions; it states that *the Brownian particle starts from scratch at time t_1*; in more

precise language, it states that, *conditional on the present* $b = x(t_1)$, *the future* $x(t+t_1)$ $(t \geqq 0)$ *is a Brownian motion starting at* b, *independent of the past* $x(t)$ $(t < t_1)$; *it is this starting afresh which the adjective Markovian describes.*

Consider the operator $\mathfrak{G} u = \frac{1}{2} u''$ acting on $D(\mathfrak{G}) = C^2(R^1)$.

Given $f \in C(R^1)$, the (bounded) solution of

$$\partial u/\partial t = \mathfrak{G} u \qquad t > 0 \tag{29}$$

$$u(0+, \cdot) = f$$

is

$$u = u(t, a) = E_a[f(x_t)] = \int_{R^1} \frac{e^{-(b-a)^2/2t}}{\sqrt{2\pi t}} f\, db,^* \tag{30}$$

and its LAPLACE transform $\hat{u} = \int_0^{+\infty} e^{-\alpha t} u\, dt\, (\alpha > 0)$ is the (bounded) solution

$$\hat{u} = \hat{u}(\alpha, a) = E_a\left[\int_0^{+\infty} e^{-\alpha t} f(x_t)\, dt\right] = \int_{R^1} \frac{e^{-\sqrt{2\alpha}|b-a|}}{\sqrt{2\alpha}} f\, db^{**} \tag{31}$$

of

$$(\alpha - \mathfrak{G})\, \hat{u} = f; \tag{32}$$

in operator language, if $e^{t\mathfrak{G}} f = u(t, \cdot)$ $(t > 0)$ and $G_\alpha f = \hat{u}$ $(\alpha > 0)$, then $e^{t\mathfrak{G}}$ maps $C(R^1)$ into $C^\infty(R^1)$, $(\partial/\partial t - \mathfrak{G})\, e^{t\mathfrak{G}} = 0$, $G_\alpha = \int_0^{+\infty} e^{-\alpha t} e^{t\mathfrak{G}}\, dt$ maps $C(R^1)$ 1 : 1 onto $C^2(R^1)$, and $G_\alpha^{-1} = \alpha - \mathfrak{G}$.

$\mathfrak{G}$ *is said to generate the Brownian motion.* G_α is the so-called GREEN operator of the BROWNian motion; see 2.4 for applications.

Problem 1 after Y. KOMATSU [*1*]. Check that

$$2\left[\sqrt{a^2+4} + a\right]^{-1} \leqq e^{a^2/2} \int_a^{+\infty} e^{-b^2/2}\, db \leqq 2\left[\sqrt{a^2+2} + a\right]^{-1} \quad \text{for} \quad a \geqq 0.$$

[Writing g_- for the under-estimate, g_+ for the over-estimate, and g for the modified error integral in the middle, a simple computation gives $g_-' \geqq a g_- - 1$, $g' = a g - 1$, $g_+' \leqq a g_+ - 1$; therefore $(g - g_-)' \leqq$ $\leqq a(g - g_-)$, $(g_+ - g)' \leqq a(g_+ - g)$, and using $g_-, g, g_+ \leqq a^{-1}$, it follows that $g - g_-$, $g_+ - g$ cannot become negative.]

---

* See I. PETROVSKY [*2*: 241—244].

** $\dfrac{e^{-\sqrt{2\alpha}|b-a|}}{\sqrt{2\alpha}} = \int_0^{+\infty} e^{-\alpha t} \dfrac{e^{-(b-a)^2/2t}}{\sqrt{2\pi t}}\, dt$; see A. ERDÉLYI [*1* (1): 146 (27)].

Problem 2. $P_0\left[\lim_{t\downarrow 0} t x(1/t) = 0\right] = 1$.

Problem 3 after P. LÉVY [*3*: 246]. Consider a standard BROWNian motion starting at 0; the problem is to show that

$$a(t) = 0 \qquad t = 0$$
$$= t x(1/t) \qquad t > 0$$

and

$$b(t) = c\, x(t/c^2) \qquad t \geqq 0 \quad (c > 0)$$

are likewise BROWNian motions starting at 0.

[Both these motions are GAUSSian with mean 0, and

$$E_0[x(s)\, x(t)] = E_0[s\, x(1/s)\, t\, x(1/t)] = E_0[c\, x(s/c^2)\, c\, x(t/c^2)] = s \wedge t.]$$

Problem 4. The BROWNian motion is differential in the sense that its increments $x[a_i, b_i) = x(b_i) - x(a_i)$ $(i \leqq n)$ over disjoint intervals $[a_i, b_i)$ $(i \leqq n)$ of $[0, +\infty)$ are independent.

Problem 5. Given $f = c_0 + c_1 x + c_2 x^2 + \cdots + c_n x^n$, check that

$$u(t, a) = e^{t\mathfrak{G}} f = \sum_{n \geqq 0} \frac{t^n (\mathfrak{G}^n f)(a)}{n!}$$

is the solution of

$$\frac{\partial u}{\partial t} = \mathfrak{G} u \qquad u(0+, \cdot) = f,$$

and deduce, for the standard BROWNian motion,

$$E_0(x_t^{2n}) = 1 . 3 \ldots (2n-1)\, t^n \qquad n \geqq 1.$$

Problem 6. Give a direct proof that 31) is the bounded solution of 32) and use this to deduce the formula

$$\frac{e^{-\sqrt{2\alpha}|b-a|}}{\sqrt{2\alpha}} = \int_0^{+\infty} e^{-\alpha t} \frac{e^{-(b-a)^2/2t}\, dt}{\sqrt{2\pi t}}$$

of the footnote to 31).

Problem 7. Give a proof that almost all BROWNian paths are nowhere differentiable. R. PALEY, N. WIENER and A. ZYGMUND [*1*] discovered this; the proof suggested below is due to A. DVORETSKI, P. ERDÖS, and S. KAKUTANI [*3*].

[If the BROWNian path is differentiable at some point $0 \leqq s \leqq 1$, then

$$|x(t) - x(s)| < l(t - s) \quad s < t < s + \frac{5}{n}, \qquad n \geqq m$$

for some $l \geqq 1$ and some $m \geqq 1$. But this event is part of the event

$$\bigcup_{l \geqq 1} \bigcup_{m \geqq 1} \bigcap_{n \geqq m} \bigcup_{0 < i \leqq n+2} \bigcap_{i < k \leqq i+3} \left(w : \left| x\left(\frac{k}{n}\right) - x\left(\frac{k-1}{n}\right)\right| < \frac{7l}{n}\right)$$

and

$$P_0\left[\bigcup_{0<i\leq n+2}\bigcap_{i<k\leq i+3}\left(w:\left|x\left(\frac{k}{n}\right)-x\left(\frac{k-1}{n}\right)\right|<\frac{7l}{n}\right)\right]$$
$$\leq (n+2)\,P_0\left[\left|x\left(\frac{1}{n}\right)\right|<\frac{7l}{n}\right]^3$$
$$=(n+2)\left(\int\limits_{|\xi|<\frac{7l}{\sqrt{n}}}\frac{e^{-\xi^2/2}}{\sqrt{2\pi}}\,d\xi\right)^3<\frac{\text{constant}}{\sqrt{n}}\downarrow 0\qquad (n\uparrow+\infty).]$$

1.5. P. Lévy's construction

P. LÉVY [*3*: 15–20] gave another construction of the standard BROWNian motion; the use of HAAR functions below is adopted from Z. CIESIELSKI [*1*].

Bring in the HILBERT space $L^2[0,1]$ with an orthonormal basis defined by the HAAR functions

1 a) $$f_0(t)=1\qquad 0\leq t\leq 1$$

1 b) $$f_{k2^{-n}}(t)=+2^{(n-1)/2}\qquad (k-1)\,2^{-n}\leq t<k\,2^{-n}$$
$$=-2^{(n-1)/2}\qquad k\,2^{-n}\leq t<(k+1)\,2^{-n}$$
$$k \text{ odd} < 2^n,\ n\geq 1.$$

Given a standard BROWNian path $x(t)$ $(t\leq 1)$ with formal derivative $x^\bullet$ (= white noise), the formal FOURIER coefficients

2 a) $$g_0=\int_0^1 f_0\,x^\bullet\,dt=x(1)$$

2 b) $$g_{k2^{-n}}=\int_0^1 f_{k2^{-n}}\,x^\bullet\,dt\qquad k \text{ odd}<2^n,\ n>1$$
$$=2^{(n-1)/2}[2x(k2^{-n})-x((k-1)\,2^{-n})-x((k+1)\,2^{-n})]$$

should be *Gaussian*:

3) $$P(g<a)=\int_{-\infty}^{a}\frac{e^{-b^2/2}}{\sqrt{2\pi}}\,db,$$

perpendicular:

4) $$E(g_{k2^{-n}}\,g_{j2^{-m}})=\int_0^1 f_{k2^{-n}}\,f_{j2^{-m}}\,dt=0\qquad (k2^{-n}\neq j2^{-m})$$

and hence *independent* (see prerequisites), and the integrated expansion

5) $$g_0\int_0^t f_0\,ds+\sum_{n=1}^{\infty}\ \sum_{k \text{ odd}<2^n} g_{k2^{-n}}\int_0^t f_{k2^{-n}}\,ds$$

should converge to $x(t)$ $(t \leqq 1)$; the idea is to use 5) to *define* the BROWNian motion.

Consider for this purpose, the algebra $\mathsf{B}[0, 1]$ of BOREL subsets of $[0, 1]$ with the classical BOREL measure P attached, let $0 \leqq u \leqq 1$ be expanded as

$$u = 2^{-1} u_1 + 2^{-2} u_2 + 2^{-3} u_3 + \cdots = .u_1 u_2 u_3 \ldots,$$

let h be the function inverse to $\int_{-\infty}^{a} e^{-b^2/2}\, db/\sqrt{2\pi}$, and defining

$$\begin{aligned} v_1 &= .u_1 u_2 u_4 u_7 \ldots \\ v_2 &= 0\, .u_3 u_5 u_8 \ldots \\ v_3 &= 0\,0\, .u_6 u_9 \ldots \\ v_4 &= 0\,0\,0\, .u_{10} \ldots \\ &\text{etc.,} \end{aligned}$$

let

$$\begin{array}{cccc} & g_0 = h(v_1) & & \\ & & g_{1/2} = h(v_2) & \\ & g_{1/4} = h(v_3) & g_{3/4} = h(v_4) & \\ g_{1/8} = h(v_5) & g_{3/8} = h(v_6) & g_{5/8} = h(v_7) & g_{7/8} = h(v_8) \\ \text{etc.,} & & & \end{array}$$

and note that the g's are independent with common distribution 3).

Consider now the sum

6) $$e_m = g_0 \int_0^t f_0\, ds + \sum_{n \leqq m} \sum_{\text{odd } k < 2^n} g_{k2^{-n}} \int_0^t f_{k2^{-n}}\, ds.$$

Because

7) $$\max_{t \leqq 1} \sum_{\text{odd } k < 2^n} \int_0^t f_{k2^{-n}}\, ds = 2^{-(n+1)/2}$$

as is clear from a picture,

8) $$\| e_n - e_{n-1} \|^* \leqq 2^{-(n+1)/2} \max_{\text{odd } k < 2^n} |g_{k2^{-n}}|,$$

so that

9) $$P\left[\| e_n - e_{n-1} \| > \sqrt{2 \cdot 2^{-n} \lg 2^n}\right] \leqq 2^n \int_{2\sqrt{n \lg 2}}^{+\infty} e^{-b^2/2} \frac{db}{\sqrt{2\pi}} < \frac{2^{-n-1}}{\sqrt{n \lg 2}}.$$

Using the first BOREL-CANTELLI lemma and the estimate

$$\sum_{m > n} \sqrt{2 \cdot 2^{-m} \lg 2^m} < \frac{3}{\lg 2} \sqrt{2 \cdot 2^{-n} \lg 2^n},$$

* $\|f\| = \max_{t \leqq 1} |f(t)|$.

it is immediate that e_n converges as $n \uparrow +\infty$ to a continuous function e_∞ and that

10) $$P\left[\|e_\infty - e_n\| \leq \frac{3}{\lg 2}\sqrt{2 \cdot 2^{-n} \lg 2^n},\ n \uparrow +\infty\right] = 1.$$

Because e_n was GAUSSian, so is e_∞; since $E(e_n)$ was 0, so is $E(e_\infty)$; and to complete the proof that $[e_\infty : 0 \leq t \leq 1, \mathsf{B}, P]$ is a BROWNian motion, it suffices to check that

11) $$E[e_\infty(s)\, e_\infty(t)] = E_0[x(s)\, x(t)],$$

i. e.,

12) $$\int_0^s f_0 \int_0^t f_0 + \sum_{n=1} \sum_{\text{odd } k < 2^n} \int_0^s f_{k2^{-n}} \int_0^t f_{k2^{-n}} = s \wedge t.$$

But this is obvious from the PARSEVAL relation.

Problem 1 after P. LÉVY [*3*: 19]. Use the BOREL-CANTELLI lemmas to find the lower bound of the values of c for which $\|e_n - e_{n-1}\| < c\sqrt{2^{-n} \lg 2^n}$ as $n \to +\infty$.

[$c = 1$].

Problem 2. Use the result of problem 1 to check that

$$P\left[\limsup_{\substack{|t_2 - t_1| = \varepsilon \downarrow 0 \\ 0 \leq t_1 < t_2 \leq 1}} \frac{e_\infty(t_2) - e_\infty(t_1)}{\sqrt{\varepsilon \lg(1/\varepsilon)}} \geq 1\right] = 1$$

(see 1.9 for more information on this point).

$$[\tfrac{1}{2}[e_\infty(m2^{-n}) - e_\infty((m-1)\,2^{-n})] + \tfrac{1}{2}[e_\infty(m2^{-n}) - e_\infty((m+1)\,2^{-n})]$$
$$= e_n(m2^{-n}) - e_{n-1}(m2^{-n})$$

for $m = 1, 3, \ldots, 2^n - 1$; therefore $\max_{m \leq 2^n} [e_\infty(m\,2^{-n}) - e_\infty((m-1)\,2^{-n})] \geq \|e_n - e_{n-1}\|$; now use the answer to problem 1.]

Problem 3. WIENER's construction of the standard BROWNian motion; see R. PALEY and N. WIENER [*1*: 140–162]. Let g_n, $n \geq 0$, be independent and GAUSS distributed:

$$P(g_n \leq a) = \int_{-\infty}^{a} \frac{1}{\sqrt{2\pi}} e^{-b^2/2}\, db.$$

Then

$$x(t) = \frac{t}{\sqrt{\pi}} g_0 + \sum_{n \geq 1} \sum_{k=2^{n-1}}^{2^n - 1} \sqrt{\frac{2}{\pi}} \frac{\sin k t}{k} g_k$$

converges uniformly in $0 \leq t \leq \pi$ with probability 1 and is a standard BROWNian motion.

[Put

$$s_{mn}(t) = \sum_m^{n-1} \frac{\sin k t}{k} g_k \qquad t_{mn} = \max_{0 \leq t \leq \pi} |s_{mn}(t)|;$$

then

$$t_{mn}^2 \leqq \max_{t \leqq \pi} \left| \sum_m^{n-1} \frac{e^{ikt}}{k} g_k \right|^2 \leqq \sum_m^{n-1} \frac{g_k^2}{k^2} + 2 \sum_{l=1}^{n-m-1} \left| \sum_{j=m}^{n-l-1} \frac{g_j g_{j+l}}{j(j+l)} \right|,$$

$$(E(t_{mn}))^2 \leqq E(t_{mn}^2) \leqq \sum_m^{n-1} \frac{1}{k^2} + 2 \sum_{l=1}^{n-m-1} \left(E\left(\left| \sum_{j=m}^{n-l-1} \frac{g_j g_{j+l}}{j(j+l)} \right|^2 \right) \right)^{\frac{1}{2}}$$

$$\leqq \sum_m^{n-1} \frac{1}{k^2} + 2 \sum_{l=1}^{n-m-1} \left(\sum_{j=m}^{n-l-1} \frac{1}{j^2(j+l)^2} \right)^{\frac{1}{2}}$$

$$\leqq \frac{n-m}{m^2} + 2(n-m) \left(\frac{n-m}{m^4} \right)^{\frac{1}{2}},$$

$$E(t_{m,2m}) \leqq \sqrt{3}\, m^{-1/4},$$

$$E\left[\sum_{n \geqq 1} t_{2^{n-1},2^n} \right] \leqq \sum_{n \geqq 1} E(t_{2^{n-1},2^n}) < +\infty,$$

$$P\left[\sum_{n \geqq 1} t_{2^{n-1},2^n} < +\infty \right] = 1,$$

and therefore our series converges uniformly with probability 1, so that almost all sample paths $x(t)$ $(t \leqq \pi)$ are continous. It is now enough to check that

$$E[x(t)\, x(s)] = \frac{ts}{\pi} + \frac{2}{\pi} \sum_{k \geqq 1} \frac{\sin kt \sin ks}{k^2} = s \wedge t.]$$

1.6. Strict Markov character

Consider the standard BROWNian motion D, let w_s^+ be the *shifted* path $x_t(w_s^+) = x_{t+s}(w)$ $(t \geqq 0)$, and, as before, let B_s be the smallest BOREL subalgebra of B including all the events $(w : a \leqq x(t) < b)$ $(t \leqq s)$. D is MARKOVian in the sense of

1) $$P_\bullet[w_s^+ \in B \,|\, \mathsf{B}_s] = P_b(B) \qquad b = x(s),\, B \in \mathsf{B}$$

[see 1.4.28)], i.e., *the Brownian traveller starts from scratch at each time* $t = s \geqq 0$.

BROWNian motion enthusiasts are familiar with the fact that *the Brownian traveller also starts afresh at certain random times, such as the passage time* $\mathfrak{m}_1 = \min(t : x(t) = 1)$*.

G. HUNT [*1*] discovered the complete statement of this feature of the BROWNian motion.

Consider a MARKOV time $0 \leqq \mathfrak{m} = \mathfrak{m}(w)$ $(\leqq +\infty)$, *i.e.*, let

2) $$(\mathfrak{m} < t) \in \mathsf{B}_t\text{**} \qquad t \geqq 0,$$

and define $\mathsf{B}_{\mathfrak{m}+}$ to be the class of sets $B \in \mathsf{B}$ such that

3) $$B \cap (\mathfrak{m} < t) \in \mathsf{B}_t \qquad t \geqq 0.$$

---

* $\min(t : x(t) = 1) = +\infty$ if the path never meets 1.

** $(\mathfrak{m} < t)$ is short for $(w : \mathfrak{m} < t)$.

$\mathsf{B}_{\mathfrak{m}+}$ is a BOREL algebra and $(\mathfrak{m} < t) \in \mathsf{B}_{\mathfrak{m}+}$ for each $t > 0$; in addition

4) $$\mathsf{B}_{\mathfrak{m}+} \supset \bigcap_{\varepsilon>0} \mathsf{B}[x(t \wedge (\mathfrak{m}+\varepsilon)) : t \geqq 0],$$

as will be seen below. $\mathsf{B}_{\mathfrak{m}+}$ *is to be thought of as measuring the Brownian path up to time* $t = \mathfrak{m}+$.

HUNT's statement is that, *conditional on the present position* $b = x(\mathfrak{m})$, *the future path* $x(t+\mathfrak{m})$ $(t \geqq 0)$ *is a standard Brownian motion starting at* b, *and this Brownian motion is independent of* $\mathsf{B}_{\mathfrak{m}+}$; in the language of conditional probabilities,

5 a) $$P_{\bullet}[w_{\mathfrak{m}}^{+} \in B \mid \mathsf{B}_{\mathfrak{m}+}] = P_b(B) \qquad \mathfrak{m} < +\infty, b = x(\mathfrak{m}), B \in \mathsf{B},$$

i. e.,

5 b) $$P_{\bullet}[A, w_{\mathfrak{m}}^{+} \in B, \mathfrak{m} < +\infty] = E_{\bullet}[A \cap (\mathfrak{m} < +\infty), P_{x(\mathfrak{m})}(B)] \qquad A \in \mathsf{B}_{\mathfrak{m}+}, B \in \mathsf{B};$$

in other words, *if* $\mathfrak{m}$ *is a Markov time and if* $\mathfrak{m}(w) < +\infty$, *then the Brownian traveller starts afresh at time* $t = \mathfrak{m}(w)$.

Before proving 5), it is convenient to examine the MARKOV times $\mathfrak{m}$ and the algebras $\mathsf{B}_{\mathfrak{m}+}$ in a little detail.

Given $t \geqq 0$, $\mathfrak{m} \equiv t$ is a MARKOV time, as is clear from 2), and $\mathsf{B}_{\mathfrak{m}+} = \mathsf{B}_{t+} = \bigcap_{s>t} \mathsf{B}_s$; indeed, if $B \in \mathsf{B}_{\mathfrak{m}+}$ and if $s > t$, then $B = B \cap (\mathfrak{m} < s) \in \mathsf{B}_s$, while, if $B \in \bigcap_{s>t} \mathsf{B}_s$, then

$$\begin{aligned} B \cap (\mathfrak{m} < s) &= B & s > t, \\ &= \emptyset & s \leqq t, \end{aligned}$$

and $B \in \mathsf{B}_{\mathfrak{m}+}$.

Given MARKOV times $\mathfrak{m}_1 \leqq \mathfrak{m}_2$, if $B \in \mathsf{B}_{\mathfrak{m}_1+}$, then $B \cap (\mathfrak{m}_2 < t) = B \cap (\mathfrak{m}_1 < t) \cap (\mathfrak{m}_2 < t) \in \mathsf{B}_t$ for each $t \geqq 0$, *i.e.*,

6) $$\mathsf{B}_{\mathfrak{m}_1+} \subset \mathsf{B}_{\mathfrak{m}_2+}.$$

Given MARKOV times $\mathfrak{m}$ and $\mathfrak{m}_1 \geqq \mathfrak{m}_2 \geqq$ *etc.* $\downarrow \mathfrak{m}$, if $B \in \bigcap_{n\geqq 1} B_{\mathfrak{m}_n+}$, then $B \cap (\mathfrak{m} < t) = \bigcup_{n\geqq 1} B \cap (\mathfrak{m}_n < t) \in \mathsf{B}_t$ for each $t \geqq 0$, *i.e.*, using 6),

7) $$\bigcap_{n\geqq 1} \mathsf{B}_{\mathfrak{m}_n+} = \mathsf{B}_{\mathfrak{m}+}.$$

Coming to the proof of 4), if $\mathfrak{m}$ is MARKOV, then $\mathfrak{m} \wedge t$ is measurable B_t for each $t \geqq 0$ as is clear from 2), and, using the fact that $x(t \wedge s, w)$ is a $\mathsf{B}[0, +\infty) \times \mathsf{B}_t$ measurable function of the pair (s, w) *, it follows that $x(\mathfrak{m} \wedge t \wedge s)$ is measurable B_t for each $s \geqq 0$. But then

$$(w : x(\mathfrak{m} \wedge s) < b) \cap (\mathfrak{m} < t) = (w : x(\mathfrak{m} \wedge t \wedge s) < b) \cap (\mathfrak{m} < t) \in \mathsf{B}_t$$

* $x(t) = \lim_{n\uparrow+\infty} x(2^{-n}[2^n t])$ is used for the proof of this point.

$(t \geqq 0)$, *i. e.*, $(w: x(\mathfrak{m} \wedge s) < b) \in \mathsf{B}_{\mathfrak{m}+}$ for each $s \geqq 0$, and, using 7) and the fact that $\mathfrak{m} + \varepsilon$ is MARKOV for each $\varepsilon > 0$, $\bigcap_{\varepsilon > 0} \mathsf{B}[x(t \wedge (\mathfrak{m} + \varepsilon))\ t \geqq 0]$ is found to be a part of $\bigcap_{\varepsilon > 0} \mathsf{B}_{(\mathfrak{m}+\varepsilon)+} = \mathsf{B}_{\mathfrak{m}+}$, as stated in 4).

Coming to the proof of 5), it is enough to show that, for each B-measurable e $(0 \leqq e \leqq 1)$,

8) $$E_{\bullet}[B, e(w_{\mathfrak{m}}^{+}), \mathfrak{m} < +\infty] = E_{\bullet}[B, E_{x(\mathfrak{m})}(e), \mathfrak{m} < +\infty] \quad B \in \mathsf{B}_{\mathfrak{m}+}.$$

Because the tame functions

9) $$e(w) = f[x(t_1), x(t_2), \ldots, x(t_n)]$$
$$0 < t_1 < t_2 < \cdots < t_n, \ 0 \leqq f \leqq 1, \ f \in C(R^n)$$

generate the class of B measurable functions, it is enough to prove 8) for them.

Given e as in 9),

10) $$\lim_{s \downarrow t} e(w_s^{+}) = e(w_t^{+}) \qquad t \geqq 0,$$

and

11) $$E_a(e) = \int_{R^n} g(t_1, a, b_1)\, g(t_2 - t_1, b_1, b_2) \ldots g(t_n - t_{n-1}, b_{n-1}, b_n)$$
$$f(b_1, b_2, \ldots, b_n)\, db_1\, db_2 \ldots db_n$$
$$\in C(R^1),$$

and so, in this special case,

12) $$\begin{aligned} &E_{\bullet}[B, e(w_{\mathfrak{m}}^{+}), \mathfrak{m} < +\infty] \\ &= \lim_{n \uparrow +\infty} E_{\bullet}(B, e(w_{\mathfrak{m}_n}^{+}), \mathfrak{m} < +\infty] \qquad \mathfrak{m}_n = 2^{-n}([2^n \mathfrak{m}] + 1) \ (n \geqq 1) \\ &= \lim_{n \uparrow +\infty} \sum_{k \geqq 1} E_{\bullet}[B \cap (\mathfrak{m}_n = k 2^{-n}), e(w_{k2^{-n}}^{+})] \\ &= \lim_{n \uparrow +\infty} \sum_{k \geqq 1} E_{\bullet}[B \cap ((k-1)\, 2^{-n} \leqq \mathfrak{m} < k 2^{-n}), e(w_{k2^{-n}}^{+})] \\ &= \lim_{n \uparrow +\infty} \sum_{k \geqq 1} E_{\bullet}[B \cap ((k-1)\, 2^{-n} \leqq \mathfrak{m} < k 2^{-n}), E_{x(k2^{-n})}(e)]\, ^{*} \\ &= \lim_{n \uparrow +\infty} E_{\bullet}[B, E_{x(\mathfrak{m}_n)}(e), \mathfrak{m} < +\infty] \\ &= E_{\bullet}[B, E_{x(\mathfrak{m})}(e), \mathfrak{m} < +\infty], \end{aligned}$$

as desired.

Problem 1. Give an example of a MARKOV time such that

$$(\mathfrak{m} \leqq t) \in \mathsf{B}_t \qquad t \geqq 0.$$

Give also a MARKOV time for which this is false.

$$[\mathfrak{m}_1 = \min(t : x(t) = 1); \ \mathfrak{m}_{0+} = \inf(t : x(t) > 0).]$$

---

* $B \cap ((k-1) 2^{-n} \leqq \mathfrak{m} < k 2^{-n}) \in \mathsf{B}_{k2^{-n}}$; now use 1).

Problem 2. BLUMENTHAL's 01 law (see R. BLUMENTHAL [1]). B_{0+} is trivial in the sense that

$$P_{\bullet}(B) = 0 \text{ or } 1 \qquad B \in \mathsf{B}_{0+}.$$

$$[P_{\bullet}(B) = E_{\bullet}(B, P_{\bullet}(B \mid \mathsf{B}_{0+})) = E_{\bullet}(B, P_{x(0)}(B)) = P_{\bullet}(B)^2.]$$

1.7. Passage times for the standard Brownian motion

The most important MARKOV times are the *passage times:*

1) $$\mathfrak{m}_a = \min(t : x_t = a) \qquad a \in R^1.$$

P. LÉVY [3: 221–223] has shown that $[\mathfrak{m}_a : a \geq 0,\ P_0]$ *is the one-sided stable process with exponent* $\frac{1}{2}$ *and rate* $\sqrt{2}$, *i.e., it is differential and homogeneous, with law*

2) $$P_0[\mathfrak{m}_b - \mathfrak{m}_a \leq t] = P_0[\mathfrak{m}_{b-a} \leq t] = \int_0^t \frac{b-a}{\sqrt{2\pi s^3}} e^{-(b-a)^2/2s} \, ds \qquad b \geq a,\ t \geq 0$$

(see note 1).

Beginning with the proof of 2), if $a \geq 0$ and if f is the indicator function of $[a, +\infty)$, then, using 1.6.5b) and the fact that $e^{-\alpha\mathfrak{m}} = 0$ if $\mathfrak{m} = +\infty$,

3) $$\begin{aligned}\int_0^{+\infty} e^{-\alpha t} P_0(x_t \geq a)\, dt &= E_0\left(\int_0^{+\infty} e^{-\alpha t} f(x_t)\, dt\right) \\ &= E_0\left(\int_{\mathfrak{m}}^{+\infty} e^{-\alpha t} f(x_t)\, dt\right) \qquad (\mathfrak{m} = \mathfrak{m}_a) \\ &= E_0\left[e^{-\alpha\mathfrak{m}} E_0\left(\int_0^{+\infty} e^{-\alpha t} f[x(t, w_{\mathfrak{m}}^+)]\, dt \mid \mathsf{B}_{\mathfrak{m}+}\right)\right]^* \\ &= E_0(e^{-\alpha\mathfrak{m}_a}) E_a\left(\int_0^{+\infty} e^{-\alpha t} f(x_t)\, dt\right) \\ &= E_0(e^{-\alpha\mathfrak{m}_a}) \int_0^{+\infty} e^{-\alpha t} P_a(x_t \geq a)\, dt \\ &= (2\alpha)^{-1} E_0(e^{-\alpha\mathfrak{m}_a}) \\ &= (2\alpha)^{-1} \int_0^{+\infty} e^{-\alpha t} P_0(\mathfrak{m}_a \in dt) \\ &= \tfrac{1}{2} \int_0^{+\infty} e^{-\alpha t} P_0(\mathfrak{m}_a \leq t)\, dt,\end{aligned}$$

---

* $\mathfrak{m}$ is measurable $\mathsf{B}_{\mathfrak{m}+}$.

proving the celebrated *reflection principle* of D. ANDRÉ:

4) $$P_0(\mathfrak{m}_a \leqq t) = 2P_0(x_t \geqq a) = 2\int_a^{+\infty} \frac{e^{-b^2/2t}}{\sqrt{2\pi t}}\,db \qquad t, a \geqq 0.$$

2) is now immediate from the formula:

$$2\int_a^{+\infty} \frac{e^{-b^2/2t}}{\sqrt{2\pi t}}\,db = \int_0^t \frac{a}{\sqrt{2\pi s^3}} e^{-a^2/2s}\,ds \qquad t, a \geqq 0,$$

and, at no extra cost, it is found that $P_a(\mathfrak{m}_b < +\infty) = 1$.

Because of 4),

5) $$E_0(e^{-\alpha \mathfrak{m}_a}) = 2\alpha \int_0^{+\infty} e^{-\alpha t}\,dt \int_a^{+\infty} \frac{e^{-b^2/2t}}{\sqrt{2\pi t}}\,db = 2\alpha \int_a^{+\infty} \frac{e^{-\sqrt{2\alpha}\,b}}{\sqrt{2\alpha}}\,db = e^{-\sqrt{2\alpha}\,a},$$

using the footnote to 1.4.31); more directly, it is clear from 2) and the first integral in 5) that $g \equiv E_0(e^{-\alpha \mathfrak{m}_a})$ is the solution of

6) $$g'' = 2\alpha g, \quad g(0) = 1, \quad g(+\infty) = 0,$$

and solving this gives 5).

To complete the proof that $\mathfrak{m}_a$ $(a \geqq 0)$ is a one-sided stable process, it is now sufficient to check its differential character. Given $0 < a < b$,

7) $$\begin{aligned} \mathfrak{m}_b &= \min(t : x(t) = b) \\ &= \mathfrak{m} + \min(t : x(t+\mathfrak{m}) = b) \qquad \mathfrak{m} = \mathfrak{m}_a < +\infty \\ &= \mathfrak{m} + \min(t : x(t, w_\mathfrak{m}^+) = b) \\ &= \mathfrak{m} + \mathfrak{m}_b(w_\mathfrak{m}^+), \end{aligned}$$

so that, using 1.6.5b) and the fact that $e^{-\alpha\mathfrak{m}} = 0$ if $\mathfrak{m} = +\infty$,

8) $$\begin{aligned} &E_0[e^{-\alpha_1 \mathfrak{m}_a} e^{-\alpha_2(\mathfrak{m}_b - \mathfrak{m}_a)}] \\ &= E_0\left[e^{-\alpha_1 \mathfrak{m}_a} E_0\left(e^{-\alpha_2 \mathfrak{m}_b(w_\mathfrak{m}^+)} \mid \mathsf{B}_{\mathfrak{m}+}\right)\right] \\ &= E_0(e^{-\alpha_1 \mathfrak{m}_a})\, E_a(e^{-\alpha_2 \mathfrak{m}_b}) \\ &= E_0(e^{-\alpha_1 \mathfrak{m}_a})\, E_0\left[e^{-\alpha_2 \mathfrak{m}_b(w_\mathfrak{m}^+)}\right] \\ &= E_0(e^{-\alpha_1 \mathfrak{m}_a})\, E_0[e^{-\alpha_2(\mathfrak{m}_b - \mathfrak{m}_a)}]. \end{aligned}$$

A similar computation shows that for

$$\alpha_1, \alpha_2, \ldots, \alpha_n > 0 \quad \text{and} \quad 0 < a_1 < a_2 < \cdots < a_n,$$

9) $$E_0\left[e^{-\alpha_1 \mathfrak{m}_{a_1}} e^{-\alpha_2(\mathfrak{m}_{a_2} - \mathfrak{m}_{a_1})} \cdots e^{-\alpha_n(\mathfrak{m}_{a_n} - \mathfrak{m}_{a_{n-1}})}\right] = E_0\left[e^{-\alpha_1 \mathfrak{m}_{a_1}}\right] E_0\left[e^{-\alpha_2(\mathfrak{m}_{a_2} - \mathfrak{m}_{a_1})}\right] \cdots E_0\left[e^{-\alpha_n(\mathfrak{m}_{a_n} - \mathfrak{m}_{a_{n-1}})}\right],$$

which completes the proof.

Coming to the actual sample paths $\mathfrak{m}_a$ $(a \geq 0)$, it is clear from

10) $$E_0(e^{-\alpha \mathfrak{m}_a}) = e^{-\sqrt{2\alpha}\, a} = e^{-a \int_0^{+\infty} (1 - e^{-\alpha l}) \frac{dl}{\sqrt{2\pi l^3}}} \qquad a \geq 0$$

that

11) $$\mathfrak{m}_a = \int_0^{+\infty} l\, \mathfrak{p}([0, a) \times dl),$$

where

12) $\mathfrak{p}(da \times dl) =$ *the number of jumps of magnitude* $l \in dl$ *that* $\mathfrak{m}_b$ $(b \geq 0)$ *suffers for* $b \in da$

is the POISSON measure with mean $da \times \dfrac{dl}{\sqrt{2\pi l^3}}$ (see note 1); in particular, $\mathfrak{m}_b$ is a sum of positive jumps. Note that $[0, a)$ is used in 11) instead of $[0, a]$ as in note 1; this is because $\mathfrak{m}_a$ is *left* continuous.

The maximum function $\mathfrak{t}^-(t) = \max_{s \leq t} x(s)$ $(t \geq 0)$ is the inverse function of the passage time process $\mathfrak{m}$; it is continuous, and, as such, it looks something like the standard CANTOR function (see 2.2 and problem 5 for additional information on this point).

Problem 1 after P. LÉVY [*3*: 211]. Give a proof of the joint law

$$P_0\left[x_t \in da, \max_{s \leq t} x_s \in db\right] = \left(\frac{2}{\pi t^3}\right)^{1/2} (2b - a)\, e^{-(2b-a)^2/2t} da\, db \quad t > 0, 0 \leq b, b \geq a.$$

[Given $b \geq a$,

$$\int_0^{+\infty} e^{-\alpha t} dt\, P_0\left[x_t \leq a, \max_{s \leq t} x_s \geq b\right] = E_0\left[\int_{\mathfrak{m}_b}^{+\infty} e^{-\alpha t} f(x_t)\, dt\right] = E_0(e^{-\alpha \mathfrak{m}_b})\, E_b\left(\int_0^{+\infty} e^{-\alpha t} f\, dt\right) = e^{-\sqrt{2\alpha}\, b} \int_{-\infty}^{a} \frac{e^{-\sqrt{2\alpha}|b-c|}}{\sqrt{2\alpha}} dc,$$

where f is the indicator of $(-\infty, a]$; thus,

$$\int_0^{+\infty} e^{-\alpha t}\, dt\, P_0\Big[x_t \in da, \max_{s \le t} x_s \in db\Big] = 2 e^{-\sqrt{2\alpha}(2b-a)}\, da\, db;$$

now use 2) and 6).]

Problem 2. Use the result of problem 1.4.3 to give a new derivation of the passage time law 5).

[Given $c > 0$, $\min(t : c\, x(t/c^2) = b) = c^2\, \mathfrak{m}_{b/c}$ and $\mathfrak{m}_b$ are identical in law, and, using the homogeneous differential character of $\mathfrak{m}_b$ $(b \geqq 0)$, it follows that $E_0(e^{-\alpha \mathfrak{m}_b})$ is an exponential $e^{-b g(\alpha)}$, where $g(c^2\alpha)/c = g(\alpha)$ $(c > 0)$, *i.e.*, $g(\alpha) = g(1)\sqrt{\alpha}$ $(\alpha > 0)$; now compute the constant $g(1)$.]

Problem 3. Check the arc sine law of P. LÉVY [3: 216]: that, for the greatest root $\mathfrak{z}_- < t$ of $x(s) = 0$,

$$P_0(\mathfrak{z}_- \leqq s) = \frac{1}{\pi}\int_0^{s/t} \frac{dl}{\sqrt{l(1-l)}} = \frac{2}{\pi} \arcsin \sqrt{s/t} \qquad t \geqq s \geqq 0;$$

compute also the distribution of the smallest root $\mathfrak{z}_+ > t$ of $x(s) = 0$ conditional on $\mathfrak{z}_-$.

$$[P_0(\mathfrak{z}_- \leqq s) = P_0[\mathfrak{m}_0(w_s^+) > t - s] = \int_{-\infty}^{+\infty} P_0[x(s) \in da]\, P_a[\mathfrak{m}_0 > t - s]$$

$$= 2\int_0^{+\infty} \frac{e^{-a^2/2s}}{\sqrt{2\pi s}}\, da \int_{t-s}^{+\infty} \frac{a}{\sqrt{2\pi l^3}}\, e^{-a^2/2l}\, dl$$

$$= \frac{1}{\pi}\int_0^{s/t} \frac{dl}{\sqrt{l(1-l)}}.$$

By a similar computation, $P_0[\mathfrak{z}_+ \in db \mid \mathfrak{z}_- = a] = \sqrt{(t-a)/(b-a)^3}\, db/2\, (b > t)$.]

Problem 4. Use the formula $(2\pi i)^{-1} \int_{1-i\infty}^{1+i\infty} e^{t\alpha} e^{-\sqrt{2\alpha}}\, d\alpha$ for the inverse LAPLACE transform to check $e^{-\sqrt{2\alpha}} = \int_0^{+\infty} e^{-\alpha t} \frac{e^{-1/2t}}{\sqrt{2\pi t^3}}\, dt$.

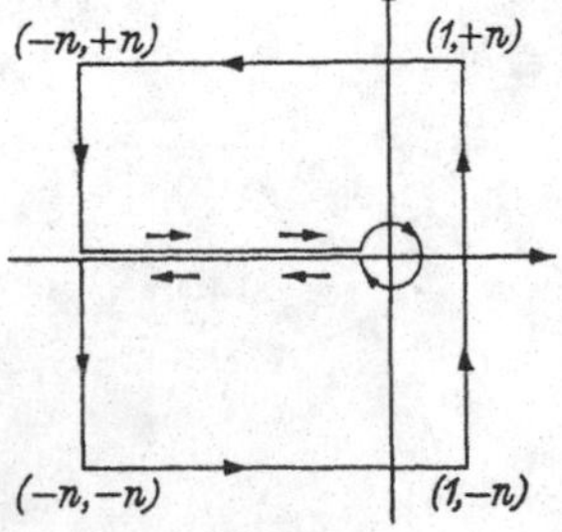

Diagram 1

[Working on the branch $\sqrt{\alpha} = \sqrt{|\alpha|}\, e^{i\varphi/2}$ $(\alpha = |\alpha|\, e^{i\varphi})$, if Γ is the closed curve pictured in the diagram, then $(2\pi i)^{-1}\int_{\Gamma} e^{t\alpha}\, e^{-\sqrt{2\alpha}}\, d\alpha = 0$, and, letting $n \uparrow +\infty$, it results that

$$(2\pi i)^{-1}\int_{1-i\infty}^{1+i\infty} e^{t\alpha}\, e^{-\sqrt{2\alpha}}\, d\alpha$$
$$= (2\pi i)^{-1}\int_{-\infty}^{0} e^{t\alpha + i\sqrt{2|\alpha|}}\, d\alpha + (2\pi i)^{-1}\int_{0}^{-\infty} e^{t\alpha - i\sqrt{2|\alpha|}}\, d\alpha$$
$$= \frac{1}{\pi}\int_{0}^{+\infty} e^{-t\alpha}\sin\sqrt{2\alpha}\, d\alpha = \frac{\sqrt{2}}{\pi t}\int_{0}^{+\infty} e^{-t\alpha^2}\cos\sqrt{2}\,\alpha\, d\alpha = \frac{e^{-1/2t}}{\sqrt{2\pi t^3}}.\Big]$$

Problem 5. The set $\mathfrak{Z}$ of roots of $x(t) = 0$ is a topological CANTOR set (closed, uncountable, of topological dimension 0, having no isolated point)[1]; its LEBESGUE measure is 0.

[$\mathfrak{Z}$ is closed because the BROWNian sample path is continuous, of LEBESGUE measure 0 because

$$E_0\,[\text{measure}(\mathfrak{Z})] = \int_0^{+\infty} dt\, P_0[x(t) = 0] = 0,$$

and of topological dimension 0 for the same reason. Because $t + \mathfrak{m}_0(w_t^+) \in \uparrow$ as a function of t,

$$\lim_{t\downarrow 0} E_0\left[e^{-\alpha(t+\mathfrak{m}_0(w_t^+))}\right] = \lim_{t\downarrow 0} E_0\left[e^{-\sqrt{2\alpha}|x(t)|}\right] = 1$$

implies that $P_0\left[\lim_{t\downarrow 0}(t + \mathfrak{m}_0(w_t^+)) = 0\right] = 1$, and, seeing as $t + \mathfrak{m}_0(w_t^+) \in \mathfrak{Z}$ if it is $< +\infty$, $\mathfrak{Z}$ is infinite. Define $\mathfrak{m} = t_1 + \mathfrak{m}_0(w_{t_1}^+)$. Given $t_2 > t_1$,

$$P_0[x(s) \textit{ has just one root between } t_1 \textit{ and } t_2]$$
$$\leq P_0\left[\mathfrak{m} < t_2,\ \lim_{\varepsilon\downarrow 0}(\varepsilon + \mathfrak{m}_0(w_{\varepsilon+\mathfrak{m}}^+)) > 0\right] \leq P_0\left[\lim_{\varepsilon\downarrow 0}\mathfrak{m}_0(w_\varepsilon^+) > 0\right] = 0,$$

i.e., $\mathfrak{Z}$ has no isolated point; it follows from this that $\mathfrak{Z}$ is uncountable, and the proof is complete.]

Problem 6. Given $a < \xi < b$,

$$E_\xi\left[e^{-\alpha\mathfrak{m}_b},\ \mathfrak{m}_b < \mathfrak{m}_a\right] = \frac{\sinh\sqrt{2\alpha}(\xi - a)}{\sinh\sqrt{2\alpha}(b - a)},$$

$$E_\xi\left[e^{-\alpha\mathfrak{m}_a},\ \mathfrak{m}_a < \mathfrak{m}_b\right] = \frac{\sinh\sqrt{2\alpha}(b - \xi)}{\sinh\sqrt{2\alpha}(b - a)},$$

$$P_\xi[\mathfrak{m}_b < \mathfrak{m}_a] = \frac{\xi - a}{b - a}, \qquad P_\xi[\mathfrak{m}_a < \mathfrak{m}_b] = \frac{b - \xi}{b - a},$$

$$E_\xi\left[e^{-\alpha\,\mathfrak{m}_a\wedge\mathfrak{m}_b}\right] = \frac{\cosh\sqrt{2\alpha}\, d}{\cosh\sqrt{2\alpha}(b - a)/2},$$

---

[1] See ALEKSANDROV-HOPF [*1*: 45].

and

$$E_\xi[e^{\alpha \mathfrak{m}_a \wedge \mathfrak{m}_b}] = \frac{\cos\sqrt{2\alpha}\, d}{\cos\sqrt{2\alpha}\,(b-a)/2} \qquad \alpha < \pi^2/2(b-a)^2,$$

where d is the distance between ξ and the midpoint of $[a, b]$.

[Given $a < \xi < b$, if $\mathfrak{m} = \mathfrak{m}_a \wedge \mathfrak{m}_b$, then

$$e^{-\sqrt{2\alpha}(\xi-a)}$$
$$= E_\xi[e^{-\alpha \mathfrak{m}_a}] = E_\xi[e^{-\alpha\mathfrak{m}}, \mathfrak{m}_a < \mathfrak{m}_b] + E_\xi[e^{-\alpha\mathfrak{m}}, \mathfrak{m}_b < \mathfrak{m}_a]\, e^{-\sqrt{2\alpha}(b-a)}$$

and

$$e^{-\sqrt{2\alpha}(b-\xi)}$$
$$= E_\xi[e^{-\alpha \mathfrak{m}_b}] = E_\xi[e^{-\alpha\mathfrak{m}}, \mathfrak{m}_a < \mathfrak{m}_b]\, e^{-\sqrt{2\alpha}(b-a)} + E_\xi[e^{-\alpha\mathfrak{m}}, \mathfrak{m}_b < \mathfrak{m}_a];$$

now solve for $E_\xi[e^{-\alpha\mathfrak{m}}, \mathfrak{m}_a < \mathfrak{m}_b]$ and $E_\xi[e^{-\alpha\mathfrak{m}}, \mathfrak{m}_b < \mathfrak{m}_a]$.]

Problem 7. Prove

$$P_a(x_t \in db, \mathfrak{m}_0 > t) = \frac{1}{\sqrt{2\pi t}}\left[e^{-\frac{(b-a)^2}{2t}} - e^{-\frac{(b+a)^2}{2t}}\right] db \qquad a\, b > 0.$$

[Use problem 1.]

Problem 8. Show that, for the standard BROWNian motion,

$$P_a[x(t) \in db, t < \mathfrak{m}_0 \wedge \mathfrak{m}_1] = \sum_{|n| \geq 0} (-)^n g(t, a, b_n)\, db \qquad t > 0, \quad 0 < a, \quad b < 1,$$

where $g(t, a, b)$ is the usual GAUSSian kernel and $b_{2n} = b + 2n$, $b_{2n-1} = -b + 2n$.

[Given a BOREL subset B of $[0, 1]$, if $B_n = (b_n : b \in B)$, then

$$P_a[x(t) \in B, t < \mathfrak{m}] \qquad 0 < a < 1, \qquad \mathfrak{m} = \mathfrak{m}_0 \wedge \mathfrak{m}_1$$
$$= \sum_n (-)^n P_a[x(t) \in B_n, t < \mathfrak{m}]$$
$$= \sum_n (-)^n P_a[x(t) \in B_n] - \sum_n (-)^n P_a[x(t) \in B_n, t \geq \mathfrak{m}];$$

now use the symmetry of the standard BROWNian motion to show that

$$\sum_n (-)^n P_a[x(t) \in B_n, t \geq \mathfrak{m}]$$
$$= \int_0^t P_a[\mathfrak{m}_0 < \mathfrak{m}_1, \mathfrak{m}_0 \in ds] \sum_{|n| \geq 0} (-)^n P_0[x(t-s) \in B_n]$$
$$+ \int_0^t P_a[\mathfrak{m}_1 < \mathfrak{m}_0, \mathfrak{m}_1 \in ds] \sum_{|n| \geq 0} (-)^n P_1[x(t-s) \in B_n]$$
$$= 0.$$

$P_a[x(t)\in db,\ t<\mathfrak{m}]$ can also be expressed as $g^\bullet(t,a,b)\,db$, $g^\bullet(t,a,b)$ being the fundamental solution

$$2\sum_{n=1}^{\infty} e^{-n^2\pi^2 t/2}\sin n\pi a\sin n\pi b$$

of

$$\frac{\partial u}{\partial t}=\mathfrak{G}u \qquad \mathfrak{G}=\frac{1}{2}\frac{d^2}{db^2}$$

$$u(t,0+)=u(t,1-)=0,$$

(see 4.11) and this can be transformed into $\sum_{|n|\geq 0}(-)^n g(t,a,b_n)$ with the aid of one of JACOBI's theta function identities.]

Note 1. Homogeneous differential processes with increasing paths. A stochastic process with sample paths $p(t)$ ($t\geq 0$, $p(0)=0$) is said to be *differential* if its increments $p[t_1,t_2)=p(t_2)-p(t_1)$ over disjoint intervals $[t_1,t_2)$ are independent, *homogeneous* if the law of $p[t_1+s,\ t_2+s)$ is independent of $s\,(\geq 0)$, and *increasing* if $p(t_1)\leq p(t_2)$ $(t_1\leq t_2)$.

P. LÉVY [*1*: 173–180] showed that if, in addition, $p(t+)=p(t)$ $(t\geq 0)$, then

1) $$E[e^{-\alpha p(t)}]=e^{-t\psi(\alpha)} \qquad \alpha>0,$$

where

2) $$\psi(\alpha)=m\alpha+\int_{0+}^{+\infty}[1-e^{-\alpha l}]\,n(dl)$$

$$m\geq 0,\ n(dl)\geq 0,\ \int_{0+}^{+\infty}[1-e^{-l}]\,n(dl)<+\infty;$$

$n(dl)$ is the so-called LÉVY measure of the process.

Corresponding to this formula, LÉVY [*1*: 173–180][1] decomposed the sample path into *a linear part plus an integral of differential Poisson processes*:

3) $$p(t)=mt+\int_{0+}^{+\infty} l\,\mathfrak{p}([0,t]\times dl) \qquad t\geq 0$$

$\mathfrak{p}(dt\times dl)$ *being the Poisson measure with mean* $dt\times n(dl)$, *i.e.*, $\mathfrak{p}$ *is Poisson distributed*:

4) $$P[\mathfrak{p}(B)=n]=\frac{\beta^n}{n!}e^{-\beta} \quad n\geq 0,\ \beta=\int_B dt\,n(dl),\ B\in\mathsf{B}([0,+\infty)\times(0,+\infty)),^*$$

differential in the sense that the masses $\mathfrak{p}(B)$ attached to disjoint $B\in\mathsf{B}[0,+\infty)\times(0,+\infty)$ are independent, and *additive* in the sense that $\mathfrak{p}\left(\bigcup_{n\geq 1}B_n\right)=\sum_{n\geq 1}\mathfrak{p}(B_n)$ for disjoint B_1, B_2, *etc.* $\in\mathsf{B}([0,+\infty)\times(0,+\infty))$.

[1] See also K. ITÔ [*1*].

* $P[\mathfrak{p}(B)=+\infty]=1$ if $\beta=+\infty$.

$\mathfrak{p}([t_1, t_2) \times [l_1, l_2))$ is just *the number of jumps of* $p(t)$; $t_1 \leqq t < t_2$ of magnitude $l_1 \leqq l < l_2$, as is clear from 3).

Here is the proof of 1) and 2).

Given $\alpha > 0$, the differential character of p implies that $e(t) = E[e^{-\alpha p(t)}]$ is a solution of

5) $$e(t) = e(t-s)\,e(s) \quad t \geqq s, \quad 0 < e \leqq 1,$$

proving

6) $$e(t) = e^{-t\psi(\alpha)} \qquad \alpha \geqq 0, \quad t \geqq 0, 0 \leqq \psi < +\infty.$$

Because

7) $$\psi'(\alpha)\,e^{-t\psi(\alpha)} = -t^{-1}\frac{\partial}{\partial\alpha}e^{-t\psi} = t^{-1}E[p(t)\,e^{-\alpha p(t)}],*$$

8) $$\psi'(\alpha) = \lim_{t \downarrow 0} t^{-1}\int_0^{+\infty} l\,e^{-\alpha l}\,P[p(t) \in dl]$$

is the LAPLACE transform of a non-negative BOREL measure $m(dl)$ on $[0, +\infty)$, and using $\psi(0) = 0$, it is seen that

9) $$\psi(\alpha) = \int_0^\alpha \psi'(\beta)\,d\beta = \int_0^\alpha d\beta \int_0^{+\infty} e^{-\beta l}\,m(dl) = \alpha\,m(0) + \int_{0+}^{+\infty}[1 - e^{-\alpha l}]\,l^{-1}\,m(dl);$$

in other words,

10) $$\psi(\alpha) = \alpha\,m + \int_{0+}^{+\infty}[1 - e^{-\alpha l}]\,n(dl)$$

with $m = m(0)$, $dn = l^{-1}\,dm\,(l > 0)$, and $\int_{0+}^{+\infty}(1 - e^{-l})\,dn \leqq \psi(1) < +\infty$ as desired.

Coming to 3), if $\mathfrak{p}(dt \times dl)$ is the POISSON measure with mean $dt \times n(dl)$, then

11) $$p(t) = t\,m + \int_{0+}^{+\infty} l\,\mathfrak{p}([0, t] \times dl)$$

is differential and homogeneous; in addition,

12) $$\begin{aligned} E[e^{-\alpha p(t)}] &= e^{-\alpha t m}\,E\left[e^{-\alpha\int_{0+}^{+\infty} l\mathfrak{p}([0,t]\times dl)}\right] \\ &= e^{-\alpha t m}\lim_{n \uparrow +\infty} E\left[e^{-\alpha\int_{0+}^{+\infty} 2^{-n}[2^n l]\mathfrak{p}([0,t]\times dl)}\right] \\ &= e^{-\alpha t m}\lim_{n \uparrow +\infty} e^{-t\int_{0+}^{+\infty}\left(1 - e^{-\alpha 2^{-n}[2^n l]}\right)n(dl)} \\ &= e^{-t[m\alpha + \int_{0+}^{+\infty}(1 - e^{-\alpha l})\,n(dl)]}, \end{aligned}$$

* $p\,e^{-\alpha p}$ is a bounded function of $p \geqq 0$ if $\alpha > 0$.

and it is immediate from this that 11) and 3) are identical in law as stated.

Call $p(t): t \geqq 0$ *one-sided stable* if to each $\alpha > 0$ corresponds a $\beta = \beta(\alpha) > 0$ such that $\alpha\, p(t/\beta): t \geqq 0$ is identical in law to $p(t): t \geqq 0$; in this case,

13) $$e^{-t\beta(\alpha\gamma)\psi(1)} = E[e^{-p(\beta(\alpha\gamma)t)}] = E[e^{-\alpha\gamma p(t)}] = E[e^{-\alpha p(t\beta(\gamma))}] = E[e^{-p(t\beta(\alpha)\beta(\gamma))}] = e^{-t\beta(\alpha)\beta(\gamma)\psi(1)},$$

i.e.,

14) $$\beta(\alpha\gamma) = \beta(\alpha)\,\beta(\gamma),$$

and, if $\psi(1) > 0$ ($p \equiv 0$ in the opposite case), then $\psi(\alpha)$ is a constant multiple $c = \psi(1)$ of $\beta = \alpha^{\varepsilon}$ with $\varepsilon > 0$ because $\psi \in \uparrow$ and $\varepsilon \leqq 1$ because $\psi' = m + \int^{+\infty} l\, e^{-\alpha l}\, dn \in \downarrow$. ε is the *exponent* of the stable process; the constant is its *rate*. $p = ct$ if $\varepsilon = 1$; $p = mt + \int_{0+}^{+\infty} l\, p([0,t] \times dl)$ and $n(dl) = \frac{c\,\varepsilon\, dl}{\Gamma(1-\varepsilon)\, l^{1+\varepsilon}}$ if $\varepsilon < 1$.*

1.8. Kolmogorov's test and the law of the iterated logarithm

Given a standard BROWNian motion starting at 0 and a positive function $h \in C(0, 1]$, the event $(w: x(t) < h(t),\ t \downarrow 0) \in \mathsf{B}_{0+}$, and an application of BLUMENTHAL's 01 law implies that $P_0[x(t) < h(t),\ t \downarrow 0] = 0$ or 1; h is said to belong to the *upper class* if this probability is 1 and to the *lower class* otherwise.

A. HINČIN's celebrated *local law of the iterated logarithm* [*1*: 72–75] states that $h(t) = (1+\varepsilon)\sqrt{2t \lg_2 \frac{1}{t}}$** *belongs to the upper class if* $\varepsilon > 0$ *and to the lower class if* $\varepsilon < 0$; *i.e.*,

1) $$P_0\left[\overline{\lim_{t\downarrow 0}} \frac{x(t)}{\sqrt{2t \lg_2 \frac{1}{t}}} = 1\right] = 1,$$

as will be proved below.

Kolmogorov's test states that *if* $h \in \uparrow$ *and if* $t^{-1/2} h \in \downarrow$ *for small* $t > 0$, *then h belongs to the upper or to the lower class according as* $\int_{0+} t^{-3/2} h\, e^{-h^2/2t}\, dt$ *converges or diverges.*

---

* See P. LÉVY [*1*: 198–204] and S. BOCHNER [*1*] for additional information on stable processes.

** $\lg_2 = \lg(\lg)$.

1) is a special case of this; see problem 2 for additional examples.

KOLMOGOROV's proof is unpublished, but I. PETROVSKI [1] gave a proof based on PERRON's ideas about the DIRICHLET problem (see also P. ERDÖS [1] and W. FELLER [2]).

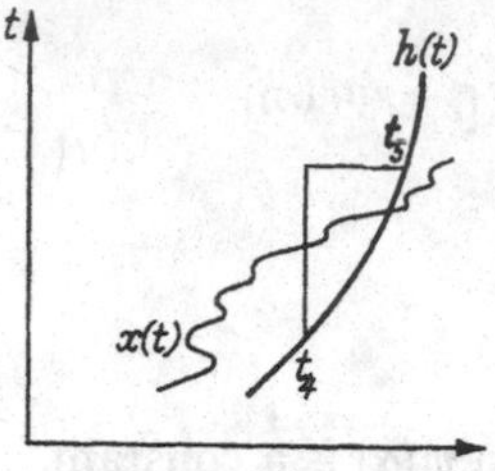

Diagram 1

We will use the passage time distribution

2) $$P_0(\mathfrak{m}_a \leqq t) = \int_0^t \frac{a}{\sqrt{2\pi s^3}} e^{-a^2/2s} ds \quad a \geqq 0$$

[see 1.8.2)] to establish the cheap half of KOLMOGOROV's test: that *if* $h \in \uparrow$ *and* $t^{-1/2} h \in \downarrow$ *for small* t, *then the convergence of the integral* $\int_{0+} t^{-3/2} h e^{-h^2/2t} dt$ *implies* $P_0[x(t) < h(t), t \downarrow 0] = 1$; the other half will be proved in 4.12 using the elegant method of M. MOTOO [1].

Given $0 < a = t_1 < t_2 < \cdots < t_n = b \leqq 1$ such that $h \in \uparrow$ for $t \leqq b$, it is clear from the diagram that

3) $$\begin{aligned} &P_0[x(t) \geqq h(t) \quad \textit{for some} \quad a \leqq t \leqq b] \\ &\quad \leqq P_0[\mathfrak{m}_{h(a)} \leqq a] + \sum_{m \geqq 2} P_0[t_{m-1} < \mathfrak{m}_{h(t_{m-1})} \leqq t_m] \\ &\quad = \int_0^a \frac{h(a)}{\sqrt{2\pi t^3}} e^{-h(a)^2/2t} dt + \sum_{m \geqq 2} \int_{t_{m-1}}^{t_m} \frac{h(t_{m-1})}{\sqrt{2\pi t^3}} e^{-h(t_{m-1})^2/2t} dt, \end{aligned}$$

and, as the subdivision becomes dense in $[a, b]$, this goes over into

4) $$\begin{aligned} &P_0[x(t) \geqq h(t) \quad \textit{for some} \quad a \leqq t \leqq b] \\ &\quad \leqq \int_0^{a/h(a)^2} \frac{e^{-1/2t} dt}{\sqrt{2\pi t^3}} + \int_a^b \frac{h(t)}{\sqrt{2\pi t^3}} e^{-h(t)^2/2t} dt. \end{aligned}$$

Because $t^{-1/2} h \in \downarrow$ for small t, the convergence of $\int_{0+} t^{-3/2} h e^{-h^2/2t} dt$ implies that $\lim_{t \downarrow 0} t^{-1/2} h = +\infty$, and letting $a \downarrow 0$ in 4), one finds that

5) $$\begin{aligned} &P_0[x(t) \geqq h(t) \quad \textit{for some} \quad 0 < t \leqq b] \\ &\quad \leqq \int_{0+}^b \frac{h}{\sqrt{2\pi t^3}} e^{-h^2/2t} dt. \end{aligned}$$

But this overestimate decreases to 0 as $b \downarrow 0$, so the proof is complete.

Consider the special case $h(t) = (1 + \varepsilon) \sqrt{2t \lg_2 \frac{1}{t}}$ $(\varepsilon > 0)$; then $h \in \uparrow$ for $t < e^{-1}$, $t^{-1/2} h \in \downarrow$ for $t \leqq 1$, and $\int_{0+} t^{-3/2} h e^{-h^2/2t} dt < +\infty$, proving

6) $$P_0\left[\overline{\lim_{t\downarrow 0}} \frac{x(t)}{\sqrt{2t \lg_2 \frac{1}{t}}} \leq 1\right] = 1;$$

the proof of 1) can be completed as follows.

Given $0 < \varepsilon < 1$, and ignoring the exceptional class of BROWNian paths for which $\overline{\lim_{t\downarrow 0}} x(t) \Big/ \sqrt{2t \lg_2 \frac{1}{t}} > 1$, if

$$A_n: x(\varepsilon^n) < (1 - 3\sqrt{\varepsilon}) \sqrt{2\varepsilon^n \lg_2 \varepsilon^{-n}}$$

as $n \uparrow +\infty$, then

$$B_n: x(\varepsilon^n) - x(\varepsilon^{n+1}) < (1 - 3\sqrt{\varepsilon}) \sqrt{2\varepsilon^n \lg_2 \varepsilon^{-n}} + \frac{3}{2} \sqrt{2\varepsilon^{n+1} \lg_2 \varepsilon^{-n-1}}$$
$$< (1 - \sqrt{\varepsilon}) \sqrt{2\varepsilon^n \lg_2 \varepsilon^{-n}}$$

as $n \uparrow +\infty$.

But

7) $$1 - P_0(B_n) = \int\limits_{\theta\sqrt{2\lg_2 \varepsilon^{-n}}} \frac{e^{-b^2/2} \, db}{\sqrt{2\pi}} \sim \text{constant} \frac{n^{-\theta^2}}{\sqrt{\lg n}}$$

$$n \uparrow +\infty, \ \theta = \frac{1 - \sqrt{\varepsilon}}{\sqrt{1 - \varepsilon}} < 1,$$

and, using the independence of the events B_n $(n \geq 1)$ and the divergence of the sum $\sum_{n\geq 2} \frac{n^{-\theta^2}}{\sqrt{\lg n}}$, it is found that

8) $$P_0\left[\bigcup_{n\geq 1} \bigcap_{m\geq n} A_m\right] \leq P_0\left[\bigcup_{n\geq 1} \bigcap_{m\geq n} B_m\right] = \lim_{n\uparrow +\infty} P_0\left[\bigcap_{m\geq n} B_m\right] = 0,$$

proving

9) $$1 = P_0[x(\varepsilon^n) \geq (1 - 3\sqrt{\varepsilon}) \sqrt{2\varepsilon^n \lg_2 \varepsilon^{-n}} \ i.o., \ n \uparrow +\infty]$$
$$\leq P_0\left[\overline{\lim_{t\downarrow 0}} \frac{x(t)}{\sqrt{2t \lg_2 \frac{1}{t}}} \geq 1 - 3\sqrt{\varepsilon}\right].$$

1) is now immediate on letting $\varepsilon \downarrow 0$.

Problem 1. Prove that if h belongs to the upper class then

$$\lim_{t\downarrow 0} t^{-1/2} h = +\infty.$$

$$\left[\lim_{t\downarrow 0} \int\limits_{-\infty}^{t^{-1/2}h} \frac{e^{-b^2/2} \, db}{\sqrt{2\pi}} = \lim_{t\downarrow 0} P_0[x(t) < h(t)] = 1.\right]$$

Problem 2 after P. ERDÖS [1]. Use KOLMOGOROV's test to show that

$$h(t) = \sqrt{2t\left[\lg_2\frac{1}{t} + \frac{3}{2}\lg_3\frac{1}{t} + \lg_4\frac{1}{t} + \cdots + \lg_{n-1}\frac{1}{t} + (1+\varepsilon)\lg_n\frac{1}{t}\right]}^{\,*}$$

belongs to the upper or to the lower class according as $\varepsilon > 0$ or $\varepsilon \leqq 0$; show also that

$$h(t) = \sqrt{2t\left[\lg_2\frac{1}{t} + \frac{3}{2}\lg_3\frac{1}{t} + \sum_{n\geqq 4}\lg_n^+\frac{1}{t}\right]}^{\,**}$$

belongs to the lower class.

Problem 3. Use KOLMOGOROV's test to prove that if $h \in C[1, +\infty)$ is positive and if $t^{-1} h \in \downarrow$ and $t^{-1/2} h \in \uparrow$ for large t, then

$$P_0[x(t) < h(t), t \uparrow +\infty] = 0 \textit{ or } 1 \textit{ according as } \int^{+\infty} t^{-3/2} h\, e^{-h^2/2t}\, dt \gtreqless +\infty.$$

Hinčin's global law of the iterated logarithm:

$$P_0\left[\varlimsup_{t\uparrow+\infty} \frac{x(t)}{\sqrt{2t\lg_2 t}} = 1\right] = 1$$

is a special case of problem 3; see A. HINČIN [1].

[Use the fact that $[t\,x(1/t) : t > 0, P_0]$ is a standard BROWNian motion starting at 0; see problem 1.4.3.]

1.9. P. Lévy's Hölder condition

K. L. CHUNG, P. ERDÖS, and T. SIRAO [1] showed that if $h \in C(0, 1]$ is positive and if $h \in \uparrow$ and $t^{-1/2} h \in \downarrow$ for small t, then

1) $$P_0\left[\max_{\substack{t_2 - t_1 = \varepsilon \\ 0 \leqq t_1 < t_2 \leqq 1}} |x(t_2) - x(t_1)| < h(\varepsilon), \varepsilon \downarrow 0\right] = 0 \textit{ or } 1 \textit{ according as}$$

$$\int_{0+} t^{-7/2} h^3 e^{-h^2/2t}\, dt \gtreqless +\infty;$$

for example if $h(t) = (1+\delta)\sqrt{2t\lg\frac{1}{t}}$, then $h \in \uparrow$ for $t \leqq e^{-1}$, $t^{-1/2} h \in \downarrow$ for $t \leqq 1$, and

2) $$\int_{0+} t^{-7/2} h^3 e^{-h^2/2t}\, dt = 2\sqrt{2}(1+\delta)^3 \int_{0+} |\lg t|^{3/2}\, t^{(1+\delta)^2 - 2}\, dt \gtreqless +\infty \quad \delta \lesseqgtr 0,$$

establishing P. LÉVY's HÖLDER condition:

3) $$P_0\left[\varlimsup_{\substack{t_2 - t_1 = \varepsilon \downarrow 0 \\ 0 \leqq t_1 < t_2 \leqq 1}} \frac{|x(t_2) - x(t_1)|}{\sqrt{2\varepsilon \lg \frac{1}{\varepsilon}}} = 1\right] = 1.$$

---

* $\lg_n = \lg(\lg_{n-1})$ $(n \geqq 3)$.

** $\lg_1^+ t = \lg(t \vee 1)$, $\lg_n^+ t = \lg^+(\lg_{n-1}^+ t)$ $(n \geqq 2)$.

LEVY's proof of 3) [1: 168—172] is a model of its kind and is presented here.

Because $1 - s < e^{-s}$, if $0 < \delta < 1$, then

4) $$P_0\left[\max_{k \leqq 2^n} [x(k2^{-n}) - x((k-1)2^{-n})] \leqq (1-\delta)\sqrt{2 \cdot 2^{-n} \log 2^n}\right]$$
$$= \left[1 - \int_{(1-\delta)\sqrt{2 \lg 2^n}} \frac{e^{-b^2/2}\,db}{\sqrt{2\pi}}\right]^{2^n}$$
$$< \left[1 - \frac{2^{-n(1-\delta)^2}}{n}\right]^{2^n}$$
$$< e^{-2^n 2^{-n(1-\delta)^2} n^{-1}}$$
$$\downarrow 0 \qquad n \uparrow +\infty$$

and, letting $\delta \downarrow 0$, it follows that

5) $$P_0\left[\overline{\lim_{\substack{t_2 - t_1 = \varepsilon \downarrow 0 \\ 0 \leqq t_1 < t_2 \leqq 1}}} \frac{|x(t_2) - x(t_1)|}{\sqrt{2\varepsilon \lg \frac{1}{\varepsilon}}} \geqq 1\right] = 1.$$

Consider, for the proof of

6) $$P_0\left[\overline{\lim_{\substack{t_2 - t_1 = \varepsilon \downarrow 0 \\ 0 \leqq t_1 < t_2 \leqq 1}}} \frac{|x(t_2) - x(t_1)|}{\sqrt{2\varepsilon \lg \frac{1}{\varepsilon}}} \leqq 1\right] = 1,$$

the function $h(\varepsilon) = \sqrt{2\varepsilon \lg \frac{1}{\varepsilon}}$ and a number $0 < \delta < 1$, and using the estimate

7) $$P_0\left[\max_{\substack{j = j_2 - j_1 \leqq 2^{n\delta} \\ 0 \leqq j_1 < j_2 \leqq 2^n}} h(j2^{-n})^{-1} |x(j_2 2^{-n}) - x(j_1 2^{-n})| \geqq 1 + \delta\right]$$
$$\leqq 2^{(1+\delta)n} \times 2 \int_{(1+\delta)\sqrt{2 \lg 2^n}} \frac{e^{-b^2/2}\,db}{\sqrt{2\pi}}$$
$$< 2^{-\delta(1+\delta)n} \qquad n \uparrow +\infty,$$

conclude from the first BOREL-CANTELLI lemma that

8) $$|x(j_2 2^{-n}) - x(j_1 2^{-n})| < (1+\delta)\, h(j2^{-n})$$
$$0 \leqq j_1 < j_2 < 2^n, \qquad j = j_2 - j_1 \leqq 2^{n\delta}, \qquad n \geqq n(w),$$

where, for later convenience, $n(w)$ is chosen so as to have

9a) $$2^{(n+1)\delta - 1} > 2,$$

9b) $$2^{-n(1-\delta)} < e^{-1},$$

and

9c) $$\sum_{m > n} h(2^{-m}) = h(2^{-n}) \sum_{m \geqq 1} \sqrt{2^{-m} \frac{\lg 2^{n+m}}{\lg 2^n}} < \delta\, h(2^{-(n+1)(1-\delta)})$$

for $n \geqq n(w)$.

Given $0 \leqq t_1 < t_2 \leqq 1$ such that $\varepsilon = t_2 - t_1 < 2^{-n(w)(1-\delta)}$, choose $n \geqq n(w)$ so as to have $2^{-(n+1)(1-\delta)} \leqq \varepsilon < 2^{-n(1-\delta)}$ and consider the expansions

10a) $$t_1 = j_1 2^{-n} - 2^{-n_1} - 2^{-n_2} - \textit{etc.} \qquad n < n_1 < n_2 < \textit{etc.}$$

10b) $$t_2 = j_2 2^{-n} + 2^{-m_1} + 2^{-m_2} + \textit{etc.} \qquad n < m_1 < m_2 < \textit{etc.}$$

Because

$$j = j_2 - j_1 \geqq 2^n \varepsilon - 2 \geqq 2^n 2^{-(n+1)(1-\delta)} - 2 = 2^{(n+1)\delta - 1} - 2 > 0$$

and

$$j \leqq 2^n \varepsilon < 2^n 2^{-n(1-\delta)} = 2^{n\delta},$$

8) justifies

11) $$\begin{aligned} &|x(t_2) - x(t_1)| \\ &\leqq |x(j_1 2^{-n}) - x(t_1)| + |x(t_2) - x(j_2 2^{-n})| + |x(j_2 2^{-n}) - x(j_1 2^{-n})| \\ &< 2 \sum_{m>n} (1+\delta)\, h(2^{-m}) + (1+\delta)\, h(j 2^{-n}), \end{aligned}$$

and, seeing as $h \in \uparrow$ for $t \leqq \varepsilon \leqq 2^{-n(1-\delta)} < e^{-1}$ [use 9b)], 9c) implies

12) $$\begin{aligned} &|x(t_2) - x(t_1)| \\ &\quad < 2(1+\delta)\,\delta h(2^{-(n+1)(1-\delta)}) + (1+\delta)\, h(j 2^{-n}) \\ &\quad < (1 + 4\delta)\, h(\varepsilon), \end{aligned}$$

and 6) follows on letting $\delta \downarrow 0$.

Problem. Use the CHUNG-ERDÖS-SIRAO test to prove that if

$$h(t) = \sqrt{2t\left[\lg\frac{1}{t} + \frac{5}{2}\lg_2\frac{1}{t} + \lg_3\frac{1}{t} + \cdots + \lg_{n-1}\frac{1}{t} + (1+\delta)\lg_n\frac{1}{t}\right]},$$

then

$$P_0\left[\max_{\substack{t_2 - t_1 = \varepsilon \\ 0 \leqq t_1 < t_2 \leqq 1}} |x(t_2) - x(t_1)| < h(\varepsilon),\ \varepsilon \downarrow 0\right] = 0 \text{ or } 1 \textit{ according as } \delta \lesseqgtr 0$$

and compute this probability for

$$h(t) = \sqrt{2t\left[\lg\frac{1}{t} + \frac{5}{2}\lg_2\frac{1}{t} + \sum_{n\geqq 3}\lg_n^+\frac{1}{t}\right]}$$

also.

[*ans.* 0]

1.10. Approximating the Brownian motion by a random walk

Given a standard BROWNian path $x(t)$ starting at 0 and an integer $l \geqq 1$, introduce the exit times

1) $$\mathrm{e}_0 = 0,\ \mathrm{e}_n = \min\left(t : |x(t) - x(\mathrm{e}_{n-1})| = \frac{1}{\sqrt{l}}\right) \qquad (n \geqq 1)$$

and let

2) $$s_l(n) = \sqrt{l}\, x(e_n) \qquad (n \geqq 1).$$

Because the BROWNian path starts a fresh at each time e, it is clear that $[s_l(n) : n \geqq 0, P_0]$ is a standard random walk.

F. KNIGHT [*1*] proved that if $x_l(t) : t \leqq 1$ is the polygonal line joining the points

$$t = k/l, \quad a = s_l(k)/\sqrt{l} \qquad (k \leqq l),$$

as in the diagram, then

3) $$P_0\left[\lim_{l\uparrow\infty} \max_{t\leqq 1} |x_l(t) - x(t)| = 0\right] = 1;$$

the proof follows.

Given $\frac{k-1}{l} \leqq t < \frac{k}{l}$ $(k \leqq l)$, an application of P. LÉVY's HÖLDER condition implies that for large l,

4) $$\begin{aligned} &|x_l(t) - x(t)| \\ &\leqq |x_l(t) - x_l(k/l)| + |x_l(k/l) - x(k/l)| + |x(k/l) - x(t)| \\ &\leqq |x(e_k) - x(e_{k-1})| + |x(e_k) - x(k/l)| + |x(k/l) - x(t)| \\ &< l^{-1/3} + \text{constant} \times \sqrt{\delta \lg(1/\delta)} \qquad \delta = \left|e_k - \frac{k}{l}\right|, \end{aligned}$$

provided, for example, that $e_l < 2$, and so, using the BOREL-CANTELLI lemma, it is enough to prove that the *time lag* $\max_{k\leqq l}\left|e_k - \frac{k}{l}\right|$ is small; for example, it would be enough to check that

$$p_l = P_0\left[\max_{k\leqq l}\left|e_k - \frac{k}{l}\right| > l^{-1/5}\right]$$

is the general term of a convergent sum.

But e_k is the sum of k independent copies of e_1, and (see problem 1.7.6)

5a) $$E_0(e_1) = l^{-1}$$

5b) $$E_0(|e_1 - l^{-1}|^2) = \text{constant} \times l^{-2}$$

5c) $$E_0(|e_1 - l^{-1}|^4) = \text{constant} \times l^{-4}.$$

Because of 5a), $e_k - k/l$ is a martingale so that $(e_k - k/l)^4$ is a submartingale, and now DOOB's submartingale inequality (see note 2.5.1) implies the desired bound:

6) $$\begin{aligned} &P_0\left[\max_{k\leqq l}\left|e_k - \frac{k}{l}\right| > l^{-1/5}\right] \\ &\leqq l^{4/5} E_0(|e_l - 1|^4) \\ &< l^{4/5} \times \text{constant } l^{-2} \\ &= \text{constant} \times l^{\,6/5}. \end{aligned}$$

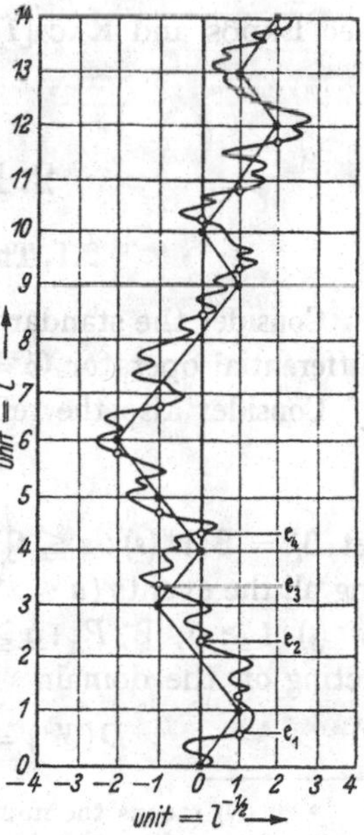

Diagram 1

KNIGHT has used his method to prove the existence (but not the smoothness) of the standard BROWNian local times of 2.8) and to discuss the time substitution of 5.2).

DONSKER's extension [1] of the DE MOIVRE-LAPLACE limit theorem follows at no extra cost: *if f is a Borel function on the space of continuous paths $x(t): t \leqq 1$ and if it is continuous in the topology of uniform convergence at all but a negligible class of Brownian paths, then*

7) $$\lim_{n\uparrow+\infty} P_0\left[a \leqq f\left(\frac{s([n t])}{\sqrt{n}}\right): t \leqq 1) < b\right] = P_0[a \leqq f(x(t): t \leqq 1) < b],$$

where $[n t]$ is the greatest integer $\leqq n t$, $s([n t])/\sqrt{n}: t \leqq 1$ is to be interpreted as the polygonal interpolation between the points $(k/n, s(k)/\sqrt{n}): k \leqq \leqq n$, and P_0 refers to the standard random walk in the first instance and to the standard BROWNian motion in the second.

Problem 1. Use 7) to give a new proof of the result of problem 1.3.1.

Problem 2. Use 7) and the evaluation

$$P_0[\text{measure}(t: x(t) > 0, t \leqq 1) < \theta] = \frac{2}{\pi} \sin^{-1} \sqrt{\theta} \qquad 0 \leqq \theta \leqq 1$$

[see 2.6.13)] to prove the arcsine law for the standard random walk:

$$\lim_{n\uparrow+\infty} P_0\left[\frac{1}{n} \# (k: s(k) > 0, k \leqq n) < \theta\right] = \frac{2}{\pi} \sin^{-1} \sqrt{\theta}*$$

(see ERDÖS and KAC [1], P. LÉVY [2], and 2.6.17)).

2. Brownian local times

2.1. The reflecting Brownian motion

Consider the standard BROWNian motion D and recall the associated differential operator $\mathfrak{G} = D^2/2$ acting on $D(\mathfrak{G}) = C^2(R^1)$ (see 1.4).

Consider also the reflected BROWNian path

1) $$x^+(t) = |x(t)| \qquad t \geqq 0,$$

let $\mathsf{B}_t^+ = \mathsf{B}[x^+(s): s \leqq t]$ be the smallest BOREL subalgebra of B including all the events $(a \leqq x^+(s) < b)$ $(s \leqq t)$, let D^+ be the stochastic motion $[x^+(t): t \geqq 0, \mathsf{B}, P_a: a \geqq 0]$, let $\mathfrak{G}^+$ be the differential operator $\mathfrak{G}$ acting on the domain

2) $$D(\mathfrak{G}^+) = C^2[0, +\infty) \cap (u: u^+(0) = 0),**$$

* $\#(Q)$ means the number of objects in the class Q.

** $u^+(0) = \lim_{\varepsilon\downarrow 0} \varepsilon^{-1}[u(\varepsilon) - u(0)]$.

and introduce the fundamental solution

3) $$g^+(t, a, b) = g(t, -a, b) + g(t, +a, b) \qquad t > 0, a, b \geqq 0$$

of $\partial u / \partial t = \mathfrak{G}^+ u$.

Given $t > s \geqq 0$,

4) $$P_a[x^+(t) \in db \mid \mathsf{B}_s]$$
$$= P_a[x^+(t - s, w_s^+) \in db \mid \mathsf{B}_s]$$
$$= P_{x(s)}[x^-(t - s) \subset db]$$
$$= g^+(t - s, x^+(s), b) \, db \qquad\qquad a, b \geqq 0,$$

and, using the fact that $\mathsf{B}_s^+ \subset \mathsf{B}_s$, it is found that D^+ is a MARKOV process.

D^+ *is the reflecting Brownian motion; it stands in the same relation to $\mathfrak{G}^+$ as the standard Brownian motion does to $\mathfrak{G}$, i.e., the fundamental solution g^+ of $\partial u / \partial t = \mathfrak{G}^+ u$ is the transition density of D^+, just as the fundamental solution $g \,(= Gauss kernel)$ of $\partial u / \partial t = \mathfrak{G} u$ was the transition density of D.*

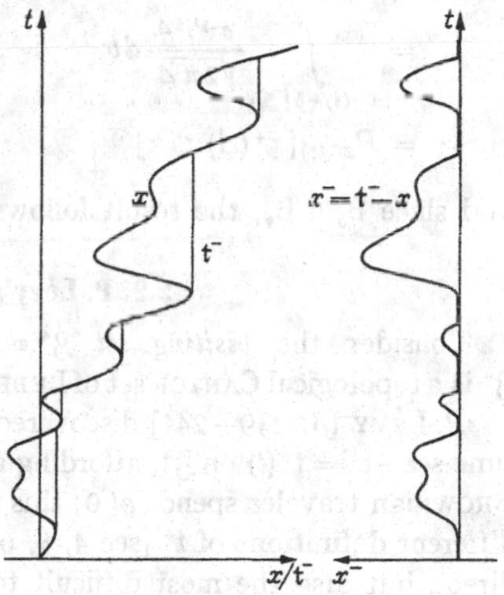

Given a standard BROWNIAN path starting at $x(0) = a \geqq 0$, introduce, next, the new sample paths

5) $x^-(t) = x(t) \qquad t \leqq \mathfrak{m}$
$\qquad = \mathfrak{t}^-(t) - x(t) \quad t > \mathfrak{m},$

Diagram 1

where $\mathfrak{m} = \mathfrak{m}_0 \equiv \min(s : x(s) = 0)$ and $\mathfrak{t}^-(t) = \max(x(s) : \mathfrak{m} \leqq s \leqq t)$, let $\mathsf{B}_t^- = \mathsf{B}[x^-(s) : s \leqq t]$, and let D^- be the motion $[x^-(t) : t \geqq 0, \mathsf{B}, P_a : a \geqq 0]$.

D^- *is identical in law to the reflecting Brownian motion as will now be proved*[1].

Because $x^-(t) = x^+(t)$ up to the MARKOV time $t = \mathfrak{m}$, it is enough to check that $[x^-(t) : t \geqq 0, \mathsf{B}, P_0]$ and $[x^+(t) : t \geqq 0, \mathsf{B}, P_0]$ are identical in law.

But, using

6) $P_0(x(t) \in da, \mathfrak{t}^-(t) \in db)$
$$= (2/\pi t^3)^{1/2}(2b - a) \, e^{-(2b-a)^2/2t} \, da \, db \qquad t > 0, a \leqq b, b \geqq 0,*$$

---

[1] P. LÉVY [3: 234].   * P. LÉVY [3: 211]; see problem 1.7.1.

it is clear that

$$7)\quad P_0[x^-(t) \leqq c \mid \mathsf{B}_s]$$
$$= P_0\Big[\mathfrak{t}^-(s) - x(t) \leqq c, \max_{\theta \leqq t-s} [x(\theta, w_s^+)] - x(t) \leqq c \mid \mathsf{B}_s\Big]$$
$$= P_0\Big[x^-(s) - [x(t) - x(s)] \leqq c, \max_{\theta \leqq t-s} [x(\theta, w_s^+) - x(s)] - - [x(t) - x(s)] \leqq c \mid \mathsf{B}_s\Big]$$
$$= \int_{\substack{x^-(s)-a \leqq c \\ b-a \leqq c \\ -\infty < a \leqq b \\ b \geqq 0}} P_0[x(\Delta) \in da, \mathfrak{t}^-(\Delta) \in db] \qquad \Delta = t - s$$
$$= \left(\int_0^{x^-(s)} db \int_{x^-(s)-c}^{b} da + \int_{x^-(s)}^{+\infty} db \int_{b-c}^{b} da\right) \sqrt{\frac{2}{\pi \Delta^3}}\,(2b - a)\, e^{-(2b-a)^2/2\Delta}$$
$$= \left(2\int_0^{x^-(s)} - \int_{-x^-(s)+c}^{x^-(s)+c} + 2\int_{x^-(s)}^{+\infty} - \int_{x^-(s)+c}^{+\infty}\right) \frac{e^{-b^2/2\Delta}}{\sqrt{2\pi\Delta}}\, db$$
$$= \int_{|x^-(s)+b| \leqq c} \frac{e^{-b^2/2\Delta}}{\sqrt{2\pi\Delta}}\, db$$
$$= P_{x^-(s)}[x^+(\Delta) \leqq c] \qquad t > s,\, c \geqq 0$$

and since $\mathsf{B}_s^- \subset \mathsf{B}_s$, the result follows.

2.2. P. Lévy's local time

Consider the *visiting set* $\mathfrak{Z}^+ \equiv (t : x^+(t) = 0) = \mathfrak{Z} \equiv (t : x(t) = 0)$. $\mathfrak{Z}^+$ is a topological CANTOR set of LEBESGUE measure 0 (see problem 1.7.5).

P. LÉVY [3: 239—241] discovered that it is possible to define a new time scale $\mathfrak{t}^+ = \mathfrak{t}^+(t)$ on $\mathfrak{Z}^+$, affording a non-trivial measure of the time the BROWNian traveler spends *at* 0; this $\mathfrak{t}^+$ is *local time*. P. LÉVY gave several different definitions of $\mathfrak{t}^+$ [see 4, 5, 6, 7)]; of these, the last is the most direct, but also the most difficult to establish.

Consider the sample path $x^-(t) : t \geqq 0$ and recall the decomposition of the passage time $\mathfrak{m}_a = \min(t : x(t) = a)$ $(a \geqq 0)$:

$$1)\qquad \mathfrak{m}_a = \int_{0+}^{+\infty} l\, \mathfrak{p}([0, a) \times dl),$$

where $\mathfrak{p}(da \times dl)$ is the POISSON measure with mean $da \times (2\pi l^3)^{-1/2}\, dl$ (see 1.7).

Because $\mathfrak{p}([0, a) \times [l, +\infty))$ is differential in l, it follows from

$$2)\qquad E_0[\mathfrak{p}([0, a) \times [\varepsilon, +\infty))] = a \times \sqrt{2/\pi\varepsilon}$$

and the strong law of large numbers that

3) $$P\left[\lim_{\varepsilon\downarrow 0}\frac{\mathfrak{p}([0,a)\times[\varepsilon,+\infty))}{\sqrt{2/\pi\varepsilon}}=a,\ a\geqq 0\right]=1,$$

and, substituting $\mathfrak{t}^-(t)=\max_{s\leqq t}x(s)$ for a, since the difference of $\mathfrak{p}([0,\mathfrak{t}^-(t))\times[\varepsilon,+\infty))$ and the number of flat stretches of the graph of $\mathfrak{t}^-(s): s\leqq t$ of length $\geqq\varepsilon$ is not greater than 1, it is seen that

4) $$P_0\left[\lim_{\varepsilon\downarrow 0}\sqrt{\frac{\pi\varepsilon}{2}}\times \textit{the number of flat stretches of } \mathfrak{t}^-(s): s\leqq t \textit{ of length}\geqq\varepsilon=\mathfrak{t}^-(t),\ t\geqq 0\right]=1.^*$$

4) shows that $\mathfrak{t}^-=\mathfrak{t}^-(t)$ is a function of $\mathfrak{Z}^-=(t:x^-(t)=0)$; in addition, it is clear that $\mathfrak{t}^-$ grows on $\mathfrak{Z}^-$ and is flat outside. Because D^+ is identical in law to D^-, there is a corresponding function $\mathfrak{t}^+=\mathfrak{t}^+(t)$ of $\mathfrak{Z}^+=(t:x^+(t)=0)$, growing on $\mathfrak{Z}^+$ and flat outside.

$\mathfrak{t}^+$ is *local time*; $\sqrt{2/\pi}\,\mathfrak{t}^+$ is the *mesure du voisinage* of P. LÉVY [*3*: 228].

Because $\mathfrak{t}^+$ and $\mathfrak{t}^-$ are identical in law and the flat stretches of the graph of $\mathfrak{t}^+$ are just the (open) intervals $\mathfrak{Z}_n\ (n\geqq 1)$ of the complement of $\mathfrak{Z}^+$, it is clear from 4) that

5) $$P_0\left[\lim_{\varepsilon\downarrow 0}\sqrt{\frac{\pi\varepsilon}{2}}\times \textit{the number of intervals } \mathfrak{Z}_n\subset[0,t] \textit{ of length}\geqq\varepsilon=\mathfrak{t}^+(t),\ t\geqq 0\right]=1,^{**}$$

and a little algebra converts 5) into

6) $$P_0\left[\lim_{\varepsilon\downarrow 0}\sqrt{\frac{\pi}{2\varepsilon}}\times \textit{the total length of the intervals } \mathfrak{Z}_n\subset[0,t] \textit{ of length}<\varepsilon=\mathfrak{t}^+(t),\ t\geqq 0\right]=1;^{***}$$

thus $\mathfrak{t}^+$ *is a function of the fine structure of* $\mathfrak{Z}^+$.

P. LÉVY [*3*: 239—241] also proved

7) $$P_0\left[\lim_{\varepsilon\downarrow 0}(2\varepsilon)^{-1}\text{ measure }(s: x_s^+\leqq\varepsilon,\ s\leqq t)=\mathfrak{t}^+(t),\ t\geqq 0\right]=1.$$

Here is his proof.

Given $\mathfrak{Z}^+$, the excursions $\mathfrak{e}_n\equiv[x(t): t\in\mathfrak{Z}_n]$ are independent, and, since $\mathfrak{Z}_n$ enters into the distribution of $\mathfrak{e}_n$ via its length $|\mathfrak{Z}_n|$ alone,[1]

---

* P. LÉVY [*3*: 224 — 225]

** P. LÉVY [*3*: 224 — 225].

*** P. LÉVY [*3*: 224 — 225]. $\sqrt{2\varepsilon/\pi}=E_0\left[\int_{0+}^{\varepsilon} l\,\mathfrak{p}([0,1)\times dl)\right]$ = the expected sum of jumps of magnitude $\leqq\varepsilon$ figuring in $\mathfrak{m}_1$.

[1] P. LÉVY [*3*: 233 — 236]; see also 2.9.

the strong law of large numbers indicates that, for $\varepsilon \downarrow 0$,

8a) $$\text{measure } (s : x_s^+ \leqq \varepsilon, s \leqq t) = \sum_{n \geqq 1} \text{measure } (s : x_s^+ \leqq \varepsilon, s \in \mathfrak{Z}_n \cap [0, t))$$

can be replaced by

8b) $$\sum_{\mathfrak{Z}_n \subset [0,t)} E_0 [\text{measure } (s : x_s^+ \leqq \varepsilon, s \in \mathfrak{Z}_n) | \mathfrak{Z}^+] = \sum_{\mathfrak{Z}_n \subset [0,t)} |\mathfrak{Z}_n| (1 - e^{-2\varepsilon^2/|\mathfrak{Z}_n|})^* = \int_0^{+\infty} l(1 - e^{-2\varepsilon^2/l})\, \mathfrak{p}([0, \mathfrak{t}^+(t)) \times dl),$$

where $\mathfrak{p}(dt \times dl)$ is the POISSON measure for the (one-sided stable) inverse function of $\mathfrak{t}^+$, $\mathfrak{p}([0, \mathfrak{t}^+(t)) \times [l, +\infty))$ being within 1 of the number of flat stretches of $\mathfrak{t}^+(s) : s \leqq t$ of length $\geqq l$. Now use 4) to conclude that

9) $$\int_0^{+\infty} l(1 - e^{-2\varepsilon^2/l})\, \mathfrak{p}([0, \mathfrak{t}^+(t)) \times dl) = \int_0^{+\infty} \mathfrak{p}([0, \mathfrak{t}^+(t)) \times [l, +\infty))\, d[l(1 - e^{-2\varepsilon^2/l})] \sim \int_0^{+\infty} \sqrt{\frac{2}{\pi l}}\, \mathfrak{t}^+(t)\, d[l(1 - e^{-2\varepsilon^2/l})] = 2\varepsilon\, \mathfrak{t}^+(t) \qquad \varepsilon \downarrow 0.$$

The difficult point of this proof is the jump between 8a) and 8b); although the meaning is clear, the complete justification escapes us (see, however, 2.8 for results of G. BOYLAN, F. KNIGHT, D. B. RAY and H. TROTTER that include 7) as a special case).

Problem 1.

$$\mathfrak{t}^+(t_2) > \mathfrak{t}^+(t_1) \qquad (t_1, t_2) \cap \mathfrak{Z}^+ \neq \emptyset, \qquad 0 \leqq t_1 < t_2$$

for almost all BROWNian paths.

[Because $\mathfrak{t}^+$ is identical in law to $\mathfrak{t}^-$, $P_0[\mathfrak{t}^+(t) > 0, t > 0] = 1$, and, if the statement were false, then $\varepsilon \equiv P_0[\mathfrak{t}^+(t_2) = \mathfrak{t}^+(t_1), (t_1, t_2) \cap \mathfrak{Z}^+ \neq \emptyset]$ would be >0 for some $t_1 < t_2$. But $\varepsilon = P_0[\mathfrak{t}^+(t_2 - \mathfrak{m}, w_{\mathfrak{m}}^+) = 0$, $\mathfrak{m} = t_1 + \mathfrak{m}_0(w_{t_1}^+) < t_2] = 0$.]

Problem 2. Prove that for almost all BROWNian paths starting at 0, $x(s) : s \leqq t$ cannot touch the line $l = \mathfrak{t}^-(t)$ more than 2 times for any $t > 0$.

[$x(s) = \mathfrak{t}^-(t)$ for 3 different times $s \leqq t$ implies that $\mathfrak{t}$ is flat on an open interval that meets $\mathfrak{Z}^-$; now use the result of problem 1, and the fact that $\mathfrak{t}^-$ is local time on $\mathfrak{Z}^-$.]

---

* P. LÉVY [3 : 238].

Problem 3. Use the result of problem 1.7.1 to deduce the joint law

$$P_0[x^+(t) \in da, \mathfrak{t}^+(t) \in db] = 2\frac{a+b}{\sqrt{2\pi t^3}} e^{-(b+a)^2/2t} da\, db \quad a, b \geqq 0.$$

Problem 4. Use the fact that $\mathfrak{t}^+$ and $\mathfrak{t}^-$ are identical in law to prove that the HAUSDORFF-BESICOVITCH dimension of $\mathfrak{Z}^+$ is $\geqq 1/2$ for almost all sample paths (see note 2.5.2 for the definition of HAUSDORFF-BESICOVITCH dimension).

[Because $\mathfrak{t}^-$ satisfies the HÖLDER condition

$$\mathfrak{t}^-(t_2) - \mathfrak{t}^-(t_1) < \sqrt{3\varepsilon|\lg\varepsilon|} \qquad 0 \leqq t_1 \leqq t_2 \leqq 1, \quad \varepsilon = t_2 - t_1 \downarrow 0$$

(see 1.9), if $[{}_nt_1, {}_nt_2] : n \geqq 1$ is a closed cover of $\mathfrak{Z}^+ \cap [0, 1]$ and if each ${}_nt_2 - {}_nt_1$ is small enough, then

$$\mathfrak{t}^+(1) \leqq \sum_{n\geqq 1} \mathfrak{t}^+({}_nt_2) - \mathfrak{t}^+({}_nt_1) <$$

$$< \sum_{n\geqq 1} \sqrt{3({}_nt_2 - {}_nt_1)|\lg({}_nt_2 - {}_nt_1)|};$$

now use the definition of dimension number and the fact that $P_0[\mathfrak{t}^+(1) > 0] = 1$.]

A. S. BESICOVITCH and S. J. TAYLOR [*1*] proved $\dim(\mathfrak{Z}^+) \leqq 1/2$ and, in a different place, S. J. TAYLOR [*1*] proved $\dim(\mathfrak{Z}^+) \geqq 1/2$ (see 2.5 for a proof of $\dim(\mathfrak{Z}^+) \leqq 1/2$ and also problem 2.8.2).

2.3. Elastic Brownian motion

Given $\gamma > 0$, we use the reflecting BROWNian local time $\mathfrak{t}^+$ of 2.2 to construct the elastic BROWNian motion associated with the differential operator

$$\mathfrak{G}^\bullet = \frac{1}{2} D^2$$

acting on the domain

$$D(\mathfrak{G}^\bullet) = C^2[0, +\infty) \cap (u : \gamma u(0) = u^+(0)).^*$$

Consider, for this purpose, the enlarged sample space $W \times [0, +\infty)$ of points $(w, \mathfrak{m}_\infty)$ and the corresponding BOREL algebra $\mathsf{B} \times \mathsf{B}[0, +\infty)$, and extend $P_\bullet$ to $\mathsf{B} \times \mathsf{B}[0, +\infty)$ according to the rule

1) $$P_\bullet(\mathfrak{m}_\infty > t \mid \mathsf{B}) = e^{-\gamma \mathfrak{t}^+(t)} \qquad t \geqq 0.$$

Adjoining to $[0, +\infty]$ an additional point ∞, define

2) $$P_\infty[\mathfrak{m}_\infty = 0] = 1$$

and introduce the stochastic process $\mathsf{D}^\bullet$ with sample paths

3) $$x^\bullet(t) = x^+(t) \quad (t < \mathfrak{m}_\infty) \qquad = \infty \quad (t \geqq \mathfrak{m}_\infty).$$

* $u^+(0) = \lim_{\varepsilon \downarrow 0} \varepsilon^{-1}[u(\varepsilon) - u(0)]$.

$\mathsf{D}^\bullet$ is a (simple) MARKOV process as will be proved in a moment; in addition, its transition density $g^\bullet(t, a, b)$:

$$\begin{aligned} g^\bullet(t, a, b)\, db &= P_a[x^\bullet(t) \in db] \\ &= P_a[x^+(t) \in db, t < \mathfrak{m}_\infty] \\ &= E_a[x^+(t) \in db, e^{-\gamma \mathfrak{t}^+(t)}] \\ &= E_a[e^{-\gamma \mathfrak{t}^+(t)} \mid x^+(t) = b]\, g^+(t, a, b)\, db \quad t > 0, \quad a, b \geqq 0 \end{aligned}$$

is the fundamental solution of the heat flow problem

4a) $$\frac{\partial u}{\partial t} = \frac{1}{2} \frac{\partial^2 u}{\partial b^2} \qquad t > 0, \quad b > 0,$$

4b) $$\gamma\, u(t, 0) = u^+(t, 0) \qquad t > 0,$$

i.e., $\mathsf{D}^\bullet$ *stands in the same relation to* $\mathfrak{G}^\bullet$ *as the reflecting Brownian motion* D^+ *does to* $\mathfrak{G}^+$.

$\mathsf{D}^\bullet$ *is the so-called elastic Brownian motion.*

Because $\mathfrak{t}^+$ is flat off $\mathfrak{Z}^+$ and $P_\bullet[\mathfrak{m}_\infty \in dt \mid \mathsf{B}, \mathfrak{m}_\infty > t] = \gamma\, \mathfrak{t}^+(dt)$, $\mathsf{D}^\bullet$ can be described as the reflecting BROWNian motion killed at $b = 0$ at the rate $\gamma\, \mathfrak{t}^+(dt) : dt$.

Beginning with the simple MARKOVian character of $\mathsf{D}^\bullet$, if e $(0 \leqq e \leqq 1)$ is a BOREL function on the space of paths: $t \to x(t) \in Q$ $(t \leqq s$, $Q = [0, +\infty) \cup \infty)$, if $e^+ = e$ evaluated at the path $x^+(t) : t \leqq s$, and if $e^\bullet = e$ evaluated at the path $x^\bullet(t) : t \leqq s$, then for $a, b \geqq 0$ and $s \leqq t$,

5) $$\begin{aligned} E_a[e^\bullet, x^\bullet(t) \in db] &= E_a[e^+, x^+(t) \in db, t < \mathfrak{m}_\infty] \\ &= E_a[e^+, x^+(t) \in db, e^{-\gamma \mathfrak{t}^+(t)}] \\ &= E_a[e^+, x^+(t - s, w_s^+) \in db, e^{-\gamma \mathfrak{t}^+(s)}\, e^{-\gamma \mathfrak{t}^+(t-s, w_s^+)}] \\ &= E_a[e^+ e^{-\gamma \mathfrak{t}^+(s)}\, E_{x^+(s)}[x^+(t - s) \in db, e^{-\gamma \mathfrak{t}^+(t-s)}]] \\ &= E_a[e^+ e^{-\gamma \mathfrak{t}^+(s)}\, P_{x^+(s)}[x^\bullet(t - s) \in db]] \\ &= E_a[e^+, s < \mathfrak{m}_\infty, P_{x^+(s)}[x^\bullet(t - s) \in db]] \\ &= E_a[e^\bullet\, P_{x^\bullet(s)}[x^\bullet(t - s) \in db]],* \end{aligned}$$

and the proof is complete.

As to the identification of $\mathsf{D}^\bullet$ as the elastic BROWNian motion, it is enough to show that

6) $$u = E_\bullet\left[\int_0^{\mathfrak{m}_\infty} e^{-\alpha t} f(x_t)\, dt\right] = E_\bullet\left[\int_0^{+\infty} e^{-\alpha t} e^{-\gamma \mathfrak{t}^+} f(x_t^+)\, dt\right] \quad \alpha > 0, f \in C(R^1)$$

---

* $P_\infty[x^\bullet(t - s) \in db] = 0$.

is a solution of

7a) $$u \in D(\mathfrak{G}^\bullet) = C^2[0, +\infty) \cap (u : \gamma u(0) = u^+(0))$$

7b) $$(\alpha - \mathfrak{G}^\bullet)\, u = f$$

and to invert the LAPLACE transform.

But

8) $$u(a) = E_a\left[\int_0^{\mathfrak{m}_0} e^{-\alpha t} f(x_t^+)\, dt\right] +$$

$$+ E_a\left[e^{-\alpha \mathfrak{m}_0} \int_0^{+\infty} e^{-\alpha t} e^{-\gamma \mathfrak{t}^+(t, w_{\mathfrak{m}}^+)} f[x^+(t, w_{\mathfrak{m}}^+)]\, dt\right] \qquad (\mathfrak{m} = \mathfrak{m}_0)$$

$$= E_a\left[\int_0^{+\infty} e^{-\alpha t} f(x_t^+)\, dt\right] - E_a\left[\int_{\mathfrak{m}_0}^{+\infty} e^{-\alpha t} f(x_t^+)\, dt\right] +$$

$$+ E_a[e^{-\alpha \mathfrak{m}_0}]\, u(0)$$

$$= E_a\left[\int_0^{+\infty} e^{-\alpha t} f(x_t^+)\, dt\right] + E_a[e^{-\alpha \mathfrak{m}_0}]\left(u(0) - E_0\left[\int_0^{+\infty} e^{-\alpha t} f(x_t^+)\, dt\right]\right)$$

$$= \int_0^{+\infty} \frac{e^{-\sqrt{2\alpha}\,|b-a|} + e^{-\sqrt{2\alpha}\,|b+a|}}{\sqrt{2\alpha}} f(b)\, db +$$

$$+ e^{-\sqrt{2\alpha}\,a}\left[u(0) - 2\int_0^{+\infty} \frac{e^{-\sqrt{2\alpha}\,b}}{\sqrt{2\alpha}} f(b)\, db\right],$$

from which it is immediate that

9) $$\alpha u - \frac{1}{2} u'' = f,$$

and, using the fact that $[x^-(t) = \mathfrak{t}^-(t) - x(t) : t \geq 0, P_0]$ is a reflecting BROWNian motion with local time $\mathfrak{t}^-$ to compute $u^+(0)$ from 8), one finds

10) $$-u^+(0) + 2\int_0^{+\infty} e^{-\sqrt{2\alpha}\,b} f(b)\, db$$

$$= \sqrt{2\alpha}\, u(0)$$

$$= \sqrt{2\alpha}\, E_0\left[\int_0^{+\infty} e^{-\alpha t} e^{-\gamma \mathfrak{t}^-(t)} f[\mathfrak{t}^-(t) - x(t)]\, dt\right]$$

$$= \sqrt{2\alpha} \int_0^{+\infty} e^{-\alpha t}\, dt \int_0^{+\infty} e^{-\gamma b}\, db \int_{-\infty}^{b} da\, 2 \cdot \frac{2b - a}{\sqrt{2\pi t^3}}\, e^{-(2b-a)^2/t} f(b - a)$$

$$= \sqrt{2\alpha} \int_0^{+\infty} e^{-\gamma b}\, db \int_{-\infty}^{b} da\, 2 e^{-\sqrt{2\alpha}(2b-a)} f(b - a)$$

$$= \frac{\sqrt{2\alpha}}{\gamma + \sqrt{2\alpha}}\, 2\int_0^{+\infty} e^{-\sqrt{2\alpha}\,b} f(b)\, db,$$

i.e.,

$$\text{11)} \qquad u^+(0) = \frac{2\gamma}{\gamma + \sqrt{2\alpha}} \int_0^{+\infty} e^{-\sqrt{2\alpha}b} f(b)\, db = \gamma\, u(0)$$

as desired.

2.4. $\mathfrak{t}^+$ and down-crossings

Consider the number of times $d_\varepsilon(t)$ that the reflecting Brownian path x^+ crosses down from $\varepsilon > 0$ to 0 before time t and let us check that

$$\text{1)} \qquad P_0\left[\lim_{\varepsilon \downarrow 0} \varepsilon\, d_\varepsilon(t) = \mathfrak{t}^+(t),\, t \geqq 0\right] = 1$$

as P. Lévy conjectured [*6*: 171], supposing for the proof that

$$\text{2)} \qquad \mathfrak{t}^+(t) = \lim_{\varepsilon \downarrow 0} (2\varepsilon)^{-1} \text{ measure } \left(s : x^+(s) < \varepsilon,\, s \leqq t\right),$$

as was mentioned in 2.2 and will be proved in 2.8.

Consider for the proof, the successive returns $\mathfrak{r}_1 < \mathfrak{r}_2 <$ *etc.* of x^+ to the origin via $\varepsilon > 0$, *i.e.*, let $\mathfrak{r}_0 = 0$, $\mathfrak{r}_1 = \mathfrak{m}_\varepsilon + \mathfrak{m}_0(w^+_{\mathfrak{m}_\varepsilon})$, and $\mathfrak{r}_n = \mathfrak{r}_{n-1} + \mathfrak{r}_1(w^+_{\mathfrak{r}_{n-1}})$ $(n > 1)$, and note that for $0 < \delta < \varepsilon$,

$$\text{3)} \qquad d_\delta(\mathfrak{m}_1) + 1 \leqq \sum_{k \leqq d_\varepsilon(\mathfrak{m}_1)+1} d_\delta\left(\mathfrak{r}_1(w^+_{\mathfrak{r}_{k-1}}),\, w^+_{\mathfrak{r}_{k-1}}\right),$$

the summands being independent, identical in law, and independent of $d_\varepsilon(\mathfrak{m}_1)$. Because of

$$\text{4)} \qquad P_0[d_\delta(\mathfrak{r}_1) = n] = \left(1 - \frac{\delta}{\varepsilon}\right)^n \frac{\delta}{\varepsilon} \qquad (n \geqq 0)$$

and the above independence,

$$\begin{aligned} \text{5)} \qquad & E_0[\delta(d_\delta(\mathfrak{m}_1) + 1) \mid d_\varepsilon(\mathfrak{m}_1)] \\ &= \delta(d_\varepsilon(\mathfrak{m}_1) + 1)\, E_0[d_\delta(\mathfrak{r}_1) + 1] \\ &= \delta(d_\varepsilon(\mathfrak{m}_1) + 1)\left[\frac{\varepsilon}{\delta}\left(1 - \frac{\delta}{\varepsilon}\right) + 1\right] \\ &= \varepsilon(d_\varepsilon(\mathfrak{m}_1) + 1) \qquad (\delta < \varepsilon), \end{aligned}$$

i.e., $\left[\varepsilon(d_\varepsilon(\mathfrak{m}_1) + 1) : \varepsilon < 1,\, P_0\right]$ is a positive backwards martingale, and since

$$\text{6)} \qquad E_0[\varepsilon(d_\varepsilon(\mathfrak{m}_1) + 1)] = \varepsilon\left[\sum_{n=0}^{\infty} n(1-\varepsilon)^n \varepsilon + 1\right] = 1,$$

Doob's martingale convergence theorem (see note 2.5.1) tells us that $\lim_{\varepsilon \downarrow 0} \varepsilon\, d_\varepsilon(\mathfrak{m}_1)$ exists.

Next, this limit is identified as $\mathfrak{t}^+(\mathfrak{m}_1)$ with the help of 2).

Using d_n in place of d_ε for $\varepsilon = 2^{-n}$, it is desired to estimate

7) $$\begin{aligned} \Delta_n &= 2^{n-1} \text{ measure } (s : x^+(s) < 2^{-n}, s < \mathfrak{m}_1) - 2^{-n} d_n(\mathfrak{m}_1) \\ &= 2^{n-1} \mathfrak{m}^+_{2^{-n}} {}^* \quad \text{if} \quad d_n(\mathfrak{m}_1) = 0 \\ &= 2^{n-1} \sum_{k \leq m} [\text{measure}(s : x^+ < 2^{-n}, \mathfrak{t}_{k-1} \leq s < \mathfrak{t}_k) - 2^{-2n+1}] \\ &\qquad\qquad \text{if} \quad d_n(\mathfrak{m}_1) = m \geq 1 . \end{aligned}$$

Here the summands are independent, identical in law, and independent of $d_n(\mathfrak{m}_1) = m$, with common mean

8) $$\begin{aligned} &\int_0^{+\infty} ds\, P_0[x^+(s) < 2^{-n}, s < \mathfrak{t}_1] \\ &\qquad = E_0(\mathfrak{m}^+_{2^{-n}}) + \int_0^{+\infty} ds\, P_{2^{-n}}[x(s) < 2^{-n}, s < \mathfrak{m}_0] \\ &\qquad = 2 \cdot 2^{-2n} {}^{**} \end{aligned}$$

and common variance

9) $$\begin{aligned} &E_0[(\text{measure}(s : x^+(s) < 2^{-n}, s < \mathfrak{t}_1) - 2^{-2n+1})^2] \\ &\qquad = \text{constant } \times 2^{-4n} \end{aligned}$$

as a simple application of the scaling $x^+(t) \to 2^{-n} x^+(2^{2n} t)$ verifies, so

10) $$\begin{aligned} &E_0(\Delta_n^2) \\ &= 2^{2n-2} E_0[(\mathfrak{m}^+_{2^{-n}})^2]\, P_0(d_n(\mathfrak{m}_1) = 0) \\ &+ 2^{2n-2} E_0[(\text{measure } (s : x^+(s) < 2^{-n}, s < \mathfrak{t}_1) - 2^{-2n+1})^2] \\ &\qquad\qquad \sum_{m=1}^{\infty} m\, P_0(d_n(\mathfrak{m}_1) = m) \\ &= 2^{2n-2} \frac{4}{3} 2^{-4n} \cdot 2^{-n} {}^{***} + 2^{2n-2} \cdot \text{constant} \times 2^{-4n} \sum_{m=1}^{\infty} (1 - 2^{-n})^m 2^{-n} \\ &< \text{constant} \times 2^{-n}, \end{aligned}$$

and combining this bound with 2) and the BOREL-CANTELLI lemma, it results that $\lim_{\varepsilon \downarrow 0} \varepsilon\, d_\varepsilon(\mathfrak{m}_1) = \mathfrak{t}^+(\mathfrak{m}_1)$.

But now, selecting $t \geq 0$, if $t < \mathfrak{m}_1$, and if $\mathfrak{r} = \min(s : x^+(s) = 0, s \geq t)$, then

11a) $$\mathfrak{t}^+(\mathfrak{m}_1) = \mathfrak{t}^+(t) + 0 \quad \text{or} \quad \mathfrak{t}^+(\mathfrak{m}_1(w^+_{\mathfrak{r}}), w^+_{\mathfrak{r}})$$

11b) $$d_\varepsilon(\mathfrak{m}_1) = d_\varepsilon(t) + 0 \quad \text{or} \quad d_\varepsilon(\mathfrak{m}_1(w^+_{\mathfrak{r}}), w^+_{\mathfrak{r}})$$

---

* $\mathfrak{m}^+_a$ denotes the reflecting BROWNian passage time to a.

** See problem 1.7.6 and 1.7.7. *** See problem 1.7.6.

according as $\mathfrak{r} \geqq \mathfrak{m}_1$ or not, with an error of 1 at most in 11b); thus,

12) $$P_0\left[\lim_{\varepsilon\downarrow 0}\varepsilon\, d_\varepsilon(t) = \mathfrak{t}^+(t) \mid t < \mathfrak{m}_1\right] = 1 \quad (t \geqq 0),$$

and to complete the proof of 1), it is enough to consider d_ε and $\mathfrak{t}^+$ between the successive passage times of x^+ to 1 via 0 and to use the fact that d_ε is an increasing function of time, while $\mathfrak{t}^+$ is both increasing and continuous.

2.5. $\mathfrak{t}^+$ as Hausdorff-Besicovitch $\frac{1}{2}$-dimensional measure[1]

Perhaps the most striking feature of the local time $\mathfrak{t}^+(t)$ is the fact that if $\bigcup_{k\geqq 1} {}_k\mathfrak{Z}_n$ is the covering

1) $${}_k\mathfrak{Z}_n = [(k-1)\,2^{-n}, k\,2^{-n}] \cap \mathfrak{Z}^+ \qquad k \geqq 1$$

of $\mathfrak{Z}^+$ and if $|{}_k\mathfrak{Z}_n|$ is the diameter of ${}_k\mathfrak{Z}_n$, then

2) $$P_0\left[\lim_{n\uparrow+\infty}\sqrt{\frac{\pi}{2}}\sum_{k2^{-n}\leqq t}|{}_k\mathfrak{Z}_n|^{1/2} = \mathfrak{t}^+(t),\ t \geqq 0\right] = 1.$$

$\mathfrak{t}^+$ is therefore similar to the HAUSDORFF-BESICOVITCH $\frac{1}{2}$-dimensional measure $\Lambda^{\frac{1}{2}}$; in particular,

3) $$P_0\left[\Lambda^{\frac{1}{2}}(\mathfrak{Z}\cap[0,t)) \leqq \sqrt{\frac{2}{\pi}}\,\mathfrak{t}^+(t),\ t \geqq 0\right] = 1,$$

proving that the HAUSDORFF-BESICOVITCH dimension number of $\mathfrak{Z}$ is $\leqq \frac{1}{2}$, and, using problem 2.2.4, it results that the dimension number of $\mathfrak{Z}$ is not smaller than and therefore $= \frac{1}{2}$.

Given $t_2 \geqq t_1 \geqq 0$, if $\mathfrak{z}_1\,(\mathfrak{z}_2) = \min\,(\max)\,\mathfrak{Z}\cap[t_1, t_2]$,* then, as will be proved below,

4a) $$E_0[\mathfrak{t}^+(t_2) - \mathfrak{t}^+(t_1) \mid x_s^+ : s \leqq t_1, \mathfrak{z}_1, \mathfrak{z}_2, x_s^+ : s \geqq t_2]$$
$$= 0 \qquad \mathfrak{z}_1 = \mathfrak{z}_2$$
$$= \sqrt{\frac{\pi}{2}(\mathfrak{z}_2 - \mathfrak{z}_1)} \qquad \mathfrak{z}_1 < \mathfrak{z}_2.$$

Because

$$\mathsf{C}_n = \mathsf{B}[x^+(l\,2^{-n}), \min {}_l\mathfrak{Z}_n, \max {}_l\mathfrak{Z}_n, l \geqq 1]$$

is a subalgebra of

$$\mathsf{B}[x_s^+ : s \leqq (k-1)\,2^{-n}, \min {}_k\mathfrak{Z}_n, \max {}_k\mathfrak{Z}_n, x_s^+ : s \geqq k\,2^{-n}],$$

4a) implies

4b) $$E_0[\mathfrak{t}^+(k\,2^{-n}) - \mathfrak{t}^+((k-1)\,2^{-n}) \mid \mathsf{C}_n] = \sqrt{\frac{\pi}{2}|{}_k\mathfrak{Z}_n|} \qquad k \geqq 1,$$

---

[1] See note 2 for the definition of the $\frac{1}{2}$-dimensional HAUSDORFF measure.

* $\mathfrak{z}_1 = \mathfrak{z}_2 = 0$ if $\mathfrak{Z}\cap[t_1, t_2] = \emptyset$.

and, noting that C_n increases to $\mathsf{B}^+ = \mathsf{B}[x_t^+ : t \geq 0]$ as $n \uparrow +\infty$, the martingale convergence theorem of J. DOOB (see note 1) implies that, for each separate $t \geq 0$,

$$5) \quad \sqrt{\frac{\pi}{2}} \sum_{k2^{-n} \leq t} |{}_k\mathfrak{z}_n|^{\frac{1}{2}} = \sum_{k2^{-n} \leq t} E_0[\mathfrak{t}^+(k2^{-n}) - \mathfrak{t}^+((k-1)2^{-n}) \mid \mathsf{C}_n]$$
$$= E_0[\mathfrak{t}^+([2^n t]\, 2^{-n}) \mid \mathsf{C}_n]$$
$$\to E_0[\mathfrak{t}^+(t) \mid \mathsf{B}^+] = \mathfrak{t}^+(t) \qquad n \uparrow +\infty.$$

2) is immediate from this because $\sum_{k2^{-n} \leq t} |{}_k\mathfrak{z}_n|^{\frac{1}{2}}$ and $\mathfrak{t}^+(t)$ are increasing functions of t and $\mathfrak{t}^+$ is continuous.

Coming to the proof of 4a), since $\mathfrak{z}_1$, $\mathfrak{z}_2$, and $\mathfrak{t}^+(t_2) - \mathfrak{t}^+(t_1) = \mathfrak{t}^+(t_2 - t_1, w_{t_1}^+)$ are all measurable $\mathsf{B}[x^+(s) : t_1 \leq s \leq t_2]$, the conditional expectation figuring in 4a) is a BOREL function of $t_1 < \mathfrak{z}_1 = s_1 < \mathfrak{z}_2 = s_2 < t_2$, $x^+(t_1) = a$, and $x^+(t_2) = b$ alone:

$$6) \quad e = E_0[\mathfrak{t}^+(t_2) - \mathfrak{t}^+(t_1) \mid x^+(t_1) = a,\ \mathfrak{z}_1 = s_1,\ \mathfrak{z}_2 = s_2,\ x^+(t_2) = b].$$

Actually, $\mathfrak{t}^+(t_2) - \mathfrak{t}^+(t_1) = \mathfrak{t}^+(t_2 - \mathfrak{z}_1, w_{\mathfrak{z}_1}^+)$, and since the BROWNian path begins afresh at time $\mathfrak{z}_1 = t_1 + \mathfrak{m}_0(w_{t_1}^+)$,

$$7) \quad e = E_0[\mathfrak{t}^+(t_2 - s_1) \mid \mathfrak{z}_3 = s_2 - s_1,\ x^+(t_2 - s_1) = b]$$

with $\mathfrak{z}_3 = \max(s : s < t_2 - s_1,\ x^+(s) = 0)$.

Now consider the conditional BROWNian motion $\mathsf{D}_{ab} = [x^+(s) : s \leq t,\ P_{ab}(B) = P_a(B \mid x^+(t) = b)]$ for $a, b \geq 0$ and $t > 0$ and run it backwards, *i.e.*, consider the backward motion $\mathsf{D}_{ab}^* = [x^+(t-s) : s \leq t,\ P_{ab}]$. A brief examination of the conditional law permits the identification of D_{ab}^* with D_{ba}. Because $\mathfrak{t}^+(t)$ is the same for the backward and forward motions, 7) becomes

$$8) \quad e = E_b[\mathfrak{t}^+(t_2 - s_1) \mid \mathfrak{m}_0 = t_2 - s_2,\ x^+(t_2 - s_1) = 0].$$

Using the same idea that led from 6) to 7), 8) can be simplified to

$$9) \quad e = E_0[\mathfrak{t}^+(s) \mid x^+(s) = 0] \qquad s = s_2 - s_1 = \mathfrak{z}_2 - \mathfrak{z}_1,$$

and this can now be evaluated by means of the joint law of $\mathfrak{t}^+$ and x^+ (see problem 2.2.3):

$$10) \quad e = \frac{2\displaystyle\int_0^\infty \frac{b^2 e^{-b^2/2s}}{\sqrt{2\pi s^3}}\, db}{2/\sqrt{2\pi s}} = \sqrt{\frac{\pi}{2}}\,(\mathfrak{z}_2 - \mathfrak{z}_1),$$

completing the proof apart from small technical gaps which the reader is invited to fill.[1]

Note 1. Submartingales. Consider a non-negative submartingale $e_n: n \geqq 1$ relative to the BOREL algebras $\mathsf{B}_1 \subset \mathsf{B}_2 \subset \mathsf{B}_3$ *etc.*, *i.e.*, let $0 \leqq e_m$ be measurable B_m with $E(e_m) < +\infty$ and $E(e_m | \mathsf{B}_n) \geqq e_n$ $(m \geqq n)$, and let us prove *the martingale convergence theorem* of J. DOOB [*1*: 324]: that *if* $\gamma = \sup_{n \geqq 1} E(e_n) < +\infty$, then

$$P\left[e_\infty = \lim_{n \uparrow \infty} e_n \textit{ exists}\right] = 1 \quad \textit{and} \quad E(e_\infty) \leqq \gamma.$$

Consider, for the proof, $l > 0$, $n_1 = \min(n : e_n = 0)$, $n_2 = \min(n : n > n_1, e_n \geqq l)$, $n_3 = \min(n : n > n_2, e_n = 0)$, *etc.*, let $\#$ denote the number of times $e_m : m < n$ crosses *up* from 0 to l (*i.e.*, let $\#$ be the number of integers n_2, n_4, n_6, *etc.* $< n$), and let $e = e_1 + \sum_{i=2}^{n} t_i (e_i - e_{i-1})$, where $t_i = 1$ if i lies in $[0, n_1] \cup (n_2, n_3] \cup (n_4, n_5]$ *etc.* and $t_i = 0$ otherwise. $(t_i = 1)$ is the event that $e_n : n < i$ has not completed its current down-crossing; this is measurable B_{i-1} so that

$$1) \qquad E(e) = E(e_1) + \sum_{i=2}^{n} E\left(t_i E(e_i - e_{i-1} \mid \mathsf{B}_{i-1})\right) \geqq 0,$$

and, using

$$
\begin{aligned}
2) \qquad e &= e_n & n &< n_1 \\
&= e_{n_1} = 0 \leqq e_n & n_1 &\leqq n < n_2 \\
&= e_{n_1} + e_n - e_{n_2} \leqq e_n - l & n_2 &\leqq n < n_3 \\
&= e_{n_1} + e_{n_3} - e_{n_2} = -e_{n_2} \leqq e_n - l & n_3 &\leqq n < n_4 \\
&= e_{n_1} + e_{n_3} - e_{n_2} + e_n - e_{n_4} \leqq e_n - 2l & n_4 &\leqq n < n_5 \\
&\quad \textit{etc.} \\
&\leqq e_n - \# l,
\end{aligned}
$$

one finds

$$3) \qquad l\, E(\#) \leqq E(e_n).$$

Given $l_2 > l_1 \geqq 0$ and using 3) with the non-negative submartingale $(e_n - l_1) \vee 0 : n \geqq 1$ in place of $e_n : n \geqq 1$ and $l_2 - l_1$ in place of l, it is found that *if* $\#$ *is the number of up-crossings of* $e_m : m \leqq n$ *from* l_1 *to* l_2, *then*

$$4) \qquad E(\#) \leqq (l_2 - l_1)^{-1} E[(e_n - l_1) \vee 0] \leqq (l_2 - l_1)^{-1} E(e_n).$$

Consider the case $\gamma = \sup_{n \geqq 1} E(e_n) \left(= \lim_{n \uparrow \infty} E(e_n)\right) < +\infty$ and make $n \uparrow \infty$ in 4); then one sees that $e_n : n \geqq 1$ crosses up from l_1 to l_2 just

[1] G. LOUCHARD [private communication] suggested this use of the backward motion improving on the clumsy computation of the original proof.

a finite number of times, and this implies the existence of $e_\infty = \lim_{n\uparrow\infty} e_n$. $E(e_\infty) \leqq \gamma$ is clear.

A special case is the result of P. LÉVY [*1*: 128—130]: that if e is non-negative and summable, if $\mathsf{B}_1 \subset \mathsf{B}_2 \subset \mathsf{B}_3$ *etc.*, and if B is the smallest BOREL algebra including $\bigcup_{n\geqq 1} \mathsf{B}_n$, then

$$5) \qquad \lim_{n\uparrow\infty} E(e \mid \mathsf{B}_n) = E(e \mid \mathsf{B}).$$

Here, $e_n = E(e \mid \mathsf{B}_n) : n \geqq 1$ is a non-negative martingale so that the existence of $e_\infty = \lim_{n\uparrow\infty} e_n$ is clear, and since

$$\begin{aligned}
&\sup_{n\geqq 1} E(e_n, e_n \geqq l) \\
&\quad = \sup_{n\geqq 1} E[E(e \mid \mathsf{B}_n), e_n \geqq l] \\
&\quad = \sup_{n\geqq 1} E[e, e_n \geqq l] \\
&\quad \leqq E\Big[e, \sup_{n\geqq 1} e_n \geqq l\Big] \downarrow 0 \qquad l \uparrow \infty,
\end{aligned}$$

$$6) \qquad E(e, B) = \lim_{m\uparrow\infty} E(e_m, B) = E(e_\infty, B) \qquad B \in \mathsf{B}_n, n \geqq 1,$$

completing the identification of e_∞ and $E(e \mid \mathsf{B})$.

Doob's submartingale inequality [*1*: 314]:

$$7) \qquad P\Big(\max_{k\leqq n} e_k \geqq l\Big) \leqq l^{-1} E(e_n)$$

is contained in 3) as a special case, but here is a direct proof.

Consider the event A_k that $e_i < l$ $(i < k)$ and $e_k \geqq l$; then $A_k \in \mathsf{B}_k$, $A_i \cap A_k = \varnothing$ $(i < k)$, and

$$\begin{aligned}
E(e_n) &\geqq E\Big(e_n, \bigcup_{k=1}^{n} A_k\Big) = \sum_{k=1}^{n} E(e_n, A_k) \\
&= \sum_{k=1}^{n} E(E(e_n \mid \mathsf{B}_k), A_k) \geqq \sum_{k=1}^{n} E(e_k, A_k) \\
&\geqq l \sum_{k=1}^{n} P(A_k) = l\, P\Big(\max_{k\leqq n} e_k \geqq l\Big).
\end{aligned}$$

Note 2. Hausdorff measure and dimension (see F. HAUSDORFF [*1*]). Given a continuous function $h(t) \in \uparrow$ $(t \geqq 0)$ with $h(0) = 0$, the *outer Hausdorff h-measure* of a linear BOREL set B is defined to be

$$\Lambda(B) = \lim_{\varepsilon\downarrow 0} \inf \sum_{n\geqq 1} h(l_n),$$

the infimum being taken over all coverings $\bigcup_{n\geqq 1} A_n$ of B by means of closed intervals of length $l_n = |A_n| < \varepsilon$ $(n \geqq 1)$.

Λ is a CARATHÉODORY outer measure, but, in general, R^1 is *not* the sum of a countable number of BOREL sets of finite measure.

Given $0 \leqq \alpha \leqq 1$, the $\wedge$ associated with $h(t) = t^\alpha$ is the *Hausdorff-Besicovitch α-dimensional measure* $\wedge^\alpha$; the numbers

$$\dim^-(B) = \begin{cases} 0 & \text{if } \wedge^0(B) < +\infty \\ \sup(\alpha : \wedge^\alpha(B) = +\infty) & \text{if } \wedge^0(B) = +\infty \end{cases}$$

$$\dim^+(B) = \begin{cases} 1 & \text{if } \wedge^1(B) = +\infty \\ \inf(\alpha : \wedge^\alpha(B) < +\infty) & \text{if } \wedge^1(B) < +\infty \end{cases}$$

coincide, and their common value $\dim(B)$ is the *Hausdorff-Besicovitch dimension number* of B. $\wedge^1$ is the classical BOREL measure.

$\wedge^{\lg 2/\lg 3}$ is of particular interest: the dimension number of the standard CANTOR set

$$K \equiv [0, 1] - \left(\frac{1}{3}, \frac{2}{3}\right) - \left(\frac{1}{9}, \frac{2}{9}\right) - \left(\frac{7}{9}, \frac{8}{9}\right) - etc.$$

is $\lg 2/\lg 3$ and $\wedge^{\lg 2/\lg 3}(K \cap [0, t])$ is the CANTOR function.

2.6. Kac's formula for Brownian functionals

Before the deep properties of BROWNian local times can be proved, a new method for computing probabilities is needed; this is embodied in a formula of KAC [*1*].

Given a piecewise continuous function $k \geqq 0$, let $\mathfrak{k}(t) = \mathfrak{k}(t, w)$ be the additive BROWNian functional $\int_0^t k[x(s)]\, ds$, so named because of the addition rule

1) $$\mathfrak{k}(t) = \mathfrak{k}(s) + \mathfrak{k}(t - s, w_s^+) \qquad (t \geqq s),$$

and let $\mathfrak{G}^\bullet$ be the differential operator

2a) $$(\mathfrak{G}^\bullet u)(a) = \tfrac{1}{2} u''(a\pm) - k(a\pm)\, u(a)$$

applied to the class of functions $u \in C(R^1)$ such that

2b) $$\tfrac{1}{2}[u'(b) - u'(a)] - \int_a^b k\, u\, d\xi = \int_a^b u^\bullet\, d\xi \qquad (a < b)$$

for some $u^\bullet$ $(\equiv \mathfrak{G}^\bullet u)$ belonging to $C(R^1)$.

KAC's formula states that for $\alpha > 0$ and $f \in C(R^1)$,

3) $$u = E_\bullet\left[\int_0^{+\infty} e^{-\alpha t}\, e^{-\mathfrak{k}(t)} f(x)\, dt\right]^*$$

is *the* bounded solution of

4) $$(\alpha - \mathfrak{G}^\bullet)\, u = f.$$

KAC's formula suggests that heat flow with cooling is the same as BROWNian motion with annihilation of particles, indeed, if $k(b)\, ds$ is the chance that a particle is killed in time ds having come safe to

* $f(x)$ stands for $f[x(t)]$ in such an integral.

the place $b = x(s)$, then its chance of surviving up to time $t \geqq 0$ conditional on the path $x(s): s \leqq t$ is

5) $$\bigcap_{s \leqq t} [1 - k(x(s))\, ds]^* = e^{-\mathfrak{k}(t)};$$

hence its chance of coming safe from a into db after a time t is

6) $$g^\bullet(t, a, b)\, db = E_a[e^{-\mathfrak{k}}, x(t) \in db] = E_a[e^{-\mathfrak{k}(t)} \mid x(t) = b]\, g(t, a, b)\, db,$$

while according to KAC's formula, $g^\bullet$ is the fundamental solution of

7a) $$\partial u / \partial t = \mathfrak{G}^\bullet u$$

7b) $$u(0+, \cdot) = f,$$

which describes heat flow with cooling.

As to the proof, if u is as in 3) and if G_α is the standard BROWNian GREEN operator of 1.4

8) $$(G_\alpha f)(a) = \int_{R^1} \frac{e^{-\sqrt{2\alpha}|b-a|}}{\sqrt{2\alpha}} f(b)\, db,$$

then

9) $$\begin{aligned} G_\alpha f - u &= E_\bullet\left[\int_0^{+\infty} e^{-\alpha t}(1 - e^{-\mathfrak{k}}) f(x)\, dt\right] \\ &= E_\bullet\left[\int_0^{+\infty} e^{-\alpha t} \int_0^t e^{-\mathfrak{k}(t-s, w_s^+)}\, \mathfrak{k}(ds)\, f(x)\, dt\right] \\ &= E_\bullet\left[\int_0^{+\infty} \mathfrak{k}(ds) \int_s^{+\infty} e^{-\alpha t} e^{-\mathfrak{k}(t-s, w_s^+)} f(x)\, dt\right] \\ &= E_\bullet\left[\int_0^{+\infty} e^{-\alpha s}\, \mathfrak{k}(ds) \int_0^{+\infty} e^{-\alpha t} e^{-\mathfrak{k}(t, w_s^+)} f[x(t, w_s^+)]\, dt\right] \\ &= E_\bullet\left[\int_0^{+\infty} e^{-\alpha s}\, k\, u(x)\, ds\right] \\ &= G_\alpha k u, \end{aligned}$$

the interchange of temporal integrations in lines 3 and 4 being justified by

10) $$\int_0^t e^{-\mathfrak{k}(t-s, w_s^+)}\, d\mathfrak{k} \leqq 1.$$

Using $(\alpha - \mathfrak{G})\, G_\alpha = 1$, it follows at once that u solves 4); also, the boundedness of u is clear, and since a solution of $\mathfrak{G}^\bullet v = \alpha v$ is convex

---

* $\bigcap_{s \leqq t}$ is meant to suggest a continuous product.

or concave according as it is positive or negative and so unbounded or $\equiv 0$, u is *the* bounded solution of 4).

GREEN functions provide the classical method for solving problems such as 4); the facts are indicated below in a form suitable for applications.

Given $\alpha > 0$, $\mathfrak{G}^{\bullet} g = \alpha g$ *has two independent solutions* $0 < g_1 \in \uparrow$ *and* $0 < g_2 \in \downarrow$, *their Wronskian* $B = g_1' g_2 - g_1 g_2'$ *is constant* (>0), *the Green function*

11) $$G(a, b) = G(b, a) = B^{-1} g_1(a) g_2(b) \qquad (a \leqq b)$$

satisfies $2\alpha \int G\, db \leqq 1$, *and* $u = 2 \int G f\, db$ *is the bounded solution* of 4).

Here is a rapid proof.

Choose $f(a) = 0$ $(a \leqq b)$ and $f(a) > 0$ $(a > b)$ in 3) and let $g_{ab} = E_a[e^{-\alpha \mathfrak{m}_b} e^{-\mathfrak{f}(\mathfrak{m}_b)}]$ $(a < b)$. u satisfies

12) $$u(a) = E_a\left[\int_{\mathfrak{m}_b}^{+\infty} e^{-\alpha t} e^{-\mathfrak{f}} f(x)\, dt\right] = g_{ab}\, u(b) \qquad (a < b),$$

and since $u(b) > 0$ and $(\alpha - \mathfrak{G}^{\bullet}) u = f = 0$ $(a < b)$, it follows from the multiplication rule

13) $$\begin{aligned} & g_{a\xi}\, g_{\xi b} & (a < \xi < b) \\ &= E_a[e^{-\alpha \mathfrak{m}} e^{-\mathfrak{f}(\mathfrak{m})}]\, E_\xi[e^{-\alpha \mathfrak{m}_b} e^{-\mathfrak{f}(\mathfrak{m}_b)}] & (\mathfrak{m} = \mathfrak{m}_\xi) \\ &= E_a\left(e^{-\alpha[\mathfrak{m} + \mathfrak{m}_b(w_{\mathfrak{m}}^+)]}\, e^{-[\mathfrak{f}(\mathfrak{m}) + \mathfrak{f}(\mathfrak{m}_b(w_{\mathfrak{m}}^+), w_{\mathfrak{m}}^+)]}\right) & \\ &= E_a[e^{-\alpha \mathfrak{m}_b} e^{-\mathfrak{f}(\mathfrak{m}_b)}] = g_{ab} & \end{aligned}$$

that

14) $$g_1(a) = \lim_{b \uparrow +\infty} g_{ab}/g_{0b}$$

is a positive increasing solution of $\mathfrak{G}^{\bullet} g = \alpha g$; the same method can be used to make positive decreasing solution. B satisfies

15) $$B\,\Big|_a^b = \int_a^b [g_2\, \mathfrak{G}^{\bullet} g_1 - g_1\, \mathfrak{G}^{\bullet} g_2]\, d\xi = 0 \quad (a < b),$$

so it is constant (>0),

16) $$\begin{aligned} & 2\alpha B \int G\, db \\ &= 2\alpha \left[g_2(a) \int_{-\infty}^{a} g_1\, db + g_1(a) \int_a^{+\infty} g_2\, db\right] \\ &= 2 \left[g_2(a) \int_{-\infty}^{a} \mathfrak{G}^{\bullet} g_1\, db + g_1(a) \int_a^{+\infty} \mathfrak{G}^{\bullet} g_2\, db\right] \\ &\leqq g_2(a) \int_{-\infty}^{a} g_1''\, db + g_1(a) \int_a^{+\infty} g_2''\, db \\ &\leqq B \end{aligned}$$

as desired, and now a routine differentiation shows that $u = 2\int G f\, db$ is a bounded solution of 4), completing the proof.

KAC [1] used his formula to prove P. LÉVY's arcsine law for the standard BROWNian motion [2: 323]:

17) $$P_0\,[\text{measure}\,(s : x(s) \geqq 0,\, s \leqq t) \leqq \theta] = \frac{1}{\pi}\int_0^{\theta/t} \frac{ds}{\sqrt{s(1-s)}} = \frac{2}{\pi}\sin^{-1}\sqrt{\theta/t} \qquad \theta \leqq t.^*$$

Use the additive functional $\mathfrak{t} =$ measure $(s: x(s) \geqq 0, s \leqq t)$ based on the indicator function e_0 of $[0, +\infty]$ and let $\mathfrak{G}^\bullet = \mathfrak{G} - \beta\, e_0$. KAC tells us that

18) $$u = E_\bullet\left(\int_0^{+\infty} e^{-\alpha t}\, e^{-\beta \mathfrak{t}}\, dt\right) \qquad \alpha, \beta > 0$$

is the bounded solution of

19) $$(\alpha - \mathfrak{G}^\bullet)\, u = 1,$$

so

20) $$u(0) = \int_0^{+\infty} e^{-\alpha t}\, dt \int_0^{+\infty} e^{-\beta s}\, P_0[\mathfrak{t}(t) \in ds] = 2B^{-1}\left[g_2(0)\int_{-\infty}^{0} g_1\, db + g_1(0)\int_0^{+\infty} g_2\, db\right],$$

and solving, it is found that

21 a) $$g_1(b) = e^{\sqrt{2\alpha}\, b} \qquad (b \leqq 0)$$

21 b) $$g_2(b) = e^{-\sqrt{2(\alpha+\beta)}\, b} \qquad (b \geqq 0)$$

21 c) $$B = \sqrt{2\alpha} + \sqrt{2(\alpha+\beta)}$$

21 d) $$u(0) = 1/\sqrt{\alpha(\alpha+\beta)}.$$

17) is now immediate from the formulas

22 a) $$\int_0^{+\infty} \frac{e^{-\gamma t}}{\sqrt{\pi t}}\, dt = \frac{1}{\sqrt{\gamma}}$$

and

22 b) $$\frac{1}{\sqrt{\alpha(\alpha+\beta)}} = \int_0^{+\infty} e^{-\alpha t}\, \frac{1}{\pi} \int_0^t \frac{e^{-\beta s}}{\sqrt{(t-s)s}}\, ds\, dt.$$

A beautiful application of KAC's formula is to the *WKB* approximation.

---

* See problem 1.7.3 for a different arcsine law also due to P. LÉVY.

Given k tending to $+\infty$ at the ends of the line, the spectrum of $\mathfrak{G}^\bullet$ acting on $L^2(R^1, db)$ is a simple series of eigenvalues $0 > \gamma_1 > \gamma_2 > etc.$ $\downarrow -\infty$, and under a simple additional condition the *WKB* approximation states that

23) $$\sum_{\gamma_n > \gamma} 1 \sim (2\pi)^{-1} \text{ area } \left[(a, b) : \frac{a^2}{2} + k(b) < |\gamma|\right]$$

as $\gamma \downarrow -\infty$. KAC's idea is to express the trace $\sum_{n \geqq 1} e^{\gamma_n t}$ as

24) $$\int_{R^1} E_0\left[e^{-\int_0^t k[x(s)+l]\,ds} \,\middle|\, x(t) = 0\right] dl \times (2\pi t)^{-1/2}$$

and then to check that 24) $\sim (2\pi t)^{-1/2} \int_{R_1} e^{-tk(l)}\,dl$ as $t \to 0$ (see KAC [1] for a complete proof and additional applications).

Problem 1. Check the evaluation

$$\int_0^{+\infty} e^{-\alpha t} E_0[e^{-\beta \,\text{measure}(s:x(s) \geqq 0,\, s \leqq t)} \mid x(t) = 0] \frac{dt}{\sqrt{2\pi t}}$$

$$= \frac{2}{\sqrt{2\alpha} + \sqrt{2(\alpha+\beta)}} \qquad \alpha, \beta > 0$$

and use it to prove P. LÉVY's law [2: 323]

$$P_0[a \leqq t^{-1} \text{ measure } (s : x(s) \geqq 0, s \leqq t) < b \mid x(t) = 0] = b - a$$

$$t > 0,\ 0 \leqq a < b \leqq 1.$$

[$2[\sqrt{2\alpha} + \sqrt{2(\alpha+\beta)}]^{-1}$ is the GREEN function of 19) above evaluated at $a = b = 0$; now invert the LAPLACE transforms.]

Problem 2. $E_0[e^{-\beta \,\text{measure}(s:x(s) \geqq 0,\, s \leqq \mathfrak{m}_1)}] = \dfrac{1}{\cosh \sqrt{2\beta}} \qquad (\beta > 0).$

[Use the additive functionals $\mathfrak{e}_0$ and $\mathfrak{e}_1$ based on the indicator functions e_0 and e_1 of $[0, +\infty)$ and $[1, +\infty]$, let

$$u = E_\bullet\left[\int_0^{+\infty} e^{-\alpha t - \beta \mathfrak{e}_0 - \gamma \mathfrak{e}_1} e_0(x(t))\,dt\right] \quad (\alpha, \beta, \gamma > 0),$$

and note that

$$\lim_{\alpha \downarrow 0} \lim_{\gamma \uparrow +\infty} u(0) = E_0\left[\int_0^{\mathfrak{m}_1} e^{-\beta \mathfrak{e}_0} e_0(x(t))\,dt\right] = \beta^{-1}[1 - E_0(e^{-\beta \mathfrak{e}_0(\mathfrak{m}_1)})].$$

KAC tells us that $u(0)$ can be expressed as

$$2g_1(0)\,B^{-1}\int_0^{+\infty} g_2\,db \quad (\mathfrak{G}^\bullet g = \alpha g,\ \mathfrak{G}^\bullet = \mathfrak{G} - \beta e_0 - \gamma e_1);$$

the formulas are

$$g_1(b) = e^{\sqrt{2\alpha}\, b} \qquad (b \leqq 0),$$

$$g_2(b) = e^{-\sqrt{2(\sigma+\gamma)}\, b} \qquad (b \geqq 1, \sigma \equiv \alpha + \beta)$$

$$= \cosh \sqrt{2\sigma}\,(1-b) + \sqrt{\frac{\sigma+\gamma}{\sigma}} \sinh \sqrt{2\sigma}\,(1-b) \quad (0 \leqq b < 1),$$

$$B = [\sqrt{2\alpha} + \sqrt{2(\sigma+\gamma)}] \cosh \sqrt{2\sigma} + [\sqrt{2\alpha(1+\gamma/\sigma)} + \sqrt{2\sigma}] \sinh \sqrt{2\sigma},$$

$$B\, u(0)/2 = \frac{\sinh \sqrt{2\sigma}}{\sqrt{2\sigma}} + \sqrt{\frac{\sigma+\gamma}{\sigma}} \frac{\cosh \sqrt{2\sigma} - 1}{\sqrt{2\sigma}} + \frac{e^{-\sqrt{2(\sigma+\gamma)}}}{\sqrt{2(\sigma+\gamma)}}.]$$

2.7. Bessel processes

We prepare the so-called BESSEL process with a view to a better understanding of BROWNian local times (see 2.8) and BROWNian excursions (see 2.9 and 2.10).

Consider the standard d-dimensional BROWNian motion $\mathsf{D} = [\mathfrak{x}(t) = (x_1(t), x_2(t), \ldots, x_d(t)) : t \geqq 0, P_\bullet]$, in which the coordinates of the sample path are independent standard 1-dimensional BROWNian motions and $P_\bullet(B)$ is the probability of B as a function of the starting point $\mathfrak{x}(0)$ of the d-dimensional BROWNian path.

Given $t > s$, if $\mathsf{B}_s = \mathsf{B}[\mathfrak{x}(\theta) : \theta \leqq s]$, then

1)
$$P_\bullet[\mathfrak{x}(t) \in d\mathfrak{b} \mid \mathsf{B}_s] = P_{\mathfrak{a}}[\mathfrak{x}(t-s) \in d\mathfrak{b}] \qquad \mathfrak{a} = \mathfrak{x}(s)$$

$$= \frac{e^{-|\mathfrak{b}-\mathfrak{a}|^2/2(t-s)}}{(2\pi(t-s))^{d/2}}\, d\mathfrak{b}$$

$$\mathfrak{a} = (a_1, a_2, \ldots, a_d), \qquad \mathfrak{b} = (b_1, b_2, \ldots, b_d),$$

$$|\mathfrak{b} - \mathfrak{a}| = \sqrt{(b_1-a_1)^2 + (b_2-a_2)^2 + \cdots + (b_d-a_d)^2},$$

$$d\mathfrak{b} = db_1\, db_2 \ldots db_d$$

because of the independence and MARKOVian character of the coordinate (1-dimensional) BROWNian motions. Because the d-dimensional GAUSS kernel $g(t, \mathfrak{a}, \mathfrak{b}) = (2\pi t)^{-(d/2)} e^{-|\mathfrak{b}-\mathfrak{a}|^2/2t}$ is the fundamental solution of

$$\frac{\partial u}{\partial t} = \mathfrak{G} u \qquad \mathfrak{G} = \frac{1}{2}\Delta = \frac{1}{2}\left(\frac{\partial^2}{\partial b_1^2} + \frac{\partial^2}{\partial b_2^2} + \cdots + \frac{\partial^2}{\partial b_d^2}\right),$$

D *is a simple Markov process, standing in the same relation to* $\mathfrak{G} = \Delta/2$ *as in the 1-dimensional case* (see 1.4; for additional information about the d-dimensional BROWNian motion, see chapter 7).

Coming to the BESSEL process $D^+ = [r(t) = |\mathfrak{x}(t)| : t \geqq 0, P_\bullet]$ (= the reflecting BROWNian motion if $d = 1$), it is clear from 1) that for $t > s$,

2)
$$\begin{aligned} P_\bullet[r(t) < b \mid B_s] &= P_{\mathfrak{a}}[r(t-s) < b] \qquad \mathfrak{a} = \mathfrak{x}(s) \\ &= \int_{|\mathfrak{b}|<b} \frac{e^{-|\mathfrak{b}-\mathfrak{a}|^2/2(t-s)}}{(2\pi(t-s))^{d/2}}\, d\mathfrak{b} \end{aligned}$$

is a function of $r(s) = |\mathfrak{x}(s)|$ alone, and noting the inclusion $R_s \equiv B[r(\theta): \theta \leqq s] \subset B_s$, it is found that

3) $P_\bullet[r(t) \in db \mid R_s] = P_{\mathfrak{a}}[r(t-s) \in db] = g^+(t-s, a, b)\, db$

where $\qquad a = r(s) = |\mathfrak{a}|,$

4)
$$\begin{aligned} g^+(t, a, b) &= \int_{S^{d-1}} (2\pi t)^{-\frac{d}{2}} e^{-|b\mathfrak{o} - a(1,0,\ldots,0)|^2/2t}\, do\, b^{d-1} * \\ &= (2\pi t)^{-\frac{d}{2}} e^{-\frac{a^2+b^2}{2t}} \int_{S^{d-1}} e^{-\frac{ab}{t}\cos\theta}\, do\, b^{d-1} ** \\ &= t^{-1} e^{-\frac{a^2+b^2}{2t}} (ab)^{1-\frac{d}{2}} I_{\frac{d}{2}-1}\left(\frac{ab}{t}\right) b^{d-1} \qquad t > 0,\ a, b > 0 *** \end{aligned}$$

is the fundamental solution of

5a) $$\frac{\partial u}{\partial t} = \frac{1}{2}\left(\frac{\partial^2 u}{\partial b^2} + \frac{d-1}{b}\frac{\partial u}{\partial b}\right) \qquad b > 0$$

5b) $$\lim_{b \downarrow 0} b^{d-1} \frac{\partial u}{\partial b} = 0$$

(see problem 1); in brief, *the Bessel motion is a simple Markov process, standing in the same relation to the Bessel operator*

6a) $\mathfrak{G}^+ = \frac{1}{2}\left(\frac{d^2}{db^2} + \frac{d-1}{b}\frac{d}{db}\right) =$ *the radial part of* $\mathfrak{G} = (1/2)\Delta$

acting on the domain

6b) $D(\mathfrak{G}^+) = C[0, +\infty) \cap (u : \mathfrak{G}^+ u \in C[0, +\infty),\ \lim_{b\downarrow 0} b^{d-1} u'(b) = 0)$ †

as in the 1-dimensional case (see 2.1).

---

* $\mathfrak{o} \in S^{d-1}$, the unit sphere. do is the element of surface area on S^{d-1}.

** θ is the angle between $\mathfrak{o}$ and $(1, 0, \ldots, 0)$.

*** $I_{(d/2)-1}$ is the usual modified BESSEL function.

† $\lim_{b\downarrow 0} b^{d-1} u'(b) = 0$ is automatic if $d \geqq 2$; see 4.6.

6a) suggests that the d-dimensional BESSEL process r should be a standard BROWNian motion b subject to a drift at rate $(d-1)/2r$, *i.e.*, that it should be a solution of $r(t) = b(t) + \frac{d-1}{2}\int_0^t ds/r(s)$; this conjecture is correct for $d \geqq 2$ (see H. P. McKEAN, JR. [3] for the proof and additional information), and it is also true for $d = 1$ if $\lim_{\varepsilon\downarrow 0} (2\varepsilon)^{-1}$ measure $(s: r(s) < \varepsilon,\ s \leqq t)$ is used in place of $\frac{d-1}{2}\int_0^t ds/r(s)$ (see H. P. McKEAN, JR. [6]).

A. YA. HINČIN's local law of the iterated logarithm

$$\text{7a)} \qquad P_0\left[\overline{\lim_{\delta\downarrow 0}}\frac{|\mathfrak{x}(\delta)|}{\sqrt{2\delta \lg_2 1/\delta}} = 1\right] = 1$$

(see 1.8) and P. LÉVY's HÖLDER condition

$$\text{7b)} \qquad P_0\left[\overline{\lim_{\substack{|t-s|=\delta\downarrow 0\\ s<t\leqq 1}}}\frac{|\mathfrak{x}(t)-\mathfrak{x}(s)|}{\sqrt{2\delta \lg 1/\delta}} = 1\right] = 1$$

(see 1.9) also hold for the d-dimensional BROWNian motion as one can easily prove, using the fact that the projection of the d-dimensional motion on a line in R^d is a standard 1-dimensional BROWNian motion and noting the obvious estimate $|\mathfrak{x}| = \sup_{n\geqq 1}|(\mathfrak{x}, \mathfrak{e}_n)|$ for $\mathfrak{x}\in R^d$ and a countable dense set of points $\mathfrak{e}_n$ $(n \geqq 1)$ on the surface of the unit ball. Because the angular distribution of the large changes in $\mathfrak{x}(t)$ $(t \leqq 1)$ is isotropic, the same laws hold for the BESSEL process as well:

$$\text{8a)} \qquad P_0\left[\overline{\lim_{\delta\downarrow 0}}\frac{|r(t+\delta)-r(t)|}{\sqrt{2\delta \lg_2 1/\delta}} = 1\right] = 1 \qquad (t \geqq 0)$$

$$\text{8b)} \qquad P_0\left[\overline{\lim_{\substack{|t-s|=\delta\downarrow 0\\ s<t\leqq 1}}}\frac{|r(t)-r(s)|}{\sqrt{2\delta \lg 1/\delta}} = 1\right] = 1.$$

A final point needed below is that the $d \geqq 2$ dimensional BESSEL path does not touch $r = 0$ at a positive time:

$$\text{9)} \qquad P_0[r(t) > 0,\ t > 0] = 1;$$

it is enough to discuss the 2-dimensional case.

Consider for the proof two circles $|\mathfrak{a}| = a$ and $|\mathfrak{b}| = b > a$ centered at 0, take a point $\mathfrak{x}$ in the shell between them $(a < |\mathfrak{x}| < b)$, draw about it a small circle $|\mathfrak{y} - \mathfrak{x}| = c$ contained in the shell, and let $\mathfrak{e}$ be the exit time $\min(t: |\mathfrak{x}(t) - \mathfrak{x}| = c)$ for BROWNian paths starting at

$\mathfrak{x}(0)=\mathfrak{x}$. Because $P_{\mathfrak{x}}[\mathfrak{x}(\mathfrak{e})\in d\mathfrak{y}]$ is invariant under rotations about $\mathfrak{x}$, it must be the uniform angular distribution do on $|\mathfrak{y}-\mathfrak{x}|=c$, and since the BROWNian traveller starts afresh at time $\mathfrak{e}$, the radial function

$$10)\quad p(|\mathfrak{x}|)=P_{\mathfrak{x}}(\mathfrak{m}_a<\mathfrak{m}_b)\quad \mathfrak{m}_r=\min(t:|\mathfrak{x}(t)|=r)=\min(t:r(t)=r)$$

satisfies

$$11)\qquad p(|\mathfrak{x}|)=P_{\mathfrak{x}}[\mathfrak{m}_a(w_{\mathfrak{e}}^+)<\mathfrak{m}_b(w_{\mathfrak{e}}^+)]=\int p(|\mathfrak{y}|)\,do,$$

i.e., it is harmonic ($\Delta p=0$) in the shell. But the projection argument used above shows $p(a+)=1$ and $p(b-)=0$, so

$$12)\qquad p(r)=\frac{\lg b/r}{\lg b/a}\qquad a<r<b,$$

and now 9) follows from

$$13)\qquad \begin{aligned} &P_{\mathfrak{x}}(\mathfrak{m}_{0+}<+\infty)\\ &=\lim_{b\uparrow+\infty}\lim_{a\downarrow 0}P_{\mathfrak{x}}(\mathfrak{m}_a<\mathfrak{m}_b)\\ &=\lim_{b\uparrow+\infty}\lim_{a\downarrow 0}\frac{\lg b/r}{\lg b/a}\qquad (|\mathfrak{x}|=r>0)\\ &=0\end{aligned}$$

(see problem 3 below and also problem 4.6.3 for other methods).

Problem 1. Give the details of the evaluation of $g^+(t,a,b)$ in 4) and also a direct proof that $g^+(t,a,b)$ is the fundamental solution of 5) (see KARLIN and McGREGOR [*5*: 9—11] and problem 4.11.2 for more information about this kernel).

Problem 2. Give complete proofs of 7) and 8) using the indications of the text.

Problem 3. Give a direct proof of 9) for $d\geqq 3$ similar to the DVORETSKY-ERDÖS-KAKUTANI proof that the 1-dimensional BROWNian path is non-differentiable (see problem 1.4.7).

$$\Big[P_0\big(r(t)=0\ \text{ has a root at some time }\ 1\leqq t<2\big)$$
$$\leqq P_0\Big(\bigcup_{m\geqq 1}\bigcap_{n\geqq m}\bigcup_{n\leqq k<2n}\bigcap_{i\leqq d}\Big(\Big|x_i\Big(\frac{k}{n}\Big)\Big|<\sqrt{\frac{3}{n}\lg n}\Big)\Big)$$
$$\leqq \lim_{n\uparrow+\infty} n\Big(\frac{3}{n\pi}\lg n\Big)^{d/2}=0.\Big]$$

2.8. Standard Brownian local time

Given a standard BROWNian motion x and a point $a \in R^1$, let x_a^+ be the reflecting BROWNian motion $x_a^+(t) = |x(t) - a|$, let $\mathfrak{t}_a^+$ be its local time, and let $\mathfrak{t}(t, a) \equiv \mathfrak{t}(t, a, w)$ be the *standard Brownian local time at* a:

1) $$\mathfrak{t}(t, a) = 0 \qquad t < \mathfrak{m}_a$$
$$= \tfrac{1}{2}\mathfrak{t}_a^+(t - \mathfrak{m}_a) \qquad t \geqq \mathfrak{m}_a.$$

$\mathfrak{t}(t, a)$ is non-negative, continuous, increasing, and flat outside the visiting set $\mathfrak{Z}_a = (t : x = a)$ for each separate a.

As will be proved below, $\mathfrak{t}$ *can be modified so as to be continuous in the pair* $(t, a) \in [0, +\infty) \times R^1$, *and with this modification,*

2a) $$\mathfrak{t}(t, a) = \lim_{b \downarrow a} \frac{\text{measure}\,(s : a \leqq x(s) < b,\, s \leqq t)}{2(b - a)} \qquad t \geqq 0,\, a \in R^1,$$

2b) $$2\int_a^b \mathfrak{t}(t, \xi)\, d\xi = \text{measure}\,(s : a \leqq x(s) < b,\, s \leqq t) \qquad t \geqq 0,\, a < b,$$

and

2c) $$\mathfrak{t}(t, a) = \mathfrak{t}(s, a) + \mathfrak{t}(t - s, a, w_s^+) \qquad s \leqq t,\, a \in R^1,$$

for all but a negligeable class of Brownian paths, but first a provisional result in the direction of 2a):

3) $$P_0\left[\left|\frac{\text{measure}\,(s : a \leqq x < b,\, s \leqq t)}{2(b - a)} - \mathfrak{t}(t, a)\right| > \delta\right]$$
$$< \text{constant} \times \varepsilon \sqrt{t}/\delta^2$$
$$t \geqq 0,\, \varepsilon = b - a,\, a < b.$$

Considering the case $a = 0$ only (the general case is left to the reader as an exercise), it suffices to check the bound

4) $$D(t) = E_0[(2\varepsilon)^{-1}\,\text{measure}\,(s : 0 \leqq x < \varepsilon,\, s \leqq t) - \mathfrak{t}(t, 0)|^2]$$
$$\leqq \text{constant} \times \varepsilon \sqrt{t}$$

and then to use ČEBYŠEV's inequality. Now

5a) $$E_0[\mathfrak{t}(t, 0)^2] = \tfrac{1}{4} E_0[\mathfrak{t}^-(t)^2] = t/4 \equiv A,$$

5b) $$(2\varepsilon)^{-2} E_0[\text{measure}\,(s : 0 \leqq x < \varepsilon,\, s \leqq t)^2]$$
$$= (2\varepsilon)^{-2}\, 2 \int_0^t ds \int_0^s d\sigma \int_0^\varepsilon da \int_0^\varepsilon db \frac{e^{-a^2/2\sigma}}{\sqrt{2\pi\sigma}} \frac{e^{-(b-a)^2/2(s-\sigma)}}{\sqrt{2\pi(s - \sigma)}}$$
$$\leqq \frac{1}{4\pi} \int_0^t ds \int_0^s \frac{d\sigma}{\sqrt{\sigma(s - \sigma)}} = \frac{t}{4}\frac{1}{\pi} \int_0^1 \frac{d\sigma}{\sqrt{\sigma(1 - \sigma)}} = t/4 \equiv B,$$

and

5c) $2(2\varepsilon)^{-1} E_0[\mathfrak{t}(t, 0) \text{ measure } (s: 0 \leqq x < \varepsilon, s \leqq t)]$

$$= \varepsilon^{-1} \int_0^t ds\, E_0[\mathfrak{t}_0^+(t), 0 \leqq x(s) < \varepsilon]$$

$$= \varepsilon^{-1} \int_0^t ds\, E_0[\mathfrak{t}_0^+(s), 0 \leqq x < \varepsilon]$$

$$+ \varepsilon^{-1} \int_0^t ds\, E_0[0 \leqq x(s) < \varepsilon, E_{x(s)}(\mathfrak{t}_0^+(t-s))]$$

$$= \varepsilon^{-1} \int_0^t ds \int_0^\varepsilon da \int_0^{+\infty} b\, db \frac{a+b}{\sqrt{2\pi s^3}} e^{-(a+b)^2/2s} {}^*$$

$$+ \varepsilon^{-1} \int_0^t ds \int_0^\varepsilon \frac{e^{-a^2/2s}}{\sqrt{2\pi s}} da \int_0^{t-s} \frac{a}{\sqrt{2\pi\sigma^3}} e^{-a^2/2\sigma} d\sigma \times$$

$$\times 2 \int_0^{+\infty} b\, db \frac{e^{-b^2/2(t-s-\sigma)}}{\sqrt{2\pi(t-s-\sigma)}}$$

$$\equiv C.$$

$\Delta \equiv A + B - C$ bounds D above and is an increasing function of t as can be seen from its LAPLACE transform

6) $$\int_0^{+\infty} e^{-\alpha t} \Delta\, dt = \frac{1}{4\varepsilon\alpha^2} \int_0^\varepsilon (1 - e^{-\sqrt{2\alpha}\,a})(2 + e^{-\sqrt{2\alpha}\,a})\, da,$$

so

7) $$D \leqq \Delta \leqq \frac{e}{t} \int_t^{+\infty} e^{-s/t} \Delta\, ds \leqq 3e\sqrt{2t}\,\varepsilon/8,$$

and this completes the proof. Now it is permissible to identify $\mathfrak{t}(t, a)$ with

$$\lim_{\substack{b = a + 2^{-n} \\ b \downarrow a}} 2^{n-1} \text{ measure } (s: a \leqq x(s) < b, s \leqq t),$$

and 2b) follows, first for each separate $t > 0$ and $a < b$, and then for all $t > 0$ and $a < b$ at once, because $\int_a^b \mathfrak{t}(t, \xi)\, d\xi$ is monotone and measure $(s: a \leqq x < b, s \leqq t)$ is monotone and continuous in the triple

* See problem 2.2.3.

$(t, a, b) \in [0, +\infty) \times R^2$; in particular, supposing that $\mathfrak{t}(t, \xi)$ is a continuous function of ξ, 2a) und 2c) follow and also

8) $$P_0\Big[\lim_{\varepsilon\downarrow 0} (2\varepsilon)^{-1} \text{ measure } (s : x^+ < \varepsilon, s \leqq t) = \mathfrak{t}^+(t), t \geqq 0\Big] = 1$$

as needed for the down-crossing lemma of 2.4 and confirming 2.2.7).

H. TROTTER [1] gave the first proof that $\mathfrak{t}(t, \xi)$ is continuous on $[0, +\infty) \times R^1$; he obtained

9a) $$P_0\left[\overline{\lim_{\substack{b-a=\delta\downarrow 0\\ a<b}}} \frac{|\mathfrak{t}(t, b) - \mathfrak{t}(t, a)|}{\sqrt{\delta \lg 1/\delta}} = 0, \quad t \geqq 0\right] = 1$$

and

9b) $$P_0\left[\overline{\lim_{\substack{t-s=\delta\downarrow 0\\ s<t\leqq 1\\ a\in R^1}}} \frac{\mathfrak{t}(t, a) - \mathfrak{t}(s, a)}{\sqrt[3]{\delta(\lg 1/\delta)^2}} = 0\right] = 1$$

using KAC's formula, and later E. BOYLAN [1] simplified the estimates with the same result. H. P. MCKEAN, JR. [4] next obtained

10a) $$P_0\left[\overline{\lim_{\delta\downarrow 0}} \frac{|\mathfrak{t}(t, \delta) - \mathfrak{t}(t, 0)|}{\sqrt{\delta \lg_2 1/\delta}} \leqq 2\sqrt{\mathfrak{t}(t, 0)}\right] = 1$$

and

10b) $$P_0\left[\overline{\lim_{\substack{b-a=\delta\downarrow 0\\ a<b}}} \frac{|\mathfrak{t}(t, b) - \mathfrak{t}(t, a)|}{\sqrt{\delta \lg 1/\delta}} \leqq 2\sqrt{\max_{a\in R^1} \mathfrak{t}(t, a)}\right] = 1$$

for each separate $t \geqq 0$, using a stochastic integral for $\mathfrak{t}$ due to H. TANAKA, and a little afterwards D. B. RAY [4] proved that 10) *is best possible (i. e., that the = sign holds in both the inside inequalities)*. RAY observed that if x is a standard BROWNian path starting at 0, stopped at its passage time $\mathfrak{m}_1$ to 1, and if $\#(t, [a, b))$ is the number of times it crosses down from b to $a < b$ before time $t \leqq \mathfrak{m}_1$, then for $2^n \geqq j > k > 0$, $\#(\mathfrak{m}_1, [(k-1) \cdot 2^{-n}, k2^{-n}))$ is the sum of $\#(\mathfrak{m}_1, [(k-1)2^{-n}, j2^{-n})) + 1$ independent copies of $\#(\mathfrak{m}_{j2^{-n}}, [(k-1)2^{-n}, k2^{-n}))$, these copies are also independent of $\#(\mathfrak{m}_1, [(j-1)2^{-n}, j2^{-n}))$ and so $[\#(\mathfrak{m}_1, [(k-1)2^{-n}, k2^{-n})) : 0 < k \leqq 2^n, P_0]$ is a MARKOV chain. But according to the down-crossing lemma of 2.4, $2^{-n}\#(\mathfrak{m}_1, [(k-1)2^{-n}, k2^{-n}))$ should tend to $\mathfrak{t}(\mathfrak{m}_1, a)$ $(k = [2^n a], n \uparrow +\infty)$, so $[\mathfrak{t}(\mathfrak{m}_1, a) : 0 \leqq a \leqq 1, P_0]$ should also be MARKOVian, and indeed, RAY proved that it can be expressed in terms of the 2 and 4-dimensional BESSEL processes of 2.7 as follows:

11a) $$\mathfrak{t}(\mathfrak{m}_1, a) = 0 \qquad a \leqq b = \min_{t\leqq \mathfrak{m}_1} x(t),$$

11b) $$P_0(b \leqq c) = \frac{1}{1-c} \qquad c \leqq 0,$$

11c) $$\mathfrak{t}(\mathfrak{m}_1, a) = r_4(a-b)^2/2 \qquad b \leqq a \leqq 0,$$

where r_4 is a 4-dimensional BESSEL process starting at 0, independent of b,

11d) $$t(m_1, a) = r_2(1-a)^2/2 \qquad 0 \leq a \leq 1,$$

where r_2 is a 2-dimensional BESSEL process starting at 0, conditioned to land at $r_4(-b)$ at time 1, but otherwise independent of r_4 and of b. The strong laws

12a) $$P_0\left[\overline{\lim_{\delta\downarrow 0}} \frac{|t(m_1,\delta) - t(m_1,0)|}{\sqrt{\delta \lg_2 1/\delta}} = 2\sqrt{t(m_1,0)}\right] = 1$$

and

12b) $$P_0\left[\overline{\lim_{\substack{b-a=\delta\downarrow 0\\ a<b}}} \frac{|t(m_1,b) - t(m_1,a)|}{\sqrt{\delta \lg 1/\delta}} = 2\sqrt{\max_{a\leq 1} t(m_1,a)}\right] = 1$$

follow at once from 2.7.8) and the above description. F. B. KNIGHT [2] obtained a similar result using the inverse local time $t^{-1}(t) = \min(s : t(s,0) = t)$ in place of m_1; this proof is left aside, but the reader is invited to prove 11d) in problems 5 and 6 below.

10) (with = in place of $\leq$) and an improvement of 9b) are proved below using a second result of D. B. RAY [4]; this states that if e is an exponential holding time with conditional law $P_\bullet(e > t \mid B) = e^{-t/2}$ then for each $a, b \in R^1$, the *conditional local times* $[t(e,\xi) : \xi \in R^1, P_{ab}(B) = P_a(B \mid x(e) = b)]$ can be described as follows:

13a) $$t(e,\xi) = 0 \qquad \xi \leq c \equiv \min_{t\leq e} x(t),$$

13b) $$P_{ab}(c < \xi) = e^{-2(a\wedge b-\xi)} \qquad \xi \leq a\wedge b,$$

13c) $$t(e,\xi) = e^{-2\xi} r_4(e^{2\xi} - e^{2c})^2/4 \qquad c \leq \xi \leq a\wedge b,$$

where r_4 is a 4-dimensional BESSEL process starting at 0, independent of c,

13d) $$t(e,\xi) = e^{-2\xi} r_2(e^{2\xi} - e^{2a\wedge b})^2/4 \quad \textit{for } \xi \textit{ between } a \textit{ and } b,$$

where r_2 is a 2-dimensional BESSEL process starting at $r_4(e^{2a\wedge b} - e^{2c})$ but otherwise independent of r_4 and c,

13e) $$t(e,\xi) = e^{2\xi} r_4(e^{-2\xi} - e^{-2d})^2/4 \qquad a\vee b \leq \xi \leq d,$$

where r_4 is another 4-dimensional BESSEL process starting at 0, conditioned so as to make $t(t,\xi)$ match at $\xi = a\vee b\,\pm$, but otherwise independent of $d \equiv \max_{t\leq e} x(t)$, r_2, c, and the old r_4,

13f) $$P_{ab}(d > \xi) = e^{-2(\xi - a\vee b)} \qquad \xi \geq a\vee b,$$

13g) $$t(e,\xi) = 0 \qquad \xi \geq d.$$

Here the MARKOVian nature of $[t(e,\xi) : \xi \in R^1, P_{ab}]$ can be traced to the fact that stopping at an exponential holding time brings to the distributions a LAPLACE transform which changes a) convolutions into simple products and b) the GAUSS kernel into a GREEN function which is also a product.

The first step is to check that if $\mathfrak{f}$ is the additive functional based on a non-negative piecewise continuous function f, then

14a) $$E_{ab}[e^{-\mathfrak{f}(e)} \mid t(e, \xi) = 0] = \frac{2e^{|b-a|}}{1 - e^{|b-a|-|b-\xi|-|\xi-a|}} \times \left[G(a, b) - \frac{G(a, \xi)\, G(\xi, b)}{G(\xi, \xi)}\right]$$

for ξ not between a and b, and

14b) $$E_{ab}[e^{-\mathfrak{f}(e)} \mid t(e, \xi) = t] = e^{|b-\xi|+|\xi-a|} \frac{G(a, \xi)\, G(\xi, b)}{G(\xi, \xi)^2} e^{-t[G(\xi,\xi)^{-1}-2]} \qquad (t > 0)$$

for all ξ, where $G(a, b) = G(b, a) = g_1(a)\, g_2(b)$ $(a \leq b)$ is the GREEN function of $1/2 - \mathfrak{G}^\bullet$ $(\mathfrak{G}^\bullet = \mathfrak{G} - f)$.

KAC tells us that if $e_{-\infty b}$ is the indicator of $a < b$, then

15) $$\begin{aligned} u(a) &= E_a[e^{-\mathfrak{f}(e)}, x(e) < b] \\ &= \tfrac{1}{2} E_a\left[\int_0^{+\infty} e^{-t/2}\, e^{-\mathfrak{f}(t)}\, e_{-\infty b}(x)\, dt\right] \\ &= \int_{c<b} G(a, c)\, dc, \end{aligned}$$

and since

16) $$P_a[x(e) \in db] = \frac{1}{2}\int_0^{+\infty} e^{-t/2}\, dt \frac{e^{-(b-a)^2/2t}}{\sqrt{2\pi t}}\, db = e^{-|b-a|}\, db/2,$$

it follows that

17) $$E_{ab}[e^{-\mathfrak{f}(e)}] = E_a[e^{-\mathfrak{f}(e)} \mid x(e) = b] = 2e^{|b-a|}\, G(a, b).$$

Given $\xi \in R^1$ and $\gamma > 0$, add to f the indicator function of $[\xi, \eta)$ $(\eta > \xi)$ times $\gamma/2(\eta - \xi)$ and let $\eta \downarrow \xi$; then the modified left hand side of 17) tends to $E_{ab}[e^{-\mathfrak{f}(e) - \gamma t(e, \xi)}]$ in view of 3), while the modified GREEN function tends to $G(a, b) - \gamma\, G(a, \xi)\, G(\xi, b)\, (1 + \gamma\, G(\xi, \xi))^{-1}$*, proving the formula

18) $$E_{ab}[e^{-\mathfrak{f}(e) - \gamma t(e, \xi)}] = 2e^{|b-a|}\left[G(a, b) - \gamma \frac{G(a, \xi)\, G(\xi, b)}{1 + \gamma\, G(\xi, \xi)}\right] \qquad \gamma > 0.$$

* G_*, the modified GREEN function, satisfies

$$G_*(a, b) = G(a, b) - (\gamma/\eta - \xi) \int_\xi^\eta G_*(a, c)\, G(c, b)\, dc;$$

making $\eta \downarrow \xi$ and solving gives the stated result.

As $\gamma \uparrow +\infty$, 18) tends to

19a) $$E_{ab}[e^{-\mathfrak{f}(e)}, \mathfrak{t}(e, \xi) = 0] = 2e^{|b-a|}\left[G(a, b) - \frac{G(a, \xi)\, G(\xi, b)}{G(\xi, \xi)}\right],$$

and subtracting 19a) from 18) and inverting as a LAPLACE transform in γ gives

19b) $$E_{ab}[e^{-\mathfrak{f}(\cdot)}, \mathfrak{t}(e, \xi) > t] = 2e^{|b-a|}\,\frac{G(a, \xi)\, G(\xi, b)}{G(\xi, \xi)}\, e^{-t/G(\xi, \xi)};$$

in the special case $f = 0$, 19) becomes

20a) $$P_{ab}[\mathfrak{t}(e, \xi) = 0] = 1 - e^{|b-a|-|b-\xi|-|\xi-a|}$$
$$P_{ab}[\mathfrak{t}(e, \xi) > t] = e^{|b-a|-|b-\xi|-|\xi-a|}\, e^{-2t},$$

and 14) follows.

Because $\mathfrak{f}(e) = 2\int \mathfrak{t}(e, \xi)\, f(\xi)\, d\xi$, *it is enough for the proof that the conditional local times are Markovian, to check that the additive functionals* $\mathfrak{f}_\pm$ *based on*

$$f_- = f \quad (\eta \leqq \xi) \quad \text{and} \quad f_+ = 0 \quad (\eta < \xi)$$
$$= 0 \quad (\eta > \xi) \qquad\qquad = f \quad (\eta \geqq \xi)$$

are independent, conditional on $\mathfrak{t}(e, \xi)$, and since the BROWNian motion is invariant under translations and reflection about its starting point, *it is permissible to take* $\xi = 0$ *and* $a < b$. Using 12) and the fact that $P_{ab}[\mathfrak{t}(e, \xi) = 0] = 0$ for ξ between a and b [see 20a)], the problem is now simplified to checking that if $G_\pm$ are the corresponding GREEN functions, then

21a) $$G(a, b) - G(a, 0)\, G(0, b)/G(0, 0) \qquad (0 \leqq a < b)$$
$$= \frac{e^{2(b-a)}}{1 - e^{-2a}}\left[\frac{G_-(a, b) - G_-(a, 0)\, G_-(0, b)}{G_-(0, 0)}\right] \times$$
$$\times \left[\frac{G_+(a, b) - G_+(a, 0)\, G_+(0, b)}{G_+(0, 0)}\right],$$

21b) $$G(a, 0)\, G(0, b)/G(0, 0)^2 = e^{|a|+|b|}\,\frac{G_-(a, 0)\, G_-(0, b)}{G_-(0, 0)^2}\,\frac{G_+(a, 0)\, G_+(0, b)}{G_+(0, 0)^2},$$

and

21c) $$G(0, 0)^{-1} = G_-(0, 0)^{-1} + G_+(0, 0)^{-1} - 2.$$

But G_- is the product of

22a−) $$g_1(\eta) \quad (\eta < 0)$$
$$g_1(0)\cosh\eta + g_1'(0)\sinh\eta \quad (\eta \geqq 0)$$

and

22b−) $g_1(0)\,g_2(\eta) - g_1(\eta)\,g_2(0) + g_1'(0)\,g_2(\eta) - g_1(\eta)\,g_2'(0) \quad (\eta < 0)$

$$e^{-\eta} \quad (\eta \geqq 0)$$

divided by their WRONSKIan

23−) $$g_1(0) + g_1'(0),$$

while G_+ is the product of

22a+) $$e^{\eta} \quad (\eta < 0)$$

$$g_1(\eta)\,g_2(0) - g_1(0)\,g_2(\eta) + g_1'(0)\,g_2(\eta) - g_1(\eta)\,g_2'(0) \quad (\eta \geqq 0)$$

and

22b+) $$g_2(0)\cosh\eta + g_2'0)\sinh\eta \quad (\eta < 0)$$

$$g_2(\eta) \quad (\eta \geqq 0)$$

divided by *their* WRONSKIan

23+) $$g_2(0) - g_2'(0),$$

and now 21 c) follows on substituting from 22) and 23). 21 b) is obvious because both sides solve $\mathfrak{G}^\bullet u = u/2$ as functions of a or of b on either side of 0. 21 a) is left to the reader.

Using 14), $E_{ab}[e^{-\gamma t(e,\eta)} \mid t(e,\xi) = t]$ is now computed for $\xi < \eta \leqq a \wedge b$ and $t = 0$ and for $\xi < \eta \leqq a \vee b$ and $t > 0$: just substitute

$$\frac{1}{2}\,e^{-|b-a|} - \frac{\gamma}{4}\left(1 + \frac{\gamma}{2}\right)^{-1} e^{-|b-\eta|-|\eta-a|}\,*$$

in place of G in 14) and obtain

24a) $E_{ab}[e^{-\gamma t(e,\eta)} \mid t(e,\xi) = 0] = \theta + (1-\theta)/\Delta \quad \xi < \eta \leqq a \wedge b$

$$\theta = \frac{1 - e^{-2(a\wedge b-\eta)}}{1 - e^{-2(a\wedge b-\xi)}}, \qquad \Delta(\xi) = 1 + \frac{\gamma}{2}\,[1 - e^{-2(\eta-\xi)}],$$

and

24b) $$E_{ab}[e^{-\gamma t(e,\eta)} \mid t(e,\xi) = t] \qquad (t > 0)$$

$$= \Delta^{-1}\,e^{-t\gamma e^{2(\xi-\eta)}/\Delta} \quad \xi < \eta \text{ between } a \text{ and } b$$

$$= \Delta^{-2}\,e^{-t\gamma e^{2(\eta-\xi)}/\Delta} \quad \xi < \eta \leqq a \wedge b.$$

Because $P_{ab}[t(e,\xi) = 0] = 0$ for ξ between a and b, it suffices for the proof of part d) of 13) to use the MARKOVian nature of the conditional local times and of the 2-dimensional BESSEL process r_2 and to

* $e^{-|b-a|}/2$ is the GREEN function of $1/2 - \mathfrak{G}$; see the line above 18) and the footnote attached to it.

check that for ξ and $\eta > \xi$ between a and b, 24b) agrees with

25) $$E_r\left[e^{-\gamma e^{-2\eta} r_2(e^{2\eta}-e^{2\xi})^2/4}\right]^{*} \qquad (r = r_2(e^{2\xi} - e^{2a\wedge b}))$$

$$= \int_{R^2} \frac{e^{-|\mathfrak{b}-\mathfrak{x}|^2/2(e^{2\eta}-e^{2\xi})}}{2\pi(e^{2\eta}-e^{2\xi})} e^{-\gamma e^{-2\eta}|\mathfrak{b}|^2/2}\, d\mathfrak{b} \quad (|\mathfrak{x}| = r)$$

$$= \frac{2e^{2\eta}/\gamma}{e^{2\eta-2\xi}+2e^{2\eta}/\gamma} e^{-r^2/(2e^{2\eta}-e^{2\xi}+2e^{2\eta}/\gamma)}.$$

As to $\mathfrak{t}(\mathfrak{e}, \xi)$ for $\xi \leq a \wedge b$, it follows from the independence of c and r_4 combined with $P_0[r_4 > 0, t > 0] = 1$ (see 2.7) that the composite motion $r_4[(e^{2\xi} - e^{2c}) \vee 0]$ is MARKOVian, and now a comparison of 24a) and 24b) for $\xi < \eta < a \wedge b$ with

26a) $$P_0[c \geq \eta \mid c \geq \xi] + E_0\left[c < \eta, e^{-\gamma e^{-2\eta} r_4(e^{2\eta}-e^{2c})^2/4} \mid c \geq \xi\right]^{**}$$

$$= \theta + \frac{2}{1-e^{-2(a\wedge b-\xi)}} \int_\xi^\eta e^{-2(a\wedge b-l)}\, dl/\Delta^2(l)$$

$$= \theta + \frac{e^{-2(a\wedge b-\eta)}}{1-e^{-2(a\wedge b-\xi)}} \frac{2}{\gamma}(1-\Delta^{-1}(\xi))$$

and

26b) $$E_r\left[e^{-\gamma e^{-2\eta} r_4(e^{2\eta}-e^{2\xi})^2/4}\right] \qquad (r = r_4(e^{2\xi} - e^{2c}))$$

$$= \int_{R^4} \frac{e^{-|\mathfrak{b}-\mathfrak{x}|^2/2(e^{2\eta}-e^{2\xi})}}{(2\pi(e^{2\eta}-e^{2\xi}))^2} e^{-\gamma e^{-2\eta}|\mathfrak{b}|^2/4}\, d\mathfrak{b} \qquad (|\mathfrak{x}| = r)$$

$$= \frac{(2e^{2\eta}/\gamma)^2}{(e^{2\eta}-e^{2\xi}+2e^{2\eta}/\gamma)^2} e^{-r^2/2(e^{2\eta}-e^{2\xi}+2e^{2\eta}/\gamma)}$$

completes the proof of 13a, 13b, and 13c). 13e, 13f, and 13g) now follow using the MARKOVian nature of the conditional local time and the obvious invariance of its law under reflection about $\xi = (a + b)/2$. Using 2.7.8),

27a) $$P_0\left[\overline{\lim_{\delta\downarrow 0}} \frac{|\mathfrak{t}(\mathfrak{e}, \delta) - \mathfrak{t}(\mathfrak{e}, 0)|}{\sqrt{\delta \lg_2 1/\delta}} = 2\sqrt{\mathfrak{t}(\mathfrak{e}, 0)}\right] = 1$$

and

27b) $$P_0\left[\overline{\lim_{\substack{b-a=\delta\downarrow 0\\ a<b}}} \frac{|\mathfrak{t}(\mathfrak{e}, b)|}{\sqrt{\delta \lg 1/\delta}} = 2\sqrt{\max_{a\in R^1} \mathfrak{t}(\mathfrak{e}, a)}\right] = 1$$

---

* E_r now stands for the 2-dimensional BESSEL expectation.

** E_r now stands for the 4-dimensional BESSEL expectation, and c is independent of r_4 with distribution 13b).

can be read off, and it follows (use FUBINI) that 10) holds (with $=$ in place of $\leqq$) at some positive time. But now the scaling $x(t) \to c\,x(t/c^2)$ $(c > 0)$ maps $\mathfrak{t}(t, a)$ into $c\,\mathfrak{t}(t/c^2, a/c)$ and changes the statements of 10) into the corresponding statements at time t/c^2, so 10) holds (with $=$ in place of $\leqq$) at all positive times, and an extra result obtained at no extra cost is that

28) $$P_0\left[\mathfrak{t}(t, \xi) > 0,\ \min_{s \leqq t} x(s) < \xi < \max_{s \leqq t} x(s),\ t > 0\right] = 1$$

as is immediate from $P_0[\mathfrak{r}_d > 0, t > 0] = 1$ $(d = 2,4)$.

Because of

29) $$\mathfrak{t}(t, a) = \lim_{\substack{b = a + 2^{-n} \\ b \downarrow a}} 2^{n-1} \text{ measure } (s : a \leqq x < b,\ s \leqq t),$$

30a) $$\mathfrak{t}(s, \xi) \leqq \mathfrak{t}(t, \xi) \qquad s \leqq t,\ \xi \in R^1;$$

also, in view of 10) and FUBINI's theorem,

30b) $\mathfrak{t}(t, \xi)$ *is a continous function of* ξ *for each* $t = k\,2^{-n} \geqq 0$,

and since $\mathfrak{t}(s, \xi)$ as a function of s is constant near $s = t$ for $\xi \neq x(t)$,

30c) $$\overline{\lim_{\substack{b - a \downarrow 0 \\ a < \xi < b}}}\ |\mathfrak{t}(t, b) - \mathfrak{t}(t, a)| \leqq \mathfrak{t}(t+, \xi) - \mathfrak{t}(t-, \xi) \qquad t \geqq 0,\ \xi \in R^1.$$

Accordingly, for *the proof that* $\mathfrak{t}(t, \xi)$ *is continuous on* $[0, +\infty) \times R^1$, it is enough to check that

31) $$P_0\left[\overline{\lim_{n \uparrow +\infty}} \max_{\substack{k < 2^n \\ \xi \in R^1}} \mathfrak{t}\,\frac{(2^{-n}, \xi, w^+_{k2^{-n}})}{2^{-n/2}\, n} \leqq \lg 4\right] = 1$$

and to use ASCOLI's theorem in conjunction with 30a); in fact 31) can be used to improve 9b) to

32) $$P_0\left[\overline{\lim_{\substack{t - s = \delta \downarrow 0 \\ s < t \leqq 1 \\ \xi \in R^1}}} \frac{\mathfrak{t}(t, \xi) - \mathfrak{t}(s, \xi)}{\sqrt{\delta}\,\lg 1/\delta} \leqq 4\right] = 1$$

as the industrious reader will check. 2a) and 2c) follow at once as noted before.

Because of the HÖLDER condition for the standard BROWNian path (1.9), it is enough for the proof of 31) to check that for $\theta > \lg 4$,

33) $$Q_n = P_0\Big[\max_{\xi \leqq 0} \mathfrak{t}(2^{-n}, \xi, w^+_{k2^{-n}}) > \theta\, 2^{-n/2}\, n,$$
$$x(2^{-n}, w^+_{k2^{-n}}) > -\sqrt{32^{-n}\, n \lg 2} \textit{ for some } k < 2^n\Big]$$

is the general term of a convergent sum, and this can be done using the standard BROWNian scaling $x(t) \longrightarrow c\,x(t/c^2)$ and RAY's description of the conditional local times as follows:

34) $$\begin{aligned} Q_n &\leqq 2^n P_0\left[\max_{\xi\leqq 0} \mathfrak{t}(2^{-n},\xi) > \theta\, 2^{-n/2} n,\, x(2^{-n}) > -2\cdot 2^{-n/2}\sqrt{n}\right] \\ &= 2^n P_0\left[\max_{\xi\leqq 0} \mathfrak{t}(1,\xi) > \theta\, n,\, x(1) > -2\sqrt{n}\right] \\ &\leqq 2^n E_0\left[\max_{\xi\leqq 0} \mathfrak{t}(1,\xi) > \theta\, n,\, e^{x(1)\wedge 0}\right] e^{2\sqrt{n}} \\ &\leqq 2^{n+2} e^{2\sqrt{n}} E_0\left[\max_{\xi\leqq 0} \mathfrak{t}(1,\xi) > \theta\, n,\, (2\sqrt{e})^{-1}\int_0^{+\infty} e^{-|b-x(1)|}\, db\right] \\ &= 2^{n+2} e^{2\sqrt{n}} P_0\left[\max_{\xi\leqq 0} \mathfrak{t}(1,\xi) > \theta\, n,\, \mathfrak{e} > 1,\, x(\mathfrak{e}) > 0\right] \\ &\leqq 2^{n+2} e^{2\sqrt{n}} P_0\left[\max_{\xi\leqq 0} \mathfrak{t}(\mathfrak{e},\xi) > \theta\, n,\, x(\mathfrak{e}) > 0\right] \\ &\leqq 2^{n+2} e^{2\sqrt{n}} P_0\left[\max_{t\leqq 1} r_4(t)^2 > 4\theta\, n\right] \\ &\leqq 2^{n+4} e^{2\sqrt{n}} P_0\left[\max_{t\leqq 1} x(t)^2 > \theta\, n\right] \\ &\leqq 2^{n+5} e^{2\sqrt{n}} P_0\left[\max_{t\leqq 1} x(t) > \sqrt{\theta\, n}\right] \\ &< \text{constant} \times 2^n e^{2\sqrt{n}} \frac{e^{-\theta n/2}}{\sqrt{\theta\, n}} \\ &< \text{constant} \times e^{-\delta n} \quad (0 < \delta < (\theta - \lg 4)/2). \end{aligned}$$

Problem 1. Check that $\int_0^1 dt/x(t) = \lim_{\varepsilon\downarrow 0} \int_{\substack{t\leqq 1\\ |x(t)|>\varepsilon}} dt/x(t)$ exists.

$\left[\int_{\substack{t\leqq 1\\ |x(t)|>\varepsilon}} dt/x(t) = 2\int_\varepsilon^{+\infty} [\mathfrak{t}(1,b) - \mathfrak{t}(1,-b)]\frac{db}{b}\right.$, and this converges as $\varepsilon\downarrow 0$, thanks to $\left.|\mathfrak{t}(1,b) - \mathfrak{t}(1,-b)| < \sqrt[3]{b}\ (b\downarrow 0).\right]$

Problem 2. Consider $\mathfrak{Z}_a = (t\colon x = a)$. Use 32) to prove that $P_0[\dim(\mathfrak{Z}_a) \geqq 1/2,\ a\in R^1] = 1$ (see note 2.5.2 for the definition of dimension; see also 2.5.3) and problem 2.2.4).

[Given $\beta < 1/2$, 32) implies that $\mathfrak{t}(t,a) - \mathfrak{t}(s,a) < \varepsilon^\beta$ $(s < t \leqq 1,$ $\varepsilon = t - s \downarrow 0,\ a\in R^1)$; this justifies

$$\Lambda^\beta \mathfrak{Z}_a \geqq \Lambda^\beta \mathfrak{Z}_a \cap [0,1) \geqq \mathfrak{t}(1,a) \qquad a\in R^1,$$

and an application of 28) completes the proof.]

Problem 3. Prove that the stochastic process $[\mathfrak{t}(\mathfrak{m}_b, 0) : b \geqq 0, P_0]$ is differential and derive the LÉVY formula

$$E_0[e^{-\alpha\,\mathfrak{t}(\mathfrak{m}_b,0)}] = (\alpha\, b + 1)^{-1} = \exp\left[-\int_0^{+\infty} [1 - e^{-\alpha l}]\, e^{-l/b}\frac{dl}{l}\right].$$

[Because $\mathfrak{t}(\mathfrak{m}_b, 0) - \mathfrak{t}(\mathfrak{m}_a, 0) = \mathfrak{t}(\mathfrak{m}_b(w^+_{\mathfrak{m}}), 0, w^+_{\mathfrak{m}})$ $(b \leqq a,\ \mathfrak{m} = \mathfrak{m}_a)$ and $\mathfrak{t}(\mathfrak{m}_c, 0) : c \leqq a$ is measurable $B_{\mathfrak{m}+}$ the differential character is clear from the fact that the BROWNian traveller starts afresh at time $\mathfrak{m}$ at the place $x(\mathfrak{m}) = a$. Consider, for the derivation of the LÉVY formula, the function

$$e(l) = e(\alpha, l) \equiv E_0[e^{-\alpha \mathfrak{t}(\mathfrak{m}_l, 0)}].]$$

Using $\mathfrak{t}(t, 0) = \mathfrak{t}(s, 0) + \mathfrak{t}(t - s, w^+_s)$ $(t \geqq s)$, deduce

$$\begin{aligned} e(l + \varepsilon) &= e(l)\,[E_l(e^{-\alpha \mathfrak{t}(\mathfrak{m}_{l+\varepsilon}, 0)}, \mathfrak{m}_0 < \mathfrak{m}_{l+\varepsilon}) + P_l(\mathfrak{m}_{l+\varepsilon} \leqq \mathfrak{m}_0)] \\ &= e(l)\left[\frac{\varepsilon}{l+\varepsilon}\, e(l+\varepsilon) + \frac{l}{l+\varepsilon}\right], \end{aligned}$$

i.e.,

$$e^+(l) \equiv \lim_{\varepsilon \downarrow 0} \frac{e(l+\varepsilon) - e(l)}{\varepsilon} = -l^{-1}\, e(l)\,[1 - e(l)],$$

and solve this to obtain

$$e(l) = [c(\alpha)\, l + 1]^{-1} \qquad c(\alpha) = \frac{e(1)}{1 - e(1)} > 0.$$

Because the scaling $x(t) \to \beta^{-1} x(\beta^2 t)$ maps $\mathfrak{m}_l$ into $\beta^{-2}\mathfrak{m}_{\beta l}$ and $\mathfrak{t}(t, l)$ into $\beta^{-1}\mathfrak{t}(\beta^2 t, \beta l)$, it is clear that $c(\alpha) = \beta c(\alpha/\beta) = \beta c(1)$ $(\beta = \alpha)$, and, to complete the proof, it is enough to evaluate $c = c(1)$. Differentiating $e = [\alpha c l + 1]^{-1}$ and letting $\alpha \downarrow 0$,

$$E_0[\mathfrak{t}(\mathfrak{m}_l, 0)] = c\, l,$$

and, using

$$\begin{aligned} E_0[\mathfrak{t}(\mathfrak{m}_1, l)] = E_0[\mathfrak{t}(\mathfrak{m}_1(w^+_{\mathfrak{m}}), l, w^+_{\mathfrak{m}})] &= E_l[\mathfrak{t}(\mathfrak{m}_1, l)] \\ &= E_0[\mathfrak{t}(\mathfrak{m}_{1-l}, 0)] \quad (\mathfrak{m} = \mathfrak{m}_l), \end{aligned}$$

one finds

$$\begin{aligned} c &= 2\int_0^1 c\, l\, dl = 2\int_0^1 E_0[\mathfrak{t}(\mathfrak{m}_l, 0)]\, dl = 2\int_0^1 E_0[\mathfrak{t}(\mathfrak{m}_1, l)]\, dl \\ &= E_0\left[2\int_0^{+\infty} \mathfrak{t}(\mathfrak{m}_1, l)\, dl\right] = E_0\,[\text{measure}\,(t : x(t) \geqq 0, t \leqq \mathfrak{m}_1)] \\ &= \int_0^{+\infty} P_0[\mathfrak{m}_1 > t, x(t) \geqq 0]\, dt \\ &= \int_0^{+\infty} P_0[\mathfrak{t}^-(t) < 1, x(t) \geqq 0]\, dt = \int_0^{+\infty} dt \int_0^1 db \int_0^b da\, 2\frac{2b - a}{\sqrt{2\pi t^3}}\, e^{-(2b-a)^2/2t} \\ &= 1, \end{aligned}$$

where problem 1.7.1 is used in the last step.]

Problem 4. Given $b > 0$, $[\mathfrak{t}(\mathfrak{t}^{-1}(t), b) : t > 0, P_0]$ is a homogeneous differential process with LÉVY measure $b^{-2} e^{-l/b} dl\,(b > 0)$; here $\mathfrak{t}^{-1}(t)$ is the inverse function of $\mathfrak{t}(t, 0)$, *i.e.*, $\mathfrak{t}^{-1}(t) = \min(s : \mathfrak{t}(s, 0) = t)$.

[Use the fact that $\mathfrak{t}^{-1}(t)$ is a MARKOV time and the addition rule $\mathfrak{t}^{-1}(t) = \mathfrak{m} + \mathfrak{t}^{-1}(t-s, w_{\mathfrak{m}}^{+})$ $(t > s,\ \mathfrak{m} = \mathfrak{t}^{-1}(s))$ to prove that $t(\mathfrak{t}^{-1}, b)$ is differential and homogeneous. As to the LÉVY measure,

$$\gamma \equiv E_0\left[\int_0^{+\infty} e^{-\alpha t} e^{-\beta t(\mathfrak{t}^{-1}(t), b)}\, dt\right]$$

$$= E_0\left[\int_0^{+\infty} e^{-\alpha \mathfrak{t}(t,0)} e^{-\beta \mathfrak{t}(t,b)}\, \mathfrak{t}(dt, 0)\right]$$

$$= E_0\left[\int_0^{\mathfrak{m}_b} e^{-\alpha \mathfrak{t}(t,0)}\, \mathfrak{t}(dt, 0)\right]$$

$$+ E_0[e^{-\alpha \mathfrak{t}(\mathfrak{m}_b, 0)}]\, E_b\left[\int_0^{+\infty} e^{-\alpha \mathfrak{t}(t,0)} e^{-\beta \mathfrak{t}(t,b)}\, \mathfrak{t}(dt, 0)\right]$$

$$= \alpha^{-1} E_0[1 - e^{-\alpha \mathfrak{t}(\mathfrak{m}_b, 0)}] + E_0[e^{-\alpha \mathfrak{t}(\mathfrak{m}_b, 0)}]\, E_b[e^{-\beta \mathfrak{t}(\mathfrak{m}_0, b)}]\,\gamma$$

and, using the result of problem 3, γ is found to be $[\alpha + \beta(\beta b + 1)^{-1}]^{-1}$. Now invert the LAPLACE transform, obtaining

$$E_0[e^{-\beta t(\mathfrak{t}^{-1}(t), b)}] = e^{-t\beta(\beta b + 1)^{-1}} = e^{-t\int_0^{+\infty} [1 - e^{-\beta l}]\, b^{-2} e^{-l/b}\, dl}$$

as stated. Because $\int_0^{+\infty} b^{-2} e^{-l/b}\, dl < +\infty$, this process has a finite number of jumps up to time $t = 1$ as is also clear from the fact that the stopped BROWNian path $x(t): t \leqq 1$ cannot come from 0 to 1 an infinite number of times.]

Problems 5 and 6 below contain the proof of RAY's description 11 d) of $[\mathfrak{t}(\mathfrak{m}_1, \xi): 0 \leqq \xi \leqq 1, P_0]$.

Problem 5. Check that

$$E_0[e^{-\gamma_1 \mathfrak{t}(\mathfrak{m}_1, \xi_1) - \gamma_2 \mathfrak{t}(\mathfrak{m}_1, \xi_2) \cdots - \gamma_n \mathfrak{t}(\mathfrak{m}_1, \xi_n)}]$$

is the reciprocal of

$$1 + \sum_{i \leqq n} \gamma_i (1 - \xi_i) + \sum_{i < j \leqq n} \gamma_i \gamma_j (\xi_j - \xi_i)(1 - \xi_j)$$
$$+ \sum_{i < j < k \leqq n} \gamma_i \gamma_j \gamma_k (\xi_j - \xi_i)(\xi_k - \xi_j)(1 - \xi_k) + \cdots$$
$$+ \gamma_1 \gamma_2 \cdots \gamma_n (\xi_2 - \xi_1)(\xi_3 - \xi_2) \cdots (\xi_n - \xi_{n-1})(1 - \xi_n)$$

for $\gamma < \gamma_1, \ldots, \gamma_n,\ 0 \leqq \xi_1, < \cdots < \xi_n < 1$, and $n \geqq 1$.

[$u(\xi) = E_\xi[e^{-\gamma_1 \mathfrak{t}(\mathfrak{m}_1, \xi_1) - etc.}]$ is bounded between 0 and 1, piecewise linear with corners at $\xi_1, \ldots, \xi_n$, and rises to 1 at $\xi = 1$. At a corner it satisfies a matching condition $u^+ - u^- = \gamma u$, for if ξ is a corner, γ the corresponding LAPLACE transform parameter, and $\mathfrak{e}$ the exit time from $|\eta - x(0)| < \delta$, then for small δ,

$$\begin{aligned} u(\xi) &= E_\xi[e^{-\gamma t(e,\xi)}\, e^{-\gamma_1 t(\mathfrak{m}_1(w_e^+),\xi_1, w_e^+)-\text{etc.}}] \\ &= E_\xi[e^{-\gamma t(e,\xi)}, x(e) = \xi - \delta]\, u(\xi - \delta) \\ &\quad + E_\xi[e^{-\gamma t(e,\xi)}, x(e) = \xi + \delta]\, u(\xi + \delta), \end{aligned}$$

and since $t(e, \xi)$ depends upon the reflecting BROWNian motion $x_\xi^+ = |x - \xi|$ alone, it follows that

$$u(\xi) = \tfrac{1}{2}[u(\xi - \delta) + u(\xi + \delta)]\, E_0[e^{-\gamma t(e,0)}].$$

But $P_0[\lim_{\delta\downarrow 0} e = 0] = 1$, so letting $\delta \downarrow 0$,

$$\begin{aligned} (u^+ - u^-)/u &= 2 \lim_{\delta\downarrow 0} \delta^{-1} E_0[1 - e^{-\gamma t(e,0)}] \\ &= c(\gamma), \end{aligned}$$

and $c(\gamma)$ can be identified as γ using the fact that the function with 1 corner $v(\xi) = E_\xi[e^{-\gamma t(\mathfrak{m}_1, 0)}]$ satisfies $v^+(0) - v^-(0) = c(\gamma)\, v(0)$, $v(0) = (1 + \gamma)^{-1}$, and $v(1) = 1$ (see problem 3). Now solve for u and use induction to obtain the stated formula.]

Problem 6. Check the evaluation

$$E_r[e^{-\gamma r_2(\xi)^2/2}] = \frac{e^{-\gamma r^2/2(1+\gamma\xi)}}{1 + \gamma\xi} \qquad (\gamma > 0)$$

and use it to prove the statement of 11d) that $[t(\mathfrak{m}_1, 1 - \xi) : 0 \leqq \xi \leqq 1, P_0]$ is identical in law to $r_2(\xi)^2/2 : \xi \leqq 1$, where r_2 is a 2-dimensional BESSEL process starting at 0.

Problem 7. Prove that $P_0[t(+\infty, \cdot) = +\infty] = 1$.

2.9. Brownian excursions

Given a standard BROWNian motion starting at 0, let $\mathfrak{Z}_n$ be the (open) interval of $[0, +\infty) - \mathfrak{Z}$ containing the first number of the list

$$\begin{matrix} 1 \\ \frac{1}{2}\ \frac{3}{2}\ 2 \\ \frac{1}{4}\ \frac{3}{4}\ \frac{5}{4}\ \frac{7}{4}\ \frac{9}{4}\ \frac{11}{4}\ 3 \end{matrix}$$

etc.

not included in $\mathfrak{Z}$ or $\bigcup_{m<n} \mathfrak{Z}_m$, introduce the *unsigned scaled excursion*

1) $$e_n(t) = \frac{|x(t|\mathfrak{Z}_n| + \inf \mathfrak{Z}_n)|}{\sqrt{|\mathfrak{Z}_n|}} \qquad 0 \leqq t \leqq 1^*$$

---

* $|\mathfrak{Z}_n|$ is the length of $\mathfrak{Z}_n$.

and the *sign*

2) $\quad e_n = -1$ *or* $+1$ *according as* $x(t) < 0$ *or* > 0 *for* $t \in \mathfrak{Z}_n$,

and note that $\mathfrak{Z}_n$, $\mathfrak{e}_n$, and e_n $(n \geq 1)$ are BOREL functions of the BROWNian path.

$\mathfrak{Z}$ is the support of the local time $\mathfrak{t}^+$; as such, *it is the closure of* $\mathfrak{t}^{-1}[0, +\infty)$, *where* $\mathfrak{t}^{-1}(t) = \min(s : \mathfrak{t}^+(s) \geq t)$ *is one-sided stable with exponent* $1/2$ and rate $\sqrt{2}$.

$\mathfrak{e}_1$, $\mathfrak{e}_2$, *etc. are independent and identical in law,* $\mathfrak{e}_1$ *being* MARKOV*ian with*

3a) $P_0[\mathfrak{e}_1(t) \in db] = h(0, 0, t, b)\, db = \dfrac{2e^{-b^2/2t(1-t)}}{\sqrt{2\pi t^3 (1-t)^3}} b^2\, db \quad 0 < t < 1$

3b) $P_0[\mathfrak{e}_1(t) \in db \mid \mathfrak{e}_1(s) = a] = h(s, a, t, b)\, db$

$$= \frac{e^{-(b-a)^2/2(t-s)} - e^{-(b+a)^2/2(t-s)}}{\sqrt{2\pi(t-s)}} \left(\frac{1-s}{1-t}\right)^{3/2} \frac{b\, e^{-b^2/2(1-t)}}{a\, e^{-a^2/2(1-s)}} db \quad 0 < s < t < 1.$$

e_1, e_2, *etc. are independent and identical in law with* $P_0[e_1 = -1] = P_0[e_1 = +1] = 1/2$, *i.e.,* e_n $(n \geq 1)$ *is a standard coin-tossing game.*

$\mathfrak{Z}$ *and* $\mathfrak{e}_n$ $(n \geq 1)$ *and* e_n $(n \geq 1)$ *are independent.*

P. LÉVY [*3*: 238] indicated a proof of these facts; here is a new one.

Consider the interval $\mathfrak{Z}_1$ that covers $t = 1$, the excursion $\mathfrak{e}_1$, and the sign e_1, and let us compute the expectation of

4) $$j_1[x(s) : s \leq \inf \mathfrak{Z}_1]\, j_2(\mathfrak{Z}_1)\, j_3(\mathfrak{e}_1)\, j_4(e_1)\, j_5(w^+_{\sup \mathfrak{Z}_1}),$$

where

a) $$j_1[x(s) : s \leq t] = j_1[x(s_1 \wedge t), x(s_2 \wedge t), \ldots, x(s_n \wedge t)]$$
$$0 < s_1 < s_2 < \cdots < s_n,$$
$$j_1[a_1, a_2, \ldots, a_n] \in C(R^n),$$

b) $j_2(\mathfrak{Z}_1) = j_2(\inf \mathfrak{Z}_1, \sup \mathfrak{Z}_1)$, $j_2(s, t) \in C[0, +\infty)^2$, and $j_2(\mathfrak{Z}_1) = 0$ for $|\mathfrak{Z}_1|$ near 0 or $+\infty$,

c) $j_3(\mathfrak{e}_1) = j_3[\mathfrak{e}_1(t_1), \mathfrak{e}_1(t_2), \ldots, \mathfrak{e}_1(t_n)]$, $\quad 0 < t_1 < t_2 < \cdots < t_n < 1$,
$$j_3[b_1, b_2, \ldots, b_n] \in C[0, +\infty)^n, \quad j_3 = 0$$
outside some compact figure of $[0, +\infty)^n$, and

d) $0 \leq j_5 \leq 1$ is a BOREL function of the BROWNian path.

$j_5(w^+_{\sup \mathfrak{Z}_1})$ is independent of the other factors of 4) because $\mathfrak{m} = \sup \mathfrak{Z}_1 = 1 + \mathfrak{m}_0(w^+_1)$ is a MARKOV time and $x(t) : t \leq \inf \mathfrak{Z}_1$, $\mathfrak{Z}_1$, $\mathfrak{e}_1$, and e_1 are measurable $\mathsf{B}_{\mathfrak{m}+}$. As to $j_1 j_2 j_3 j_4$, its integral is

5) $$\lim_{n\uparrow+\infty}\sum_{\substack{i\leq 2^n\\ k\geq 2^n}} E_0[(i-1)2^{-n}\leq \inf \mathfrak{Z}_1 < i2^{-n}, k2^{-n} < \sup\mathfrak{Z}_1 \leq (k+1)2^{-n},$$
$$j_1[x(s): s\leq (i-1)\,2^{-n}]\, j_2(i2^{-n}, k2^{-n})\times$$
$$\times j_3\left[\frac{x(t_1\Delta + i\,2^{-n})}{\sqrt{\Delta}}, \frac{x(t_2\Delta+i\,2^{-n})}{\sqrt{\Delta}}, \textit{etc.}\right] j_4(\operatorname{sign} x(i2^{-n}))]$$
$$\Delta = (k-i)\,2^{-n}$$
$$= \lim_{n\uparrow+\infty}\sum_{\substack{i\leq 2^n\\ k\geq 2^n}} \int_0^{+\infty} E_0[j_1[x(s): s\leq (i-1)\,2^{-n}], (i-1)\,2^{-n}\leq \inf\mathfrak{Z}_1 < i2^{-n},$$
$$|x(i2^{-n})|\in da]\, j_2(i2^{-n}, k2^{-n})$$
$$\times \int_{[0,+\infty)^n} g^-(t_1\Delta, a, b_1)\, g^-((t_2-t_1)\Delta, b_1, b_2)\ldots g^-((t_n-t_{n-1})\Delta, b_{n-1}, b_n)$$
$$\times j_3\left[\frac{b_1}{\sqrt{\Delta}}, \frac{b_2}{\sqrt{\Delta}}, \ldots, \frac{b_n}{\sqrt{\Delta}}\right] db_1\, db_2\ldots db_n \frac{1}{2}[j_4(-1)+j_4(+1)]$$
$$\times \int_0^{+\infty} g^-((1-t_n)\Delta, b_n, b)\, db\, P_b[\mathfrak{m}_0 < 2^{-n}],^{*}$$

and substituting $\sqrt{\Delta}\, b_1$, $\sqrt{\Delta}\, b_2$, *etc.* in place of b_1, b_2, *etc.* and introducing the kernel h of 3), 5) becomes

6) $$\lim_{n\uparrow+\infty}\sum_{\substack{i\leq 2^n\\ k\geq 2^n}} \int_0^{+\infty} E_0[j_1[x(s): s\leq (i-1)\,2^{-n}], (i-1)\,2^{-n}\leq \inf\mathfrak{Z}_1 < i2^{-n},$$
$$|x(i2^{-n})|\in da]$$
$$\int_0^{+\infty} g^-(\Delta, a, b)\, db\, P_b[\mathfrak{m}_0 < 2^{-n}]\, j_2(i2^{-n}, k2^{-n})$$
$$\int_{[0,+\infty)^n} h(0,0,t_1,b_1)\, h(t_1,b_1,t_2,b_2)\ldots h(t_{n-1},b_{n-1},t_n,b_n)\times$$
$$\times j_3[b_1, b_2, \ldots, b_n]\, db_1\, db_2\ldots db_n\cdot \tfrac{1}{2}[j_4(-1)+j_4(+1)]\times$$
$$\times\, \varepsilon(\Delta, a, b_1, b_n, b),$$

where

7) $$\varepsilon(\Delta, a, b_1, b_n, b)$$
$$= \sqrt{\Delta}\,\frac{g^-(t_1\Delta, a, \sqrt{\Delta}\, b_1)\, g^-((1-t_n)\,\Delta, \sqrt{\Delta}\, b_n, b)}{h(0,0,t_1,b_1)\, g^-(\Delta, a, b)}\left(\frac{1-t_n}{1-t_1}\right)^{3/2}\frac{b_1 e^{-b_1^2/2(1-t_1)}}{b_n e^{-b_n^2/2(1-t_n)}}$$
$$= s\left(\frac{ab}{\Delta}\right)^{-1} s\left(\frac{a b_1}{t_1\sqrt{\Delta}}\right) s\left(\frac{b_n b}{(1-t_n)\sqrt{\Delta}}\right) e^{-a^2(1-t_1)/2\Delta t_1}\, e^{-b^2 t_n/2\Delta(1-t_n)}.^{**}$$

---

* $g^-(t, a, b) = (2\pi t)^{-1/2}[e^{-(b-a)^2/2t} - e^{-(b+a)^2/2t}]$. ** $s(b) = b^{-1}\sinh b$.

Because of the definitions of j_2 and j_3, ε is bounded for the purposes of this integral; in addition, it tends to 1 as a and $b \downarrow 0$, with the result that

$$8) \quad E_0[j_1[x(s): s < \inf \mathfrak{Z}_1]\, j_2(\mathfrak{Z}_1)\, j_3(\mathfrak{e}_1)\, j_4(e_1)\, j_5(w^+_{\sup \mathfrak{Z}_1})]$$

$$= \lim_{n\uparrow+\infty} \sum_{\substack{i\leq 2^n\\ k\geq 2^n}} \int_0^{+\infty} E_0[j_1[x(s): s \leq (i-1)\,2^{-n}],\ (i-1)\,2^{-n} \leq \inf \mathfrak{Z}_1 < i2^{-n},$$

$$|x(i2^{-n})| \in da]$$

$$\times \int_0^{+\infty} g^-(\Delta, a, b)\, db\, P_b[\mathfrak{m}_0 < 2^{-n}]\, j_2(i2^{-n}, k2^{-n})$$

$$\times \int_{[0,+\infty)^n} h(0, 0, t_1, b_1) \quad \textit{etc.} \quad j_3\, db_1\, db_2 \ldots db_n$$

$$\times \tfrac{1}{2}[j_4(-1) + j_4(+1)]\, E_0(j_5)$$

$$= E_0[j_1[x(s): s \leq \inf \mathfrak{Z}_1]\, j_2(\mathfrak{Z}_1)]$$

$$\times \int_{[0,+\infty)^n} h(0, 0, t_1, b_1) \quad \textit{etc.} \quad j_3\, db_1\, db_2 \ldots db_n$$

$$\times \tfrac{1}{2}[j_4(-1) + j_4(+1)]\, E_0(j_5),$$

proving that $\mathfrak{e}_1$ and e_1 are subject to the correct law and are independent of $\mathfrak{Z}_1$ and of $x(t): t \notin \mathfrak{Z}_1$, *i.e.*, independent of $\mathfrak{Z}$ and of $\mathfrak{e}_n$, e_n $(n \geq 2)$; the reader is invited to complete the discussion.

Problem 1 after J. LAMPERTI [*1*]. Use problem 1.7.3 to prove that $[\mathfrak{z}(t) = t - \max(s; x(s) = 0,\ s \leq t), P_0]$ is MARKOVian with

$$P_0[\mathfrak{z}(t) > b] = \frac{2}{\pi} \cos^{-1} \sqrt{\frac{b}{t}} \qquad (b < t)$$

and

$$P_0[\mathfrak{z}(t) > b \mid \mathfrak{z}(s) = a] \qquad (a < s < t,\ b < t)$$

$$= \int_0^{(\Delta-b)\vee 0} \frac{\sqrt{a}\, d\theta}{2(\theta + a)^{3/2}}\, \frac{2}{\pi} \cos^{-1} \sqrt{\frac{b}{\Delta - \theta}} + \sqrt{\frac{a}{\Delta + a}} \quad \textit{or} \quad 0$$

according as $\Delta = t - s > b - a$ *or not*; observe that $\mathfrak{z}$ is the same as $t - \mathfrak{t}^{-1}(\mathfrak{t}^+)$ $(\mathfrak{t}^{-1}(t) = \min(s: \mathfrak{t}^+(s) = t))$ and hence identical in law to $t - \mathfrak{m}_a$ $(a = \mathfrak{t}^-(t) = \max_{s\leq t} x(s))$.

[Using $\mathfrak{z}_-$ = the biggest root of $x(t) = 0$ $(t < s)$ and $\mathfrak{z}_+$ = the smallest root of $x(t) = 0$ $(t > s)$, it is clear that

$$P_0(\mathfrak{z}_+ < t \mid \mathfrak{Z} \cap [0, s]) = P_0(\mathfrak{z}_+ < t \mid \mathfrak{z}_-) = P_0(\mathfrak{z}_+ < t \mid \mathfrak{z}(s)),$$

and so the MARKOVian nature of $\mathfrak{z}$ follows from

$$\begin{aligned}\mathfrak{z}(t) &= t - s + \mathfrak{z}(s) & \mathfrak{z}_+ > t\\ &= \mathfrak{z}(t - \mathfrak{z}_+, w^+_{\mathfrak{z}_+}) & \mathfrak{z}_+ < t;\end{aligned}$$

now use problem 1.7.3.]

Problem 2. Write G_ε for the BOREL algebra of events $B \in \mathsf{B}$ whose indicators are BOREL functions of the intersection of the graph of the standard BROWNian motion with the strip $[0, +\infty) \times [0, \varepsilon)$ and prove that, the *germ* $\mathsf{G}_{0+} = \bigcap_{\varepsilon > 0} \mathsf{G}_\varepsilon$ is identical to $\mathsf{B}(\mathfrak{Z}) \times \mathsf{B}(e_n : n \geqq 1) = \mathsf{B}(\mathfrak{l}) \times \mathsf{B}(e_n : n \geqq 1)$, up to sets of WIENER measure 0.

[Because G_{0+} is independent of $\mathsf{B}(e_n : n \geqq 1)$ and the universal algebra B is identical to $\mathsf{B}(\mathfrak{Z}) \times \mathsf{B}(e_n : n \geqq 1) \times \mathsf{B}(e_n : n \geqq 1)$, the identification of G_{0+} with $\mathsf{B}(\mathfrak{Z}) \times \mathsf{B}(e_n : n \geqq 1)$ follows from the obvious inclusion $\mathsf{G}_{0+} \supset \mathsf{B}(\mathfrak{Z}) \times \mathsf{B}(e_n : n \geqq 1)$.]

2.10. Application of the Bessel process to Brownian excursions

We saw in 2.9 that the scaled BROWNian excursion $\mathfrak{e}_1(t) : 0 \leqq t \leqq 1$ is MARKOV with

1a) $P_0[\mathfrak{e}_1(t) \in db] = h(0, 0, t, b)\, db = 2 \dfrac{e^{-b^2/2t(1-t)}}{\sqrt{\pi t^3 (1-t)^3}} b^2\, db \qquad 0 < t < 1$

1b) $P_0[\mathfrak{e}_1(t) \in db \mid \mathfrak{e}_1(s) = a]$

$$\begin{aligned}&\equiv h(s, a, t, b)\, db\\ &= \left[\frac{e^{-(b-a)^2/2(t-s)}}{\sqrt{2\pi(t-s)}} - \frac{e^{-(b+a)^2/2(t-s)}}{\sqrt{2\pi(t-s)}}\right]\left(\frac{1-s}{1-t}\right)^{3/2} \frac{b\, e^{-b^2/2(1-t)}}{a\, e^{-a^2/2(1-s)}}\, db\\ &= \frac{e^{-(a^2+b^2)/2(t-s)}}{(t-s)\sqrt{a b}}\, I_{1/2}\left(\frac{a b}{t-s}\right)\left(\frac{1-s}{1-t}\right)^{3/2} \frac{e^{-b^2/2(1-t)}}{e^{-a^2/2(1-s)}}\, b^2\, db^*\end{aligned}$$

$$0 < s < t < 1.$$

The BESSEL motion of 2.7 can be used to give a new interpretation of $\mathfrak{e}_1$: *if* $r_-(t)$ *and* $r_+(t)$ $(0 \leqq t \leqq \frac{1}{2})$ *are 2 independent copies of the radial* (BESSEL) *part of the 3-dimensional* BROWNian *motion subject to the conditional law*

2a) $$Q(B) = P_0(B \mid r(1/2) = b)$$

and if the terminal point b is subject to the law

2b) $$\frac{16}{\sqrt{2\pi}} e^{-2b^2} b^2\, db \qquad b > 0,$$

then the composite motion

3) $$\mathfrak{e}(t) = r_-(t)\ \left(0 \leqq t \leqq \frac{1}{2}\right) \quad = r_+(1-t)\ \left(\frac{1}{2} \leqq t \leqq 1\right)$$

is identical in law to $\mathfrak{e}_1$.

---

* $I_{1/2}(\xi) = (2/\pi\, \xi)^{1/2} \sinh \xi$.

Consider, for the proof, the BESSEL kernel 2.7.4) in the case $n = 3$:

$$4)\qquad g^+(t, a, b) = \frac{e^{-\frac{a^2+b^2}{2t}}}{t\sqrt{ab}} I_{1/2}\left(\frac{ab}{t}\right) b^2 \qquad t > 0, a, b > 0.$$

Given $1 > t > s > \frac{1}{2}$ and $0 < a, b$ and using the MARKOVian character of $r_+(1-t)$ $(t \geqq \frac{1}{2})$, one finds

$$5)\qquad P_0\left[r_+(1-t) \in db \,\middle|\, r_+(1-\theta) : \frac{1}{2} \leqq \theta \leqq s\right]$$

$$= \frac{g^+(1-t, 0, b)}{g^+(1-s, 0, a)} g^+(t-s, b, a)\, db \qquad a = r_+(1-s)$$

$$= \frac{e^{-\frac{a^2+b^2}{2(t-s)}}}{(t-s)\sqrt{ab}} I_{1/2}\left(\frac{ab}{t-s}\right)\left(\frac{1-s}{1-t}\right)^{3/2} \frac{e^{-b^2/2(1-t)}}{e^{-a^2/2(1-s)}} b^2\, db$$

$$\equiv h(s, a, t, b)\, db.$$

Because $\mathfrak{e}(\frac{1}{2})$ and $\mathfrak{e}_1(\frac{1}{2})$ are identical in law [see 1a) and 2b)], 5) implies that $\mathfrak{e}(t)$ $(t \geqq \frac{1}{2})$ is identical in law to $\mathfrak{e}_1(t)$ $(t \geqq \frac{1}{2})$, and to complete the proof, it suffices to note that $\mathfrak{e}$ and $\mathfrak{e}_1$ are symmetrical in law about $t = \frac{1}{2}$.

Here is an application to the fine structure of the standard 1-dimensional BROWNian path.

A. DVORETSKY and P. ERDÖS [1] proved that if $0 \leqq f$ and if $t^{-\frac{1}{2}} f(t) \in \uparrow$ for small t, then

$$6)\qquad P_0[r(t) > f(t), t \downarrow 0] = 0 \text{ or } 1 \textit{ according as } \int_{0+}^{1} t^{-3/2} f(t)\, dt \lesseqgtr +\infty$$

where P_0 is the WIENER measure for the 3-dimensional BROWNian motion starting at 0*. Now using the definition 2.9.1) of the scaled BROWNian excursions coupled with the fact that

$$|\mathfrak{Z}_n|^{-1/2} \int_0^1 t^{-3/2} f(|\mathfrak{Z}_n|\, t)\, dt = \int_0^{|\mathfrak{Z}_n|} t^{-3/2} f\, dt,$$

one finds

$$7)\qquad P_0[x(t, w^+_{\inf \mathfrak{Z}_n}) > f(t), t \downarrow 0, n \geqq 1] = 0 \textit{ or } 1 \textit{ according as}$$

$$\int_{0+}^{1} t^{-3/2} f(t)\, dt \lesseqgtr +\infty,$$

where P_0 is now the WIENER measure for the 1-dimensional BROWNian motion starting at 0 and the $\mathfrak{Z}_n$ $(n \geqq 1)$ are the open intervals of the

* See 4.12 for a proof of 6).

complement of $\mathfrak{Z} = (t : x(t) = 0)$; for example

$$8)\quad P_0\left[\,|x(t, w^+_{\inf \mathfrak{Z}_n})| > \frac{\sqrt{t}}{\lg\frac{1}{t}\,\lg_2\frac{1}{t}\cdots\left(\lg_n\frac{1}{t}\right)^{1+\varepsilon}},\ t\downarrow 0,\ n\geqq 1\right] = 0 \text{ or } 1$$

according as $\varepsilon \leqq 0$ *or* $\varepsilon > 0$.

According to 2.7.8a),

$$9)\quad P_0\left[\overline{\lim_{t\downarrow 0}}\,\frac{r(t)}{\sqrt{2t\lg_2\frac{1}{t}}} = 1\right] = 1,$$

from which one deduces

$$10)\quad P_0\left[\overline{\lim_{t\downarrow 0}}\,\frac{|\mathfrak{Z}_n|^{-1/2}\,|x(t\,|\mathfrak{Z}_n|, w^+_{\inf\mathfrak{Z}_n})|}{\sqrt{2t\lg_2\frac{1}{t}}} = \overline{\lim_{t\downarrow 0}}\,\frac{|x(t, w^+_{\inf\mathfrak{Z}_n})|}{\sqrt{2t\lg_2\frac{1}{t}}} = 1,\ n\geqq 1\right] = 1,$$

in contrast to 8).

2.11. A time substitution

Given a standard 1-dimensional Brownian motion starting at 0, the decomposition of 2.9 can be used to prove that *if* $\mathfrak{f}^{-1}$ *is the inverse function of* $\mathfrak{f}(t) =$ measure $(s : x(s) \geqq 0,\ s \leqq t)$, *then* $[x(\mathfrak{f}^{-1}), P_0]$ *is identical in law to the reflecting* Brownian *motion* $[x^+ = |x|, P_0]$ *of* 2.1.

Given a Brownian path starting at 0, throw out the negative excursions and shift the positive excursions down on the time scale,

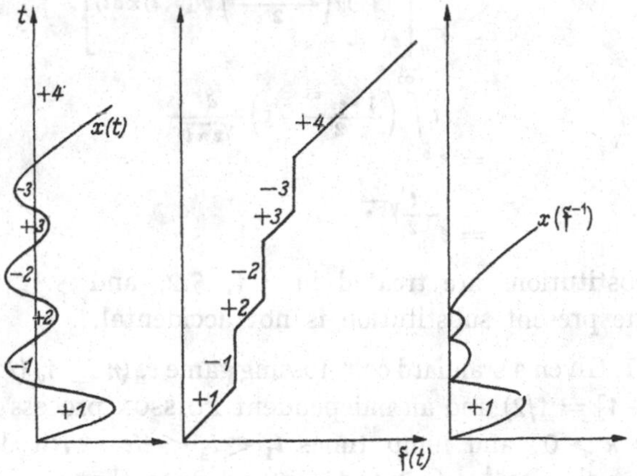

Diagram 1

closing up the gaps as indicated in the diagram; the path so defined is $x(\mathfrak{f}^{-1})$, and it is clear from the picture that, for its identification as a reflecting BROWNian motion, it suffices to show that $\mathfrak{Z}^{\bullet} = (t : x(\mathfrak{f}^{-1}(t)) = 0) = \mathfrak{f}(\mathfrak{Z}^{+})$ is identical in law to $\mathfrak{Z}^{+} = (t : x^{+}(t) = 0)$.

Because $\mathfrak{f}(\mathfrak{Z}^{+})$ is the closure of $\mathfrak{f}\,\mathfrak{t}^{-1}[0, +\infty)$, where $\mathfrak{t}^{-1}$ is the left-continuous inverse function $\min(s : \mathfrak{t}^{+}(s) \geqq t)$ of the reflecting BROWNian local time $\mathfrak{t}^{+}$, it suffices to show that $\mathfrak{f}(\mathfrak{t}^{-1})$ is a one-sided stable process with exponent $1/2$.

But $\mathfrak{f}(\mathfrak{t}^{-1})$ is $\mathfrak{t}^{-1}$ minus the jumps arising from the negative excursions:

1) $$\mathfrak{f}(\mathfrak{t}^{-1}(t)) = \sum_{\mathfrak{Z}_n \subset [0,\, \mathfrak{t}^{-1}(t))} \tfrac{1}{2}(1 + e_n)\,|\mathfrak{Z}_n|,$$

and, using the POISSON integral $\int_0^{+\infty} l\, \mathfrak{p}([0, t) \times dl)$ for $\mathfrak{t}^{-1}(t)$ (see 1.7) coupled with the fact that the coin-tossing game e_n $(n \geqq 1)$ is independen tof $\mathfrak{Z}^{+}$, it is clear that the effect on the POISSON process $\mathfrak{p}([0,t) \times dl)$ of ignoring the negative excursions is to cut its rate in half (see problem 1). $\mathfrak{f}(t^{-1})$ is then identical in law to $\mathfrak{t}^{-1}(t/2) = \int_0^{+\infty} l\,\mathfrak{p}([0, t/2) \times dl)$ $(t \geqq 0)$, and the proof is complete.

An alternative proof is to note that $\mathfrak{f}(\mathfrak{t}^{-1})$ is differential and homogeneous (see problem 2) and to use 1) to compute

2) $$\begin{aligned} &E_0[e^{-\alpha \mathfrak{f}(\mathfrak{t}^{-1}(t))}] \\ &= E_0\Big[\prod_{\mathfrak{Z}_n \subset [0,\, \mathfrak{t}^{-1}(t))} e^{-\alpha(1+e_n)|\mathfrak{Z}_n|/2}\Big] \\ &= E_0\Big[\prod_{\mathfrak{Z}_n \subset [0,\, \mathfrak{t}^{-1}(t))} \tfrac{1}{2}(1 + e^{-\alpha|\mathfrak{Z}_n|})\Big] \\ &= E_0\Big[e^{\int_0^{+\infty} \lg\left(\frac{1+e^{-\alpha l}}{2}\right) \mathfrak{p}([0,t) \times dl)}\Big] \\ &= e^{t\int_0^{+\infty}\left(\frac{1+e^{-\alpha l}}{2} - 1\right)\frac{dl}{\sqrt{2\pi l^3}}} \\ &= e^{-\frac{t}{2}\sqrt{2\alpha}} \end{aligned}$$

Time substitutions are treated in 5.1, 5.2, and 5.3; it will be seen that the present substitution is not accidental.

Problem 1. Given a standard coin-tossing game e_n $(n \geqq 1)$ $(P[e_1 = -1] = P[e_1 = +1] = 1/2)$ and an independent POISSON process with jump size 1, rate $\varkappa > 0$, and jump times $t_1 < t_2 <$ *etc.* with distribution $P[t_n - t_{n-1} > t] = e^{-\varkappa t}$ $(n \geqq 1,\ t_0 = 0)$, deduce that

$$e(t) = \sum_{t_n \leqq t} \tfrac{1}{2}(1 + e_n)$$

is POISSON with rate $\varkappa/2$.

[Given disjoint intervals Δ_1, Δ_2, *etc.* of length d_1, d_2, *etc.*,

$$P\Big[\sum_{t_n\in\Delta_1}\tfrac{1}{2}(1+e_n)=j_1,\ \sum_{t_n\in\Delta_2}\tfrac{1}{2}(1+e_n)=j_2\ \text{etc.}\Big]$$

$$=\sum_{\substack{n_1\geqq j_1\\ n_2\geqq j_2}} P\Big[\sum_{t_n\in\Delta_1}1=n_1,\ \sum_{n\leqq n_1}\tfrac{1}{2}(1+e_n)=j_1,\ \sum_{t_n\in\Delta_2}1=n_2,\ \sum_{n_1<n\leqq n_1+n_2}\tfrac{1}{2}(1+e_n)=j_2,\ \text{etc.}\Big]$$

$$=\sum_{\substack{n_1\geqq j_1\\ n_2\geqq j_2}}\frac{(\varkappa d_1)^{n_1}}{n_1!}e^{-\varkappa d_1}\binom{n_1}{j_1}2^{-n_1}\frac{(\varkappa d_2)^{n_2}}{n_2!}e^{-\varkappa d_2}\binom{n_2}{j_2}2^{-n_2}\ \text{etc.}$$

$$=\frac{(\bar\varkappa d_1)^{j_1}}{j_1!}e^{-\bar\varkappa d_1}\frac{(\bar\varkappa d_2)^{j_2}}{j_2!}e^{-\bar\varkappa d_2}\ \text{etc.}\quad\text{with}\quad \bar\varkappa=\varkappa/2.]$$

Problem 2. $\mathfrak{f}(\mathfrak{t}^{-1})$ is differential and homogeneous.

[Because of the addition rule $\mathfrak{t}^{-1}(t)=\mathfrak{t}^{-1}(s)+\mathfrak{t}^{-1}(t-s,w_{\mathfrak{m}}^{+})$ $(t>s$, $\mathfrak{m}=\mathfrak{t}^{-1}(s))$, one has $\mathfrak{f}(\mathfrak{t}^{-1}(t))-\mathfrak{f}(t^{-1}(s))=\mathfrak{f}(\mathfrak{t}^{-1}(t-s,w_{\mathfrak{m}}^{+}),w_{\mathfrak{m}}^{+})$ $(t>s)$; in addition, $\mathfrak{m}$ is a MARKOV time because $\mathfrak{m}<t$ means that $\mathfrak{t}^{+}\geqq s$ at some instant $<t$. Given $t\leqq s$,

$$(\mathfrak{f}(\mathfrak{t}^{-1}(t))<b)\cap(\mathfrak{m}<a)$$
$$\in \mathbf{B}_a,$$

i.e., $(\mathfrak{f}(\mathfrak{t}^{-1}(t))<b)\in\mathbf{B}_{\mathfrak{m}+}$ $(t\leqq s)$, and the homogeneous differential character of $\mathfrak{f}(\mathfrak{t}^{-1})$ follows on noting that the BROWNian traveller starts afresh at 0 at time $t=\mathfrak{m}$.]

3. The general 1-dimensional diffusion

3.1. Definition

Roughly speaking, a 1-dimensional diffusion is a model of the (stochastic) motion of a particle with *life time* $\mathfrak{m}_\infty\leqq+\infty$ (also called killing time), continuous path $x(t):t<\mathfrak{m}_\infty$, and *no memory*, travelling in a linear interval Q; the phrase *no memory* is meant to suggest that the motion *starts afresh* at certain (MARKOV) times including all the *constant* times $\mathfrak{m}\equiv s\geqq 0$, *i.e.*, if $\mathfrak{m}$ is a MARKOV time, then, *conditional on the present position* $x(\mathfrak{m})$, *the statistical properties of the future motion* $x(t+\mathfrak{m})$ $(t\geqq 0)$ *do not depend upon the past* $x(t)$ $(t\leqq\mathfrak{m})$.

Before stating the precise definition, introduce the following objects:

Q a subinterval of the closed line.
∞ an isolated point adjointed to Q.

w a sample path $w: [0, +\infty) \to Q \cup \infty$ with coordinates $x(t) = x(t, w) = x_t = x_t(w) \in Q \cup \infty$ and life time $\mathfrak{m}_\infty = \mathfrak{m}_\infty(w) < +\infty$ such that

$$x(t) = x(t \pm) \in Q \qquad t < \mathfrak{m}_\infty$$
$$= \infty \qquad t \geqq \mathfrak{m}_\infty,$$

esp., $x(+\infty) \equiv \infty$.

W the space of all such sample paths.

B_s the smallest BOREL algebra of subsets of W including $(w: a \leqq x(t) < b)$ for each choice of $t \leqq s$ and $a \leqq b$.

$\mathsf{B} \equiv \mathsf{B}_{+\infty}$.

$P_\bullet$ a function from Q to non-negative BOREL measures on B of total mass $+1$ such that, for each $B \in \mathsf{B}$, $P_a(B)$ is a BOREL function of $a \in Q$, and, for each $a \in Q$, $P_a[x(0) \in db]$ is the unit mass at $b = a$.

P_∞ the unit mass at the single path $x(t) \equiv \infty \quad (t \geqq 0)$.

$\mathfrak{m}$ a MARKOV time, *i.e.*, a non-negative BOREL function $\mathfrak{m} = \mathfrak{m}(w) \leqq +\infty$ of the sample path such that $(w: \mathfrak{m}(w) < t) \in \mathsf{B}_t \quad (t \geqq 0)$.

$w_\mathfrak{m}^+$ the shifted path $x(t, w_\mathfrak{m}^+) \equiv x(t + \mathfrak{m}, w) \quad (t \geqq 0)$.

$\mathsf{B}_{\mathfrak{m}+}$ the BOREL algebra of events $B \in \mathsf{B}$ such that $B \cap (\mathfrak{m} < t) \in \mathsf{B}_t$ $(t \geqq 0)$.

Q, B, $P_\bullet$ as above are said to define a *motion* D with *state interval* Q, *path space* W, and *universal Borel algebra* B. $P_a(B)$ is read: *the probability of the event B for paths starting* at a; note that $P_a[\lim_{t \downarrow 0} x(t) = x(0) = a] = P_a[\mathfrak{m}_\infty > 0] \equiv 1 \quad (a \in Q)$.

D is said to be *conservative*, if $P_a(\mathfrak{m}_\infty < +\infty) = 0 \quad (a \in Q)$.

D is said to be *simple Markov* if it *starts afresh at each constant time* $t \geqq 0$, *i.e.*, if

1a) $$P_\bullet(w_t^+ \in B \mid \mathsf{B}_t) = P_a(B) \quad a = x(t), \quad B \in \mathsf{B},$$

or, what is the same, if

1b) $$P_\bullet(x(t) \in db \mid \mathsf{B}_s) = P_a(x(t-s) \in db) \quad a = x(s), \quad t \geqq s \geqq 0.$$

D is said to be *strict Markov* or a *diffusion* if it *starts afresh at each Markov time* $\mathfrak{m}$, *i.e.*, if

2a) $$P_\bullet(w_\mathfrak{m}^+ \in B \mid \mathsf{B}_{\mathfrak{m}+}) = P_a(B) \quad a = x(\mathfrak{m}), \quad B \in \mathsf{B},$$

or, what is the same, if

2b) $$P_\bullet(x(t + \mathfrak{m}) \in db \mid \mathsf{B}_{\mathfrak{m}+}) = P_a(x(t) \in db) \quad a = x(\mathfrak{m}), \quad t \geqq 0$$

(note that $x(\mathfrak{m}) \equiv \infty$ if $\mathfrak{m} = +\infty$).

Given an exponential holding time $\mathfrak{e}$ with law $P(\mathfrak{e} > t) = e^{-t}$, if $w^\bullet$ is the sample path

3) $$x^\bullet(t) \equiv 0 \quad (t < \mathfrak{e}) \quad = t - \mathfrak{e} \quad (t \geqq \mathfrak{e})$$

and if $P_\bullet$ is chosen so as to have

4a) $$P_0(B) = P(w^\bullet \in B)$$

4b) $$P_a[x(t) \equiv a + t, t \geqq 0] = 1 \qquad a > 0,$$

then the motion so defined on $Q = [0, +\infty)$ is *simple Markov* but not *strict Markov*; indeed, $\mathfrak{m} \equiv \inf(t: x(t) > 0)$ is a MARKOV time,

$$P_0[\mathfrak{m} > 0, \mathfrak{m}(w^+_\mathfrak{m}) = 0, x(\mathfrak{m}) = 0] = 1,$$

and, should the motion start afresh at time $\mathfrak{m}$, $P_0[\mathfrak{m} > 0,\ \mathfrak{m}(w^+_\mathfrak{m}) = 0]$ would be $P_0(\mathfrak{m} > 0)\, P_0(\mathfrak{m} = 0) = 1$, which is absurd.

Given a pair of motions D and $\mathsf{D}^\bullet$, if f is a BOREL function mapping part of Q *onto* $Q^\bullet$ and if $\mathfrak{f}$ is a BOREL function mapping $(w: x(0, w) \in f^{-1}(Q^\bullet))$ *into* the dot path space such that

5a) $$f[x(0, w)] = x(0, \mathfrak{f}\, w) \qquad x(0, w) \in f^{-1}(Q^\bullet)$$

5b) $$P_a(\mathfrak{f}^{-1} B^\bullet) = P^\bullet_b(B^\bullet) \qquad a \in f^{-1}(b), \qquad B^\bullet \in \mathsf{B}^\bullet,$$

then $[\mathfrak{f}\, w, P_{f^{-1}(b)}: b \in Q^\bullet]$ is said to be a ***non-standard*** description of the dot motion as opposed to the *standard* description used before; as an example, if D is the standard 1-dimensional BROWNian motion, then

6a) $$x^+(t) = |x(t)| \qquad t \geqq 0$$

6b) $$x^-(t) = x(t) \qquad t < \mathfrak{m}_0$$

$$= \mathfrak{t}^-(t) - x(t) \qquad t \geqq \mathfrak{m}_0$$

$$\mathfrak{t}^-(t) = \max_{\mathfrak{m}_0 \leqq s \leqq t} x(s)$$

6c) $$x^\bullet(t) = x(\mathfrak{f}^{-1}) \qquad t \geqq 0$$

$$\mathfrak{f}(t) = \text{measure}(s: x(s) \geqq 0, s \leqq t)$$

are three separate non-standard descriptions of the reflecting BROWNian motion (see 2.1 and 2.12); additional examples of non-standard description will be found in 2.5 (elastic BROWNian motion), 2.10 (BESSEL motion), and problem 2 below.

Given a motion, the choice of description depends upon the problem in hand; for example, the non-standard descriptions 6a) and 6b) were useful for the discussion of the BROWNian local time $\mathfrak{t}(t, 0)$ (see 2.2) but the proof that a motion starts afresh at a MARKOV time is best done in standard description (see 3.9, 5.2, and the second part of problem 3 below).

Problem 1. Prove that 2a) includes BLUMENTHAL's 0 1 law:

$$P_\bullet(B) = 0 \text{ or } 1 \qquad B \in \mathsf{B}_{0+} \left(= \bigcap_{\varepsilon > 0} \mathsf{B}_\varepsilon \right).$$

[$\mathfrak{m} \equiv 0$ is a MARKOV time, and so, for $B \in \mathsf{B}_{0+}$,

$$P_\bullet(B) = P_\bullet(B, w^+_0 \in B) = E_\bullet(B, P_\bullet(w^+_0 \in B \mid \mathsf{B}_{0+})) = P_\bullet(B)^2.]$$

Problem 2. Explain the correspondence between the MARKOV time $\mathfrak{m}$ and BOREL algebra $\mathsf{B}_{\mathfrak{m}+}$ of

a) the standard 1-dimensional BROWNian motion

b) its non-standard description as a scaled BROWNian motion

$$[x^\bullet(t) = c\,x(t/c^2) : t \geqq 0, P_{\xi/c} : \xi \in R^1].$$

[Because $\mathsf{B}^\bullet_t \equiv \mathsf{B}[x^\bullet(s) : s \leqq t] = \mathsf{B}_{t/c^2}$, if $\mathfrak{m}^\bullet$ is MARKOV for the scaled motion, *i.e.*, if $(\mathfrak{m}^\bullet < t) \in \mathsf{B}^\bullet_t$ $(t \geqq 0)$, then $(c^{-2}\mathfrak{m}^\bullet < t) = (\mathfrak{m}^\bullet < c^2 t) \in \mathsf{B}_t$ $(t \geqq 0)$, *i.e.* $\mathfrak{m} = c^{-2}\mathfrak{m}^\bullet$ is MARKOV for the unscaled motion, and

$$\begin{aligned}\mathsf{B}^\bullet{}_{\mathfrak{m}^\bullet+} &= (B : B \cap (\mathfrak{m}^\bullet < t) \in \mathsf{B}^\bullet_t, t \geqq 0)\\ &\equiv \mathsf{B}_{\mathfrak{m}+} = (B : B \cap (\mathfrak{m} < t) \in \mathsf{B}_t, t \geqq 0).]\end{aligned}$$

Problem 3. Prove that the reflecting BROWNian motion of 2.1, the elastic BROWNian motion of 2.5, and the BESSEL motions of 2.9 are diffusions (use the standard description for the elastic BROWNian case).

3.2. Markov times

We discuss some interesting properties of MARKOV times.

Given $t \geqq 0$, define the *stopped path* $x(s, w^\bullet_t) = x(t \wedge s, w)$ $(s \geqq 0)$ and the BOREL algebra $\mathsf{B}^\bullet_t = \mathsf{B} \cap (B\colon w \in B \textit{ iff } w^\bullet_t \in B)$ and let us prove

1) $$\mathsf{B}^\bullet_t = \mathsf{B}_t.$$

Because $\mathsf{B}^\bullet_t$ contains $(w\colon x(s) < b)$ for each choice of $b \in R^1$ and $s \leqq t$, $\mathsf{B}^\bullet_t \subset \mathsf{B}_t$; as to the opposite inclusion, if $\mathsf{B}^\bullet$ is the BOREL algebra $\mathsf{B} \cap (B\colon (w\colon w^\bullet_t \in B) \in \mathsf{B}_t, t \geqq 0)$, then $(w\colon x(s) < b) \in \mathsf{B}^\bullet$ for each choice of $b \in R^1$ and $s \geqq 0$, *i.e.*, $\mathsf{B}^\bullet = \mathsf{B}$, and it follows that, if $B \in \mathsf{B}^\bullet_t$, then $B = (w : w^\bullet_t \in B) \in \mathsf{B}_t$ as desired.

A. R. GALMARINO showed us a proof of the useful fact that a *non-negative Borel function* $\mathfrak{m} = \mathfrak{m}(w) \leqq +\infty$ *is a Markov time iff*

2a) $$x_s(u) = x_s(v) \qquad s < t$$

2b) $$\mathfrak{m}(u) < t$$

implies

3) $$\mathfrak{m}(u) = \mathfrak{m}(v).$$

Consider, for the proof, a MARKOV time $\mathfrak{m}$, sample paths u and v as in 2), and an instant $s < t$. Because $B = (w : \mathfrak{m} < s) \in \mathsf{B}_s = \mathsf{B}^\bullet_s$, it is clear, using 2a), that $\mathfrak{m}(u) < s$ iff $\mathfrak{m}(u^\bullet_s) < s$ iff $\mathfrak{m}(v^\bullet_s) < s$ iff $\mathfrak{m}(v) < s$, and 3) follows on using 2b). As to the opposite implication, if $0 \leqq \mathfrak{m} \leqq +\infty$ is BOREL and if 2) implies 3), then $\mathfrak{m}(w) < t$ iff $\mathfrak{m}(w^\bullet_t) < t$, *i.e.*, $(w : \mathfrak{m} < t) \in \mathsf{B}^\bullet_t = \mathsf{B}_t$, showing that $\mathfrak{m}$ is, indeed, a MARKOV time.

Using GALMARINO's result, it is a simple matter to show that *the class of Markov times is closed under the operations*:

4a) $$\mathfrak{m}_1 \wedge \mathfrak{m}_2,$$

4b) $$\mathfrak{m}_1 \vee \mathfrak{m}_2,$$

5a) $$\mathfrak{m}_n \downarrow \mathfrak{m},$$

5b) $$\mathfrak{m}_n \uparrow \mathfrak{m},$$

6a) $$\mathfrak{m}_1 + \mathfrak{m}_2,$$

6b) $$\mathfrak{m}_1 + \mathfrak{m}_2(w^+_{\mathfrak{m}_1}),$$

6c) $$\mathfrak{m}_2(w^\bullet_{\mathfrak{m}_1}).$$

Consider, as an example, the proof that 6b) is a MARKOV time.

Given MARKOV times $\mathfrak{m}_1$ and $\mathfrak{m}_2$, if $\mathfrak{m} \equiv \mathfrak{m}_1 + \mathfrak{m}_2(w^+_{\mathfrak{m}_1})$, if $x_s(u) = x_s(v)$ $(s < t)$, and if $t > \mathfrak{m}(u)$, then $t > \mathfrak{m}_1(u)$, proving $\mathfrak{m}_1(u) = \mathfrak{m}_1(v)$; in addition,

7a) $$x_s(u^+_\theta) = x_s(v^+_\theta) \quad s < t - \theta, \quad \theta = \mathfrak{m}_1(u) = \mathfrak{m}_1(v)$$

7b) $$\mathfrak{m}_2(u^+_\theta) < t - \theta,$$

proving $\mathfrak{m}_2(u^+_\theta) = \mathfrak{m}_2(v^+_\theta)$, and adding,

8) $$\mathfrak{m}(u) = \mathfrak{m}_1(u) + \mathfrak{m}_2(u^+_\theta) = \mathfrak{m}_1(v) + \mathfrak{m}_2(v^+_\theta) = \mathfrak{m}(v),$$

completing the proof.

As in 1.7, the BOREL algebra

$$\mathsf{B}_{\mathfrak{m}+} = \mathsf{B} \cap (B : B \cap (\mathfrak{m} < t) \in \mathsf{B}_t, t \geq 0)^*$$

measures $\mathfrak{m}$ and satisfies

9a) $$\mathsf{B}_{\mathfrak{m}_1+} \subset \mathsf{B}_{\mathfrak{m}_2+} \qquad \mathfrak{m}_1 \leq \mathfrak{m}_2,$$

9b) $$\bigcap_{n \geq 1} \mathsf{B}_{\mathfrak{m}_n+} = \mathsf{B}_{\mathfrak{m}+} \qquad \mathfrak{m}_n \downarrow \mathfrak{m},$$

10a) $$\mathsf{B}_{\mathfrak{m}+} \supset \bigcap_{\varepsilon > 0} \mathsf{B}[x(t \wedge (\mathfrak{m} + \varepsilon)) : t \geq 0];$$

in fact

10b) $$\mathsf{B}_{\mathfrak{m}+} = \bigcap_{\varepsilon > 0} \mathsf{B}[x(t \wedge (\mathfrak{m} + \varepsilon)) : t \geq 0]$$

(see problem 1 below); in addition,

10c) $$\mathsf{B}_{\mathfrak{m}+} = \mathsf{B} \cap (B : w \in B \text{ iff } w^\bullet_t \in B, t > \mathfrak{m}(w))$$

as will now be proved.

Given $B \in \mathsf{B}$, if $w \in B$ iff $w^\bullet_t \in B$ for $t > \mathfrak{m}(w)$, then $B \cap (\mathfrak{m} < t) = (w : w^\bullet_t \in B) \cap (\mathfrak{m} < t)$ is a member of B_t as is clear from 1), *i.e.*, $B \in \mathsf{B}_{\mathfrak{m}+}$. As to the opposite inclusion, if $B \in \mathsf{B}_{\mathfrak{m}+}$, then $B \cap (\mathfrak{m} < t) \in \mathsf{B}^\bullet_t$, and

---

* $(\mathfrak{m} < t)$ is short for $(w : \mathfrak{m} < t)$.

using GALMARINO's theorem in case $t > \mathfrak{m}(w)$, one finds $w \in B$ iff $w \in B \cap (\mathfrak{m} < t)$ iff $w_t^\bullet \in B \cap (\mathfrak{m} < t)$ iff $w_t^\bullet \in B$, completing the proof.

Problem 1. $\mathsf{B}_{\mathfrak{m}+} = \bigcap_{\varepsilon > 0} \mathsf{B}[x(t \wedge (\mathfrak{m} + \varepsilon)) : t \geqq 0]$ (see 10b)).

[Given $B \in \mathsf{B}_{\mathfrak{m}+}$, $w \in B$ iff $w_t^\bullet \in B$ on $(\mathfrak{m} < t)$ thanks to 10c). $B \in \mathsf{B}[x(t \wedge (\mathfrak{m} + \varepsilon)) : t \geqq 0]$ can be deduced from this as in the proof of 1).]

Problem 2. Prove that the BOREL extension of $\mathsf{B}_{\mathfrak{m}+}$ and $\mathsf{B}[x(t, w_{\mathfrak{m}}^+) : \geqq 0]$ is the universal BOREL algebra B.

$$[x(t) = x(t \wedge \mathfrak{m}) \times \textit{the indicator of } (\mathfrak{m} \geqq t) + \\ + \lim_{n \uparrow +\infty} \sum_{k 2^{-n} < t} x(t - (k-1) 2^{-n}, w_{\mathfrak{m}}^+) \times \\ \times \textit{the indicator of } ((k-1) 2^{-n} \leqq \mathfrak{m} < k 2^{-n})$$

for each $t \geqq 0$; now use the fact that $\mathfrak{m}$ and $x(t \wedge \mathfrak{m})$ are measurable $\mathsf{B}_{\mathfrak{m}+}$.]

Problem 3.

$$\begin{aligned} \mathfrak{m}_\infty &= \min(t : x(t) = \infty), \\ \mathfrak{m}_{0\infty} &= \inf(t : x(t) > 0), \\ \mathfrak{m}_0 &= \min(t : x(t) = 0), \\ \mathfrak{m}_{0+} &= \lim_{\varepsilon \downarrow 0} \min(t : x(t) = \varepsilon) \end{aligned}$$

are all MARKOV times.

$$[(\mathfrak{m}_\infty < t) = \bigcup_{k 2^{-n} < t} (w : x(k 2^{-n}) = \infty) \in \mathsf{B}_t \quad (t \geqq 0), \ i.e.,$$

$\mathfrak{m}_\infty$ is MARKOV; the proof that $\mathfrak{m}_{0\infty}$ is a MARKOV time is similar. $\mathfrak{m}_0$ is MARKOV because $\mathfrak{m}_0 < t$ means that the sample path tends to 0 along some decreasing series of times $k 2^{-n} < t$. $\mathfrak{m}_{0+}$ is MARKOV because $\mathfrak{m}_{0+} < t$ means that the points $x(k 2^{-n}) : k 2^{-n} < t$ cluster to 0 from above.]

Problem 4. $P_\bullet[\mathfrak{m} > 0] = 0$ or 1 if $\mathfrak{m}$ is a MARKOV time.

[$(\mathfrak{m} > 0) = \bigcup_{m \geqq 1} \bigcap_{n \geqq m} (\mathfrak{m} \geqq n^{-1}) \in \bigcap_{m \geqq 1} \mathsf{B}_{m^{-1}} = \mathsf{B}_{0+}$; now use BLUMENTHAL's 01 law.]

Problem 5. $(\mathfrak{m}_1 < \mathfrak{m}_2) \in \mathsf{B}_{\mathfrak{m}_1+}$.

$$[(\mathfrak{m}_1 < \mathfrak{m}_2) \cap (\mathfrak{m}_1 < t) = (\mathfrak{m}_1 < \mathfrak{m}_2 < t) \cup (\mathfrak{m}_1 < t \leqq \mathfrak{m}_2) \\ = (\mathfrak{m}_1(w_t^\bullet) < \mathfrak{m}_2(w_t^\bullet) < t) \cup (\mathfrak{m}_1 < t \leqq \mathfrak{m}_2) \in \mathsf{B}_t \quad (t \geqq 0),$$

thanks to 1) and GALMARINO's theorem.]

Problem 6.

$$\mathfrak{m}_1(w_{\mathfrak{m}_2}^\bullet) \wedge \mathfrak{m}_2 = \mathfrak{m}_1 \wedge \mathfrak{m}_2.$$

[Using GALMARINO's theorem, if $\mathfrak{m}_1 < \mathfrak{m}_2$, then $\mathfrak{m}_1(w_{\mathfrak{m}_2}^\bullet) = \mathfrak{m}_1$, while, if $\mathfrak{m}_1 \geqq \mathfrak{m}_2$, then *either* $\mathfrak{m}_1(w_{\mathfrak{m}_2}^\bullet) \geqq \mathfrak{m}_2$ *or* $\mathfrak{m}_1(w_{\mathfrak{m}_2}^\bullet) < \mathfrak{m}_2$, in which case $\mathfrak{m}_1 = \mathfrak{m}_1(w_{\mathfrak{m}_2}^\bullet) < \mathfrak{m}_2$, contradicting $\mathfrak{m}_1 \geqq \mathfrak{m}_2$.]

Problem 7. Show that, for the exit time $\mathfrak{m} = \inf(t: x(t) \neq 0)$,

$$P_0[\mathfrak{m} > t] = e^{-\varkappa t} \quad 0 \leqq \varkappa \leqq +\infty, \qquad t \geqq 0$$

and

$$P_0[\mathfrak{m} = \mathfrak{m}_\infty] = 1 \qquad \varkappa < +\infty.$$

[Because $(\mathfrak{m} > s) \in \mathsf{B}_{s+}$ $(s \geqq 0)$, $\mathfrak{m}$ is a MARKOV time and $P_0[\mathfrak{m} > t + s] = P_0[\mathfrak{m} > s,\ \mathfrak{m}(w_s^+) > t] = P_0[\mathfrak{m} > s]\, P_0[\mathfrak{m} > t]$, proving $P_0[\mathfrak{m} > t] = e^{-\varkappa t}$. Given $\varkappa < +\infty$, $P_0[\mathfrak{m} > 0] = 1$, and since $\mathfrak{m} \leqq \mathfrak{m}_\infty$, it is enough to use $(\mathfrak{m} < \mathfrak{m}_\infty) \in \mathsf{B}_{\mathfrak{m}+}$ (see problem 5) to show that $P_0[\mathfrak{m} < \mathfrak{m}_\infty] = P_0[\mathfrak{m} < \mathfrak{m}_\infty,\ \mathfrak{m}(w_\mathfrak{m}^+) = 0] \leqq P_0[\mathfrak{m} = 0] = 0$.]

Problem 8 (problem 7 continued). $\mathfrak{Z} = (t: x(t) = 0)$ has no interior point in case $P_0[\mathfrak{m} = 0] = 1$ $(\varkappa = +\infty)$.

3.3. Matching numbers

Given $a \leqq b \leqq c$, if $x(0) = a$, then $\mathfrak{m}_c = \mathfrak{m}_b + \mathfrak{m}_c(w_\mathfrak{m}^+)$ $(\mathfrak{m} = \mathfrak{m}_b)$, and since $\mathfrak{m}$ is measurable $\mathsf{B}_{\mathfrak{m}+}$ and the path starts afresh at time $t = \mathfrak{m}$ at the place $x(\mathfrak{m}) = b$,

$$1) \qquad E_a(e^{-\mathfrak{m}_c}) = E_a\left[e^{-\mathfrak{m}_b} E_a\left(e^{-\mathfrak{m}_c(w_\mathfrak{m}^+)} \,\middle|\, \mathsf{B}_{\mathfrak{m}+}\right)\right] = E_a(e^{-\mathfrak{m}_b})\, E_b(e^{-\mathfrak{m}_c}).$$

$E_a(e^{-\mathfrak{m}_b})$ is therefore monotone in a and b $(a \leqq b)$, proving the existence of the limits

2a) $$e_1 = \lim_{b \downarrow a} E_a(e^{-\mathfrak{m}_b}) = E_a(e^{-\mathfrak{m}_{a+}})$$

2b) $$e_2 = \lim_{b \downarrow a} \lim_{c \downarrow a} E_c(e^{-\mathfrak{m}_b}) = \lim_{b \downarrow a} E_{a+}(e^{-\mathfrak{m}_b})$$

2c) $$e_3 = \lim_{b \downarrow a} E_b(e^{-\mathfrak{m}_a}) = E_{a+}(e^{-\mathfrak{m}_a})$$

2d) $$e_4 = \lim_{b \downarrow a} E_b(e^{-\mathfrak{m}_{a+}}),$$

in which $\mathfrak{m}_{a+} \equiv \lim_{b \downarrow a} \mathfrak{m}_b$, *i.e.*, $\mathfrak{m}_{a+} = \inf(t: x(t) > a)$ or $\min(t: \lim_{s \uparrow t} x(s) = a$ according as $x(0) \leqq a$ or $x(0) > a$.

We are going to prove that

3a) $$e_1 = 0 \text{ or } 1$$

3b) $$e_2 = 0 \text{ or } 1$$

3c) $$e_4 = 0 \text{ or } 1$$

4a) $$e_1 = e_1 e_2 \leqq e_2$$

4b) $$e_3 = e_3 e_4 \leqq e_4$$

5) $$e_1 e_3 = e_2 e_4.$$

Given a sample path starting at a, if $\mathfrak{m} = \mathfrak{m}_{a+} = \inf(t: x(t) > a)$ then, $\mathfrak{m}(w_\mathfrak{m}^+) = 0$ and $x(\mathfrak{m}) = a$ in case $\mathfrak{m} < +\infty$, so

$$6) \quad e_1 = E_a(e^{-\mathfrak{m}_{a+}}) = E_a\left[e^{-\mathfrak{m}},\ \mathfrak{m} < +\infty,\ x(\mathfrak{m}) = a,\ E_a\left(e^{-\mathfrak{m}(w_\mathfrak{m}^+)} \,\middle|\, \mathsf{B}_{\mathfrak{m}+}\right)\right] = e_1^2,$$

proving 3a); 3b) is clear from

7) $$e_2 = \lim_{b\downarrow a} E_{a+}(e^{-\mathfrak{m}_b}) = \lim_{b\downarrow a}\lim_{c\downarrow a} E_{a+}(e^{-\mathfrak{m}_c})\, E_c(e^{-\mathfrak{m}_b})$$
$$= e_2 \lim_{b\downarrow a} E_{a+}(e^{-\mathfrak{m}_b}) = e_2^2;$$

3c) from

8) $$e_4 = \lim_{b\downarrow a} E_b(e^{-\mathfrak{m}_{a+}}) = \lim_{b\downarrow a}\lim_{c\downarrow a} E_b(e^{-\mathfrak{m}_c})\, E_c(e^{-\mathfrak{m}_{a+}})$$
$$= \lim_{b\downarrow a} E_b(e^{-\mathfrak{m}_{a+}})\, e_4 = e_4^2;$$

4a) and 4b) are trivial; and, for the proof of 5), it is enough to cite the fact that, in case $x(0) > a$, $\mathfrak{m}_a > \mathfrak{m}_{a+}$ implies $+\infty = \mathfrak{m}_a > \mathfrak{m}_{a+} = \mathfrak{m}_\infty$ and to infer that

9) $$e_1 e_3 = \lim_{b\downarrow a}\lim_{c\downarrow a} E_c\left[e^{-\mathfrak{m}_a} e^{-\mathfrak{m}_b(w^+_{\mathfrak{m}_a})}\right] = \lim_{b\downarrow a}\lim_{c\downarrow a} E_c\left[e^{-\mathfrak{m}_{a+}} e^{-\mathfrak{m}_b(w^+_{\mathfrak{m}_{a+}})}\right]$$
$$= \lim_{b\downarrow a}\lim_{c\downarrow a}\lim_{d\downarrow a} E_c\left[e^{-\mathfrak{m}_d} e^{-\mathfrak{m}_b(w^+_{\mathfrak{m}_d})}\right]^*$$
$$= \lim_{b\downarrow a}\lim_{c\downarrow a}\lim_{d\downarrow a} E_c(e^{-\mathfrak{m}_d})\, E_d(e^{-\mathfrak{m}_b}) = e_2 e_4.$$

We can now prove

10a) $$e_1 = P_a(\mathfrak{m}_{a+} < +\infty) = P_a(\mathfrak{m}_{a+} = 0)$$

10b) $$e_2 = \lim_{b\downarrow a} P_{a+}(\mathfrak{m}_b < \varepsilon)$$

10c) $$e_3 = \lim_{b\downarrow a} P_b(\mathfrak{m}_a < +\infty) = \lim_{b\downarrow a} P_b(\mathfrak{m}_a < \varepsilon)$$

10d) $$e_4 = \lim_{b\downarrow a} P_b(\mathfrak{m}_{a+} < +\infty) = \lim_{b\downarrow a} P_b(\mathfrak{m}_{a+} < \varepsilon),$$

in which $\varepsilon > 0$ is chosen at pleasure; indeed 10a), 10b), and 10d) are clear from the 01 laws 3a), 3b), and 3c) and the fact that, for $a \leqq b$, $P_a(\mathfrak{m}_b < +\infty)$ and $P_a(\mathfrak{m}_b < \varepsilon)$ are monotone in a and b. 10c) follows from

11) $$0 \leqq P_b(\mathfrak{m}_a < +\infty) - E_b(e^{-\mathfrak{m}_a})$$
$$= P_b(\mathfrak{m}_a < +\infty) - E_b(e^{-\mathfrak{m}_{a+}}, \mathfrak{m}_a < +\infty)$$
$$= E_b(1 - e^{-\mathfrak{m}_{a+}}, \mathfrak{m}_a < +\infty) \leqq E_b(1 - e^{-\mathfrak{m}_{a+}}) \downarrow 0 \quad \textit{if } b\downarrow a,\ e_4 = 1$$
$$\leqq P_b(\mathfrak{m}_a < +\infty) = 0 \quad \textit{if } b > a,\ e_4 = 0$$

and

12) $$P_b(\varepsilon \leqq \mathfrak{m}_a < +\infty) \leqq P_b(\varepsilon \leqq \mathfrak{m}_{a+} < +\infty) \leqq P_b(\varepsilon \leqq \mathfrak{m}_{a+}) \downarrow 0 \quad \textit{if } b\downarrow a,\ e_4 = 1$$
$$\leqq P_b(\mathfrak{m}_{a+} < +\infty) = 0 \quad \textit{if } b > a,\ e_4 = 0.$$

---

* $$\lim_{d\downarrow a}[\mathfrak{m}_d + \mathfrak{m}_b(w^+_{\mathfrak{m}_d})] = \lim_{d\downarrow a} \min(t : x(t) = b, t > \mathfrak{m}_d) = \min(t : x(t) = b, t > \mathfrak{m}_{a+})$$
$$= \mathfrak{m}_{a+} + \mathfrak{m}_b(w^+_{\mathfrak{m}_{a+}}).$$

$2^3 = 8$ combinations of the values of e_1, e_2, e_4 are possible. 3 of these violate 4a), 4b), or 5), leaving 5 possible cases:

e_1	e_2	e_3	e_4
1	1	1	1
1	1	0	0
0	1	0	0
0	0		1
0	0	0	0

(the values of e_3 follow from 4b) and 5); the blank indicates that all values $0 \leqq e_3 \leqq 1$ are possible).

Consider, now, the corresponding *left limits* e_1^-, e_2^-, *etc.* and let $e_1 = e_1^+$, $e_2 = e_2^+$, *etc.* $5^2 = 25$ combinations of e_1^-, e_2^-, e_4^- and e_1^+, e_2^+, e_4^+ are possible. 3 combinations violate the fact that $e_1^- e_1^+ = 1$ implies $e_3^- e_3^+ = 1$ (see problem 1), leaving 22 possible cases.

Problem 1. $e_3^- e_3^+ = 1$ if $e_1^- e_1^+ = 1$.

$$\Big[1 = e_1^- = E_a(e^{-\mathfrak{m}_{a-}}) = \lim_{c\uparrow a} E_a(e^{-\mathfrak{m}_c}) = \lim_{c\uparrow a} E_a\big[e^{-\mathfrak{m}_{a+}}\, e^{-\mathfrak{m}_c(w^+_{\mathfrak{m}_{a+}})}\big]$$

$$= \lim_{c\uparrow a} \lim_{b\downarrow a} E_a\big[e^{-\mathfrak{m}_b - \mathfrak{m}_c(w^+_{\mathfrak{m}_b})}\big] \leqq \lim_{b\downarrow a} E_b(e^{-\mathfrak{m}_a}) = e_3^+ \text{ as in 9), } \textit{etc.}\Big]$$

Problem 2. Give concrete examples of the 5 possible combinations of e_1^+, e_2^+, e_3^+, e_4^+.

3.4. Singular points

Keeping the 01 law

$$1) \qquad e_1^\pm = P_\xi(\mathfrak{m}_{\xi\pm} = 0) = 0 \text{ or } 1$$

in mind, the point $\xi \in Q$ is defined to be *regular* if $e_1^- = e_1^+ = 1$, *singular* ($\xi \in K_- \cup K_+$) if $e_1^- e_1^+ = 0$, *left singular* ($\xi \in K_-$) if $e_1^+ = 0$, *right singular* ($\xi \in K_+$) if $e_1^- = 0$, a *left shunt* if $e_1^- = 1$ and $e_1^+ = 0$, a *right shunt* if $e_1^- = 0$ and $e_1^+ = 1$, a *trap* ($\xi \in K_- \cap K_+$) if $e_1^- = e_1^+ = 0$.

K_- (= the class of left singular points) *is closed from the right in* Q, meaning that if the left singular points $b_1 > b_2 > b_3$ *etc.* $\downarrow a \in Q$, then a is left singular also; for, in that case, $P_a(\mathfrak{m}_b < +\infty) = 0$ $(a < b)$ because (a, b) contains points that cannot be crossed from the left.

K_+ (= the class of right singular points) *is closed from the left in* Q, meaning that if the right singular points $b_1 < b_2 < b_3$ *etc.* $\uparrow a \in Q$, then a is right singular also; the proof is the same.

K_- and K_+ are Borel sets; for example, if the points, a_1, a_2, *etc.* are dense in K_+ and include all its left isolated points, then the fact

that K_+ is left closed in Q implies

$$K_+ = Q \cap \Big[\bigcap_{m \geqq 1} \bigcup_{n \geqq 1} [a_n, a_n + m^{-1}) \Big].$$

$Q - K_- \cup K_+$ (= the class of regular points) *is open in the line* (*i.e.*, not just relative to Q); indeed, if ξ is regular, then $e_1^- = e_1^+ = 1$ which implies $e_3^- = e_3^+ = 1$, or, what is the same, $P_a(\mathfrak{m}_b < +\infty) \cdot P_b(\mathfrak{m}_a < +\infty) > 0$ for some $a < \xi < b$, and the whole neighborhood (a, b) must then be free of singular points.

$K_- \cup K_+$ (= the class of singular points) *is then closed in* Q.

3.5. Decomposing the general diffusion into simple pieces

Given a diffusion D, if either $P_a(\mathfrak{m}_b < +\infty)$ or $P_b(\mathfrak{m}_a < +\infty)$ is positive, then a and b are said to be in *direct communication*; if there is a finite chain of points c_1, c_2, *etc.*, leading from a to b, such that c_1 is in direct communication with c_2, c_2 with c_3, *etc.*, then a and b are said to be in *indirect communication*. On Q, (indirect) communication satisfies the following rules: each point communicates with itself; if a and b communicate, then so do b and a; if a and b communicate and also b and c, then so do a and c.

Q therefore splits into *non-communicating classes Q^* with perfect (indirect) communication inside each class* and this splitting induces a splitting of D into *autonomous diffusions*

$$\mathsf{D}^* = [W^* = W(Q^*), \mathsf{B}^* = \mathsf{B}(W^*), P_a^* = P_a : a \in Q^*].$$

D^* is trivial if Q^* is a single point 0, for then

$$P_0[x(t) = 0, t < \mathfrak{m}_\infty] = 1, \; f(t) = P_0(\mathfrak{m}_\infty > t) = P_0[x(s) = 0, s \leqq t]$$

satisfies

1) $\quad f(t+s) = E_0[\mathfrak{m}_\infty > s, x(s) = 0, P_0(\mathfrak{m}_\infty(w_s^+) > t \mid \mathsf{B}_s)] = f(t)\, f(s),$

and therefore

2) $$P_0(\mathfrak{m}_\infty > t) = e^{-\varkappa t} \qquad t \geqq 0, 0 \leqq \varkappa < +\infty.$$

When Q^* is not a point, it is an interval, for if $a, b \in Q^*$, then the existing communication between them implies that each point of (a, b) communicates with a; thus, $[a, b] \subset Q^*$ and the result follows.

Consider the (open) set of regular points in Q^*, split it into disjoint open intervals, single out the *special intervals* that end in opposing shunts, and call the end points of the special intervals and the traps belonging to int Q^*, *special singular points*.

Given special singular points $c_1 > c_2 > c_3$ *etc.* $\downarrow a \in \text{int}\, Q^*$,

$$P_a(\mathfrak{m}_{a+} < +\infty) = P_{a+}(\mathfrak{m}_a < +\infty) = 0,$$

violating the fact that $Q^* \cap (b \leqq a)$ and $Q^* \cap (b > a)$ are in (indirect) communication, and, making a similar argument for special singular points $c_1 < c_2 < c_3$ *etc.* $\uparrow a \in \operatorname{int} Q^*$, it develops that the special singular points are finite in number on each closed subinterval of Q^*.

Thus, the series of special singular points cuts Q^* into non-overlapping intervals $Q^\bullet$ (closed at the cuts), each with perfect (indirect) communication and no internal special singular points.

$Q^\bullet$ contains either no internal singular points or just one kind of internal shunt: for if $Q^\bullet$ contains two opposing internal shunts $c_1 < c_2$, then there is a maximal shunt $c_1 \leqq c_3 < c_2$ of the same kind as c_1 and a minimal shunt $c_3 < c_4 \leqq c_2$ of the same kind as c_2, and either (c_3, c_4) is a special interval, violating the choice of $Q^\bullet$ or (c_3, c_4) contains singular points, violating the choice of c_3 or c_4.

Consider, for the next step, the exit time $\mathfrak{e} = \inf(t : x(t) \notin Q^\bullet)$. $\mathsf{D}^\bullet = [x(t, w_\mathfrak{e}^\bullet) : t \geqq 0, P_a^\bullet = P_a : a \in Q^\bullet]$ is, itself, a diffusion (see 3.9); moreover, a particle starting in an interval $Q^\bullet = Q_1^\bullet$ performs the corresponding diffusion $\mathsf{D}^\bullet = \mathsf{D}_1^\bullet$, up to the exit time $\mathfrak{e} = \mathfrak{e}_1$ $(\leqq +\infty)$; if $\mathfrak{e}_1 < +\infty$, it enters into a new interval $Q^\bullet = Q_2^\bullet$ (never coming back to $Q_1^\bullet - Q_2^\bullet$ at all), performing the corresponding diffusion $\mathsf{D}_2^\bullet$ up to the exit time $\mathfrak{e}_2$ $(\leqq +\infty)$, *etc.* D^* is thus decomposed into the simpler diffusions $\mathsf{D}^\bullet$. Up to a change of the sign of the scale on the intervals $Q^\bullet$ without internal left singular points, 9 separate cases arise:

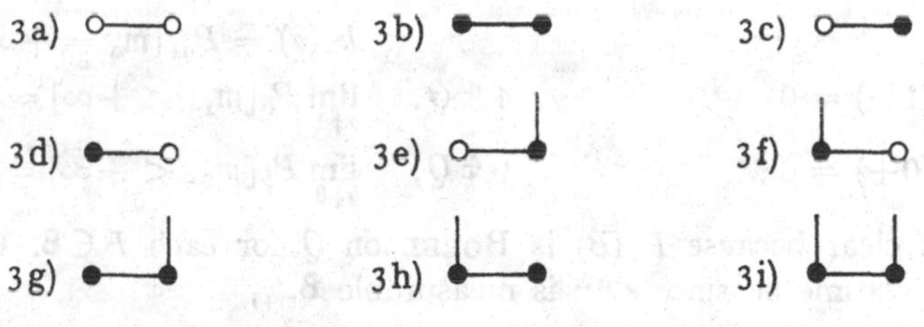

in which o— and —o mark the end points not belonging to $Q^\bullet$, •— and —• the end points belonging to $Q^\bullet - (K_- \cap K_+)$, and ↓ the end points belonging to $Q^\bullet \cap (K_- \cap K_+)$, and, in each case, $Q^\bullet$ has perfect (indirect) communication and no internal left singular points, so that if $\inf Q^\bullet < a < b < \sup Q^\bullet$, then $P_a(\mathfrak{m}_b < +\infty) > 0$, *i.e.*, *direct* communication prevails on $\operatorname{int} Q^\bullet$; indeed, if $a \in \operatorname{int} Q^\bullet$, then $0 < P_b(\mathfrak{m}_c < +\infty)$ for some $b < a < c$, because, if not, then a is a right shunt, *i.e.*, $P_a(\mathfrak{m}_{a+} = 0) = P_a(\mathfrak{m}_{a-} = +\infty) = 1$, and $P_{a-}(\mathfrak{m}_a < +\infty) = 0$ would prohibit communication between $Q^\bullet \cap (b < a)$ and $Q^\bullet \cap (b > a)$; now a simple HEINE-BOREL covering completes the proof.

In the complicated bits below, case 3g) will often be singled out, the other cases being left to the reader.

3.6. Green operators and the space D

Given a diffusion D on an interval Q with endpoints 0 and 1, let $B(Q)$ be the space of (real) BOREL functions f defined on $Q \cup \infty$ such that

1a) $$f(\infty) = 0$$

1b) $$\|f\| < +\infty,$$

and, introducing *the Green operators*

2) $$G_\alpha : f \in B(Q) \to E_\bullet\left[\int_0^{+\infty} e^{-\alpha t} f(x_t)\, dt\right] \qquad \alpha > 0,$$

let us show that $u = G_\alpha f$ satisfies

3) $u \in B(Q)$

4a) $u(a-) = u(a)$ $\qquad P_a[\mathfrak{m}_{a-} = 0] = 1$

4b) $u(a+) = u(a)$ $\qquad P_a[\mathfrak{m}_{a+} = 0] = 1$

5a) $u(a-)$ *exists* $\qquad \lim_{b\uparrow a} E_{a-}(e^{-\mathfrak{m}_b}) = 1$

5b) $u(a+)$ *exists* $\qquad \lim_{b\downarrow a} E_{a+}(e^{-\mathfrak{m}_b}) = 1$

6a) $u(a-) = [1 - k_+(a)]\, u(a)$ $\qquad \lim_{b\uparrow a} P_b[\mathfrak{m}_{a-} < +\infty] = 1,$
$\qquad k_+(a) = P_{a-}[\mathfrak{m}_a = +\infty]$

6b) $u(a+) = [1 - k_-(a)]\, u(a)$ $\qquad \lim_{b\downarrow a} P_b[\mathfrak{m}_{a+} < +\infty] = 1,$
$\qquad k_-(a) = P_{a+}[\mathfrak{m}_a = +\infty]$

7a) $u(1-) = 0$ $\qquad 1 \notin Q, \quad \lim_{b\uparrow 1} P_b[\mathfrak{m}_{1-} < +\infty] = 1$

7b) $u(0+) = 0$ $\qquad 0 \notin Q, \quad \lim_{b\downarrow 0} P_b[\mathfrak{m}_{0+} < +\infty] = 1.$

3) is clear because $P_\bullet(B)$ is BOREL on Q for each $B \in \mathsf{B}$. Given a MARKOV time $\mathfrak{m}$, since $e^{-\alpha\mathfrak{m}}$ is measurable $\mathsf{B}_{\mathfrak{m}+}$,

8) $$\begin{aligned} u = G_\alpha f &= E_\bullet\left[\int_0^{\mathfrak{m}} e^{-\alpha t} f(x_t)\, dt + e^{-\alpha\mathfrak{m}} \int_0^{+\infty} e^{-\alpha t} f(x(t, w_{\mathfrak{m}}^+))\, dt\right] \\ &= E_\bullet\left[\int_0^{\mathfrak{m}} e^{-\alpha t} f(x_t)\, dt\right] + E_\bullet\left[e^{-\alpha\mathfrak{m}} E_\bullet\left(\int_0^{+\infty} e^{-\alpha t} f(x(t, w_{\mathfrak{m}}^+))\, dt \,\Big|\, \mathsf{B}_{\mathfrak{m}+}\right)\right] \\ &= E_\bullet\left[\int_0^{\mathfrak{m}} e^{-\alpha t} f(x_t)\, dt\right] + E_\bullet[e^{-\alpha\mathfrak{m}} u(x_{\mathfrak{m}})]; \end{aligned}$$

4), 5), 6), 7) are going to follow from this.

Beginning with 4b), if $P_a[\mathfrak{m}_{a+} = 0] = 1$, then 8) implies

9) $$u(a) = \lim_{b\downarrow a}\left[E_a\left[\int_0^{\mathfrak{m}_b} e^{-\alpha t} f(x_t)\, dt\right] + E_a(e^{-\alpha\mathfrak{m}_b})\, u(b)\right] = \lim_{b\downarrow a} u(b),$$

in which the existence of the limit is part of the assertion; as to 5b, if $\lim_{b\downarrow a} E_{a+}(e^{-\mathfrak{m}_b}) = 1$, then 8) implies

$$10)\quad \overline{\lim_{c\downarrow a}}\, u(c) = \lim_{b\downarrow a}\overline{\lim_{c\downarrow a}}\left[E_c\left[\int_0^{\mathfrak{m}_b} e^{-\alpha t} f(x_t)\,dt\right] + E_c(e^{-\alpha\,\mathfrak{m}_b})\,u(b)\right] = \underline{\lim_{b\downarrow a}}\, u(b)$$

as desired; and, as to 6b), if $\lim_{b\downarrow a} P_b[\mathfrak{m}_{a+} < +\infty] = 1$ and if $P_{a+}[\mathfrak{m}_a < +\infty] = 1 - k_-(a)$, then [see 3.3.10c)] $E_{a+}(e^{-\alpha\mathfrak{m}_a}) = 1 - k_-(a)$, and, using 8) coupled with $u(\infty) = 0$, it is found that

$$\begin{aligned}
11)\quad \lim_{b\downarrow a} u(b) &= \lim_{b\downarrow a}\left(E_b\left[\int_0^{\mathfrak{m}_{a+}} e^{-\alpha t} f(x_t)\,dt\right] + E_b[e^{-\alpha\mathfrak{m}_{a+}} u(x(\mathfrak{m}_{a+}))]\right)\\
&= \lim_{b\downarrow a} E_b[e^{-\alpha\mathfrak{m}_{a+}},\ \mathfrak{m}_{a+} = \mathfrak{m}_a < +\infty]\,u(a)\\
&= \lim_{b\downarrow a} E_b(e^{-\alpha\mathfrak{m}_a})\,u(a)\\
&= [1 - k_-(a)]\,u(a).
\end{aligned}$$

Because $\mathfrak{m}_{0+} = \mathfrak{m}_\infty$, if $\mathfrak{m}_{0+} < +\infty$ and $0 \notin Q$, 7b) is a special case of 6b); the proofs of 4a, 5a, 6a, 7a) are similar (see D. B. RAY [2] for the original of this method).

A special case of 3, 4, 5, 6, 7) is the fact that *the* GREEN *operators map $B(Q)$ into the space D of functions $f \in B(Q)$ such that*

$$12\text{a})\qquad \lim_{b\uparrow a} f(b) = f(a) \quad \textit{if} \quad P_a[\mathfrak{m}_{a-} = 0] = 1$$

$$12\text{b})\qquad \lim_{b\downarrow a} f(b) = f(a) \quad \textit{if} \quad P_a[\mathfrak{m}_{a+} = 0] = 1;$$

this will be important for 3.7.

Problem 1. A simple MARKOVian motion D, defined as in 3.1, is strict MARKOV if

$$P_\bullet\left[\lim_{s\downarrow t}(G_\alpha f)(x_s) = (G_\alpha f)(x_t),\ t \geqq 0\right] = 1 \qquad f \in C(Q),\quad f(\infty) = 0.$$

[Consider a MARKOV time $\mathfrak{m}$, let $\mathfrak{m}_n = ([\mathfrak{m}2^n] + 1)2^{-n}$ $(n \geqq 1)$, let $e = e(w)$ $(0 \leqq e \leqq 1)$ be measurable $\mathbf{B}_{\mathfrak{m}+}$, and select $f \in C[0, 1]$ such that $f(\infty) = 0$. Because $(\mathfrak{m}_n = k2^{-n}) \in \mathbf{B}_{k2^{-n}}$ $(k \geqq 0)$ and $e(w^\bullet_{\mathfrak{m}_n}) = e$ (see 3.2.10c)), it is found, as in 1.7, that

$$\begin{aligned}
E_\bullet\left[e\int_0^{+\infty} e^{-\alpha t} f[x(t+\mathfrak{m})]\,dt\right] &= \lim_{n\uparrow+\infty} E_\bullet\left[e(w^\bullet_{\mathfrak{m}_n})\int_0^{+\infty} e^{-\alpha t} f[x(t+\mathfrak{m}_n)]\,dt\right]\\
&= \lim_{n\uparrow+\infty} E_\bullet(e\,G_\alpha f[x(\mathfrak{m}_n)])\\
&= E_\bullet(e\,G_\alpha f[x(\mathfrak{m})]),
\end{aligned}$$

i.e., inverting the LAPLACE transform, and expressing the result in terms of conditional probabilities,

$$P_\bullet[x(t+\mathfrak{m}) \in db \mid \mathbf{B}_{\mathfrak{m}+}] = P_a[x(t) \in db] \quad a = x(\mathfrak{m}).]$$

Problem 2. A simple MARKOVian motion D defined as in 3.1, starting afresh at the special MARKOV times

$$\mathfrak{m}_a = \min(t : x(t) = a) \qquad a \in Q$$
$$\mathfrak{m}_{a-} = \lim_{b \uparrow a} \mathfrak{m}_b \qquad a > \inf Q$$
$$\mathfrak{m}_{a+} = \lim_{b \downarrow a} \mathfrak{m}_b \qquad a < \sup Q,$$

also starts afresh at the general MARKOV time, *i.e.*, it is strict MARKOV.

[3.3, 3.4, 3.6 can be copied unchanged. Given a dense countable subset I of K_+ including all the left isolated points of K_+,

$$\begin{aligned} &P_\bullet[x_s < x_t \in K_+ \quad \text{for some} \quad s > t \geqq 0] \\ &\leqq \sum_{l \in I} P_\bullet[\mathfrak{m}_l < +\infty, x(s + \mathfrak{m}_l) < l \quad \text{for some} \quad s > 0] \\ &\leqq \sum_{l \in I} P_\bullet[\mathfrak{m}_l < +\infty]\, P_l[\mathfrak{m}_{l-} < +\infty] \\ &= 0, \end{aligned}$$

i.e., the path cannot cross down over a right singular point; the proof that it cannot cross up over a left singular point is similar. Because $G_\alpha B(Q) \subset D$, it follows that $P_\bullet\left[\lim_{s \downarrow t} (G_\alpha f)(x_s) = (G_\alpha f)(x_t), t \geqq 0\right] = 1$, $f \in B(Q)$, and an application of the result of problem 1 completes the solution.]

Problem 3. Given a diffusion D with state interval $(0, 1]$, if 0 is *entrance* in the sense that

$$\lim_{\varepsilon \downarrow 0} E_{0+}(e^{-\mathfrak{m}_\varepsilon}) = 1$$
$$\lim_{\varepsilon \downarrow 0} P_\varepsilon(\mathfrak{m}_{0+} < +\infty) = 0,$$

then it is possible to define a new diffusion $D^\bullet$ with state interval $[0, 1]$, $P_b^\bullet = P_b \quad (b > 0)$, and $P_0^\bullet\left[\mathfrak{m}_{0+} = 0, \lim_{\varepsilon \downarrow 0} \mathfrak{m}_0(w_\varepsilon^+) = +\infty\right] = 1$.

[Define $P = P_{1/2} \times P_{1/3} \times P_{1/4} \times$ *etc.*, and using $w_{1/2}, w_{1/3}$, *etc.* to denote sample paths starting at $\frac{1}{2}, \frac{1}{3}$, *etc.*, let

$$\mathfrak{m}^\bullet_{1/n} = \sum_{l \geqq n} \mathfrak{m}_{1/l}(w^+_{1/l+1}) \qquad n \geqq 2,$$

and let $w^\bullet$ be the new sample path

$$\begin{aligned} x^\bullet(t) &= 0 & & t = 0 \\ &= x(t - \mathfrak{m}^\bullet_{1/n}, w_{1/n}) & \mathfrak{m}^\bullet_{1/n} \leqq t < \mathfrak{m}^\bullet_{1/n-1}, & \quad n \geqq 3 \\ &= x(t - \mathfrak{m}^\bullet_{1/2}, w_{1/2}) & \mathfrak{m}^\bullet_{1/2} \leqq t < +\infty. & \end{aligned}$$

Because

$$\begin{aligned} P(\mathfrak{m}^\bullet_{1/n} < +\infty) \geqq E[e^{-\mathfrak{m}^\bullet_{1/n}}] &= E\left[e^{-\mathfrak{m}_{1/n}(w_{1/n+1})}\right] E\left[e^{-\mathfrak{m}_{1/n+1}(w_{1/n+2})}\right] \quad \textit{etc.} \\ &= E_{0+}\left[e^{-\mathfrak{m}_{1/n}}\right] \uparrow 1 \qquad n \uparrow +\infty, \end{aligned}$$

$P\left[\lim_{n\uparrow+\infty} \mathfrak{m}^{\bullet}_{1/n} = 0\right] = 1$, or, what interests us, $P\left[\lim_{t\downarrow 0} x^{\bullet}(t) = 0\right] = 1$, and it follows that $P^{\bullet}_0(B) = P[w^{\bullet} \in B]$ is a non-negative BOREL measure on $\mathbf{B}$ of total mass $+1$ such that $P^{\bullet}_0\left[\mathfrak{m}_{0+} = 0,\ \lim_{s\downarrow 0} \mathfrak{m}_0(w^+_s) = +\infty\right] = 1$. Choose $f \in B[0, 1]$; then

$$(G^{\bullet}_{\alpha} f)(0) = E^{\bullet}_0\left[\int_0^{+\infty} e^{-\alpha t} f(x_t)\, dt\right]$$

$$= \lim_{\varepsilon\downarrow 0} E^{\bullet}_0\left[\int_0^{+\infty} e^{-\alpha t} f(x(t + \mathfrak{m}_{\varepsilon}(w^+_{\mathfrak{m}}) + \mathfrak{m}))\, dt\right]$$

$$\mathfrak{m} = \mathfrak{m}_{1/n}, \qquad \frac{1}{n} \leqq \varepsilon < \frac{1}{n-1}$$

$$= \lim_{\varepsilon\downarrow 0} E_{1/n}\left[\int_0^{+\infty} e^{-\alpha t} f(x(t + \mathfrak{m}_{\varepsilon}))\, dt\right]$$

$$= \lim_{\varepsilon\downarrow 0} (G_{\alpha} f)(\varepsilon)$$

$$= (G^{\bullet}_{\alpha} f)(0+),$$

which implies $P^{\bullet}_{\bullet}\left[\lim_{r\downarrow t} (G^{\bullet}_{\alpha} f)(x_s) = (G^{\bullet}_{\alpha} f)(x_t),\ t \geqq 0\right] = 1$, and to complete the solution, it suffices to show that $[x(t): t \geqq 0, P^{\bullet}_0]$ is simple MARKOV (see problem 1). But, if $0 \leqq e_1 \leqq 1$ is measurable $\mathbf{B}_s$, if $\lim_{\varepsilon\downarrow 0} e_1(w^!_s) = e_1$, if $e_2 \in C(0, 1]$, and if $e_2(\infty) \equiv 0$, then, computing as above,

$$E^{\bullet}_0\left[e_1 \int_0^{+\infty} e^{-\alpha t} e_2(x_{t+s})\, dt\right]$$

$$= \lim_{n\uparrow+\infty} E^{\bullet}\left[e_1(w^+_{\mathfrak{m}}) \int_0^{+\infty} e^{-\alpha t} e_2(w_{t+s+\mathfrak{m}})\, dt\right] \qquad \mathfrak{m} = \mathfrak{m}_{1/n}$$

$$= \lim_{n\uparrow+\infty} E_{1/n}[e_1 (G_{\alpha} e_2)(x_s)]$$

$$= \lim_{n\uparrow+\infty} E^{\bullet}_0[e_1(w^+_{\mathfrak{m}}) (G_{\alpha} e_2)(x_{s+\mathfrak{m}})]$$

$$= E^{\bullet}_0[e_1 (G^{\bullet}_{\alpha} e_2)(x_s)],$$

and, inverting the LAPLACE transform, one finds

$$P^{\bullet}_0[x(t+s) \in db \mid \mathbf{B}_s] = P^{\bullet}_a[x(t) \in db] \qquad a = x(s).]$$

Problem 4. Give an explicit isomorphism of $G_1 B(Q)$ into $C[0, 1]$ in case $P_0(\mathfrak{m}_1 < +\infty) > 0$.

[Consider the mapping $u \in G_1 B(Q) \to u^{\bullet} = P_{\bullet}(\mathfrak{m}_1 < +\infty)^{-1} u$. Because $K_- \cap [0, 1) = \varnothing$, $u(a+) = u(a)$ and $\lim_{b\downarrow a} P_b(\mathfrak{m}_1 < +\infty)$ $= \lim_{b\downarrow a} P_a(\mathfrak{m}_b < +\infty) P_b(\mathfrak{m}_1 < +\infty) = P_a(\mathfrak{m}_1 < +\infty)$ $(a \leqq 1)$, *i.e.*, $u^{\bullet}(a+) = u^{\bullet}(a)$ $(a < 1)$; as to $u^{\bullet}(a-)$, $u(a-) = [1 - k_+(a)]\, u(a)$ and $\lim_{b\uparrow a} P_b(\mathfrak{m}_1 < +\infty) = \lim_{b\uparrow a} P_b(\mathfrak{m}_a < +\infty) P_a(\mathfrak{m}_1 < +\infty) = [1 - k_+(a)] \times$ $\times P_a(\mathfrak{m}_1 < +\infty)$ $(a \geqq 0)$, *i.e.*, $u^{\bullet}(a-) = u^{\bullet}(a)$ $(a > 0)$, so $u \to u^{\bullet}$ is a possible isomorphism.]

3.7. Generators

Consider the GREEN operators $G_\alpha (\alpha > 0)$ acting on the space D as defined in 3.6.

Given $\alpha, \beta > 0$,

1) $$G_\alpha - G_\beta + (\alpha - \beta) G_\alpha G_\beta = 0,$$

because

2) $$G_\beta G_\alpha f = E_\bullet\left[\int_0^{+\infty} e^{-\beta s} (G_\alpha f)(x_s)\, ds\right]$$

$$= E_\bullet\left[\int_0^{+\infty} e^{-\beta s}\, ds \int_0^{+\infty} e^{-\alpha t} f(x_{t+s})\, dt\right]$$

$$= E_\bullet\left[\int_0^{+\infty} e^{-\alpha\tau} f(x_\tau)\, d\tau \int_0^{\tau} e^{-(\beta-\alpha)\sigma}\, d\sigma\right] \qquad \tau = t + s, \sigma = s$$

$$= E_\bullet\left[\int_0^{+\infty} \frac{(e^{-\beta\tau} - e^{-\alpha\tau})}{\alpha - \beta} f(x_\tau)\, d\tau\right]$$

$$= (\alpha - \beta)^{-1} (G_\beta - G_\alpha) f,$$

and it follows that the range $D(\mathfrak{G}) \equiv G_\alpha D \subset D$ and the null-space $G_\alpha^{-1}(0) \equiv D \cap (f : G_\alpha f = 0)$ are independent of $\alpha > 0$.

Because of the definition of the space D, if $a \in Q$ and if $f \in D$, then $P_a[\lim_{t\downarrow 0} f(x_t) = f(a)] = 1$, permitting us to deduce from $f \in G_\alpha^{-1}(0)$ that

3) $$0 \equiv \lim_{\beta\uparrow+\infty} \beta G_\beta f = E_\bullet\left[\lim_{\beta\uparrow+\infty} \beta \int_0^{+\infty} e^{-\beta t} f(x_t)\, dt\right] = f.$$

$G_\alpha : D \to D(\mathfrak{G})$ is therefore invertible, and, defining

4) $$\mathfrak{G} = 1 - G_1^{-1} : D(\mathfrak{G}) \to D$$

and letting $1 - \mathfrak{G}$ act on both sides of 1) with $\alpha = 1$, it is found that

5) $$G_\beta^{-1} = (\beta - \mathfrak{G}) \qquad \beta > 0.$$

$\mathfrak{G}$ *is the so-called generator of* D.

Given a MARKOV time $\mathfrak{m}$, $\alpha > 0$, and $u \in D(\mathfrak{G})$, if $f = (\alpha - \mathfrak{G}) u$ then, as in 3.6.8),

6) $$u = G_\alpha f = E_\bullet\left[\int_0^{\mathfrak{m}} e^{-\alpha t} f(x_t)\, dt\right] + E_\bullet[e^{-\alpha\mathfrak{m}} (G_\alpha f)(x_{\mathfrak{m}})]$$

$$= E_\bullet\left[\int_0^{\mathfrak{m}} e^{-\alpha t} (\alpha - \mathfrak{G})\, u(x_t)\, dt\right] + E_\bullet[e^{-\alpha\mathfrak{m}} u(x_{\mathfrak{m}})],$$

and, for $E_\bullet(\mathfrak{m}) < +\infty$ and $\alpha \downarrow 0$*, 6) goes over into E. B. DYNKIN's formula [4]:

$$7)\quad E_\bullet[u(x_{\mathfrak{m}})] - u = E_\bullet\left[\int_0^{\mathfrak{m}} (\mathfrak{G}u)(x_t)\,dt\right] \qquad u \in D(\mathfrak{G}),\ E_\bullet(\mathfrak{m}) < +\infty.$$

$\mathfrak{G}$ is computed from 7) following E. B. DYNKIN [4, 5].

Given a *trap* $a \in Q$, $P_a[\mathfrak{m}_\infty > t] = e^{-\varkappa t}$ $(0 \leqq \varkappa < +\infty)$ (see problem 3.2.7), and, if $\varkappa = 0$, then $P_a[\mathfrak{m}_\infty = +\infty] = 1$ and $(\mathfrak{G}u)(a) = 0$, while, if $\varkappa > 0$, then $0 < E_a(\mathfrak{m}_\infty) = \varkappa^{-1} < +\infty$, and using DYNKIN's formula,

$$\begin{aligned} 8)\qquad (\mathfrak{G}u)(a)\, E_a(\mathfrak{m}_\infty) &= \lim_{n \uparrow +\infty} E_a\left[\int_0^{\mathfrak{m}_\infty \wedge n} (\mathfrak{G}u)(x_t)\,dt\right] \\ &= -\lim_{n \uparrow +\infty} P_a[\mathfrak{m}_\infty < n]\, u(a) \\ &= -u(a), \end{aligned}$$

i.e.,

$$9)\qquad (\mathfrak{G}u)(a) = -\frac{u(a)}{E_\bullet(\mathfrak{m}_\infty)} \qquad u \in D(\mathfrak{G}).$$

Given $a \in Q$ such that $(\mathfrak{G}u)(a) = 0$ for each $u \in D(\mathfrak{G})$, $\alpha(G_\alpha f)(a) = f(a)$ for each $f \in D$ and $\alpha > 0$, and taking $f \in C(Q) \subset D$ so as to have $0 \leqq f(a) < f(b)$ $(a \neq b)$, it is found that $P_a[x(t) \equiv a, t \geqq 0] = 1$, *i.e.*, *a is a trap.*

Because of this, *if $a \in Q$ is not a trap*, then $(\mathfrak{G}u)(a) > 1$ for some $u \in D(\mathfrak{G})$, and using the fact that $\mathfrak{G}u \in D$ to select a neighborhood B of a:

$$\begin{aligned} B &= (b, a] && b < a \quad \textit{if}\quad a \text{ is a left shunt} \\ &= [a, b) && b > a \quad \textit{if}\quad a \text{ is a right shunt} \\ &= (c, b) && c < a < b \quad \textit{if}\quad a \text{ is non-singular} \end{aligned}$$

such that $\mathfrak{G}u > 1$ on B and introducing the exit time $\mathfrak{m} = \min(t : x(t) \notin B)$, DYNKIN's formula applied to u and $\mathfrak{m} \wedge n$ shows that

$$10)\qquad E_a(\mathfrak{m}) \leqq \lim_{n \uparrow +\infty} E_a\left[\int_0^{\mathfrak{m} \wedge n} (\mathfrak{G}u)(x_t)\,dt\right] \leqq 2\,\|u\| < +\infty.$$

But then DYNKIN's formula can be applied to each $u \in D(\mathfrak{G})$ and exit time $\mathfrak{m}$, provided B is small enough, and it follows that

$$11)\qquad (\mathfrak{G}u)(a) = \lim_{B \downarrow a} \frac{E_\bullet[u(x_{\mathfrak{m}})] - u(a)}{E_\bullet(\mathfrak{m})} \qquad u \in D(\mathfrak{G}),$$

* f depends upon α but u does not.

i.e., making cases,

12a) $(\mathfrak{G} u)(a) = \lim_{b \uparrow a} \frac{u(b)\, P_a[\mathfrak{m}_b < \mathfrak{m}_\infty] - u(a)}{E_a(\mathfrak{m}_b \wedge \mathfrak{m}_\infty)}$ *if a is a left shunt*

12b) $(\mathfrak{G} u)(a) = \lim_{b \downarrow a} \frac{u(b)\, P_a[\mathfrak{m}_b < \mathfrak{m}_\infty] - u(a)}{E_a(\mathfrak{m}_b \wedge \mathfrak{m}_\infty)}$ *if a is a right shunt*

12c) $(\mathfrak{G} u)(a) = \lim_{\substack{c \uparrow a \\ b \downarrow a}} \frac{u(c)\, P_a[\mathfrak{m}_c < \mathfrak{m}_b] + u(b)\, P_a[\mathfrak{m}_b < \mathfrak{m}_c] - u(a)}{E_a(\mathfrak{m}_c \wedge \mathfrak{m}_b \wedge \mathfrak{m}_\infty)}$

if a is non-singular.

$\mathfrak{G}$ is then a local operator, i.e., if $a \in Q$ and if $u_1 \in D(\mathfrak{G})$ coincides with $u_2 \in D(\mathfrak{G})$ on a neighborhood B of a:

$B = a$		*if a is a trap*
$= (b, a]$	$b < a$	*if a is a left shunt*
$= [a, b)$	$b > a$	*if a is a right shunt*
$= (c, b)$	$c < a < b$	*if a is non-singular,*

then $(\mathfrak{G} u_1)(a) = (\mathfrak{G} u_2)(a)$.

Problem 1. Given $u \in D(\mathfrak{G})$ and a neighborhood B of $a \in Q$ as described above, if $u(b) \leqq u(a)$ on B and if $u(a) \geqq 0$, then $(\mathfrak{G}u)(a) \leqq 0$. [Use 3.7.12).]

W. Feller [9] proved that if $\mathfrak{G}$ is *local* and if it is also *negative* in the sense of problem 1, then it has to be a differential operator of degree $\leqq 2$ (see 4.1 for more information on this point).

3.8. Generators continued

Here is another method of computing $\mathfrak{G}$.

Consider

1) $$\mathfrak{G}_\varepsilon u = \varepsilon^{-1}(E_\bullet[u(x_\varepsilon)] - u) \qquad \varepsilon > 0$$

and define

2) $$\mathfrak{G}^\bullet = \lim_{\varepsilon \downarrow 0} \mathfrak{G}_\varepsilon,$$

where $\mathfrak{G}^\bullet$ is to be applied to the class $D(\mathfrak{G}^\bullet)$ of functions $u \in D$ such that $\mathfrak{G}^\bullet u \equiv \lim_{\varepsilon \downarrow 0} \mathfrak{G}_\varepsilon u$ exists pointwise and belongs to D and $\sup_{\varepsilon > 0} \|\mathfrak{G}_\varepsilon u\| < +\infty$; *it is to be proved that*

3) $$\mathfrak{G}^\bullet = \mathfrak{G},$$

i.e.,

4a) $$D(\mathfrak{G}^\bullet) = D(\mathfrak{G})$$

4b) $$\mathfrak{G} u = \lim_{\varepsilon \downarrow 0} \frac{E_\bullet[u(x_\varepsilon)] - u}{\varepsilon} \qquad u \in D(\mathfrak{G}).$$

4b) *is to be compared to* 3.7.11); the use of $\mathfrak{G}^\bullet$ was suggested by W. Feller [5].

DYNKIN's formula 3.7.7) applied to $u \in D(\mathfrak{G})$ and $\mathfrak{m} = \varepsilon > 0$ implies that $\mathfrak{G}^\bullet = \mathfrak{G}$ on $D(\mathfrak{G}) \subset D(\mathfrak{G}^\bullet)$ because

5a) $$\lim_{\varepsilon \downarrow 0} \mathfrak{G}_\varepsilon u = \lim_{\varepsilon \downarrow 0} \varepsilon^{-1} E_\bullet \left[\int_0^\varepsilon (\mathfrak{G} u)(x_t)\, dt \right] = \mathfrak{G} u$$

and

5b) $$\| \mathfrak{G}_\varepsilon u \| \leqq \| \mathfrak{G} u \|,$$

and, to complete the proof, it is enough to show that $D(\mathfrak{G}^\bullet) \subset D(\mathfrak{G})$.

But, if $u \in D(\mathfrak{G}^\bullet)$, then

6) $$\begin{aligned} G_\alpha(\alpha - \mathfrak{G}^\bullet) u &= \alpha G_\alpha u - \lim_{\varepsilon \downarrow 0} G_\alpha \mathfrak{G}_\varepsilon u \\ &= \alpha G_\alpha u - \lim_{\varepsilon \downarrow 0} \varepsilon^{-1} E_\bullet \left[\int_0^{+\infty} e^{-\alpha t} [u(x_{t+\varepsilon}) - u(x_t)]\, dt \right] \\ &= \alpha G_\alpha u - \lim_{\varepsilon \downarrow 0} \varepsilon^{-1} E_\bullet \left[(e^{\alpha\varepsilon} - 1) \int_\varepsilon^{+\infty} e^{-\alpha t} u(x_t)\, dt - \int_0^\varepsilon e^{-\alpha t} u(x_t)\, dt \right] \\ &= u, \end{aligned}$$

and, since $(\alpha - \mathfrak{G}^\bullet) u \in D$, $u = G_\alpha(\alpha - \mathfrak{G}^\bullet) u \in D(\mathfrak{G})$ as desired.

$\mathfrak{G}^\bullet = \mathfrak{G}$ can be used to give a new proof of the fact, noted at the end of 3.7, that $\mathfrak{G}$ is a local operator.

Because the path cannot cross up over a left singular point or down over a right singular point, it suffices to show that

7) $$\lim_{t \downarrow 0} t^{-1} P_a(\mathfrak{m}_b < t) = 0 \qquad a < b,$$

as will now be done using the method of D. B. RAY [2].

Given $a < c < b$,

8) $$\begin{aligned} P_a(\mathfrak{m}_b < t) &= P_a[\mathfrak{m} + \mathfrak{m}_b(w_\mathfrak{m}^+) < t] \qquad \mathfrak{m} = \mathfrak{m}_c \\ &\leqq P_a[\mathfrak{m} < t,\ \mathfrak{m}_b(w_\mathfrak{m}^+) < t] \\ &= E_a(\mathfrak{m}_c < t,\ P_a[\mathfrak{m}_b(w_\mathfrak{m}^+) < t \mid \mathsf{B}_{\mathfrak{m}+}]) \\ &= P_a(\mathfrak{m}_c < t)\, P_c(\mathfrak{m}_b < t) \end{aligned}$$

and

9) $$\begin{aligned} & P_c[\mathfrak{m}_a \wedge \mathfrak{m}_\infty \wedge \mathfrak{m}_b > s]\, [s/t]\, P_a(\mathfrak{m}_b < t) \\ &\leqq \sum_{n \leqq [s/t]} P_c[\mathfrak{m}_a \wedge \mathfrak{m}_\infty \wedge \mathfrak{m}_b > n t]\, P_a(\mathfrak{m}_b < t) \\ &\leqq \sum_{n \leqq [s/t]} \int_{(a,b)} P_c[\mathfrak{m}_a \wedge \mathfrak{m}_\infty \wedge \mathfrak{m}_b > n t,\ x_{nt} \in dl]\, P_l(\mathfrak{m}_b \leqq t) \\ &\leqq \sum_{n \leqq [s/t]} P_c[\mathfrak{m}_b \in (n t, (n+1) t]] \\ &\leqq 1 \qquad s > 0. \end{aligned}$$

But, since $P_c[\mathfrak{m}_a \wedge \mathfrak{m}_\infty \wedge \mathfrak{m}_b > 0] = 1$, $\limsup_{t \downarrow 0} t^{-1} P_a(\mathfrak{m}_b < t) < +\infty$, and so [use 8)] $\overline{\lim_{t\downarrow 0}}\, t^{-2} P_a(\mathfrak{m}_b < t) < +\infty$, completing the proof of 7); as a matter of fact, 8) implies $P_a(\mathfrak{m}_b < t) <$ constant $\times\, t^4, t^8$, *etc.* $t \downarrow 0)$, proving

10) $$\lim_{t\downarrow 0} t^{-n} P_a(\mathfrak{m}_b < t) = 0 \qquad a < b, \quad n \geqq 1.$$

Problem 1. Give a new proof of 10), using DYNKIN's formula to check

$$E_a(e^{-\alpha \mathfrak{m}_b}) \leqq [\alpha E_c(\mathfrak{m}_a \wedge \mathfrak{m}_\infty \wedge \mathfrak{m}_b)]^{-1} \qquad a < c < b$$

and then making use of $P_a(\mathfrak{m}_b \leqq t) \leqq e \int_0^t e^{-s/t} P_a(\mathfrak{m}_b \in ds) \leqq e\, E_a(e^{-t^{-1}\mathfrak{m}_b})$ (see problem 4.7.4 for still another proof).

[Given $a < b$ such that $P_a(\mathfrak{m}_{b+} < +\infty) > 0$, if $f \in D$ is 0 to the left of b and positive elsewhere and if $\alpha > 0$, then $\gamma = (G_\alpha f)(b) > 0$, $0 < u \equiv \gamma^{-1} G_\alpha f = E_\bullet(e^{-\alpha \mathfrak{m}_b}) \in \uparrow$ and solves $\mathfrak{G} u = \alpha u$ on $[a, b)$, and using DYNKIN's formula with $\mathfrak{m} = n \wedge \mathfrak{m}_a \wedge \mathfrak{m}_\infty \wedge \mathfrak{m}_b$, it is found that, if $a < c < b$, then $1 \geqq E_c[u(x_\mathfrak{m})] - u(c) = E_c\left(\int_0^\mathfrak{m} (\mathfrak{G} u)(x_t)\, dt\right) \geqq \alpha u(a)\, E_c(\mathfrak{m}) \uparrow \alpha u(a)\, E_c(\mathfrak{m}_a \wedge \mathfrak{m}_\infty \wedge \mathfrak{m}_b)$ as $n \uparrow +\infty$.]

3.9. Stopped diffusion

We will prove that *if $Q^\bullet$ is a subinterval of Q, closed in Q, and if* $\mathfrak{e}$ *is the exit time* $\inf(t : x(t) \notin Q^\bullet)$, *then the stopped motion* $\mathsf{D}^\bullet = [x(t \wedge \mathfrak{e}) : t \geqq 0, P_a : a \in Q^\bullet]$ *is itself a diffusion*; this fact was used in 3.5.

Consider the standard description of the stopped motion with sample paths $w^\bullet : t \to Q^\bullet$, universal BOREL algebra $\mathsf{B}^\bullet$, MARKOV times $\mathfrak{m}^\bullet = \mathfrak{m}^\bullet(w^\bullet)$, subalgebras $\mathsf{B}_{\mathfrak{m}^\bullet +}$, and probabilities

1) $$P_a^\bullet(B^\bullet) = P_a(B) \quad a \in Q^\bullet, \quad B \equiv (w : w_\mathfrak{e}^\bullet \in B^\bullet), \quad B^\bullet \in \mathsf{B}^\bullet;$$

the problem is to show that, if $B^\bullet \in \mathsf{B}_{\mathfrak{m}^\bullet +}$, then

2) $$P_\bullet^\bullet[B^\bullet, x(t + \mathfrak{m}^\bullet) \in db] = E_\bullet^\bullet[B^\bullet, P_{x(\mathfrak{m}^\bullet)}^\bullet(x(t) \in db)].$$

But, if $\mathfrak{m}^\bullet = \mathfrak{m}^\bullet(w^\bullet)$ is a MARKOV time for the stopped motion (in standard description), then ignorning paths not starting from $Q^\bullet$, $\mathfrak{m}(w) \equiv \mathfrak{m}^\bullet(w_\mathfrak{e}^\bullet)$ is a MARKOV time for the original motion because

3) $$\begin{aligned}(w : \mathfrak{m} < t) &= (w : \mathfrak{m}^\bullet(w_\mathfrak{e}^\bullet) < t) \\ &= (w : w_\mathfrak{e}^\bullet \in B^\bullet) \qquad B^\bullet = (w^\bullet : \mathfrak{m}^\bullet < t) \in \mathsf{B}_t^\bullet \\ &= (w : (w_\mathfrak{e}^\bullet)_t^\bullet \in B^\bullet) \\ &= (w : w_{\mathfrak{e} \wedge t}^\bullet \in B^\bullet) \\ &\in \mathsf{B}_t \qquad\qquad t \geqq 0;^*\end{aligned}$$

* Use 3.2.1) and $\mathfrak{e} \wedge t = \mathfrak{e}(w_t^\bullet) \wedge t$ (see problem 3.2.6).

in addition, if $B^\bullet \in \mathsf{B}_{\mathfrak{m}^\bullet +}$, then $B \equiv (w : w^\bullet_{\mathfrak{e}} \in B^\bullet) \in \mathsf{B}_{(\mathfrak{m}\wedge\mathfrak{e})+}$ as is clear from

4)
$$\begin{aligned} B \cap (w : \mathfrak{m} \wedge \mathfrak{e} < t) &= (w^\bullet_{\mathfrak{e}} \in B^\bullet) \cap (\mathfrak{m}^\bullet (w^\bullet_{\mathfrak{e}}) < t) \cap (\mathfrak{m} < \mathfrak{e}) \\ &+ (w^\bullet_{\mathfrak{e}} \in B^\bullet) \cap (\mathfrak{e} < t) \cap (\mathfrak{m} \geqq \mathfrak{e}) \\ &= ((w^\bullet_{\mathfrak{e}})^\bullet_t \in B^\bullet) \cap (\mathfrak{m} < t) \cap (\mathfrak{m} < \mathfrak{e})^* \\ &+ ((w^\bullet_{\mathfrak{e}})^\bullet_t \in B^\bullet) \cap (\mathfrak{e} < t) \cap (\mathfrak{m} \geqq \mathfrak{e}) \\ &= (w^\bullet_{\mathfrak{e}\wedge t} \in B^\bullet) \cap (\mathfrak{m} \wedge \mathfrak{e} < t) \\ &\in \mathsf{B}_t \qquad\qquad t \geqq 0. \end{aligned}$$

Using

5)
$$\mathfrak{e} = \mathfrak{m} \wedge \mathfrak{e} + \mathfrak{e}(w^+_{\mathfrak{m}\wedge\mathfrak{e}}),$$

6)
$$\begin{aligned} x(t + \mathfrak{m}, w^\bullet_{\mathfrak{e}}) &= x((t + \mathfrak{m} \wedge \mathfrak{e}) \wedge (\mathfrak{e}(w^+_{\mathfrak{m}\wedge\mathfrak{e}}) + \mathfrak{m} \wedge \mathfrak{e})) \\ &= x(t \wedge \mathfrak{e}(w^+_{\mathfrak{m}\wedge\mathfrak{e}}), w^+_{\mathfrak{m}\wedge\mathfrak{e}}), \end{aligned}$$

and 1), it is found that

7)
$$\begin{aligned} P^\bullet_\bullet[B^\bullet, x(t + \mathfrak{m}^\bullet) \in db] &= P_\bullet[B, x(t + \mathfrak{m}, w^\bullet_{\mathfrak{e}}) \in db] \\ &= P_\bullet[B, x(t \wedge \mathfrak{e}(w^+_{\mathfrak{m}\wedge\mathfrak{e}}), w^+_{\mathfrak{m}\wedge\mathfrak{e}}) \in db] \\ &= E_\bullet[B, P_{x(\mathfrak{m}\wedge\mathfrak{e})}(x(t \wedge \mathfrak{e}) \in db)] \\ &= E_\bullet[B, P^\bullet_{x(\mathfrak{m}^\bullet(w^\bullet_{\mathfrak{e}}), w^\bullet_{\mathfrak{e}})}(x(t) \in db)] \\ &= E^\bullet_\bullet[B^\bullet, P^\bullet_{x(\mathfrak{m}^\bullet)}(x(t) \in db)] \end{aligned}$$

as desired.

Consider the generators $\mathfrak{G}$ and $\mathfrak{G}^\bullet$ of D and $\mathsf{D}^\bullet$.

$\mathfrak{G}$ is a contraction of $\mathfrak{G}^\bullet$ in the sense that, if $u \in D(\mathfrak{G})$, then

8a) *the restriction $u^\bullet$ of u to $Q^\bullet$ belongs to $D(\mathfrak{G}^\bullet)$*

and

8b) *$\mathfrak{G}^\bullet u^\bullet = \mathfrak{G} u$ on* int $Q^\bullet$ *and at those endpoints of $Q^\bullet$ at which*
$$P^\bullet_\bullet[\mathfrak{e} > 0] = 1,$$

as will now be proved.

Given $u \in D(\mathfrak{G})$, if $f = (1 - \mathfrak{G})u$, then

9a)
$$f \in D$$

9b)
$$u = G_1 f = E_\bullet\left[\int_0^{\mathfrak{e}} e^{-t} f(x_t)\, dt\right] + E_\bullet[e^{-\mathfrak{e}} u(x_{\mathfrak{e}})],$$

and, if

10)
$$\begin{aligned} f^\bullet(a) &= f(a) & a &\in Q^\bullet, & P_a[\mathfrak{e} > 0] &= 1 \\ &= (G_1 f)(a) & a &\in Q^\bullet, & P_a[\mathfrak{e} = 0] &= 1 \\ &= 0 & a &= \infty, \end{aligned}$$

---

* Use the fact that $w^\bullet \in B^\bullet$ if and only if $(w^\bullet)^\bullet_t \in B^\bullet$ for each $t > \mathfrak{m}^\bullet(w^\bullet)$; see 3.2.10c).

then

11a) $$f^\bullet \in D^\bullet,$$

$D^\bullet$ being the space D attached to $\mathsf{D}^\bullet$, and on $Q^\bullet$,

11b) $$\begin{aligned} G^\bullet f^\bullet &= E^\bullet_\bullet\left[\int_0^{+\infty} e^{-t} f^\bullet(x)\,dt\right] \\ &= E_\bullet\left[\int_0^{+\infty} e^{-t} f^\bullet[x(t\wedge \mathfrak{e})]\,dt\right] \\ &= E_\bullet\left[\int_0^{\mathfrak{e}} e^{-t} f^\bullet(x_t)\,dt\right] + E_\bullet\left[\int_{\mathfrak{e}}^{+\infty} e^{-t}\,dt\, f^\bullet(x_{\mathfrak{e}})\right] \\ &= E_\bullet\left[\int_0^{\mathfrak{e}} e^{-t} f(x_t)\,dt\right] + E_\bullet[e^{-\mathfrak{e}}(G_1 f)(x_{\mathfrak{e}})] \\ &= G_1 f = u. \end{aligned}$$

8a) is clear from 11); as to 8b), 11b) implies that on $Q^\bullet$,

12) $$\mathfrak{G}^\bullet u^\bullet = (G_1^\bullet - 1) f^\bullet = G_1 f - f^\bullet = \mathfrak{G} u + f - f^\bullet = \mathfrak{G} u$$
$$P_\bullet[\mathfrak{e} > 0] = 1,$$

and 8b) follows.

$\mathfrak{G}$ is now seen to be *local* in the sense that, if A is the closure of a neighborhood B of $a \in Q$:

$$\begin{aligned} B &= (b, a] \quad b < a && \textit{if } a \textit{ is a left singular point} \\ &= [a, b) \quad b > a && \textit{if } a \textit{ is a right singular point} \\ &= (c, b) \quad c < a < b && \textit{if } a \textit{ is non-singular,} \end{aligned}$$

then $\mathfrak{G}$ agrees on B with the generator $\mathfrak{G}^\bullet$ of the stopped motion

$$[x(t\wedge\mathfrak{e}) : t \geqq 0,\ P_\xi : \xi \in A] \qquad \mathfrak{e} = \inf(t : x(t) \notin A);$$

this usage of *local* is to be compared with the usage in 3.7 and 3.8.

Problem 1. What step of the proof that $\mathsf{D}^\bullet$ is a diffusion fails if $\mathfrak{e}$ is a general Markov time? Give a concrete example.

[5) is false; indeed, if D is the standard Brownian motion and if $f \in C(R^1)$ is positive, then $\mathfrak{e} = \min\left(t : \int_0^t f(x_s)\,ds = 1\right)$ is a Markov time, and if 5) held and $\mathfrak{m} = t$ were $< \mathfrak{e}$, we should have

$$1 = \int_0^{\mathfrak{e}} f(x_s)\,ds = \int_0^t f(x_s)\,ds + \int_0^{\mathfrak{e}(w_t^+)} f(x(s, w_t^+))\,ds = \int_0^t f(x_s)\,ds + 1,$$

which is absurd.]

Problem 2. Given $\mathfrak{e} = \inf(t: x(t) \notin Q^\bullet)$, show that, if $P_\bullet(\mathfrak{e} < +\infty) > 0$ on $Q^\bullet$, then the stopped conditional motion $\mathsf{D}^\bullet = [x(t \wedge \mathfrak{e}): t \geqq 0, P_a(B \mid \mathfrak{e} < +\infty): a \in Q^\bullet]$ is a diffusion (see 4.3 for an application).

[Given $\mathfrak{m}^\bullet$, $\mathfrak{m}$, $B^\bullet$, B as above, if $P^\bullet_\bullet(B^\bullet) = P_\bullet(B \mid \mathfrak{e} < +\infty)$ on $Q^\bullet$ and if $db \subset Q^\bullet$, then, as in 7),

$$
\begin{aligned}
&P^\bullet_\bullet[B^\bullet, x(t + \mathfrak{m}^\bullet) \in db] P_\bullet(\mathfrak{e} < +\infty) \\
&\quad = P_\bullet[B, x(t \wedge \mathfrak{e}(w^+_{\mathfrak{m}\wedge\mathfrak{e}}), w^+_{\mathfrak{m}\wedge\mathfrak{e}}) \in db, \mathfrak{m} \wedge \mathfrak{e} + \mathfrak{e}(w^+_{\mathfrak{m}\wedge\mathfrak{e}}) < +\infty] \\
&\quad = E_\bullet[B, P_{x(\mathfrak{m}\wedge\mathfrak{e})}(x(t \wedge \mathfrak{e}) \in db, \mathfrak{e} < +\infty), \mathfrak{m} \wedge \mathfrak{e} < +\infty] \\
&\quad = E_\bullet[B, P^\bullet_{x(\mathfrak{m}\wedge\mathfrak{e})}(x(t) \in db)\, P_{x(\mathfrak{m}\wedge\mathfrak{e})}(\mathfrak{e} < +\infty)] \\
&\quad = E_\bullet[B, \mathfrak{m} \wedge \mathfrak{e} + \mathfrak{e}(w^+_{\mathfrak{m}\wedge\mathfrak{e}}) < +\infty, P^\bullet_{x(\mathfrak{m}\wedge\mathfrak{e})}(x(t) \in db)] \\
&\quad = E_\bullet[B, \mathfrak{e} < +\infty, P^\bullet_{x(\mathfrak{m}\wedge\mathfrak{e})}(x(t) \in db)] \\
&\quad = E^\bullet_\bullet[B^\bullet, P^\bullet_{x(\mathfrak{m}^\bullet)}(x(t) \in db)] P_\bullet(\mathfrak{e} < +\infty),
\end{aligned}
$$

completing the solution.]

4. Generators

4.1. A general view

A particle starts at time $t = 0$ at $-1 \leqq l < 0$, moving at speed $+1$ until it hits $l = 0$; at that moment, it begins a reflecting Brownian motion on $[0, +\infty)$, stopping at the passage time $\mathfrak{m}_1$ to $l = 1$, waiting at that place for an exponential holding time $\mathfrak{e}$ with mean $\frac{1}{3}$, and jumping at time $\mathfrak{m}_1 + \mathfrak{e}$ to the point ∞.

E. B. Dynkin's formula 3.7.11) can be used to check that $\mathfrak{G}u$ coincides with

$$
\begin{aligned}
1) \quad (\mathfrak{G}^\bullet u)(l) &= u^+(l) && -1 \leqq l < 0 \\
&= \lim_{\varepsilon \downarrow 0} \frac{u(\varepsilon) - u(0)}{\varepsilon^2} && l = 0 \\
&= \tfrac{1}{2} u''(l) && 0 < l < 1 \\
&= -3u(1) && l = 1
\end{aligned}
$$

on its domain $\mathsf{D}(\mathfrak{G}) = \mathsf{D}(\mathfrak{G}^\bullet)$ comprising all functions $u \in C[-1, 1] \cap C^1[-1, 0) \cap C^2[0, 1)$ such that

$$
\begin{aligned}
&2a) && u^-(0) = \tfrac{1}{2} u''(0+) \\
&2b) && u^+(0) = 0^* \\
&2c) && \tfrac{1}{2} u''(1-) = -3u(1),
\end{aligned}
$$

* $u^+(0) = 0$ is automatic from the second line of 1 because $\varepsilon^{-1}[u(\varepsilon) - u(0)]$ is bounded, as $\varepsilon \downarrow 0$.

especially, $u \to (\mathfrak{G}u)(l)$ is a differential operator of degree 2,1, or 0 according as l is a non-singular point, a shunt, or a trap.

Given a reflecting BROWNian motion x^+ as described in 2.1 with passage time $\mathfrak{m}_1 = \min(t: x^+ = 1)$ and an independent exponential holding time $\mathfrak{e}$ with law $P_\bullet(\mathfrak{e} > t) = e^{-3t}$, then

$$
\begin{aligned}
3)\quad x^\bullet(t) &= t + l && 0 \leqq t < -l \\
&= x^+(t + l) && -l \leqq t \leqq -l + \mathfrak{m}_1 \\
&= 1 && -l + \mathfrak{m}_1 \leqq t < -l + \mathfrak{m}_1 + \mathfrak{e} \\
&= \infty && t \geqq -l + m_1 + \mathfrak{e}
\end{aligned}
$$

is a (non-standard) description of the motion starting at $-1 \leqq l < 0$.

$\mathfrak{G}$, $D(\mathfrak{G})$, and the associated sample paths admit a similar description in the general case as will be explained below. W. FELLER discovered an expression for $\mathfrak{G}$ as a differential operator $\mathfrak{G}^\bullet$ in the case that the GREEN operators map $C(Q)$ into itself and discussed it in a series of basic papers [*5, 7, 9, 10, 11*]. E. B. DYNKIN [*4, 5, 8*] found an elegant probabilistic method based on 3.7.12) of deriving FELLER's $\mathfrak{G}^\bullet$ in the *conservative* case, at the same time giving it a more probabilistic expression. The expression for $\mathfrak{G}^\bullet$ on the shunts, the properties of killing measures, and the concrete expression of the sample path in terms of $\mathfrak{G}^\bullet$ and a standard BROWNian motion, as described in 5.1, etc., are published here for the first time.

Begin with the non-singular case:

$$4\text{a})\qquad Q = [0, 1]$$

$$4\text{b})\qquad P_a[\mathfrak{m}_{a-} = \mathfrak{m}_{a+} = 0] = 1 \qquad 0 < a < 1$$

$$5\text{a})\qquad E_0(e^{-\mathfrak{m}_{0+}}) = \lim_{b \downarrow 0} E_{0+}(e^{-\mathfrak{m}_b}),\ E_{0+}(e^{-\mathfrak{m}_0}) = \lim_{b \downarrow 0} E_b(e^{-\mathfrak{m}_{0+}})$$

$$5\text{b})\qquad E_1(e^{-\mathfrak{m}_{1-}}) = \lim_{b \uparrow 1} E_{1-}(e^{-\mathfrak{m}_b}),\ E_{1-}(e^{-\mathfrak{m}_1}) = \lim_{b \uparrow 1} E_b(e^{-\mathfrak{m}_{1-}}).$$

4b) states that $(0, 1)$ contains no singular points; see 4.5 for a discussion of 5).

Given $0 < a < b < 1$, the *hitting probabilities* and the *mean exit time*

$$6\text{a})\qquad p_{ab}(\xi) = P_\xi[\mathfrak{m}_a < \mathfrak{m}_b]$$

$$6\text{b})\qquad p_{ba}(\xi) = P_\xi[\mathfrak{m}_b < \mathfrak{m}_a] \qquad a < \xi < b$$

$$6\text{c})\qquad e_{ab}(\xi) = E_\xi(\mathfrak{m}_a \wedge \mathfrak{m}_b \wedge \mathfrak{m}_\infty)$$

can be used to introduce a (continuous, increasing) *scale* $s = s_{ab}$, a (non-negative) *killing measure* $k = k_{ab}$, and a (positive) *speed mea-*

sure $m = m_{ab}$ according to the formulas

7a) $$s(d\xi) = p_{ab}(\xi)\, p_{ba}(d\xi) - p_{ba}(\xi)\, p_{ab}(d\xi) > 0$$

7b) $$k(d\xi) = \frac{p^+_{ab}(d\xi)}{p_{ab}(\xi)} = \frac{p^+_{ba}(d\xi)}{p_{ba}(\xi)} \geq 0$$

7c) $$m(d\xi) = -[e^+_{ab}(d\xi) - e_{ab}(\xi)\, k_{ab}(d\xi)] > 0 \qquad a < \xi < b,^*$$

in terms of which $\mathfrak{G}$ can be expressed as a differential operator

8a) $$(\mathfrak{G}^\bullet u)(\xi) = \frac{u^+(d\xi) - u(\xi)\, k(d\xi)}{m(d\xi)} \qquad a < \xi < b,$$

the precise meaning of 8a) being

8b) $$\int_{[c,d)} (\mathfrak{G}\, u)(\xi)\, m(d\xi) = u^+(d) - u^+(c) - \int_{[c,d)} u(\xi)\, k(d\xi) \quad a < c < d < b.$$

Given $0 < a^\bullet < a < b < b^\bullet < 1$, the new $s_{a^\bullet b^\bullet}$, $k_{a^\bullet b^\bullet}$, $m_{a^\bullet b^\bullet}$ are related to the old according to the rules

9a) $$s_{a^\bullet b^\bullet}(d\xi) = B\begin{pmatrix} a^\bullet & b^\bullet \\ a & b \end{pmatrix} s_{ab}(d\xi)$$

9b) $$k_{a^\bullet b^\bullet}(d\xi) = B\begin{pmatrix} a^\bullet & b^\bullet \\ a & b \end{pmatrix}^{-1} k_{ab}(d\xi) \qquad a < \xi < b,$$

9c) $$m_{a^\bullet b^\bullet}(d\xi) = B\begin{pmatrix} a^\bullet & b^\bullet \\ a & b \end{pmatrix}^{-1} m_{ab}(d\xi)$$

where

$$B\begin{pmatrix} a^\bullet & b^\bullet \\ a & b. \end{pmatrix} = \begin{vmatrix} p_{b^\bullet a^\bullet}(b) & p_{b^\bullet a^\bullet}(a) \\ p_{a^\bullet b^\bullet}(b) & p_{a^\bullet b^\bullet}(a) \end{vmatrix} > 0,$$

and, using this rule, it is a simple matter to introduce universal scale, killing measure, and speed measure such that 8) holds for all $0 < \xi < 1$.

In case $E_0(e^{-\mathfrak{m}_{0+}}) = 1$,

10a) $$k(0) = \lim_{b \downarrow 0} \frac{P_0(\mathfrak{m}_b = +\infty)}{s(b) - s(0)} < +\infty$$

10b) $$m(0) = \lim_{b \downarrow 0} \frac{E_0(\mathfrak{m}_b \wedge \mathfrak{m}_\infty)}{s(b) - s(0)} < +\infty$$

11a) $$\int_0^{1/2} k[0, \xi)\, s(d\xi) < +\infty$$

11b) $$\int_0^{1/2} m[0, \xi)\, s(d\xi) < +\infty,$$

and

12a) $$(\mathfrak{G}\, u)(0)\, m(0) = u^+(0) - u(0)\, k(0) \qquad u \in D(\mathfrak{G}),$$

---

* $e^+(\xi) = \lim_{\eta \downarrow \xi} [s(\eta) - s(\xi)]^{-1} [e(\eta) - e(\xi)]$ $(e = p_{ab}, p_{ba}, e_{ab})$ and $e^+(d\xi)$ is the Borel measure induced by the interval function $e^+(\xi, \eta] = e^+(\eta) - e^+(\xi)$ $(\xi < \eta)$.

where

12b) $$u^+(0+) = u^+(0) = \lim_{b\downarrow 0} \frac{u(b) - u(0)}{s(b) - s(0)},$$

with the understanding that $k(0) = m(0) = u^+(0) \equiv 0$ in case $s(0) \equiv s(0+) = -\infty$; similar results hold at 1 in case $E_1(e^{-\mathfrak{m}_{1-}}) = 1$.

$D(\mathfrak{G})$ can now be identified as the class $D(\mathfrak{G}^\bullet)$ of functions $u \in D$ with

13a) $$u(0+) = u(0) \quad \textit{if} \quad E_{0+}(e^{-\mathfrak{m}_0}) = 1$$

13b) $$u(1-) = u(1) \quad \textit{if} \quad E_{1-}(e^{-\mathfrak{m}_1}) = 1$$

such that, for some $u^\bullet \in D$,

14) $$u^\bullet(\xi)\, m(d\xi) = u^+(d\xi) - u(\xi)\, k(d\xi) \qquad 0 < \xi < 1,$$

15a) $$u^\bullet(0)\, m(0) = u^+(0) - u(0)\, k(0) \qquad \textit{if} \quad E_0(e^{-\mathfrak{m}_{0+}}) = 1$$

15b) $$u^\bullet(1)\, m(1) = -u^-(1) - u(1)\, k(1) \qquad \textit{if} \quad E_1(e^{-\mathfrak{m}_{1-}}) = 1,$$

16a) $$u^\bullet(0) = -\varkappa(0)\, u(0) \quad \textit{if} \quad E_0(e^{-\mathfrak{m}_{0+}}) = 0, \quad \varkappa(0) = E_0(\mathfrak{m}_\infty)^{-1}$$

16b) $$u^\bullet(1) = -\varkappa(1)\, u(1) \quad \textit{if} \quad E_1(e^{-\mathfrak{m}_{1-}}) = 0, \quad \varkappa(1) = E_1(\mathfrak{m}_\infty)^{-1},$$

and $\mathfrak{G}$ can be identified as the global differential operator $\mathfrak{G}^\bullet$ mapping $u \in D(\mathfrak{G}^\bullet) \to u^\bullet$ in agreement with 9). $\mathfrak{G}^\bullet$ is unambiguous because $u^\bullet \in D$ and $0 < m[a, b)$ $(0 < a < b < 1)$.

$\mathfrak{G}$ *at* 0 depends in part upon s, m, and k *near* 0 as epitomised in the table below; in this table, 0 is said to be

$$\textit{an exit point if } +\infty > \int_0^{\frac{1}{2}} k(\xi, \tfrac{1}{2}]\, s(d\xi) + \int_0^{\frac{1}{2}} m(\xi, \tfrac{1}{2}]\, s(d\xi)$$

$$\textit{an entrance point if } +\infty > \int_0^{\frac{1}{2}} k(0, \xi]\, s(d\xi) + \int_0^{\frac{1}{2}} m(0, \xi]\, s(d\xi),$$

and $(\mathfrak{G}u)(0) = m(0)^{-1}[u^+(0) - u(0)\, k(0)]$ stands for $u^+(0) = u(0)\, k(0)$ in case $m(0) = 0$ this classification is due to W. FELLER [*4*].

	exit and entrance	exit not entrance	entrance not exit	neither exit nor entrance
$P_0[\mathfrak{m}_{0+} = 0] =$	0 *or* 1	0	1	0
$E_{0+}(e^{-\mathfrak{m}_{0+}}) =$	1	1	0	0
$(\mathfrak{G}u)(0) =$	$\dfrac{u^+(0) - u(0)\, k(0)}{m(0)}$ *if* $P_0[\mathfrak{m}_{0+} = 0] = 1$; $-\varkappa u(0)$ *if* $P_0[\mathfrak{m}_{0+} = 0] = 0$	$-\varkappa u(0)$	$u^+(0) = 0$	$-\varkappa u(0)$

$\mathfrak{G}$ *at* 1 depends upon s, m, and k *near* 1 in the same manner.

Consider, as a second example, the singular case:

17a) $$Q = [0, 1]$$

17b) $$P_0[\mathfrak{m}_1 < +\infty] > 0$$

17c) $$P_1[\mathfrak{m}_{1-} \wedge \mathfrak{m}_\infty = +\infty] = 1$$

17d) $$E_0(\mathfrak{m}_1 \wedge \mathfrak{m}_\infty) < +\infty;$$

the content of 17b) is that $[0, 1)$ contains no left singular point.

Q breaks up into the *shunts*

18) $$K_+ \cap [0, 1) = [0, 1) \cap (a : P_a[\mathfrak{m}_{a-} = +\infty] = 1),$$

the *non-singular intervals*

19) $$Q_n = [l_n, r_n) \qquad E_{l_n}(e^{-\mathfrak{m}_{l_n+}}) = 1 \qquad n \geqq 1,$$

and

20) *a single isolated trap at* 1,

and on $\bigcup_{n\geqq 1} [l_n, r_n)$, it is possible to express $\mathfrak{G}$ as

21a) $$(\mathfrak{G}\, u)(\xi)\, m(d\xi) = u^+(d\xi) \quad u(\xi)\, k(d\xi) \qquad l_n < \xi < r_n$$

21b) $$(\mathfrak{G}\, u)(l_n)\, m(l_n) = u^+(l_n) - u(l_n)\, k(l_n),$$

where, for $l_n < a < b < r_n$ and some positive constant B depending upon a and b,

22a) $$s(d\xi) = B[p_{ab}(\xi)\, p_{ba}(d\xi) - p_{ba}(\xi)\, p_{ab}(d\xi)] > 0$$

22b) $$k(d\xi) = \frac{p^+_{ab}(d\xi)}{p_{ab}(\xi)} = \frac{p^+_{ba}(d\xi)}{p_{ba}(\xi)} \geqq 0$$

22c) $$m(d\xi) = -[e^+_{ab}(d\xi) - e_{ab}(\xi)\, k(d\xi)] > 0^*$$

as in the non-singular case described above. At the left end,

23a) $$u^+(l_n+) = u^+(l_n) = \lim_{b \downarrow l_n} \frac{u(b) - u(l_n)}{s(b) - s(l_n)}$$

23b) $$k(l_n) = \lim_{b \downarrow l_n} \frac{P_{l_n}[\mathfrak{m}_b = +\infty]}{s(b) - s(l_n)}$$

23c) $$m(l_n) = \lim_{b \downarrow l_n} \frac{E_{l_n}(\mathfrak{m}_b \wedge \mathfrak{m}_\infty)}{s(b) - s(l_n)}$$

with the understanding that $u^+(l_n) = k(l_n) = m(l_n) \equiv 0$ in case $s(l_n) = s(l_n+) = -\infty$, and the additional condition

24) $$\sum_{n \geqq 1} \int_{Q_n} k[l_n, \xi)\, s(d\xi) + \int_{Q_n} m[l_n, \xi)\, s(d\xi) < +\infty$$

holds.

---

* p^+_{ab}, p^+_{ba}, *etc.* are computed using the *universal scale* s of 22a), not the *local scale* $s_{ab}(d\xi) = p_{ab}(\xi)\, p_{ba}(d\xi) - p_{ba}(\xi)\, p_{ab}(d\xi)$.

Besides this, the hitting probabilities and mean exit time

$$\text{25a)}\quad p_b(a) = P_a(\mathfrak{m}_b < +\infty) \qquad a \leqq b$$

$$\text{25b)}\quad e_b(a) = E_a(\mathfrak{m}_b \wedge \mathfrak{m}_\infty)$$

can be used to introduce a *shunt killing measure* k_+ and a *shunt scale* s_+:

$$\text{26a)}\quad 0 \leqq k_+(d\xi) = \frac{p_b(d\xi)}{p_b(\xi)} \quad (\xi < b) \qquad = \frac{p_1(d\xi)}{p_1(\xi)} \quad (\xi < 1)$$

$$k_+(0) = 0, \quad 0 \leqq k_+(1) = P_{1-}[\mathfrak{m}_1 = +\infty]$$

$$\text{26b)}\quad s_+(d\xi) = -[e_b(d\xi) - e_b(\xi)\, k_+(d\xi)] \qquad (\xi < b)$$

$$= -[e_1(d\xi) - e_1(\xi)\, k_+(d\xi)] \qquad (\xi < 1)$$

such that

$$\text{27a)}\quad s_+(b) \equiv \int_0^b s_+(d\xi) \quad \textit{is continuous}$$

$$\text{27b)}\quad s_+(a) < s_+(b) \qquad (a < b)$$

$$\text{27c)}\quad s_+(1) + k_+[0, 1) < +\infty$$

$$\text{28a)}\quad \int_{[a,b)\cap K_+} (\mathfrak{G} u)(\xi)\, s_+(d\xi) = \int_{[a,b)\cap K_+} [u(d\xi) - u(\xi)\, k_+(d\xi)]$$

$$a < b, \; u \in D(\mathfrak{G}),$$

and, for $l_n < \xi < r_n$,

$$\text{29a)}\quad k_+(d\xi) = \frac{s(d\xi)}{p(\xi)} \int_{[l_n, \xi)} p(\eta)\, k(d\eta)$$

$$\text{29b)}\quad s_+(d\xi) = \frac{s(d\xi)}{p(\xi)} \int_{[l_n, \xi)} p(\eta)\, m(d\eta),$$

p being the solution of

$$\text{30a)}\quad p^+(d\xi) = p(\xi)\, k(d\xi) \qquad l < \xi < r$$

$$\text{30b)}\quad p^+(l_n) = p(l_n)\, k(l_n)$$

$$\text{30c)}\quad p(l_n+) = p(l_n) = 1.$$

28a) can be put in the alternative form

$$\text{28b)}\quad (\mathfrak{G} u)(a) = \lim_{b \downarrow a} \frac{u(b) - u(a) - \left[\left[\bigcap_{a<\xi\leqq b} [1 - k_+(d\xi)]\right]^{-1} - 1\right] u(a)}{s_+(b) - s_+(a)}$$

$$a \in K_+ \cap [0, 1), \quad u \in D(\mathfrak{G}),$$

$\bigcap_{a<\xi\leqq b} [1 - k_+(d\xi)]$ being meant as a suggestive notation for $e^{-j_+(a,b]} \times (1 - \varkappa_1)(1 - \varkappa_2)$ *etc.*, where j_+ is the continuous part of k_+ and $\varkappa_1$, $\varkappa_2$, *etc.* ($\leqq 1$) are the jumps of k_+ for $a < \xi \leqq b$.

$D(\mathfrak{G})$ can now be described as the class $D(\mathfrak{G}^\bullet)$ of functions $u \in D$ such that, for some $u^\bullet \in D$,

31 a) $$\int_{[a,b)} u^\bullet(\xi)\, m(d\xi) = \int_{[a,b)} [u^+(d\xi) - u(\xi)\, k(d\xi)] \qquad l_n < a < b < r_n$$

31 b) $$u^\bullet(l_n)\, m(l_n) = u^+(l_n) - u(l_n)\, k(l_n),$$

32) $$\int_{[a,b)\cap K_+} u^\bullet(\xi)\, s_+(d\xi) = \int_{[a,b)\cap K_+} [u(d\xi) - u(\xi)\, k_+(d\xi)] \qquad 0 \leqq a < b \leqq 1^*,$$

33) $$u^\bullet(1) = 0,$$

and $\mathfrak{G}$ can be identified as the global differential operator $\mathfrak{G}^\bullet$ mapping $u \in D(\mathfrak{G}^\bullet) \to u^\bullet$. $\mathfrak{G}^\bullet$ is unambiguous because $u^\bullet \in D$, $0 < m[a, b)$ $(l_n \leqq a < b < r_n)$, and $s_+(a) < s_+(b)$ $(a < b)$.

A similar description of $\mathfrak{G}$ holds in the general case; the expressions become a bit complicated but nothing new happens and the proofs can be carried through using the special case explained above as a model and appealing to the splitting of 3.5 and the stopped motions of 3.9.

The reader may wish to skip the *singular* case at a first reading; this means that he should read 4.2—4.7 and 5.1—5.9 and then come back and read 4.8—4.10, 5.10, and 5.11.

4.2. $\mathfrak{G}$ as local differential operator: conservative non-singular case

Consider the non-singular conservative case:

1) $$P_a(\mathfrak{m}_\infty = +\infty) = 1 \qquad 0 \leqq a \leqq 1,$$

2a) $$P_a(\mathfrak{m}_{a+} = 0) = 1 \qquad 0 < a < 1$$

2b) $$P_a(\mathfrak{m}_{a-} = 0) = 1 \qquad 0 < a < 1,$$

suppose, in addition, that

3a) $$P_a(\mathfrak{m}_0 < +\infty) > 0 \qquad a < 1$$

3b) $$P_a(\mathfrak{m}_1 < +\infty) > 0 \qquad a > 0,$$

4a) $$P_0(\mathfrak{m}_{0+} = +\infty) = 1$$

4b) $$P_1(\mathfrak{m}_{1-} = +\infty) = 1,$$

5) $$E_a(\mathfrak{m}_0 \wedge \mathfrak{m}_1) < +\infty \qquad 0 \leqq a \leqq 1,$$

and let us construct a *scale* (*i.e.*, a continuous function with $s(a) < s(b)$ $(a < b)$) and a *speed measure* m (*i.e.*, a non-negative BOREL measure with $m[a, b) > 0$ $(a < b)$) such that

6) $$(\mathfrak{G} u)(\xi) = \frac{u^-(d\xi)}{m(d\xi)} = \frac{u^+(d\xi)}{m(d\xi)} \qquad u \in D(\mathfrak{G}),\quad 0 < \xi < 1,$$

* 32) contains the matching condition $u(b-) = [1 - k_+(b)]\, u(b)$ $(0 \leqq b < 1)$.

where u^- and u^+ are the one-sided scale derivatives of u:

$$u^-(a) = \lim_{b \uparrow a} \frac{u(a) - u(b)}{s(a) - s(b)}, \qquad u^+(a) = \lim_{b \downarrow a} \frac{u(b) - u(a)}{s(b) - s(a)},$$

and 6) is short for

7a) $$u^-(b) - u^-(a) = \int_{[a,b)} \mathfrak{G} u \, dm$$

7b) $$u^+(b) - u^+(a) = \int_{(a,b]} \mathfrak{G} u \, dm \qquad 0 < a < b < 1.$$

Define

8) $$s(\xi) = P_\xi(\mathfrak{m}_1 < +\infty) = P_\xi(\mathfrak{m}_1 < \mathfrak{m}_0) \qquad 0 < \xi < 1,$$

let $s(0) = 0$, $s(1) = 1$, and let e_1 be the indicator of the point 1; then

9) $$u = E_\cdot(e^{-\varepsilon \mathfrak{m}_1}) = \varepsilon G_\varepsilon e_1 \uparrow s \qquad \varepsilon \downarrow 0$$

and making $\varepsilon \downarrow 0$ in

10) $$\varepsilon \times [G_\alpha e_1 - G_\varepsilon e_1 + (\alpha - \varepsilon) G_\alpha G_\varepsilon e_1] = 0$$

proves

11) $$\alpha G_\alpha s = s,$$

or, what interests us,

12a) $$s \in D(\mathfrak{G})$$

12b) $$\mathfrak{G} s = 0.$$

Because $s(a) \leq s(b)$ $(a < b)$, if $s(a) = s(b)$ for some $a < b$, then *either* $s(b) = 0$ contradicting 3b), *or* $s(a)/s(b) = P_a(\mathfrak{m}_b < +\infty) = 1$, and

13) $$0 < E_a(e^{-\mathfrak{m}_0}) = E_a(e^{-\mathfrak{m}_0}, \mathfrak{m}_b < \mathfrak{m}_0) = E_a(e^{-\mathfrak{m}_b}) E_b(e^{-\mathfrak{m}_0}) < E_a(e^{-\mathfrak{m}_0}),$$

which is absurd; in brief, s *is a scale.*

Because of 5), if $a < \xi < b$, then

14) $$1 = P_\xi(\mathfrak{m}_a \wedge \mathfrak{m}_b < +\infty) = P_\xi(\mathfrak{m}_a < \mathfrak{m}_b) + P_\xi(\mathfrak{m}_b < \mathfrak{m}_a)$$

and, solving 14) and

15) $$\begin{aligned} s(\xi) &= P_\xi(\mathfrak{m} < +\infty, \mathfrak{m}_1(w^+_{\mathfrak{m}}) < +\infty) \qquad \mathfrak{m} = \mathfrak{m}_a \wedge \mathfrak{m}_b \\ &= P_\xi(\mathfrak{m}_a < \mathfrak{m}_b)\, s(a) + P_\xi(\mathfrak{m}_b < \mathfrak{m}_a)\, s(b) \end{aligned}$$

for $P_\xi(\mathfrak{m}_a < \mathfrak{m}_b)$ and $P_\xi(\mathfrak{m}_b < \mathfrak{m}_a)$ in terms of the scale,

16) $$P_\xi(\mathfrak{m}_a < \mathfrak{m}_b) = \frac{s(b) - s(\xi)}{s(b) - s(a)}, \qquad P_\xi(\mathfrak{m}_b < \mathfrak{m}_a) = \frac{s(\xi) - s(a)}{s(b) - s(a)},$$
$$a < \xi < b.$$

Defining e_{ab} to be the *mean exit time*

17) $$e_{ab}(\xi) = E_\xi(\mathfrak{m}_a \wedge \mathfrak{m}_b) \qquad a < \xi < b,$$

16) implies that, if $a < \xi < b$ and if $\mathfrak{m} = \mathfrak{m}_a \wedge \mathfrak{m}_b$, then

18) $$\begin{aligned} e(\xi) &= e_{01}(\xi) = E_\xi[\mathfrak{m} + \mathfrak{m}_0(w_\mathfrak{m}^+) \wedge \mathfrak{m}_1(w_\mathfrak{m}^+)] \\ &= e_{ab}(\xi) + \frac{s(b) - s(\xi)}{s(b) - s(a)} e(a) + \frac{s(\xi) - s(a)}{s(b) - s(a)} e(b), \end{aligned}$$

i.e., e is a concave function of the scale.

e is also a member of $D(\mathfrak{G})$; for, if f is the indicator of $(0, 1)$, then

$$G_\varepsilon f = E_\bullet\left[\int_0^{\mathfrak{m}_0 \wedge \mathfrak{m}_1} e^{-\varepsilon t}\, dt\right] \uparrow e \qquad \varepsilon \downarrow 0,$$

and letting $\varepsilon \downarrow 0$ in

19) $$[G_\alpha - G_\varepsilon + (\alpha - \varepsilon) G_\alpha G_\varepsilon] f = 0,$$

it is found that $e = G_\alpha[f + \alpha e]$, or what interests us,

20a) $$e \in D(\mathfrak{G})$$

20b) $$-\mathfrak{G} e = f = 1 \quad (0 < \xi < 1) \qquad = 0 \quad (\xi = 0, 1);$$

especially, e is continuous on $[0, 1]$ because of 2) and 3) (see 3.6).

Now, in general, if e is concave and if

$$e(0) = e(0+) = e(1) = e(1-) = 0,$$

then

21) $$e(a) = -\int_0^1 G(a, b)\, e^\pm(db) \qquad 0 < a < 1,$$

where G is the *Green function*

22) $$G(a, b) = G(b, a) = \frac{[s(a) - s(0)][s(1) - s(b)]}{s(1) - s(0)} \qquad a \leqq b$$

and $e^-(db) = e^+(db)$ is the non-positive BOREL measure induced by the interval functions

23a) $$e^-[a, b) = e^-(b) - e^-(a) \qquad e^-(a) = \lim_{b \uparrow a} \frac{e(a) - s(b)}{s(a) - s(b)}$$

23b) $$e^+(a, b] = e^+(b) - e^+(a) \qquad e^+(a) = \lim_{b \downarrow a} \frac{e(b) - e(a)}{s(b) - s(a)}.$$

Here is the proof.

Because e is concave,

24a) $$\frac{e(b) - e(a)}{s(b) - s(a)} \uparrow e^+(a)\ (b \downarrow a), \quad e^+(a) < +\infty\ (a > 0), \quad e^+ \in \downarrow$$

24b) $$\frac{e(a) - e(b)}{s(a) - s(b)} \downarrow e^-(a)\ (b \uparrow a), \quad e^-(a) > -\infty\ (a < 1), \quad e^- \in \downarrow,$$

and

25a) $$e^-(a-) = e^+(a-) = e^-(a) \qquad a < 1$$

25b) $$e^-(a+) = e^+(a+) = e^+(a) \qquad a > 0,$$

and it follows that the interval functions 23) induce the *same* non-positive BOREL measure $e^-(db) = e^+(db)$. Because of $e^\pm \in \downarrow$ and

$$e(0) = e(0+) = e(1) = e(1-) = 0,$$

26a) $$e(b) - e(0) \geqq e^\pm(b)\,[s(b) - s(0)] \geqq e^\pm(\tfrac{1}{2})\,[s(b) - s(0)] \qquad b \leqq \tfrac{1}{2}$$

26b) $$e(1) - e(b) \leqq e^\pm(b)\,[s(1) - s(b)] \leqq e^\pm(\tfrac{1}{2})\,[s(1) - s(b)] \qquad b \geqq \tfrac{1}{2}$$

and using this, it develops that

27) $$\int_0^1 G(a, b)\, e^+(db) \qquad 0 < a < 1$$

$$= \frac{s(1) - s(a)}{s(1) - s(0)} \int_{0<b\leqq a} [s(b) - s(0)]\, e^\pm(db) +$$

$$+ \frac{s(a) - s(0)}{s(1) - s(0)} \int_{a<b<1} [s(1) - s(b)]\, e^\pm(db)$$

$$= \frac{s(1) - s(a)}{s(1) - s(0)} \left[[s(a) - s(0)]\, e^+(a) - \int_0^a e^\pm(b)\, s(db) \right] +$$

$$+ \frac{s(a) - s(0)}{s(1) - s(0)} \left[-[s(1) - s(a)]\, e^+(a) + \int_a^1 e^\pm(b)\, s(db) \right] = -e(a)$$

as stated in 21).

21) is now applied to $e = E_\bullet(\mathfrak{m}_0 \wedge \mathfrak{m}_1)$, with the result that $e = \int_0^1 G\,dm$, where m is the *speed measure*

28) $$m(db) = -e^\pm(db) \geqq 0.$$

Because the e_{ab} in 18) is positive, $m[a, b) > 0$ $(a < b)$.

6) can now be proved in a few lines.

Given $u \in D(\mathfrak{G})$, DYNKIN's formula 3.7.7) implies

29) $$\frac{s(\xi) - s(0)}{s(1) - s(0)} u(1) + \frac{s(1) - s(\xi)}{s(1) - s(0)} u(0) - u(\xi)$$

$$= E_\xi[u(x_{\mathfrak{m}})] - u(\xi) \qquad \mathfrak{m} = \mathfrak{m}_0 \wedge \mathfrak{m}_1$$

$$= E_\xi\left[\int_0^{\mathfrak{m}} (\mathfrak{G}\,u)\,(x_t)\,dt\right] \qquad 0 < \xi < 1,$$

and to establish 6) it is enough to prove

30) $$E_\bullet\left[\int_0^{\mathfrak{m}} f(x_t)\,dt\right] = \int_0^1 G\,f\,dm \qquad f \in D$$

and to differentiate the resulting formula

$$31)\quad \frac{s(\xi)-s(0)}{s(1)-s(0)}u(1)+\frac{s(1)-s(\xi)}{s(1)-s(0)}u(0)-u(\xi) = \int_0^1 G(\xi,\eta)\,(\mathfrak{G}u)(\eta)\,m(d\eta)$$

Given $B\subset[0,1]$ with indicator f, $u_B = E_{\bullet}\left[\int_0^{\mathfrak{m}} f(x_t)\,dt\right]$ is concave because

$$32)\quad u_B(\xi) = E_\xi\left[\int_0^{\mathfrak{m}_a\wedge\mathfrak{m}_b} f(x_t)\,dt\right] + \frac{s(b)-s(\xi)}{s(b)-s(a)}u_B(a) + \frac{s(\xi)-s(a)}{s(b)-s(a)}u_B(b)$$

$$a<\xi<b,$$

$u_B(0+) = u_B(1-) = 0$ because $0\leqq u_B\leqq e_{01}$, and, choosing $B=(b,1]$ $(0<b<1)$, an application of 21) coupled with

$$33)\quad u(\xi) = e_{b1}(\xi) + \frac{s(1)-s(\xi)}{s(1)-s(b)}u(b)\qquad b\leqq\xi,\quad u = u_{(b,1]},\ e_{01}$$

implies

$$34a)\quad u_{(b,1]} \geqq \int_{(b,1]} G\,dm$$

$$34b)\quad u_{[b,1]} = \lim_{a\uparrow b} u_{[a,1]} \geqq \int_{[b,1]} G\,dm.$$

But now

$$34c)\quad u_{[0,b)} \geqq \int_{[0,b)} G\,dm$$

for the same reasons, and adding 34b) and 34c), it appears that

$$e = u_{[0,b)} + u_{[b,1)} \geqq \int_{[0,b)} G\,dm + \int_{[b,1]} G\,dm = \int_0^1 G\,dm = e,$$

i.e.,

$$35)\quad u_{(b,1]} = \int_{(b,1]} G\,dm\qquad 0\leqq b\leqq 1,$$

leading at once to 30).

Problem 1. Consider the splitting of the standard Brownian motion into independent $\mathfrak{Z} = (t: x(t)=0)$, excursions $\mathfrak{e}_n: n\geqq 1$, and signs $e_n: n\geqq 1$ (see 2.9); the problem is to compute the scale $s(d\xi)$ and the speed measure $m(d\xi)$ for the *skew* Brownian motion made up from standard Brownian $\mathfrak{Z}$ and $\mathfrak{e}_n: n\geqq 1$ and *skew* signs $e_n = \pm 1$,

$P_\bullet[e_n = +1] = \alpha, n \geqq 1, 0 < \alpha < 1$. ($\alpha = \frac{1}{2}$ for the standard BROWNian motion.)

$$[s(d\xi) = (1-\alpha)^{-1} d\xi, \; m(d\xi) = 2(1-\alpha)\, d\xi \quad \text{for } \xi \leqq 0,$$
$$s(d\xi) = \alpha^{-1} d\xi, \qquad m(d\xi) = 2\alpha\, d\xi \quad \text{for } \xi \geqq 0;$$

in fact, if $\mathfrak{Z}_n: n \geqq 1$ is a (BOREL) numbering of the open intervals of the complement of $\mathfrak{Z}$ and if $\mathfrak{m}^*_{-1}$, $\mathfrak{m}^*_{+1}$ are the skew BROWNian passage times to ± 1, then

$$\frac{s(+1) - s(0)}{s(+1) - s(-1)} = P_0[\mathfrak{m}^*_{-1} < \mathfrak{m}^*_{+1}] = \sum_{n \geqq 1} P_0[e_n = -1, \mathfrak{m}_{-1} \wedge \mathfrak{m}_{+1} \in \mathfrak{Z}_n]$$
$$= (1-\alpha) \sum_{n \geqq 1} P_0[\mathfrak{m}_{-1} \wedge \mathfrak{m}_{+1} \in \mathfrak{Z}_n] = 1 - \alpha,$$

or, what is the same, $\dfrac{s(+1) - s(0)}{s(0) - s(-1)} = \dfrac{\alpha}{1-\alpha}$, and for the rest, it is sufficient to point out that the skew and standard BROWNian motions agree up to the passage time to 0.]

Problem 2. Consider a particle moving among the integers Z^1 according to the rule: wait at $x(0) = l \in Z^1$ for an exponential holding time $\mathfrak{e}_1 > 0$, at time $\mathfrak{e}_1 -$ make a unit step (± 1) with probabilities depending upon l:

$$0 < p_\pm = P_l[x(\mathfrak{e}_1) = l \pm 1] \quad p_- + p_+ = 1,$$

and then start afresh, waiting at $x(\mathfrak{e}_1) = l \pm 1$ for an independent exponential holding time $\mathfrak{e}_2 > 0$, etc. W. FELLER [*12*] pointed out that the generator $\mathfrak{G}$ of such a *birth and death process* can be expressed in terms of a distance (scale) and positive weights (speed measure):

$$\mathfrak{G}\, u = \frac{u^+ - u^-}{m},$$

where

$$u^-(l) = \frac{u(l) - u(l-1)}{s(l) - s(l-1)}$$
$$u^+(l) = \frac{u(l+1) - u(l)}{s(l+1) - s(l)}$$
$$P_l[\mathfrak{m}_a < \mathfrak{m}_b] = \frac{s(b) - s(l)}{s(b) - s(a)} \qquad a < l < b$$
$$0 < m(l) = \frac{s(l+1) - s(l-1)}{[s(l) - s(l-1)][s(l+1) - s(l)]} E_l(\mathfrak{e}_1).$$

Give a complete proof using the method presented above.

4.3. $\mathfrak{G}$ as local differential operator: general non-singular case

Now consider the general non-singular diffusion D with state interval $= [0, 1]$ and let us prove the existence of a *scale* s, a non-negative *killing measure* k and a positive *speed measure* m defined on $(0, 1)$

such that, for each $u \in D(\mathfrak{G})$,

1) $$(\mathfrak{G} u)(\xi)\, m(d\xi) = u^-(d\xi) - u(\xi)\, k(d\xi) = u^+(d\xi) - u(\xi)\, k(d\xi) \qquad 0 < \xi < 1.$$

Given $0 < a < b < 1$, it is clear that, up to the obvious change of scale, the conditional diffusion (see problem 3.9.2)

2) $$\mathsf{D}^* = [W, \mathsf{B}, P^*_\xi(B) = P_\xi(w^\bullet_{\mathfrak{m}} \in B \mid \mathfrak{m} < +\infty), a \leqq \xi \leqq b] \qquad \mathfrak{m} = \mathfrak{m}_a \wedge \mathfrak{m}_b$$

satisfies all the conditions of the preceding section, and we conclude that, for the *scale*

3) $$s^*(\xi) = P^*_\xi(\mathfrak{m}_b < +\infty)$$

the *speed measure*

4) $$m^*(d\xi) = -e^{*+}(d\xi)^* \qquad e^*(\xi) = E^*_\xi(\mathfrak{m}_b),$$

and the appropriate *Green function* G^*,

5) $$E^*_\xi\left[\int_0^{\mathfrak{m}} f(x_t)\, dt\right] = \int_a^b G^*(\xi, \eta)\, f(\eta)\, m^*(d\eta) \qquad a \leqq \xi \leqq b, \quad f \in C(a, b)$$

Define $f^* = f/P_\bullet(\mathfrak{m} < +\infty)$ on $[a, b]$.

Given $u \in D(\mathfrak{G})$, it is clear from DYNKIN's formula that, for $a \leqq \xi \leqq b$,

6) $$\begin{aligned}
[E^*_\xi[u^*(x_{\mathfrak{m}})] - u^*(\xi)]\, P_\xi(\mathfrak{m} < +\infty)
&= E_\xi[u(x_{\mathfrak{m}}), \mathfrak{m} < +\infty] - u(\xi) \\
&= E_\xi\left[\int_0^{\mathfrak{m} \wedge \mathfrak{m}_\infty} (\mathfrak{G} u)(x_t)\, dt\right] \\
&= \int_0^{+\infty} dt\, E_\xi[(\mathfrak{G} u)^*(x_t)\, P_{x_t}(\mathfrak{m} < +\infty), t < \mathfrak{m} \wedge \mathfrak{m}_\infty] \\
&= \int_0^{+\infty} dt\, E^*_\xi[(\mathfrak{G} u)^*(x_t), t < \mathfrak{m}]\, P_\xi(\mathfrak{m} < +\infty) \\
&= E^*_\xi\left[\int_0^{\mathfrak{m}} (\mathfrak{G} u)^*(x_t)\, dt\right] P_\xi(\mathfrak{m} < +\infty),
\end{aligned}$$

and recalling 5), it results from 6) that

7) $$\begin{aligned}
-u^*(\xi) + \frac{s^*(b) - s^*(\xi)}{s^*(b) - s^*(a)}\, u^*(a) + \frac{s^*(\xi) - s^*(a)}{s^*(b) - s^*(a)}\, u^*(b) \\
= \int_b^a G^*(\xi, \eta)\, (\mathfrak{G} u)^*\, m^*(d\eta) \qquad a \leqq \xi \leqq b,
\end{aligned}$$

---

* $-$ and $+$ applied to functions e^* (with a star) indicate left and right differentiation with respect to s^*.

or, what interests us, that

8) $(\mathfrak{G}\,u)^*(\xi)\, m^*(d\xi) = u^{*-}(d\xi) = u^{*+}(d\xi) \qquad a < \xi < b, \quad u \in D(\mathfrak{G})$.

Consider the new scale

9) $s(d\xi) = p(\xi)^2 s^*(d\xi) \quad p(\xi) = P_\xi(\mathfrak{m} < +\infty),\ a \leqq \xi \leqq b.$

Given $\alpha > 0$, $G_\alpha 1 \in D(\mathfrak{G})$, $(\mathfrak{G} G_\alpha 1)^* = (\alpha G_\alpha 1 - 1)^* \leqq 0$, and using 7), it results that $(G_\alpha 1)^*$ is a concave function of s^*; thus, $1^* = \lim_{\alpha \uparrow +\infty} \alpha (G_\alpha 1)^*$ is a concave function of s^*, and computing -1^{*+} in terms of s, we find that

10) $$-1^{*+}(\xi) = p^+(\xi) \in \uparrow.$$

10) permits the introduction of the non-negative *killing measure*

11) $$k(d\xi) = p(\xi)^{-1} p^+(d\xi) = p(\xi)^{-1} p^-(d\xi)$$

and defining also the positive *speed measure*

12) $$m(d\xi) = p(\xi)^{-2} m^*(d\xi),$$

it results from 8) that, for $u \in D(\mathfrak{G})$,

13) $$\begin{aligned}(\mathfrak{G}\,u)(\xi)\, m(d\xi) &= p(\xi)^{-1} (\mathfrak{G}\,u)^*(\xi)\, m^*(d\xi) = p(\xi)^{-1} u^{*+}(d\xi)\\ &= p(\xi)^{-1} (p^2 (u\,p^{-1})^+)(d\xi) = p(\xi)^{-1}(u^+ p - u\,p^+)(d\xi)\\ &= u^+(d\xi) - u(\xi)\, p(\xi)^{-1} p^+(d\xi) = u^+(d\xi) - u(\xi)\, k(d\xi)\\ &= u^-(d\xi) - u(\xi)\, k(d\xi) \qquad\qquad a < \xi < b,\end{aligned}$$

as in 1).

13) is a local formula; the global formula 1) will now be proved. Given some new $s^\bullet$, $k^\bullet$, $m^\bullet$ such that 13) holds, and defining

14a) $$g_1(\xi) = E_\xi(e^{-\alpha \mathfrak{m}_b})$$

14b) $$g_2(\xi) = E_\xi(e^{-\alpha \mathfrak{m}_a}),$$

if $u(\xi) = G_\alpha[(a - \xi) \vee 0]$, then $u(a) > 0$, $g_2 = u(a)^{-1} u$ $(a < \xi < b)$, and using 13) with $s^\bullet$, $k^\bullet$, $m^\bullet$, one finds

15a) $$\alpha\, g_2(\xi)\, m^\bullet(d\xi) = g_2^{\bullet +}(d\xi) - g_2(\xi)\, k^\bullet(d\xi) \quad a < \xi < b^{**}$$

and likewise

15b) $$\alpha\, g_1(\xi)\, m^\bullet(d\xi) = g_1^{\bullet +}(d\xi) - g_1(\xi)\, k^\bullet(d\xi).$$

Because $g_1 \in \uparrow$, $g_2 \in \downarrow$, and $0 = g_1^{\bullet +}(d\xi)\, g_2(\xi) - g_1(\xi)\, g_2^{\bullet +}(d\xi)$, the WRONSKIAN $g_1^{\bullet +} g_2 - g_1 g_2^{\bullet +}$ is constant (> 0). $s^\bullet(d\xi)$ is therefore determined up to a constant factor (> 0) and, fixing $s^\bullet(d\xi)$, 15) deter-

* $-$ and $+$ applied to functions e (without the *) indicate left and right differentiation with respect to s.

** $\bullet +$ applied to function g indicates right differentiation with respect to $s^\bullet$.

mines $k^{\bullet}(d\xi)$ and $m^{\bullet}(d\xi)$:

16) $s^{\bullet}(d\xi) = l\, s(d\xi), \quad k^{\bullet}(d\xi) = l^{-1} k(d\xi), \quad m^{\bullet}(d\xi) = l^{-1} m(d\xi)$

$$l = \frac{s^{\bullet}(a, b)}{s(a, b)} > 0.$$

Now each point $0 < \xi < 1$ belongs to some interval (a, b) on which are defined s, k, and m such that 13) holds. Given two such intervals the corresponding s, k, m match on the overlap as in 16), and using this, it is a simple matter to piece together these local scales, speed measures, and killing measures so as to obtain global ones for which 1) holds.

4.4. A second proof

4.3.1) can be proved using an elaboration of the method of 4.2; the conditional diffusion of 4.3 is thus avoided.

Given $0 < a < b < 1$,

1) $$p_{ba}(\xi) = P_{\xi}[\mathfrak{m}_b < \mathfrak{m}_a] \qquad a \leqq \xi \leqq b$$

satisfies

2a) $$p_{ba}(\xi) < p_{ba}(\eta) \qquad \xi < \eta$$

2b) $$p_{ba} \in C[a, b]$$

2c) $$p_{ba}(a) = 0, \quad p_{ba}(b) = 1.$$

2c) is automatic; as to 2a), if $p_{ba}(\eta) = 0$ for some $a < \eta < b$, then

$$0 < E_{\eta}[e^{-\mathfrak{m}_b}] = E_{\eta}[e^{-\mathfrak{m}_b}, \mathfrak{m}_a < \mathfrak{m}_b] < E_a[e^{-\mathfrak{m}_b}],$$

which is impossible, and since $p_{\eta a}(\xi) < 1$ $(\xi < \eta)$ for similar reasons,

$$p_{ba}(\xi) = p_{\eta a}(\xi)\, p_{ba}(\eta) < p_{ba}(\eta) \qquad \xi < \eta$$

as desired. Because $p_{\eta a}(\xi)$ is continuous in η $(\eta > \xi)$, $p_{ba}(\eta)$ is continuous $(\eta > a)$, and the estimate

$$p_{\eta a}(a+) \leqq P_{\eta}[\mathfrak{m}_b < \varepsilon] \downarrow 0 \qquad (\varepsilon \downarrow 0, a < \eta < b)$$

completes the proof of 2b).

Changing the roles of a and b,

3) $$p_{ab}(\xi) = P_{\xi}[\mathfrak{m}_a < \mathfrak{m}_b] \qquad a \leqq \xi \leqq b$$

satisfies the analogous

4a) $$p_{ab}(\xi) > p_{ab}(\eta) \qquad \xi < \eta$$

4b) $$p_{ab} \in C[a, b]$$

4c) $$p_{ab}(a) = 1, \quad p_{ab}(b) = 0.$$

p_{ab} *is a convex function* of $t = p_{ba}$; for, if $a < a^\bullet < \xi < b^\bullet < b$, then

5a) $$p_{ab}(\xi) = p_{a^\bullet b^\bullet}(\xi)\, p_{ab}(a^\bullet) + p_{b^\bullet a^\bullet}(\xi)\, p_{ab}(b^\bullet)$$

5b) $$p_{ba}(\xi) = p_{a^\bullet b^\bullet}(\xi)\, p_{ba}(a\ ) + p_{b^\cdot a^\cdot}(\xi)\, p_{ba}(b\ ),$$

and, solving 5b) and

6) $$p(\xi) = p_{a^\bullet b^\bullet}(\xi) + p_{b^\bullet a^\bullet}(\xi) \qquad (\leqq 1)$$

in terms of p and $t = p_{ba}$ for

7a) $$p_{b^\bullet a^\bullet}(\xi) = \frac{t(\xi) - p(\xi)\, t(a^\bullet)}{t(b^\bullet) - t(a^\bullet)} = \frac{t(\xi) - t(a^\bullet)}{t(b^\bullet) - t(a^\bullet)} + \frac{[1 - p(\xi)]\, p_{ba}(a^\bullet)}{t(b^\bullet) - t(a^\bullet)}$$

7b) $$p_{a^\bullet b^\bullet}(\xi) = \frac{p(\xi)\, t(b^\bullet) - t(\xi)}{t(b^\bullet) - t(a^\bullet)} = \frac{t(b^\bullet) - t(\xi)}{t(b^\bullet) - t(a^\bullet)} - \frac{[1 - p(\xi)]\, p_{ba}(b^\bullet)}{t(b^\bullet) - t(a^\bullet)}$$

and substituting them back into 5a), one finds

8) $$\begin{aligned} p_{ab}(\xi) &= \frac{t(b^\bullet) - t(\xi)}{t(b^\bullet) - t(a^\bullet)}\, p_{ab}(a^\bullet) + \frac{t(\xi) - t(a^\bullet)}{t(b^\bullet) - t(a^\bullet)}\, p_{ab}(b^\bullet) \\ &\quad - \frac{1 - p(\xi)}{t(b^\bullet) - t(a^\bullet)} \times [p_{ba}(b^\bullet)\, p_{ab}(a^\bullet) - p_{ba}(a^\bullet)\, p_{ab}(b^\bullet)] \\ &< \frac{t(b^\bullet) - t(\xi)}{t(b^\bullet) - t(a^\bullet)}\, p_{ab}(a^\bullet) + \frac{t(\xi) - t(a^\bullet)}{t(b^\bullet) - t(a^\bullet)}\, p_{ab}(b^\bullet) \end{aligned}$$

as desired.

But then

9) $$0 \geqq p^\bullet = \lim_{\eta \downarrow \xi} \frac{p_{ab}(\eta) - p_{ab}(\xi)}{t(\eta) - t(\xi)} \qquad \xi < b$$

satisfies

10a) $$p^\bullet \in \uparrow$$

10b) $$\lim_{\eta \downarrow \xi} p^\bullet(\eta) = p^\bullet(\xi),$$

and, defining the *scale*

11) $$s(\xi) = s_{ab}(\xi) = \int_a^\xi [p_{ab}\, dp_{ba} - p_{ba}\, dp_{ab}] = \int_a^\xi [p_{ab} - t\, p^\bullet]\, dt,$$

it is found that

12) $$t^+ = p_{ba}^+(\xi) = \lim_{\eta \downarrow \xi} \frac{t(\eta) - t(\xi)}{s(\eta) - s(\xi)} = [p_{ab} - t\, p^\bullet]^{-1} \in \uparrow$$

13) $$p_{ab}^+(\xi) = \lim_{\eta \downarrow \xi} \frac{p_{ab}(\eta) - p_{ab}(\xi)}{s(\eta) - s(\xi)} = p^\bullet\, t^+ \in \uparrow$$

and [see 11)]

14) $$p_{ba}^+\, p_{ab} - p_{ba}\, p_{ab}^+ = t^+[p_{ab} - t\, p^\bullet] = 1,$$

permitting the introduction of the *killing measure*;

15) $$0 \leqq k_{ab}(d\xi) = \frac{p^+(d\xi)}{p(\xi)} \qquad a < \xi < b, \quad p = p_{ab}, p_{ba}.$$

Given $a < a^\bullet < b^\bullet < b$, the same method applied to $p_{a^\bullet b^\bullet}$ and $p_{b^\bullet a^\bullet}$ leads to a scale $s_{a^\bullet b^\bullet}$ and a killing measure $k_{a^\bullet b^\bullet}$ for $a^\bullet < \xi < b^\bullet$, and solving the identities 5) for

16a) $$p_{a^\bullet b^\bullet}(\xi) = \frac{\begin{vmatrix} p_{ab}(\xi) & p_{ab}(b^\bullet) \\ p_{ba}(\xi) & p_{ba}(b^\bullet) \end{vmatrix}}{\begin{vmatrix} p_{ab}(a^\bullet) & p_{ab}(b^\bullet) \\ p_{ba}(a^\bullet) & p_{ba}(b^\bullet) \end{vmatrix}} = \frac{B\begin{pmatrix} a & b \\ \xi & b^\bullet \end{pmatrix}}{B\begin{pmatrix} a & b \\ a^\bullet & b^\bullet \end{pmatrix}}$$

16b) $$p_{b^\bullet a^\bullet}(\xi) = \frac{\begin{vmatrix} p_{ab}(a^\bullet) & p_{ab}(\xi) \\ p_{ba}(a^\bullet) & p_{ba}(\xi) \end{vmatrix}}{\begin{vmatrix} p_{ab}(a^\bullet) & p_{ab}(b^\bullet) \\ p_{ba}(a^\bullet) & p_{ba}(b^\bullet) \end{vmatrix}} = \frac{B\begin{pmatrix} a & b \\ a^\bullet & \xi \end{pmatrix}}{B\begin{pmatrix} a & b \\ a^\bullet & b^\bullet \end{pmatrix}},$$

a simple calculation leads to the rules

17a) $$s_{a^\bullet b^\bullet}(d\xi) = B^{-1} s_{ab}(d\xi)$$

17b) $$k_{a^\bullet b^\bullet}(d\xi) = B\, k_{ab}(d\xi) \qquad a^\bullet < \xi < b^\bullet, \quad B = B\begin{pmatrix} a & b \\ a^\bullet & b^\bullet \end{pmatrix}$$

Coming to the *speed measure*, introduce the mean exit time

18) $$e(\xi) = e_{ab}(\xi) = E_\xi[\mathfrak{m}_a \wedge \mathfrak{m}_b \wedge \mathfrak{m}_\infty] \qquad a \leqq \xi \leqq b$$

and choose (a, b) so small that $\mathfrak{G}u > 0$ $(a \leqq \xi \leqq b)$ for some $u \in D(\mathfrak{G})$ (such intervals (a, b) cover $(0, 1)$).

Given $a < \xi < b$,

19) $$E_\xi[u(x(\mathfrak{m}_a \wedge \mathfrak{m}_b \wedge \mathfrak{m}_\infty))] - u(\xi) = E_\xi\left[\int_0^{\mathfrak{m}_a \wedge \mathfrak{m}_b \wedge \mathfrak{m}_\infty} \mathfrak{G}u\, dt\right]$$
$$\geqq \min_{ab} \mathfrak{G}u \times e(\xi),$$

showing that e is bounded and that

20) $$e(a+) = e(b-) = 0;$$

in addition,

21) $$e(\xi) = e_{a^\bullet b^\bullet}(\xi) + p_{a^\bullet b^\bullet}(\xi)\, e(a^\bullet) + p_{b^\bullet a^\bullet}(\xi)\, e(b^\bullet) \quad a < a^\bullet < \xi < b^\bullet < b,$$

and, using 19) with $(a^\bullet, b^\bullet)$ in place of (a, b) to check that $\sup_{a^\bullet b^\bullet} e_{a^\bullet b^\bullet}$ is small for small $(a^\bullet, b^\bullet)$, it follows from 21) that $e_{ab} \in C[a, b]$.

Now substitute 16a) and 16b) into 21), finding

22) $$B\begin{pmatrix} a & b \\ a^\bullet & b^\bullet \end{pmatrix} e(\xi) = B\begin{pmatrix} a & b \\ a^\bullet & b^\bullet \end{pmatrix} e_{a^\bullet b^\bullet}(\xi) + B\begin{pmatrix} a & b \\ \xi & b^\bullet \end{pmatrix} e(a^\bullet) +$$
$$+ B\begin{pmatrix} a & b \\ a^\bullet & \xi \end{pmatrix} e(b^\bullet),$$

i.e., collecting the determinants,

23)
$$0 \geqq -B\begin{pmatrix} a & b \\ a^\bullet & b^\bullet \end{pmatrix} e_{a^\bullet b^\bullet}(\xi) = \begin{vmatrix} e(b^\bullet) & e(\xi) & e(a^\bullet) \\ p_{ba}(b^\bullet) & p_{ba}(\xi) & p_{ba}(a^\bullet) \\ p_{ab}(b^\bullet) & p_{ab}(\xi) & p_{ab}(a^\bullet) \end{vmatrix} \quad a < a^\bullet < \xi < b^\bullet < b.$$

21) states that $e(\xi)$ lies above the linear interpolation

$$l(\xi) = p_{a^\bullet b^\bullet}(\xi)\, e(a^\bullet) + p_{b^\bullet a^\bullet}(\xi)\, e(b^\bullet) \qquad (a^\bullet < \xi < b^\bullet),$$

and, thinking of l as a straight line because of

$$l^\pm(d\xi) - k_{ab}(d\xi)\, l(\xi) = 0,^*$$

it is natural to describe the situation of 21, 22, 23) with the phrase; *e is concave relative to the linear form*

$$u^\pm(d\xi) - k_{ab}(d\xi)\, u(\xi),^{**}$$

and one expects

24)
$$e(\xi) = e_{ab}(\xi) = \int_a^b G_{ab}(\xi, \eta)\, m_{ab}(d\eta) \qquad a < \xi < b,$$

where $m_{ab}(d\xi)$ is the (positive) measure $-[e^\pm(d\xi) - k(d\xi)\, e(\xi)]$ and G_{ab} is the GREEN function

25)
$$G_{ab}(\xi, \eta) = G_{ab}(\eta, \xi) = p_{ba}(\xi)\, p_{ab}(\eta) \qquad \xi \leqq \eta.$$

Consider, for the proof of 24), a subdivision Δ of $[a, b]$ and, between each pair of successive division points $a \leqq a^\bullet < b^\bullet \leqq b$, define the interpolation

26)
$$e_\Delta(\xi) = p_{a^\bullet b^\bullet}(\xi)\, e(a^\bullet) + p_{b^\bullet a^\bullet}(\xi)\, e(b^\bullet) \quad a^\bullet \leqq \xi \leqq b^\bullet$$

as in the diagram.

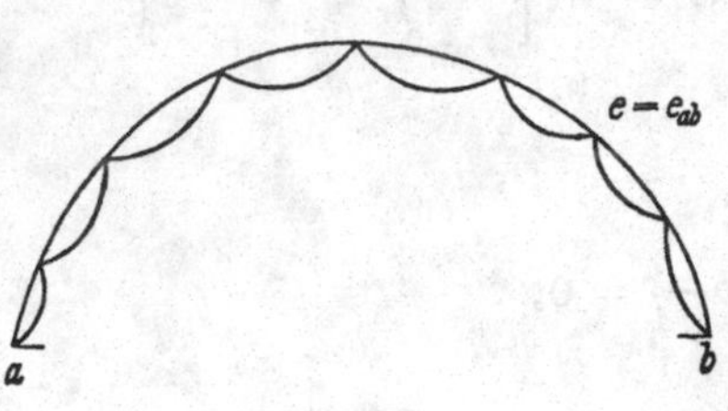

Diagramm 1

$e_\Delta \in C[a, b]$, $e_\Delta \leqq e$, $e_\Delta \uparrow e$ as the modulus of $\Delta \downarrow 0$, and $m_\Delta(d\xi) \equiv -[e_\Delta^\pm(d\xi) - k_{ab}(d\xi)\, e_\Delta(\xi)]$ $(a < \xi < b)$ is concentrated at the division points of Δ.

m_Δ *is non-negative*; for, if $a < \xi < b$ is a division point of Δ and if $a < a^\bullet < \xi < b^\bullet < b$, then [use $e_\Delta \leqq e$, $e_\Delta(\xi) = e(\xi)$, and 23)]

27)
$$\begin{vmatrix} e_\Delta(b^\bullet) & e_\Delta(\xi) & e_\Delta(a^\bullet) \\ p_{ba}(b^\bullet) & p_{ba}(\xi) & p_{ba}(a^\bullet) \\ p_{ab}(b^\bullet) & p_{ab}(\xi) & p_{ab}(a^\bullet) \end{vmatrix} \leqq \begin{vmatrix} e(b^\bullet) & e(\xi) & e(a^\bullet) \\ p_{ba}(b^\bullet) & p_{ba}(\xi) & p_{ba}(a^\bullet) \\ p_{ab}(b^\bullet) & p_{ab}(\xi) & p_{ab}(a^\bullet) \end{vmatrix} < 0,$$

---

* $l^+(\xi) = \lim_{\eta \downarrow} \dfrac{l(\eta) - l(\xi)}{s_{ab}(\eta) - s_{ab}(\xi)}$; $l^-(\xi) = \lim_{\eta \uparrow \xi} \dfrac{l(\xi) - l(\eta)}{s_{ab}(\xi) - s_{ab}(\eta)}$.

** See M. HEINS [1] for a special case of this idea.

and transforming the determinant on the left, it appears that

$$28)\qquad 0 \leqq \lim_{\substack{a^\bullet \uparrow \xi \\ b^\bullet \downarrow \xi}} \begin{vmatrix} \frac{e_\Delta(b^\bullet)-e_\Delta(\xi)}{s(b^\bullet)-s(\xi)} & e_\Delta(\xi) & \frac{e_\Delta(\xi)-e_\Delta(a^\bullet)}{s(\xi)-s(a^\bullet)} \\ \frac{p_{ba}(b^\bullet)-p_{ba}(\xi)}{s(b^\bullet)-s(\xi)} & p_{ba}(\xi) & \frac{p_{ba}(\xi)-p_{ba}(a^\bullet)}{s(\xi)-s(a^\bullet)} \\ \frac{p_{ab}(b^\bullet)-p_{ab}(\xi)}{s(b^\bullet)-s(\xi)} & p_{ab}(\xi) & \frac{p_{ab}(\xi)-p_{ab}(a^\bullet)}{s(\xi)-s(a^\bullet)} \end{vmatrix}$$

$$= \begin{vmatrix} e_\Delta^+(\xi) & e_\Delta(\xi) & e_\Delta^-(\xi) \\ p_{ba}^+(\xi) & p_{ba}(\xi) & p_{ba}^-(\xi) \\ p_{ab}^+(\xi) & p_{ab}(\xi) & p_{ab}^-(\xi) \end{vmatrix} = \begin{vmatrix} e_\Delta^+(\xi) - e_\Delta^-(\xi) - k_{ab}(\xi)e_\Delta(\xi) & e_\Delta(\xi) & e_\Delta^-(\xi) \\ p_{ba}^+(\xi) - p_{ba}^-(\xi) - k_{ab}(\xi)p_{ba}(\xi) & p_{ba}(\xi) & p_{ba}^-(\xi) \\ p_{ab}^+(\xi) - p_{ab}^-(\xi) - k_{ab}(\xi)p_{ab}(\xi) & p_{ab}(\xi) & p_{ab}^-(\xi) \end{vmatrix}$$

$$= \begin{vmatrix} -m_\Delta(\xi) & e_\Delta(\xi) & e_\Delta^-(\xi) \\ 0 & p_{ba}(\xi) & p_{ba}^-(\xi) \\ 0 & p_{ab}(\xi) & p_{ab}^-(\xi) \end{vmatrix} = m_\Delta(\xi) \times [p_{ba}^- p_{ab} - p_{ba} p_{ab}^-] = m_\Delta(\xi),$$

i.e., $m_\Delta \geqq 0$ at each division point.

$e_\Delta = \int G_{ab}\, dm_\Delta$ because $e^\bullet \equiv e_\Delta - \int G_{ab}\, dm_\Delta$ is a solution of

$$29a)\qquad e^{\bullet +}(d\xi) - k_{ab}(d\xi)\, e^\bullet(\xi) = 0 \qquad a < \xi < b$$

$$29b)\qquad e^\bullet(a+) = e^\bullet(b-) = 0,$$

as a simple computation proves, and, as such, it has to be $\equiv 0$; for example, if $0 < d = \max_{[a,b]} e^\bullet$ and if $a < \eta < b$ is the least root of $e^\bullet = d$, then [use 29a)] $0 \geqq e^{\bullet -}$ on a left neighbourhood of η, contradicting $d > e^\bullet(\xi)$ $(\xi < \eta)$.

$e \geqq e_\Delta = \int G_{ab}\, dm_\Delta$ now shows that, if $a < c < b$, then the total mass $n_\Delta[a, b]$ of

$$30)\qquad n_\Delta(d\xi) = p_{ba}(\xi)\, m_\Delta(d\xi) \qquad a < \xi < c$$
$$= p_{ab}(\xi)\, m_\Delta(d\xi) \qquad b > \xi \geqq c$$

is bounded, and selecting subdivisions Δ with moduli $\downarrow 0$ such that n_Δ converges to a non-negative BOREL measure n on $[a, b]$, one finds

$$31)\qquad e_\Delta = \int_a^b G_{ab}\, dm_\Delta \uparrow e = n(a)\, p_{ab}(\xi) + n(b)\, p_{ba}(\xi) + \int_a^b G_{ab}\, dm_{ab},$$

where

$$32)\qquad m_{ab}(d\xi) = p_{ba}(\xi)^{-1}\, n(d\xi) \qquad a < \xi < c$$
$$p_{ab}(\xi)^{-1}\, n(d\xi) \qquad b > \xi \geqq c$$

and

$$33)\qquad n(a) = e(a+) = n(b) = e(b-) = 0,$$

completing the proof of 24).

m_{ab} is the *speed measure*. 21) and 17) are now used to show that m_{ab} transforms according to the rule

34) $$m_{a^\bullet b^\bullet}(d\xi) = B\begin{pmatrix} a & b \\ a^\bullet & b^\bullet \end{pmatrix} m_{ab}(d\xi)$$

and the proof of 4.3.1) is completed as in 4.2.

Problem 1. Prove that, for $a < b$,

$$p_{ab}(\xi) \leqq \frac{s(b) - s(\xi)}{s(b) - s(a)}, \quad p_{ba}(\xi) \leqq \frac{s(b) - s(\xi)}{s(b) - s(a)}$$

where $=$ holds if and only if $k(a, b) = 0$.

[$p(\xi) = p_{ab}(\xi)$ satisfies $p^+(d\xi) = p(\xi)\,k(d\xi)$, $p(a) = 1$, $p(b) = 0$, and therefore

$$p(\xi) \leqq \frac{s(b) - s(\xi)}{s(b) - s(a)} p(a) + \frac{s(\xi) - s(a)}{s(b) - s(a)} p(b) = \frac{s(b) - s(\xi)}{s(b) - s(a)} \equiv l(\xi).$$

p cannot touch l inside (a, b) unless it is a linear function of s, *i.e.* unless $p^+(d\xi) = p(\xi)\,k(d\xi) = 0$.]

Problem 2. Prove that $P_a(\mathfrak{m}_b \leqq t)$ is a convex function of $s = s(a)$ for $a \leqq b$ and for $a \geqq b$.

$$[P_a(\mathfrak{m}_b \leqq t) \leqq P_a(\mathfrak{m}_b(w^+_{\mathfrak{m}_\xi \wedge \mathfrak{m}_\eta}) \leqq t)$$
$$= p_{\xi\eta}(a)\,P_\xi(\mathfrak{m}_b \leqq t) + p_{\eta\xi}(a)\,P_\eta(\mathfrak{m}_b \leqq t)$$

for $\xi < a < \eta < b$ or $\xi > a > \eta > b$; now use the result of problem 1.]

Problem 3. Consider the class $D(\mathfrak{G}^\bullet)$ of continuous functions u such that

$$u^+(b) - u^+(a) = \int_{(a,b]} [u(\xi)\,k(d\xi) + u^\bullet(\xi)\,m(d\xi)] \qquad 0 < a < b < 1$$

for some continuous $u^\bullet$ and let $\mathfrak{G}^\bullet$ be the differential operator

$$(\mathfrak{G}^\bullet u)(\xi) \equiv u^\bullet(\xi) = \frac{u^+(d\xi) - u(\xi)\,k(d\xi)}{m(d\xi)} \qquad 0 < \xi < 1;$$

the problem is to show that $\mathfrak{G}^\bullet u \leqq 0$ at each non-negative local maximum of $u(\xi)$ $(0 < \xi < 1)$ (see problem 3.7.1).

[Given $0 < a < \xi < b < 1$ such that $u(\eta) \leqq u(\xi)$ $(\geqq 0)$ and $(\mathfrak{G}^\bullet u)(\eta) > 0$ for $a < \eta < b$, it is found that

$$0 > p_{ab}(\xi)\,u(a) + p_{ba}(\xi)\,u(b) - u(\xi) = \int_a^b G_{ab}(\xi, \eta)\,(\mathfrak{G}^\bullet u)(\eta)\,m(d\eta) > 0,$$

which is absurd.]

Problem 4. $P_a(\mathfrak{m}_b < +\infty)\,P_b(\mathfrak{m}_a < +\infty)$ is either $\equiv 1$ on $(0, 1) \times (0, 1)$ or < 1 on $(0, 1) \times (0, 1)$. D is said to be *persistent* in the first case and *transient* in the second.

[Given $0 < a < b < 1$ with $P_a(\mathfrak{m}_b < +\infty)\, P_b(\mathfrak{m}_a < +\infty) = 1$, $p_b(\xi) = P_\xi(\mathfrak{m}_b < +\infty)$ is a solution of

$$\begin{aligned} p^+(d\xi) &= p(\xi)\, k(d\xi) && 0 < \xi < b \\ p(\xi) &= 1 && \xi = a, b \end{aligned}$$

and, using $0 \leqq p_b^+ \in \uparrow$, it is found that $p_b(\xi) \equiv 1$ $(0 < \xi < b)$. But then $k(d\xi) = 0$ $(0 < \xi < b)$ and since $k(d\xi) = 0$ $(a < \xi < 1)$ for similar reasons, $p_\eta(\xi) = P_\xi(\mathfrak{m}_\eta < +\infty)$ $(1 > \eta > b)$ is a solution of

$$\begin{aligned} p^+(d\xi) &= 0 && 0 < \xi < \eta \\ p(\eta-) &= 1 && \\ p^+(\xi) &= 0 && a < \xi < b, \end{aligned}$$

i.e., $p_\eta(\xi) = 1$ $(0 < \xi < \eta)$ for each $\eta < 1$.]

4.5. $\mathfrak{G}$ at an isolated singular point

Given a non-singular diffusion D on an interval Q with endpoints 0 and 1, if $0 \notin Q$, let it be added and let $P_0[\mathfrak{m}_{0+} = \mathfrak{m}_\infty = +\infty] \equiv 1$. D *will be modified so as to have*

$$1\text{a)} \qquad e_1 = E_0(e^{-\mathfrak{m}_{0+}}) = e_2 = \lim_{b \downarrow 0} E_{0+}(e^{-\mathfrak{m}_b})$$

$$1\text{b)} \qquad e_3 = E_{0+}(e^{-\mathfrak{m}_0}) = e_4 = \lim_{b \downarrow 0} E_b(e^{-\mathfrak{m}_{0+}}).$$

Because of

$$2\text{a)} \qquad e_1, e_2, e_4 = 0 \text{ or } 1$$

$$2\text{b)} \qquad e_1 \leqq e_2, \quad e_3 \leqq e_4$$

$$2\text{c)} \qquad e_1 e_3 = e_2 e_4$$

[see 3), 4) and 5) of 3.3], *if* $e_1 < e_2$, then

$$3\text{a)} \qquad e_1 = 0, \quad e_2 = 1, \quad e_3 = e_4 = 0,$$

1b) is automatic, and (see problem 3.6.3) P_0 can be modified so as to have $P_0[\mathfrak{m}_{0+} = 0] = 1$, *i.e.*, $e_1 = 1$, while *if* $e_1 = e_2$ and if $e_3 < e_4$, then

$$3\text{b)} \qquad e_3 < 1 = e_4, \quad e_1 = e_2 = 0,$$

1a) is automatic, $\mathfrak{m}_{0+} = \mathfrak{m}_0 \wedge \mathfrak{m}_\infty$ in case $x(0) > 0$ and $\mathfrak{m}_{0+} < +\infty$, and the desired modification is $[x^\bullet(t);\, t \geqq 0, P_\bullet]$, where

$$\begin{aligned} x^\bullet(t) &= x(t) && t < \mathfrak{m}_0 \wedge \mathfrak{m}_{0+} \\ &= 0 && t \geqq \mathfrak{m}_0 \wedge \mathfrak{m}_{0+} \end{aligned}$$

D *is now modified at* 1 *also, so as to have* $1 \in Q$ and

4a) $$E_1(e^{-\mathfrak{m}_1-}) = \lim_{b\uparrow 1} E_{1-}(e^{-\mathfrak{m}_b})$$

4b) $$E_{1-}(e^{-\mathfrak{m}_1}) = \lim_{b\uparrow 1} E_b(e^{-\mathfrak{m}_1-});$$

this modification will be understood below.

Given $u \in D(\mathfrak{G})$, if $P_0(\mathfrak{m}_{0+} < +\infty) = 0$, then

5a) $$(\mathfrak{G}\, u)\,(0) = -\varkappa(0)\, u(0) \qquad \varkappa(0) = E_0(\mathfrak{m}_\infty)^{-1},^*$$

while, if $P_0(\mathfrak{m}_{0+} = 0) = 1$ and if a dependence $c_1\, u(0) + c_2(\mathfrak{G} u)(0) = 0$ prevails for each choice of $u \in D(\mathfrak{G})$, then $[c_1 + c_2\,\alpha](G_\alpha f)(0) = c_2 f(0)$ for each $f \in D$ and $\alpha > 0$, and choosing $f(0) = 0 < f(b)$ $(b > 0)$, $(G_\alpha f)(0) > 0$ implies $c_1 = c_2 = 0$; in brief, *if* $P_0[\mathfrak{m}_{0+} = 0] = 1$, *then* $u(0)$ *and* $(\mathfrak{G} u)(0)$ *are independent for* $u \in D(\mathfrak{G})$.

But if $P_0[\mathfrak{m}_{0+} = 0] = 1$, then, as will be proved below,

6) $$u^+(0) = \lim_{b\downarrow 0} \frac{u(b) - u(0)}{s(b) - s(0)} = u^+(0+)^{**}$$

exists for each $u \in D(\mathfrak{G})$, and a dependence holds between $u(0)$, $u^+(0)$, and $(\mathfrak{G} u)(0)$:

5b) $$(\mathfrak{G}\, u)\,(0)\, m(0) = u^+(0) - u(0)\, k(0) \qquad u \in D(\mathfrak{G}),$$

where

7a) $$k(0) = \lim_{b\downarrow 0} \frac{P_0(\mathfrak{m}_b = +\infty)}{s(b) - s(0)}$$

7b) $$m(0) = \lim_{b\downarrow 0} \frac{E_0(\mathfrak{m}_b \wedge \mathfrak{m}_\infty)}{s(b) - s(0)},^{***}$$

complementing 4.3.1):

$$(\mathfrak{G}\, u)\,(\xi)\, m(d\xi) = u^+(d\xi) - u(\xi)\, k(d\xi) \qquad 0 < \xi < 1.$$

Beginning with the proof of 6), if $f \in D$ is such that $f(a) = 0$ $(a > \frac{1}{2})$ then $u = G_1 f \in D(\mathfrak{G})$ satisfies $u(a) = E_a(e^{-\mathfrak{m}_{1/2}})\, u(\frac{1}{2})$ $(a \leqq \frac{1}{2})$. Because $u(\frac{1}{2}) > 0$, $0 \leqq u(a) \in \uparrow$ $(0 < a \leqq \frac{1}{2})$; because $P_0(\mathfrak{m}_{0+} = 0) = 1$, $u(0+) = u(0) > 0$; and, using $\mathfrak{G}\, u(a) = u(a)$ $(a < \frac{1}{2})$ to deduce from $0 \leqq u \in \uparrow$ that $0 \leqq u^+ \in \uparrow$ to the left of $\frac{1}{2}$, it follows that

8) $$k(0, \tfrac{1}{2}] + m(0, \tfrac{1}{2}] \leqq u(0)^{-1} \left[\int_{0+}^{\frac{1}{2}} u\, k(d\xi) + \int_{0+}^{\frac{1}{2}} u\, m(d\xi)\right]$$
$$\leqq u(0)^{-1}[u^+(\tfrac{1}{2}) - u^+(0+)] < +\infty.$$

---

* $e = E_0(\mathfrak{m}_\infty) > 0$; $e^{-1} \equiv 0$ if $e = +\infty$.

** $u^+(0) \equiv 0$ if $s(0) = -\infty$.

*** $k(0) = m(0) = 0$ if $s(0) = -\infty$.

8) implies the convergence of the integral

$$\int_{0+}^{\frac{1}{2}} u\,k(d\xi) + \int_{0+}^{\frac{1}{2}} \mathfrak{G} u\,m(d\xi) = u^+(\tfrac{1}{2}) - u^+(0+)$$

for each $u \in D(\mathfrak{G})$, proving the existence of $u^+(0+)$; as to the rest of 6), *either* $s(0) > -\infty$, in which case

$$u^+(0) = \lim_{b \downarrow 0} \frac{u(b) - u(0)}{s(b) - s(0)} = \lim_{b \downarrow 0} [s(b) - s(0)]^{-1} \int_0^b u^+ s(d\xi)$$
$$= u^+(0+),^*$$

or $s(0) = -\infty$, in which case $u^+(0) = 0$, $u^+(0+) = 0$ owing to $u(\frac{1}{2}) - u(0) = \int_{0+}^{\frac{1}{2}} u^+ ds$, and $k(0) = m(0) = 0$ in agreement with 5b).

Given $s(0) > -\infty$, DYNKIN's formula,

9) $$(\mathfrak{G} u)(0) = \lim_{b \downarrow 0} \frac{P_0(\mathfrak{m}_b < +\infty)\, u(b) - u(0)}{E_0(\mathfrak{m}_b \wedge \mathfrak{m}_\infty)},$$

implies

10) $$(\mathfrak{G} u)(0) \frac{E_0(\mathfrak{m}_b \wedge \mathfrak{m}_\infty)\, P_0(\mathfrak{m}_b < +\infty)^{-1}}{s(b) - s(0)}$$
$$= \frac{u(b) - u(0)}{s(b) - s(0)} - \frac{P_0(\mathfrak{m}_b < +\infty)^{-1} - 1}{s(b) - s(0)} u(0) + o(1) \qquad b \downarrow 0,$$

and an application of 6) coupled with the independence of $u(0)$ and $(\mathfrak{G}u)(0)$ implies the separate convergence of the coefficients

11a) $$\frac{P_0(\mathfrak{m}_b < +\infty)^{-1} - 1}{s(b) - s(0)} = \frac{P_0(\mathfrak{m}_b = +\infty)}{s(b) - s(0)} + o(1)$$

11b) $$\frac{E_0(\mathfrak{m}_b \wedge \mathfrak{m}_\infty)\, P_0(\mathfrak{m}_b < +\infty)^{-1}}{s(b) - s(0)} = \frac{E_0(\mathfrak{m}_b \wedge \mathfrak{m}_\infty)}{s(b) - s(0)} + o(1)$$

to non-negative limits $k(0)$ and $m(0) < +\infty$. 5b) is immediate from this.

$\mathfrak{G}$ satisfies similar conditions at 1; if $u \in D(\mathfrak{G})$ and if $P_1(\mathfrak{m}_{1-} = 0) = 0$ then

12a) $$(\mathfrak{G} u)(1) = -\varkappa(1)\, u(1) \qquad \varkappa(1) = E_1(\mathfrak{m}_\infty)^{-1},$$

while, if $P_1[\mathfrak{m}_{1-} = 0] = 1$, then

12b) $$(\mathfrak{G} u)(1)\, m(1) = -u^-(1) - u(1)\, k(1),$$

where

13) $$u^-(1) = \lim_{b \uparrow 1} \frac{u(1) - u(b)}{s(1) - s(b)} = u^-(1-)$$

* $P_0[\mathfrak{m}_{0+} = 0] = 1$ implies $u(0+) = u(0)$.

and

14a) $$k(1) = \lim_{b\uparrow 1} \frac{P_1(\mathfrak{m}_b = +\infty)}{s(1) - s(b)}$$

14b) $$m(1) = \lim_{b\uparrow 1} \frac{E_1(\mathfrak{m}_b \wedge \mathfrak{m}_\infty)}{s(1) - s(b)}.$$

Problem 1. Give an alternative proof of

$$(\mathfrak{G}\, u)\,(0)\, m(0) = u^+(0) - u(0)\, k(0) \qquad u \in D(\mathfrak{G})$$

in the case

$$P_0(\mathfrak{m}_1 < +\infty) > 0, \quad E_0(\mathfrak{m}_1 \wedge \mathfrak{m}_\infty) < +\infty.$$

Use

$$p_1(\xi) = P_\xi(\mathfrak{m}_1 < +\infty)$$
$$p_{01}(\xi) = P_\xi(\mathfrak{m}_0 < \mathfrak{m}_1)$$
$$e_1(\xi) = E_\xi(\mathfrak{m}_1 \wedge \mathfrak{m}_\infty)$$

as in 4.4 to define the killing and speed measures

$$k(d\xi) = \frac{p_1^+(d\xi)}{p_1(\xi)} \qquad 0 < \xi < 1$$

$$k(0) = \frac{p_1^+(0)}{p_1(0)} = \frac{p_1^+(0+)}{p_1(0)}$$

$$m(d\xi) = -[e_1^+(d\xi) - e_1(\xi)\, k(d\xi)] \qquad 0 < \xi < 1$$

$$m(0) = -[e_1^+(0) - e_1(0)\, k(0)]$$

and express

$$p_1(\xi)\, u(1) - u(\xi) = E_\xi\left[\int_0^{\mathfrak{m}_1 \wedge \mathfrak{m}_\infty} (\mathfrak{G}\, u)\,(x_t)\, dt\right] \qquad u \in D(\mathfrak{G})$$

by means of the GREEN function

$$G(\xi, \eta) = G(\eta, \xi) = \frac{p_1(\xi)\, p_{01}(\eta)}{B} \qquad \xi \leqq \eta,$$

$$B = p_1^+(\xi)\, p_{01}(\eta) - p_1(\xi)\, p_{01}^+(\eta) \qquad 0 \leqq \xi < 1.$$

4.6. Solving $\mathfrak{G}^\bullet u = \alpha u$

Consider the modified non-singular diffusion of 4.5; it is obvious from the composition rule

1) $$E_a(e^{-\alpha \mathfrak{m}_b}) = E_a(e^{-\alpha \mathfrak{m}_c})\, E_c(e^{-\alpha \mathfrak{m}_b}) \qquad a < c < b, \quad \alpha > 0,$$

that

2a) $$g_1(a) = E_a(e^{-\alpha \mathfrak{m}_{\frac{1}{2}}}) \qquad 0 \leqq a \leqq \tfrac{1}{2}$$
$$= E_{\frac{1}{2}}(e^{-\alpha \mathfrak{m}_a})^{-1} \qquad \tfrac{1}{2} < a \leqq 1$$

satisfies

3a) $$E_a(e^{-\alpha \mathfrak{m}_b}) = \frac{g_1(a)}{g_1(b)} \qquad a < b$$

4a) $$g_1(a) < g_1(b) \qquad a < b$$

5a) $$g_1(a+) = g_1(a) < +\infty \qquad 0 \leqq a < 1$$
$$g_1(a-) = g_1(a) > 0 \qquad 0 < a \leqq 1.$$

Now as in the argument before 4.5.8), if $0 < b < 1$ and if $f \in D$ satisfies $f(a) = 0$ $(a \leqq b)$ and $f(a) > 0$ $(a > b)$, then $u = G_\alpha f \in D(\mathfrak{G})$ satisfies $u(a) = E_a(e^{-\alpha \mathfrak{m}_b})\, u(b)$ $(a \leqq b)$. Because $(\mathfrak{G}u)(a) = \alpha u(a)$ $(a \leqq b)$ and $u(b) > 0$, g_1 is now seen to be a solution of

6a) $$(\mathfrak{G}^{\bullet} g_1)(\xi) = \alpha g_1(\xi) \qquad 0 < \xi < 1,$$

coupled with

7a) $$\alpha g_1(0)\, m(0) = g_1^+(0) - g_1(0)\, k(0)^* \qquad E_0(e^{-\mathfrak{m}_{0+}}) = 1$$
$$g_1(0) = g_1(0+) = 0 \qquad E_0(e^{-\mathfrak{m}_{0+}}) = 0,$$

$\mathfrak{G}^{\bullet}$ being the differential operator

$$(\mathfrak{G}^{\bullet} u)(\xi) = \frac{u^+(d\xi) - u(\xi)\, k(d\xi)}{m(d\xi)} \qquad 0 < \xi < 1.$$

A similar argument shows that

2b) $$g_2(a) = E_a\left(e^{-\alpha \mathfrak{m}_{\frac{1}{2}}}\right) \qquad \tfrac{1}{2} < a \leqq 1$$
$$= E_{\frac{1}{2}}(e^{-\alpha \mathfrak{m}_a})^{-1} \qquad 0 \leqq a \leqq \tfrac{1}{2}$$

satisfies

3b) $$E_a(e^{-\alpha \mathfrak{m}_b}) = \frac{g_2(a)}{g_2(b)} \qquad a > b$$

4b) $$g_2(a) < g_2(b) \qquad a > b$$

5b) $$g_2(a+) = g_2(a) > 0 \qquad 0 \leqq a < 1$$
$$g_2(a-) = g_2(a) < +\infty \qquad 0 < a \leqq 1$$

and

6b) $$(\mathfrak{G}^{\bullet} g_2)(\xi) = \alpha g_2(\xi) \qquad 0 < \xi < 1,$$

coupled with

7b) $$\alpha g_2(1)\, m(1) = -g_2^-(1) - g_2(1)\, k(1)^{**} \qquad E_1(e^{-\mathfrak{m}_{1-}}) = 1$$
$$g_2(1) = g_2(1-) = 0 \qquad E_1(e^{-\mathfrak{m}_{1-}}) = 0.$$

g_1 and g_2 span the solutions of

8) $$(\mathfrak{G}^{\bullet} u)(\xi) = \alpha u(\xi) \qquad 0 < \xi < 1,$$

as will now be proved.

---

* $g_1^+(0) = g_1^+(0+)$. $m(0) = k(0) = 0$ if $s(0) = -\infty$.

** $g_2^-(1) = g_2^-(1-)$. $m(1) = k(1) = 0$ if $s(1) = +\infty$.

Given $0 < a < b < 1$ and a solution u of 8), the fact that the determinant $g_1(a)\, g_2(b) - g_1(b)\, g_2(a)$ is < 0 permits the choice of constants c_{ab} and c_{ba} such that $u^\bullet = u - [c_{ab}\, g_1 + c_{ba}\, g_2] = 0$ at a and at b. Because $u^\bullet$ is a solution of 8), if $a < \xi < b$ is a positive local maximum of $u^\bullet(\xi)$, then $0 < \alpha\, u^\bullet(\xi) = (\mathfrak{G}^\bullet u^\bullet)(\xi) \leqq 0$ (see problem 4.4.3), which is absurd. $u^\bullet(\xi)$ is then $\leqq 0$ $(a < \xi < b)$, and, using $-u^\bullet$ in place of $u^\bullet$, it appears that $u^\bullet(\xi) = 0$ $(a < \xi < b)$. But then $u(\xi) = c_{ab}\, g_1(\xi) + c_{ba}\, g_2(\xi)$ $(a \leqq \xi \leqq b)$, and as the reader will check, c_{ab} and c_{ba} do not depend upon a and b; this completes the proof.

Because

$$9)\qquad g_1^+ g_2 - g_1 g_2^+ \big|_a^b = \int_a^b [g_2\, \mathfrak{G}^\bullet g_1 - g_1\, \mathfrak{G}^\bullet g_2]\, m(d\xi) \qquad 0 < a < b < 1,$$

the Wronskian

$$10)\qquad \begin{aligned} B = B[g_1, g_2] &= g_1^-(\xi)\, g_2(\xi) - g_1(\xi)\, g_2^-(\xi) \\ &= g_1^+(\xi)\, g_2(\xi) - g_1(\xi)\, g_2^+(\xi) \qquad\qquad 0 < \xi < 1 \end{aligned}$$

is constant (> 0); this will be useful below.

Given $\alpha > 0$, let $j(db) = k(db) + \alpha m(db)$ $(0 < b < 1)$, let 0 be called

an exit point if $\qquad +\infty > \int_{0+}^{\frac12} s(da) \int_a^{\frac12} j(db)$

an entrance point if $\quad +\infty > \int_{0+}^{\frac12} j(da) \int_a^{\frac12} s(db),$

note that this classification is independent of α, and let us check the statements of tables 1 and 2 below.

Table 1

	exit and entrance	exit not entrance	entrance not exit	neither exit nor entrance
$g_1(0)$	$\geqq 0$	$= 0$	> 0	$= 0$
$g_2(0)$	$< +\infty$	$< +\infty$	$= +\infty$	$= +\infty$
$g_1^+(0)$	$\geqq 0$	> 0	$= 0$	$= 0$
$-g_2^+(0)$	$< +\infty$	$= +\infty$	$< +\infty$	$= +\infty$
$\int_0^{1/2} g_1\, dj$	$< +\infty$	$< +\infty$	$< +\infty$	$< +\infty$
$\int_{0+}^{1/2} g_2\, dj$	$< +\infty$	$= +\infty$	$< +\infty$	$= +\infty$

Table 2

	exit and entrance	exit not entrance	entrance not exit	neither exit nor entrance
$P_0[\mathfrak{m}_{0+} = 0] =$	0 *or* 1	0	1	0
$E_{0+}(e^{-\mathfrak{m}_{0+}}) =$	1	1	0	0
$(\mathfrak{G} u)(0) =$	$\frac{u^+(0) - u(0)\,k(0)}{m(0)}$ *if* $P_0[\mathfrak{m}_{0+} = 0] = 1$; $-\varkappa u(0)$ *if* $P_0[\mathfrak{m}_{0+} = 0] = 0$	$-\varkappa u(0)$	$u^+(0) = 0$	$-\varkappa u(0)$

A glance at line 1 of table 2 will explain the usage of *exit* and *entrance*; in table 2, $(\mathfrak{G}u)(0) = m(0)^{-1}[u^+(0) - k(0)\,u(0)]$ means $u^+(0) = k(0)\,u(0)$ if $m(0) = 0$; in table 1, $g_2^+(0) = g_2^+(0+)$.

Endpoint 1 is said to be

$$\textit{an exit point if} \qquad +\infty > \int_{\frac{1}{2}}^{1-} s(da) \int_{\frac{1}{2}}^{a} j(db)$$

$$\textit{an entrance point if} \qquad +\infty > \int_{\frac{1}{2}}^{1-} j(da) \int_{\frac{1}{2}}^{a} s(db);$$

$$g_1(1),\, g_2(1),\, g_1^-(1) = g_1^-(1-), \quad P_1[\mathfrak{m}_{1-} = 0], \quad \lim_{b \uparrow 1} E_b(e^{-\mathfrak{m}_1}), \quad \textit{etc.}$$

can again be described as in tables 1 and 2.

We label the statements of table 1 like the entries 11, 12, ..., 64 of a 6×4 matrix; the reader is advised to take a pencil and to check off each statement as it is proved.

Because $g_1^+(0) \geqq 0$, row 5 is evident from $g_1^+(\frac{1}{2}) - g_1^+(0) = \int_{0+}^{\frac{1}{2}} g_1\, dj$, and, using $g_2^+(\frac{1}{2}) - g_2^+(0) = \int_{0+}^{\frac{1}{2}} g_2\, dj$, row 6 is seen to follow from row 4. Because $0 > g_2^+ \in \uparrow$, $g_2^+(0) = g_2^+(0+) < 0$, and so $g_2(0) < +\infty$ implies that $s(0) > -\infty$ and

$$\begin{aligned} +\infty &> g_2(0) - g_2(\tfrac{1}{2}) - [s(\tfrac{1}{2}) - s(0)]\, g_2^-(\tfrac{1}{2}) \\ &= \int_0^{\frac{1}{2}} s(d\xi) \int_\xi^{\frac{1}{2}} g_2\, dj \geqq g_2(\tfrac{1}{2}) \int_0^{\frac{1}{2}} j(\xi, \tfrac{1}{2}]\, s(d\xi), \end{aligned}$$

i.e., that 0 is an *exit* point, proving 23 and 24, from which 33 and 34 follow using the fact that $g_1^+ g_2$ is smaller than the constant $B = g_1^+ g_2 - g_1 g_2^+$. 12 and 14 follow from 42, 44, and $-g_1 g_2^+ \leqq B$; as to 42 and 44 themselves, $g_2^+(0) > -\infty$ implies that

$$+\infty > g_2^+(\tfrac{1}{2}) - g_2^+(0) = \int_{0+}^{\frac{1}{2}} g_2\, dj \geq g_2^+(\tfrac{1}{2}) \int_{0+}^{\frac{1}{2}} [s(\tfrac{1}{2}) - s(\xi)]\, dj$$

i.e., that 0 is an *entrance* point as it be.

Given that 0 is *entrance* but *not exit*, $g_1^+(0) = 0$ (see 33),

$$0 < g_1^+(\xi)/g_1(\xi) = g_1(\xi)^{-1} \int_{0+}^{\xi} g_1\, dj \leq j(0, \xi],$$

and so

$$\lg g_1 \Big|_0^{\frac{1}{2}} = \int_0^{\frac{1}{2}} \frac{g_1^+}{g_1}\, ds \leq \int_0^{\frac{1}{2}} j(0, \xi]\, ds = \int_{0+}^{\frac{1}{2}} \left[s\left(\frac{1}{2}\right) - s(\xi) \right] dj < +\infty$$

proving 13. 43 follows from $-g_1 g_2^+ \leq B$.

Given that 0 is an *exit* point,

$$0 \leq g_2(\xi)^{-1}[g_2^+(\tfrac{1}{2}) - g_2^+(\xi)] = g_2(\xi)^{-1} \int_{\xi}^{\frac{1}{2}} g_2\, dj,$$

and so

$$g_2^+(\tfrac{1}{2}) \int_0^{\frac{1}{2}} g_2^{-1}\, ds - \lg g_2 \Big|_0^{\frac{1}{2}} \leq \int_0^{\frac{1}{2}} j(\xi, \tfrac{1}{2}]\, s(d\xi) < +\infty,$$

proving 21 and 22. 32 is now evident from $g_1^+ g_2 - g_1 g_2^+ = B > 0$ and $-g_2^+(\xi)\, g_1(\xi) = -g_2^+(\xi) \int_0^{\xi} g_1^+\, ds \leq -g_1^+(\xi) \int_0^{\xi} g_2^+\, ds$, and 41 follows from 21 and $g_2^+(\tfrac{1}{2}) - g_2^+(0) = \int_{0+}^{\frac{1}{2}} g_2\, dj \leq g_2(0)\, j(0, \tfrac{1}{2}]$ (see problem 1 for additional information about 11 and 31).

Table 1 is now established. Table 2 follows at once.

Problem 1. Given that 0 is both an exit and an entrance point, construct the solutions $g_\bullet$ and $g^\bullet$ of

11) $\quad \alpha\, g(\xi) = (\mathfrak{G}^\bullet g)(\xi) \qquad 0 < \xi \leq \tfrac{1}{2}$

such that

12a) $\quad g_\bullet(0) = 0, \quad g_\bullet^+(0) > 0, \quad g_\bullet(\tfrac{1}{2}) = 1$

12b) $\quad g^\bullet(0) > 0, \quad g^{\bullet +}(0) = 0, \quad g^\bullet(\tfrac{1}{2}) = 1$

and prove that $g_\bullet (g^\bullet)$ is the smallest (greatest) positive increasing solution of 11) subject to $g(\tfrac{1}{2}) = 1$.

$$[g_\bullet = B_\bullet / B_\bullet(\tfrac{1}{2}) \qquad B_\bullet = g_1\, g_2(0) - g_1(0)\, g_2$$

and

$$g^\bullet = B^\bullet / B^\bullet(\tfrac{1}{2}) \qquad B^\bullet = g_1\, g_2^+(0) - g_1^+(0)\, g_2$$

solve 11) and 12), $0 \leq g_{\bullet} \leq g^{\bullet}$, and if $0 \leq g \in \uparrow$ is another solution of 11) with $g(\frac{1}{2}) = 1$, then the obvious independence of $g_{\bullet}$ and $g^{\bullet}$ permits the choice of $c_{\bullet}$ and $c^{\bullet}$ such that $g = c_{\bullet} g_{\bullet} + c^{\bullet} g^{\bullet}$. $g_{\bullet} \leq g \leq g^{\bullet}$ is now clear from $g_{\bullet} \leq g^{\bullet}$ and

$$g(0)/g^{\bullet}(0) = c^{\bullet} \geq 0, \quad g^{+}(0)/g_{\bullet}^{+}(0) = c_{\bullet} \geq 0,$$
$$g(\tfrac{1}{2}) = 1 = c_{\bullet} + c^{\bullet}.]$$

Problem 2. Compute the increasing and decreasing solutions g_1 and g_2 and the WRONSKIAN $g_1^{+} g_2 - g_1 g_2^{+}$ for $\mathfrak{G}^{\bullet} = \frac{1}{2}|b|^{-\gamma} D^2$ $(-1 < \gamma < +\infty)$* (see 6.8 for an application).

[$g_1(\alpha, a) = g_2(\alpha, -a)$; setting $\beta = (\gamma + 2)^{-1}$, $g_2(\alpha, a)$

$$= \sqrt{a}\, K_{\beta}(2\beta \sqrt{2\alpha}\, a^{1/2\beta})$$

and

$$g_1(\alpha, a) = \Gamma(1-\beta)\, \Gamma(\beta) \sqrt{a}\, I_{\beta}(2\beta \sqrt{2\alpha}\, a^{1/2\beta}) + g_2 \quad \text{for} \quad a \geq 0;$$

the WRONSKIAN is $\Gamma(1-\beta)\,\Gamma(\beta)/2\beta$.]

Problem 3. Determine the scale and speed measure for the BESSEL motion with generator $\mathfrak{G}^{\bullet} = \frac{1}{2}\left(\frac{d^2}{dr^2} + \frac{d-1}{r}\frac{d}{dr}\right)$ and check that

$$P_{\bullet}\left[\lim_{t \uparrow +\infty} r(t) = +\infty\right] = 0 \textit{ or } 1 \textit{ according as } d = 2 \textit{ or } d \geq 3;$$

prove also that $P_{\bullet}[r > 0, t > 0] = 1$ for $d \geq 2$ (see 2.7).

$$[s(dr) = r^{1-d}\, dr, \quad m(dr) = 2r^{d-1}\, dr, \quad s(+\infty) \gtreqless +\infty \; (d \lesseqgtr 2),$$

$$\int_1^{+\infty} s\, dm = \int_1^{+\infty} m\, ds = +\infty.]$$

Problem 4. Compute the increasing and decreasing solutions g_1 and g_2 and the WRONSKIAN for the BESSEL operator $\mathfrak{G}^{\bullet} = \frac{1}{2}\left(\frac{d^2}{dr^2} + \frac{1}{r}\frac{d}{dr}\right)$ $(r > 0)$, and use g_2 to prove that

$$\lim_{\varepsilon \downarrow 0} E_1\left(e^{-\alpha \varepsilon^{\gamma} m_{\varepsilon}}\right) = \frac{\gamma}{\gamma + 2} \qquad \gamma > 0,$$

What is the meaning of the disappearance of the α?[1]

$$[g_1(\alpha, \xi) = I_0(\xi \sqrt{2\alpha}), \quad g_2(\alpha, \xi) = K_0(\xi \sqrt{2\alpha}), \quad g_1^{+} g_2 - g_1 g_2^{+} = 1,$$
$$E_1\left(e^{-\alpha m_{\varepsilon}}\right) = K_0(\sqrt{2\alpha})/K_0(\varepsilon \sqrt{2\alpha}), \quad K_0(\xi) \sim -\lg \xi \quad (\xi \downarrow 0).]$$

* F. B. HILDEBRAND [1; 161—167] is helpful.

[1] See problem 6.8.4 for a related result.

Problem 5. Give a new proof of D. RAY's estimate

$$\lim_{t\downarrow 0} t^{-n} P_a(\mathfrak{m}_b \leqq t) = 0 \qquad a < b, \quad n \geqq 1$$

for the general non-singular diffusion, using

$$\mathfrak{G}^{\bullet} g_1 = \alpha g_1$$

to check

$$e^{-1} P_a(\mathfrak{m}_b \leqq t) \leqq g_1(a)/g_1(b) \leqq [1 + \alpha p_2 + \alpha^2 p_4 + \alpha^3 p_6 + \dots]^{-1}$$
$$\alpha = t^{-1},$$

where

$$p_{2n} = \int_a^b s(d\xi_1) \int_{a+}^{\xi_1} m(d\xi_2) \dots \int_a^{\xi_{2n-2}} s(d\xi_{2n-1}) \int_{a+}^{\xi_{2n-1}} m(d\xi_{2n}) \qquad n \geqq 1$$

(see 3.8.8) and problem 3.8.1).

Problem 6. A non singular diffusion D is persistent if

13) $k = 0$

14a) $m(0, \frac{1}{2}] < +\infty$, $(\mathfrak{G} u)(0)\, m(0) = u^+(0)$ *in case* $s(0) > -\infty$

14b) $m[\frac{1}{2}, 1) < +\infty$, $(\mathfrak{G} u)(1)\, m(1) = u^+(1)$ *in case* $s(1) < +\infty$

and transient otherwise (see problem 4.4.4 for the definition of persistence).

[Given a persistent diffusion, if $-s(0) < +\infty = m(0, \frac{1}{2}]$ and if $0 < a < b$, then

$$P_a(\mathfrak{m}_b < +\infty) = \lim_{\alpha\downarrow 0} \frac{g_1(a)}{g_1(b)} \leqq \lim_{\alpha\downarrow 0} \frac{g_1^+(a)}{g_1(b)} [s(a) - s(0)]$$
$$\leqq \frac{s(a) - s(0)}{s(b) - s(a)} \downarrow 0 \qquad a \downarrow 0,$$

which is impossible; in brief, $s(0) > -\infty$ implies $m(0, \frac{1}{2}] < +\infty$. But then $E_{0+}(e^{-\mathfrak{m}_0}) = 1$, because of persistence $E_0(e^{-\mathfrak{m}_{0+}})$ is likewise 1, and now the fact that $P_\xi(\mathfrak{m}_b < +\infty)$ is a solution of

$$0 = p^+(d\xi) - p(\xi)\, k(d\xi) \qquad 0 < \xi < b$$
$$0 = p^+(0) - p(0)\, k(0) \qquad p(0) = 1, \quad E_0(e^{-\mathfrak{m}_{0+}}) = 1$$

implies 13) and 14a). The proof of the converse is similar: using 13) and 14a), $p(\xi) = P_\xi(\mathfrak{m}_b < +\infty)$ is found to be the solution of

$$p^+(d\xi) = 0 \qquad 0 < \xi < b$$
$$p^+(0) = 0 \qquad E_0(e^{-\mathfrak{m}_{0+}}) = 1$$
$$p(b) = 1;$$

as such, $p(\xi) = 1 + \text{constant} \times [s(b) - s(\xi)]$, the constant has to be 0 either because $s(0) = -\infty$ and $0 \leqq p \leqq 1$ or because $s(0) > -\infty$ and $p^+(0) = 0$, *etc.*]

Problem 7 after J. L. DOOB [3]. Given a *conservative* diffusion with $s(0) > -\infty$ and $m(0, \frac{1}{2}] = +\infty$, $\lim_{b\downarrow 0} P_b[\lim_{t\uparrow+\infty} x(t) = 0] = 1$ even if 0 is not *an exit point.*

[As in the solution of problem 5, $s(0) > -\infty$ and $m(0, \frac{1}{2}] = +\infty$ implies $P_{0+}(\mathfrak{m}_b < +\infty) = 0$ $(b > 0)$.

$$P_\xi[\mathfrak{m}_a \wedge \mathfrak{m}_{\frac{1}{2}} < +\infty] \equiv 1 \qquad (a \leqq \xi \leqq \tfrac{1}{2})$$

because it is a solution of

$$(\mathfrak{G}^\bullet u)(\xi) = 0 \qquad (a \leqq \xi \leqq \tfrac{1}{2})$$

$$u(a+) = u(\tfrac{1}{2}-) = 1,$$

and so

$$\lim_{\varepsilon\downarrow 0} P_\varepsilon\Big[\lim_{t\uparrow+\infty} x(t) = 0\Big] \geqq \lim_{\varepsilon\downarrow 0}\lim_{b\downarrow 0}\lim_{a\downarrow 0} P_\varepsilon[\mathfrak{m}_a < +\infty, \mathfrak{m}_b(w^+_{\mathfrak{m}_a}) = +\infty]$$

$$= \lim_{\varepsilon\downarrow 0}\lim_{b\downarrow 0}\lim_{a\downarrow 0} P_\varepsilon[\mathfrak{m} < +\infty] P_a[\mathfrak{m}_b = +\infty] = \lim_{\varepsilon\downarrow 0}\lim_{a\downarrow 0} P_\varepsilon[\mathfrak{m}_a < +\infty]$$

$$\geqq \lim_{\varepsilon\downarrow 0}\lim_{a\downarrow 0} P_\varepsilon[\mathfrak{m}_a \wedge \mathfrak{m}_{\frac{1}{2}} < +\infty, \mathfrak{m}_{\frac{1}{2}} = +\infty] = 1.]$$

Problem 8. Prove that a non-singular diffusion in the natural scale on $Q = R^1$ with $P_a(x(t) \leqq a) = P_a(x(t) > a) = 1/2$ for each $t > 0$ and $a \in R^1$ is a standard BROWNIAN motion up to a constant scale factor.

[Derive first the D. ANDRÉ reflection principles $P_a(\mathfrak{m}_b < t) = 2P_a(x(t) > b)$ $(a < b)$ and $P_a(\mathfrak{m}_b < t) = 2P_a(x(t) < b)$ $(a > b)$ as in 1.7, then take their LAPLACE transforms and deduce that $-g_1 g_2^+ = g_1^+ g_2 = 1/2$ $(g_1^+ g_2 - g_1 g_2^+ = 1)$ and hence $g_1 g_2 =$ constant; the rest is plain sailing using $\mathfrak{G} g = \alpha g$.]

4.7. $\mathfrak{G}$ as global differential operator: non-singular case

$\mathfrak{G}$ can now be identified as a global differential operator $\mathfrak{G}^\bullet$.

$D(\mathfrak{G}^\bullet)$ shall be the class of functions $u \in D$ such that, for some $u^\bullet \in D$,

1) $u^\bullet(\xi)\, m(d\xi) = u^+(d\xi) - u(\xi)\, k(d\xi)$ *in case* $0 < \xi < 1$

2a) $u^\bullet(0)\, m(0) = u^+(0) - u(0)\, k(0)$, $u^+(0) = u^+(0+)$ *in case* $E_0(e^{-\mathfrak{m}_{0+}}) = 1$

2b) $u^\bullet(1)\, m(1) = -u^-(1) - u(1)\, k(1)$, $u^-(1) = u^-(1-)$ *in case* $E_1(e^{-\mathfrak{m}_{1-}}) = 1$

3a) $u^\bullet(0) = -\varkappa(0)\, u(0)$ *in case* $E_0(e^{-\mathfrak{m}_{0+}}) = 0$

3b) $u^\bullet(1) = -\varkappa(1)\, u(1)$ *in case* $E_1(e^{-\mathfrak{m}_{1-}}) = 0$

4a) $u(0) = u(0+)$ *in case* $\int_0^{\frac{1}{2}} j(b, \tfrac{1}{2}]\, s(db) < +\infty$

4b) $u(1) = u(1-)$ *in case* $\int_{\frac{1}{2}}^{0} j[\tfrac{1}{2}, b)\, s(db) < +\infty$

and $\mathfrak{G}^\bullet$ shall be the map $u \in D(\mathfrak{G}^\bullet) \to u^\bullet$.

Because $m[a, b) > 0$ ($a < b$) and $D(\mathfrak{G}^\bullet) \subset D$, $\mathfrak{G}^\bullet u = u^\bullet$ is unambiguous. $\mathfrak{G}^\bullet \supset \mathfrak{G}$ as is clear on using 4.5 [see 3.6.6)] and entries 21 and 22 of table 4.6.2 in connection with 4)), and, for the identification of $\mathfrak{G}$ and $\mathfrak{G}^\bullet$, it suffices to show that $D(\mathfrak{G}^\bullet) \subset D(\mathfrak{G})$.

But, if $u \in D(\mathfrak{G}^\bullet)$, then $(1 - \mathfrak{G}^\bullet) u \in D$, $u^\bullet \equiv u - G_1(1 - \mathfrak{G}^\bullet) u \in D(\mathfrak{G}^\bullet)$ is a solution of $\mathfrak{G}^\bullet u^\bullet = u^\bullet$, and, as such, it has to be $\equiv 0$; for example, if $P_0[\mathfrak{m}_{0+} = 0] = 1$ and if 1 is neither an exit nor an entrance point, then $u^\bullet(1) = 0$ because $(\mathfrak{G}^\bullet u^\bullet)(1) = -\varkappa(1) u^\bullet(1) = u^\bullet(1)$, $u^\bullet(0+) = u^\bullet(0)$, $u^\bullet(\xi) = \text{constant} \times g_2(\xi)$ $(0 \leqq \xi < 1)$ because g_1 is unbounded near 1, and the constant has to be 0 because

$$0 < \mathfrak{G}^\bullet g_2(0)\, m(0) - [g_2^+(0) - g_2(0)\, k(0)];$$

the proof in the other cases is similar.

$\mathfrak{G}^\bullet$ stands for a *local or a global differential* operator below, the letter $\mathfrak{G}$ being kept for the *global* operator; for example, $(\mathfrak{G}^\bullet u)(\xi) = u^\bullet(\xi)$ $(0 \leqq \xi < \frac{1}{2})$ is short for the statement that both $u(\xi)$ and $u^\bullet(\xi) = [u^+(d\xi) - u(\xi)\, k(d\xi)]/m(d\xi)$ are continuous $(0 < \xi < \frac{1}{2})$, coupled with

5a) $u(0) = u(0+)$, $\quad u^\bullet(0) = u^\bullet(0+)$, $\quad u^+(0) = u^+(0+)$,
$u^\bullet(0)\, m(0) = u^+(0) - u(0)\, k(0)$ *in case* $E_0(e^{-\mathfrak{m}_{0-}}) = 1$,

5b) $u(b)$ *and* $u^\bullet(b)$ *bounded near* $b = 0$, $u^\bullet(0) = -\varkappa(0)\, u(0)$
in case $E_0(e^{-\mathfrak{m}_{0+}}) = 0$,

and 4a). As an illustration, 4.6.6a) and 4.6.7a) can be summed up as

$$(\mathfrak{G}^\bullet g_1)(\xi) = \alpha\, g_1(\xi) \qquad (0 \leqq \xi < 1),$$

and 4.6.6b) and 4.6.7b) as

$$(\mathfrak{G}^\bullet g_2)(\xi) = \alpha\, g_2(\xi) \qquad (0 < \xi \leqq 1).$$

4.8. $\mathfrak{G}$ on the shunts

Given a diffusion with state interval $Q = [0, 1]$, if

1a) $$P_0(\mathfrak{m}_1 < +\infty) > 0$$

1b) $$P_1(\mathfrak{m}_{1-} = \mathfrak{m}_\infty = +\infty) = 1$$

1c) $$E_0(\mathfrak{m}_1 \wedge \mathfrak{m}_\infty) < +\infty,$$

then it is possible to define a *shunt scale* $s_+(\xi)$ and a *shunt killing measure* $k_+(d\xi)$ with $k_+(0) = 0$ such that, for $u \in D(\mathfrak{G})$,

2a) $$\int_{[a,b) \cap K_+} (\mathfrak{G} u)(\xi)\, s_+(d\xi) = \int_{[a,b) \cap K_+} u(d\xi) - \int_{[a,b) \cap K_+} u(\xi)\, k_+(d\xi) \qquad a \leqq b < 1,$$

2b) $$u(1-) = [1 - k_+(1)]\, u(1) \qquad k_+(1) = P_{1-}[\mathfrak{m}_1 = +\infty]$$

2c) $$(\mathfrak{G} u)(a) = \lim_{b \downarrow a} \frac{u(b) - u(a) - \left[\left(\bigcap_{a < \xi \leqq b} [1 - k_+(d\xi)]\right)^{-1} - 1\right] u(a)}{s_+(b) - s_+(a)} \qquad a \in K_+ \cap [0, 1),$$

where $\bigcap_{a<\xi\leq b} [1 - k_+(d\xi)]$ is meant as a suggestive notation for $e^{-j_+(a,b]} \times$ $\times (1-\varkappa_1)(1-\varkappa_2)$ *etc.*, $j_+(d\xi)$ being the continuous part of $k_+(d\xi)$ and $\varkappa_1, \varkappa_2$, *etc.* (≤ 1) the jumps of $k_+(d\xi)$ $(a < \xi \leq b)$.

Q splits into K_+ plus the non-singular intervals (l_n, r_n), $P_{l_n}(\mathfrak{m}_{l_n} = 0) = 1$, and, on $Q_n \equiv [l_n, r_n)$, it is possible to define s, k, m such that, for $u \in D(\mathfrak{G})$,

3a) $$(\mathfrak{G} u)(\xi)\, m(d\xi) = u^+(d\xi) - u(\xi)\, k(d\xi) \qquad l_n < \xi < r_n$$

3b) $$(\mathfrak{G} u)(l_n)\, m(l_n) = u^+(l_n) - u(l_n)\, k(l_n)$$

$$u^+(l_n) = \lim_{b \downarrow l_n} \frac{u(b) - u(l_n)}{s(b) - s(l_n)}$$

as in 4.3.1) and 4.5.5b).

Given $n \geq 1$,

4a) $$k_+(d\xi) = \frac{s(d\xi)}{p(\xi)} \int_{[l_n, \xi)} p(\eta)\, k(d\eta) \qquad l_n < \xi < r_n$$

4b) $$s_+(d\xi) = \frac{s(d\xi)}{p(\xi)} \int_{[l_n, \xi)} p(\eta)\, m(d\eta) \qquad l_n < \xi < r_n,$$

p being the solution of

5a) $$p^+(db) = p(b)\, k(db) \qquad l_n < b < r_n$$

5b) $$p^+(l_n) = p(l_n)\, k(l_n), \quad p(l_n +) = p(l_n) = 1;$$

a special case of 4) is

6a) $$k(l_n) = \lim_{b \downarrow l_n} \frac{k_+(l_n, b]}{s(b) - s(l_n)}$$

6b) $$m(l_n) = \lim_{b \downarrow l_n} \frac{s_+(b) - s_+(l_n)}{s(b) - s(l_n)};$$

the global condition

7a) $$k_+[0, 1) \leq \sum_{n \geq 1} \int_{Q_n} k[l_n, \xi)\, s(d\xi) + k_+(K_+) < +\infty$$

7b) $$s_+(1) - s_+(0) \leq \sum_{n \geq 1} \int_{Q_n} m[l_n, \xi)\, s(d\xi) + s_+(K_+) < +\infty$$

has also to be met.

Here is the proof.

Given $a \leq b$, if

8a) $$p_b(a) = P_a(\mathfrak{m}_b < +\infty)$$

8b) $$e_b(a) = E_a(\mathfrak{m}_b \wedge \mathfrak{m}_\infty),$$

then [see 1)]

9a) $$p_b(a) \geq p_1(0) > 0$$

9b) $$e_b(a) \leq e_1(0) < +\infty,$$

and using

10a) $$\lim_{a\uparrow b} p_b(a) = P_{b-}(\mathfrak{m}_b < +\infty) = 1 - k_+(b) \qquad b > 0$$

10b) $$\lim_{b\downarrow a} p_b(a) = P_a(\mathfrak{m}_{a+} < +\infty) = 1 \qquad a < 1^*$$

and the composition rules

11a) $$p_b(a) = p_c(a)\, p_b(c)$$

11b) $$e_b(a) = e_c(a) + p_c(a)\, e_b(c) \qquad a \leqq c \leqq b,$$

it is a simple matter to show that

12a) $$p_b(a+) = p_b(a) \qquad a < b$$

12b) $$p_b(a-) = p_b(a)\,[1 - k_+(a)] \qquad a \leqq b$$

and

13a) $$e_b(a+) = e_b(a) \qquad a < b$$

13b) $$e_b(a-) = e_b(a)\,[1 - k_+(a)] \qquad a \leqq b.$$

12) is clear; as to 13a), let $c \downarrow a$ in 11b) and use 10b) and $\lim_{c\downarrow a} e_c(a) = E_a(\mathfrak{m}_{a+} \wedge \mathfrak{m}_\infty) = 0$; as to 13b), let $c \uparrow a$ in

$$e_b(c) = e_a(c) + p_a(c)\, e_b(a) \qquad (c < a \leqq b)$$

and use

$$\begin{aligned}\lim_{c\uparrow a} e_a(c) &= \lim_{c\uparrow a} p_c(0)^{-1} E_0(\mathfrak{m}_a \wedge \mathfrak{m}_\infty - \mathfrak{m}_c \wedge \mathfrak{m}_\infty,\, \mathfrak{m}_c < +\infty)\\ &= \lim_{c\uparrow a} p_1(0)^{-1} E_0(\mathfrak{m}_a \wedge \mathfrak{m}_\infty - \mathfrak{m}_c \wedge \mathfrak{m}_\infty) = 0.\end{aligned}$$

Because of the composition rule 11a), the *shunt killing measure* $k_+(da)$ as defined in

14) $$k_+(da) = \frac{p_b(da)}{p_b(a)} \qquad a < b^{**}$$

does not depend upon b. 10a) is consistent with the present definition. $k_+(0) = 0$, and $k_+[0, 1) < +\infty$ because $p_1(0) > 0$.

Coming to the definition of the shunt scale,

15) $$v_b(a) = \frac{e_b(a)}{p_b(a)} \qquad a \leqq b,$$

is continuous [see 12) and 13)] and, using 11), if $\xi < \eta \leqq b$, then

16) $$\begin{aligned}v_b(\eta) - v_b(\xi) &= p_b(\xi)^{-1}[p_\eta(\xi)\, e_b(\eta) - e_b(\xi)]\\ &= -\frac{e_\xi(\eta)}{p_b(\xi)} < 0,\end{aligned}$$

---

* See 3.3.10c) and 3.6.6a). ** $k_+(da) = d\,\lg p_1(a)$ if p_1 is continuous.

so that, taking differentials and using 11b) and 14), the (continuous) *shunt scale* $s_+(\xi)$ as defined in

17)
$$\begin{aligned} 0 &< -\int_\xi^\eta p_b(a)\, v_b(da) \\ &= -\int_\xi^\eta [e_b(da) - v_b(a)\, p_b(da)] \\ &= -\int_\xi^\eta [e_b(da) - e_b(a)\, k_+(da)] \\ &= s_+(\eta) - s_+(\xi) \qquad \xi < \eta < b \end{aligned}$$

does not depend upon b. Because $p_1(0) > 0$ and $e_1(0) < +\infty$, $s_+(1) - s_+(0) < +\infty$; in addition (use $\lim_{a\uparrow b} v_b(a) = 0$),

18)
$$\begin{aligned} e_b(a) &= p_b(a)\, v_b(a) = p_b(a) \int_a^b \frac{s_+(d\xi)}{p_b(\xi)} \\ &= \int_a^b p_\xi(a)\, s_+(d\xi) \qquad a \leq b \end{aligned}$$

2c) can now be proved as follows. Given $u \in D(\mathfrak{G})$, if $a \in K^+ \cap [0, 1)$, then DYNKIN's formula implies

19)
$$\begin{aligned} (\mathfrak{G}\, u)\,(a) &= \lim_{b\downarrow a} \frac{p_b(a)\, u(b) - u(a)}{E_a(\mathfrak{m}_b \wedge \mathfrak{m}_\infty)} \\ &= \lim_{b\downarrow a} \frac{u(b) - u(a) - [p_b(a)^{-1} - 1]\, u(a)}{e_b(a)/p_b(a)} \end{aligned}$$

as in 4.5. Because of 18), $e_b(a)/p_b(a) = s_+(b) + s_+(a) + o(1)$ $(b \downarrow a)$, and it is a simple matter to conclude from 14) that $\bigcap_{a<\xi\leq b} [1 - k_+(d\xi)]$ as described below 2b) is identical to $p_b(a)$; this completes the proof.

Consider the indicator f of the point 1; $f \in D$ [see 1b)], $\varepsilon G_\varepsilon f = E_\bullet(e^{-\varepsilon \mathfrak{m}_1})\, \varepsilon\, (G_\varepsilon f)\,(1)$ increases to $p_1 \in D$ as $\varepsilon \downarrow 0$, and so, letting $\varepsilon \downarrow 0$ in

20)
$$\varepsilon \times [G_\alpha - G_\varepsilon + (\alpha - \varepsilon)\, G_\alpha\, G_\varepsilon]\, f = 0,$$

it is found that

21)
$$\alpha\, G_\alpha\, p_1 = p_1,$$

i.e.,

22a)
$$p_1 \in D(\mathfrak{G})$$

22b)
$$\mathfrak{G}\, p_1 = 0$$

as in 4.2.12). An application of the composition rule 11a) coupled with 3) now implies

23a) $$p_b^+(da) = p_b(a)\,k(da) \qquad l_n < a < r_n, \quad a \leqq b$$

23b) $$p_b^+(l) = p_b(l_n)\,k(l_n) \qquad l_n \leqq b,$$

from which 4a) and 5) are immediate.

As to 4b), if f is the indicator of $[0, 1)$, then $G_\varepsilon f = E_\bullet\left[\int_0^{\mathfrak{m}_\infty \wedge \mathfrak{m}_1} e^{-\varepsilon t} f(x_t)\,dt\right]$ increases to $e_1 \in D$ as $\varepsilon \downarrow 0$, and so, letting $\varepsilon \downarrow 0$ in

24) $$[G_\alpha - G_\varepsilon + (\alpha - \varepsilon)\, G_\alpha G_\varepsilon]\, f = 0,$$

it is found that

25) $$e_1 = G_\alpha[f + \alpha\, e_1]$$

i.e.,

26a) $$e_1 \in D(\mathfrak{G})$$

26b) $$(\mathfrak{G}\, e_1)\,(b) = -1 \qquad b < 1$$

as in 4.2.20). An application of the composition rule 11b) now implies

27a) $$e_b^+(da) - e_b(a)\,k(da) = -m(da) \qquad l_n < a < r_n,$$

27b) $$e_b^+(l_n) - e_b(l_n)\,k(l_n) = -m(l_n) \qquad l_n \leqq b,$$

from which 4b) is immediate on using 17).

7a) and 7b) are clear from 4a) and 4b).

Coming to the proof of 2a), if $J_+ = K_+ - \bigcup_{n \geqq 1} l_n$, then

28) $$E_\xi[\text{measure}\,(t : x(t) \in [a, 1) \cap J_+, t \leqq \mathfrak{m}_1 \wedge \mathfrak{m}_\infty)] = \int_{[a \vee \xi, 1) \cap K_+} p_\eta(\xi)\, s_+(d\eta),$$

as will now be proved.

Because a sample path starting at ξ cannot enter $[0, \xi) \cap J_+$ without crossing down over a point of $\bigcup_{n \geqq 1} l_n$,

29) $$\begin{aligned} &E_\xi[\text{measure}\,(t : x(t) \in [0, 1) \cap J_+, t < \mathfrak{m}_1 \wedge \mathfrak{m}_\infty)] \\ &= p_b(\xi)\, E_b[\text{measure}\,(t : x(t) \in J_+, t \leqq \mathfrak{m}_1 \wedge \mathfrak{m}_\infty)] \qquad b = \xi \vee a \\ &= p_b(\xi)\, e_1'(b) - p_b(\xi) \sum_{n \geqq 1} E_b[\text{measure}\,(t : l_n \leqq x(t) < r_n, t < \mathfrak{m}_1 \wedge \mathfrak{m}_\infty)] \\ &= p_b(\xi)\, e_1(b) - p_b(\xi) \sum_{r_n > b} p_{b \vee l_n}(b)\, e_{r_n}(b \vee l_n) \\ &= \int_\xi^1 p_\eta(\xi)\, s_+(d\eta) - p_b(\xi) \sum_{r_n > b} p_{b \vee l_n}(b) \int_{b \vee l_n}^{r_n} p_\eta(b \vee l_n)\, s_+(d\eta) \\ &= \int_\xi^1 p_\eta(\xi)\, s_+(d\eta) - \sum_{r_n > b} \int_{b \vee l_n}^{r_n} p_\eta(\xi)\, s_+(d\eta) \\ &= \int_{[\xi, 1) \cap K_+} p_\eta(\xi)\, s_+(d\eta), \end{aligned}$$

as stated.

2a) can now be proved in a few lines.

Given $u \in D(\mathfrak{G})$, if $e(d\eta) \equiv -[u(d\eta) - u(\eta)\, k_+(d\eta)]$, then

30) $$p_b(a)\, u(b) - u(a) = \int\limits_{[a,b)} p_\eta(a)\, e(d\eta) \qquad a \leq b,$$

the proof being identical to that of 18), and now an application of DYNKIN's formula coupled with 28) implies

31) $$\begin{aligned}
&\int\limits_{[\xi,1)} p_\eta(\xi)\, e(d\eta) \\
&= p_1(\xi)\, u(1) - u(\xi) \\
&= E_\xi\left[\int\limits_0^{\mathfrak{m}_1 \wedge \mathfrak{m}_\infty} (\mathfrak{G} u)(x_t)\, dt\right] \\
&= E_\xi\left[\int\limits_{\substack{t \leq \mathfrak{m}_1 \wedge \mathfrak{m}_\infty \\ x(t) \in J_+}} (\mathfrak{G} u)(x_t)\, dt\right] + \sum_{r_n > \xi} p_{l_n \vee \xi}(\xi)\, E_{l_n \vee \xi}\left[\int\limits_0^{\mathfrak{m}_{r_n} \wedge \mathfrak{m}_\infty} (\mathfrak{G} u)(x_t)\, dt\right] \\
&= \int\limits_{[\xi,1) \cap K_+} p_\eta(\xi)\, (\mathfrak{G} u)(\eta)\, s_+(d\eta) + \sum_{r_n > \xi} p_{l_n \vee \xi}(\xi) \int\limits_{[l_n \vee \xi,\, r_n)} p_\eta(l_n \vee \xi)\, e(d\eta) \\
&= \int\limits_{[\xi,1) \cap K_+} p_\eta(\xi)\, (\mathfrak{G} u)(\eta)\, s_+(d\eta) + \int\limits_{\bigcup_{r_n > \xi} [l_n \vee \xi,\, r_n)} p_\eta(\xi)\, e(d\eta),
\end{aligned}$$

i.e.,

32) $$\int\limits_{[\xi,1) \cap K_+} p_\eta(\xi)\, (\mathfrak{G} u)(\eta)\, s_+(d\eta) = \int\limits_{[\xi,1) \cap K_+} p_\eta(\xi)\, e(d\eta).$$

2a) is immediate from this.

Problem 1. If $K_+ = [0, 1)$, then the motion starting at $a < 1$ is just translation at speed 1 in the scale s_+:

$$\begin{aligned}
x(t) &= s_+^{-1}(t + s_+(0, a]) && t < \mathfrak{m}_\infty \wedge s_+(a, 1] \\
&= \infty && t \geq \mathfrak{m}_\infty < s_+(a, 1] \\
&= 1 && t \geq s_+(a, 1] < \mathfrak{m}_\infty,
\end{aligned}$$

where s_+^{-1} is the inverse function of $s_+(0, a]$.

[$x^{-1}(b) = \mathfrak{m}_b$ if $\mathfrak{m}_b < +\infty$, and using the obvious

$$\alpha\, p(\xi)\, s_+(d\xi) = p(d\xi) - p(\xi)\, k_+(d\xi) \quad \xi < b,\ p = E_\bullet(e^{-\alpha \mathfrak{m}_b}),$$

one finds

$$\begin{aligned}
\frac{p(a)}{p(b)} &= e^{-\alpha s_+(a, b]} \bigcap_{0 < \xi \leq b} [1 - k_+(d\xi)] \\
&= E_a(e^{-\alpha \mathfrak{m}_b}) \\
&= E_a[e^{-\alpha x^{-1}(b)},\ \mathfrak{m}_b < +\infty] \qquad a < b;
\end{aligned}$$

now the result is evident.]

4.9. $\mathfrak{G}$ as global differential operator: singular case

Continuing with the case

1a) $$P_0(\mathfrak{m}_1 < +\infty) > 0$$

1b) $$P_1(\mathfrak{m}_{1-} \wedge \mathfrak{m}_\infty = +\infty) = 1$$

1c) $$E_0(\mathfrak{m}_1 \wedge \mathfrak{m}_\infty) < +\infty,$$

$\mathfrak{G}$ *can now be identified* as *a global differential operator* $\mathfrak{G}^\bullet$.

$D(\mathfrak{G}^\bullet)$ shall be the class of functions $u \in D$ such that, for some $u^\bullet \in D$,

2a) $$\int_{[a,b)} u^\bullet(\xi)\, m(d\xi) = \int_{[a,b)} u^+(d\xi) - \int_{[a,b)} u(\xi)\, k(d\xi)$$
$$l_n < a < b < r_n, \quad n \geqq 1$$

2b) $$u^\bullet(l_n)\, m(l_n) = u^+(l_n) - u(l_n)\, k(l_n) \qquad n \geqq 1$$

3a) $$\int_{[a,b)\cap K_+} u^\bullet(\xi)\, s_+(d\xi) = \int_{[a,b)\cap K_+} u(d\xi) - \int_{[a,b)\cap K_+} u(\xi)\, k_-(d\xi) \quad a < b < 1$$

3b) $$u(1-) = [1 - k_+(1)]\, u(1)$$

4) $$u^\bullet(1) = 0,$$

and $\mathfrak{G}^\bullet$ shall be the map $u \in D(\mathfrak{G}^\bullet) \to u^\bullet$.

Because $m[a,b) > 0$ $(a < b)$ on Q_n and $s_+(a) < s_+(b)$ $(a < b)$, $\mathfrak{G}^\bullet: u \to u^\bullet$ is unambiguous, and using 4.8, it is clear that for the identification of $\mathfrak{G}$ and $\mathfrak{G}^\bullet$, it suffices to prove $D(\mathfrak{G}^\bullet) \subset D(\mathfrak{G})$.

But, if $u \in D(\mathfrak{G}^\bullet)$, then $(1 - \mathfrak{G}^\bullet)\, u \in D$,

5a) $$u^\bullet \equiv u - G_1(1 - \mathfrak{G}^\bullet)\, u \in D(\mathfrak{G}^\bullet)$$

is a solution of

5b) $$\mathfrak{G}^\bullet u^\bullet = u^\bullet,$$

and, as will be proved below, $0 < d \equiv \sup_{0 \leqq \xi \leqq 1} u^\bullet(\xi)$ leads to a contradiction. $u^\bullet(\xi)$ is then $\leqq 0$ $(0 \leqq \xi \leqq 1)$, and using $-u^\bullet$ in place of $u^\bullet$, it develops that $u^\bullet(\xi) \equiv 0$ $(0 \leqq \xi \leqq 1)$, *i.e.*, $u = G_1(1 - \mathfrak{G}^\bullet)\, u \in D(\mathfrak{G})$.

Given $d > 0$, if $\lim_{n\uparrow+\infty} b_n = a$ and if $u(b_n) \uparrow d$ $(n \uparrow +\infty)$, then

$$\begin{aligned} 0 < d &= \lim_{n\uparrow+\infty} u^\bullet(b_n) \\ &= u^\bullet(a) && \textit{in case } b_n \geqq a \quad \textit{i.o.} \\ &= P_{a-}[\mathfrak{m}_a < +\infty]\, u^\bullet(a) && \textit{in case } b_n < a \quad \textit{i.o.}, \end{aligned}$$

and, using $0 < u^\bullet(a) \leqq d$ in the third line, $d = u^\bullet(a)$ is seen to hold in that case also.

But, *if a is non-singular*, then (see problem 4.4.3), $u^\bullet(a) = (\mathfrak{G}^\bullet u)(a) \leqq 0$, contradicting $d > 0$; *if* $a = 1$, then $0 < d = u^\bullet(1) = (\mathfrak{G}^\bullet u)(1) = 0$;

and, *if* $a \in K_+ \cap [0, 1)$, then choosing $b > a$ such that $u^\bullet(\xi) > d/2 > 0$ $(a \leqq \xi \leqq b)$ and noting

$$\begin{aligned} u^{\bullet +}(\xi) &= u^{\bullet +}(l_n) + \int_{(l_n, \xi]} [(\mathfrak{G}^\bullet u^\bullet)(\eta)\, m(d\eta) + u^\bullet(\eta)\, k(d\eta)] \\ &> u^{\bullet +}(l_n) \\ &= (\mathfrak{G}^\bullet u^\bullet)(l_n)\, m(l_n) + u^\bullet(l_n)\, k(l_n) \\ &\geqq 0 \qquad\qquad a \leqq l_n \leqq \xi < r_n \wedge b \end{aligned}$$

it develops that

$$\begin{aligned} u^\bullet(b) &= u^\bullet(a) + \sum_{a \leqq l_n < b} \int_{l_n}^{r_n \wedge b} u^+ \, ds + \\ &\qquad + \int_{K_+ \cap [a, b)} [(\mathfrak{G}^\bullet u^\bullet)(\eta)\, s_+(d\eta) + u^\bullet(\eta)\, k_+(d\eta)] \\ &> u^\bullet(a) = d, \end{aligned}$$

which is also absurd.

Problem 1. Consider the differential operator

$$(\mathfrak{G}^\bullet u)(b) = \frac{u^+(db)}{m(db)} \qquad b \in R^\bullet \equiv R^1 - 0,$$

where $u^+(a) = \lim_{b \downarrow a} [b - a]^{-1} [u(b) - u(a)]$, $m(db)$ is a speed measure on $R^\bullet$, and $u \in D(\mathfrak{G}^\bullet) = C(R^\bullet) \cap (u : \mathfrak{G}^\bullet u \in C(R^\bullet))$; the problem is to determine all generators $\mathfrak{G}$ of conservative diffusions on R^1 such that $\mathfrak{G}^\bullet \supset \mathfrak{G}$.

[One possible choice is to declare $(\mathfrak{G} u)(0) = 0$ for each $u \in D(\mathfrak{G})$. $D(\mathfrak{G})$ is the class of functions $u \in D(\mathfrak{G}^\bullet)$, left continuous at 0 if $\int_{-1}^{0} m[-1, \xi)\, d\xi < +\infty$, and right continuous at 0 if $\int_0^1 m(\xi, 1]\, d\xi < < +\infty$. If $m(0, +1] < +\infty$, it is possible to introduce a right shunt at 0, *i.e.*, to define

$$(\mathfrak{G} u)(0) = \lim_{\varepsilon \downarrow 0} \frac{u(\varepsilon) - u(0)}{s_+(\varepsilon) - s_+(0)}$$

with $s_+(a) = \int_0^a m[0, b)\, db$ and $m(0) \geqq 0$. $D(\mathfrak{G})$ is now the class of functions $u \in D(\mathfrak{G}^\bullet)$, left continuous at 0 if $\int_{-1}^{0} m[-1, \xi)\, d\xi < +\infty$, with $u(0+) = u(0)$ and $(\mathfrak{G} u)(0+) = (\mathfrak{G} u)(0)$. If $m[-1, 0) < +\infty$, a left shunt can be introduced at 0, while if $m[-1, 0) + m(0, +1] < < +\infty$, it is possible to define

$$(\mathfrak{G} u)(0) = \lim_{\substack{b \downarrow 0 \\ a \uparrow 0}} \frac{u^+(b) - u^+(a)}{m(a, b]}$$

with $m(0) \geqq 0$ and $D(\mathfrak{G})$ the class of functions $u \in D(\mathfrak{G}^\bullet)$ with $u(0\pm) = u(0)$ and $(\mathfrak{G} u)(0\pm) = (\mathfrak{G} u)(0)$.]

4.10. Passage times

Consider, as in 4.8, the case

1 a) $$P_0[\mathfrak{m}_1 < +\infty] > 0$$

1 b) $$E_0[\mathfrak{m}_1 \wedge \mathfrak{m}_\infty] < +\infty,$$

introduce the letter $Q^\bullet = \bigcup_{n\geq 1} [l_n, r_n)$, and let us show that $[\mathfrak{m}_b: 0 \leq b \leq 1, P_0]$ is identical in law to

2) $$s_+([0,b) \cap K_+) + \int_{(0,+\infty)} l\, \mathfrak{p}([0,b)\times dl) + \infty \times \mathfrak{p}((0,b]\times +\infty) \qquad 0 \leq b \leq 1,$$

where $+\infty \times 0 \equiv 0$ and $\mathfrak{p}(db \times dl)$ $(0 \leq b \leq 1,\ 0 < l \leq +\infty)$ is the POISSON measure with mean $n(db \times dl)$:

3 a) $$e^{-n((a,b]\times+\infty)} = P_a[\mathfrak{m}_b < +\infty] = \bigcap_{a<\xi\leq b} [1 - k_+(d\xi)]^* \qquad a < b$$

3 b) $$n(K_+ \times dl) = 0 \qquad l < +\infty$$

3 c) $$n(db \cap Q^\bullet \times dl) = s(db) \lim_{a\uparrow b} \frac{P_\bullet(\mathfrak{m}_b \in dl)}{s(b) - s(a)} \qquad l > 0.$$

$[\mathfrak{m}_b: 0 \leq b \leq 1, P_0(B \mid \mathfrak{m}_1 < +\infty)]$ is differential, $P_0\left[\lim_{a\uparrow b} \mathfrak{m}_a = \mathfrak{m}_b,\ 0 < b \leq 1 \mid \mathfrak{m}_1 < +\infty\right] = 1$, and $P_0\left[\lim_{a\downarrow b} \mathfrak{m}_a = \mathfrak{m}_b\right] = 1$ $(0 \leq b \leq 1)$, so $\mathfrak{m}_b$ can be expressed as

4) $$t(b) + \int_{(0,+\infty)} l\, \mathfrak{p}([0,b)\times dl) \qquad 0 \leq b \leq 1$$

relative to the *conditional* law $P_0(B \mid \mathfrak{m}_1 < +\infty)$, where $t \in C[0,1]$ is independent of the sample path and $\mathfrak{p}(db \times dl)$ is a POISSON measure.[1] Taking $\mathfrak{p}((a,b] \times +\infty)$ to be independent of 4), differential, and POISSON distributed with mean $n((a,b]\times+\infty) \equiv -\lg P_a[\mathfrak{m}_b < +\infty]$ $(0 \leq a < b \leq 1)$ as in 3 a), $\mathfrak{m}_b$ can now be expressed as

5) $$t(b) + \int_{(0,+\infty)} l\, \mathfrak{p}([0,b)\times dl) + \infty \times \mathfrak{p}((0,b]\times+\infty) \qquad 0 \leq b \leq 1$$

relative to the *unconditional* law P_0.

t and n have now to be identified with the expressions in 2) and 3).

Given $a \leq b$ inside some interval of $Q^\bullet$, $P_a(\mathfrak{m}_b \geq l)$ is a concave function of $s(a)$ (see problem 4.4.2), permitting the definition of $n^\bullet(b \times dl)$ as in

6) $$0 \leq [s(b) - s(a)]^{-1} P_a(\mathfrak{m}_b \geq l) \uparrow n^\bullet(b \times [l, +\infty)) \qquad a \uparrow b, \quad l > 0$$

if $l > 0$ is not a jump of $n^\bullet(b \times dl)$. Using the solution $g_1 = E_\bullet(e^{-\alpha \mathfrak{m}_1})$ of

7) $$(\mathfrak{G}^\bullet g_1)(b) = \alpha g_1(b) \qquad 0 \leq b < 1,$$

* See the explanation below 4.8.2).

[1] See the note placed at the end of this section.

it is immediate from

8) $$\alpha \int_0^{+\infty} e^{-\alpha l} n^\bullet (b \times [l, +\infty))\, dl$$
$$= \lim_{a \uparrow b} \alpha \int_0^{+\infty} e^{-\alpha l} [s(b) - s(a)]^{-1} P_a(\mathfrak{m}_b \geqq l)\, dl$$
$$= \lim_{a \uparrow b} [s(b) - s(a)]^{-1} \int_{(0, +\infty]} [1 - e^{-\alpha l}]\, P_a(\mathfrak{m}_b \in dl)$$
$$= \lim_{a \uparrow b} g_1(b)^{-1} \frac{g_1(b) - g_1(a)}{s(b) - s(a)}$$
$$= \frac{g_1^-(b)}{g_1(b)}$$

that

9) $$\int_{(0, +\infty]} [1 - e^{-\alpha l}]\, n^\bullet (b \times dl) = \frac{g_1^-(b)}{g_1(b)}$$

and

10) $$\int_{[a, b)} s(d\xi)\, n^\bullet (\xi \times +\infty) = \lim_{\alpha \downarrow 0} \lg \frac{g_1(b)}{g_1(a)} = -\lg P_a[\mathfrak{m}_b < +\infty].$$

But now, using 5) to check

11) $$e^{-\alpha [t(b) - t(a)] - \int_{(0, +\infty)} [1 - e^{-\alpha l}] n([a, b) \times dl) - n((a, b] \times +\infty)}$$
$$= E_a(e^{-\alpha \mathfrak{m}_b}) = \frac{g_1(a)}{g_1(b)} \qquad 0 \leqq a < b \leqq 1,$$

it appears from 7), 9), and 10) that

12) $$\alpha [t(b) - t(a)] + \int_{(0, +\infty)} [1 - e^{-\alpha l}]\, n([a, b) \times dl) + n((a, b] \times +\infty)$$
$$= \lg \frac{g_1(b)}{g_1(a)}$$
$$= \lg \frac{u(b)}{u(a)} - \lg P_a[\mathfrak{m}_b < +\infty] \qquad u = p^{-1} g_1 \in C[0, 1],$$
$$= \int_{[a, b)} \frac{u(d\xi)}{u(\xi)} - \lg P_a[\mathfrak{m}_b < +\infty] \qquad p = P_\bullet[\mathfrak{m}_1 < +\infty]$$
$$= \int_{[a, b)} \left[\frac{g_1(d\xi)}{g_1(\xi)} - \frac{p(d\xi)}{p(\xi)} \right] - \lg P_a[\mathfrak{m}_b < +\infty]$$
$$= \int_{[a, b) \cap K_+} g_1(\xi)^{-1} [(\mathfrak{G}^\bullet g_1)(\xi)\, s_+(d\xi) + g_1(\xi)\, k_+(d\xi)] - k_+[a, b)$$
$$+ \int_{[a, b) \cap Q^\bullet} \frac{g_1^-(\xi)}{g_1(\xi)} s(d\xi) - \lg P_a[\mathfrak{m}_b < +\infty]$$
$$= \alpha s_+([a, b) \cap K_+) - k_+([a, b) \cap Q^\bullet) + \int_{[a, b) \cap Q^\bullet} s(d\xi) \int_{(0, +\infty]} [1 - e^{-\alpha l}] n^\bullet (\xi \times dl)$$
$$- \lg P_a[\mathfrak{m}_b < +\infty]$$
$$= \alpha s_+([a, b) \cap K_+) + \int_{[a, b) \cap Q^\bullet} s(d\xi) \int_{(0, +\infty)} [1 - e^{-\alpha l}]\, n^\bullet (\xi \times dl)$$
$$- \lg P_a[\mathfrak{m}_b < +\infty].$$

$t(b) - t(a) = s_+([a, b) \cap K_+)$ and 3) follow, and the proof is complete.

Note 1. Differential processes with increasing paths. Given a differential process $p(t)$ $(t \geqq 0)$ such that $p(0) = 0, p(t) \in \uparrow, p(t) = p(t+) < +\infty$ $(t \geqq 0)$, and $E[e^{-p(t)}]$ is continuous $(t \geqq 0)$, P. LÉVY [*1*: 173—180] proved that

$$1)\qquad p(t) = c(t) + \int_{0+}^{+\infty} \mathfrak{p}([0, t] \times dl) \qquad t \geqq 0,$$

where $c(t)$ is continuous and independent of the sample path, and

2) $\mathfrak{p}(dt \times dl) =$ *the number of jumps of* p *of magnitude* $l \in dl$ *in time* dt

is a POISSON measure with mean $n(dt \times dl)$ such that

$$3)\qquad \int_{0+}^{+\infty} (1 - e^{-l})\, n((0, t] \times dl) < +\infty \qquad t > 0$$

(see the note placed at the end of 1.8 for the definition of POISSON measure).

P. LÉVY's formula [*1*: 173—180]:

$$4)\quad E[e^{-\alpha(p(t_2) - p(t_1))}] = \exp\Big[-\alpha[c(t_2) - c(t_1)] - \int_{0+}^{+\infty} (1 - e^{-\alpha l})\, n((t_1, t_2] \times dl)\Big] \qquad \alpha > 0,\ t_1 \leqq t_2$$

is immediate from 1), and 3) is immediate from the fact that the left side of 4) is > 0.

The following proof of 1) is adapted from K. ITÔ [*1*]. Given $\mathfrak{p}$ as in 2),

$$5)\qquad c(t) \equiv p(t) - \int_{0+}^{+\infty} l\, \mathfrak{p}([0, t] \times dl) \qquad t \geqq 0$$

is continuous, non-negative, increasing, and differential, and, as such, it has to be independent of the sample path as will now be proved in the case $t = 1$.

Because

$$6)\qquad \frac{1}{2} \max_{k \leqq n} E(l_{nk}) \qquad l_{nk} = \left[c\left(\frac{k}{n}\right) - c\left(\frac{k-1}{n}\right)\right] \wedge 1, \quad k \leqq n, n \geqq 1$$

$$\leqq \max_{k \leqq n} E[1 - e^{-l_{nk}}]$$

$$\leqq \max_{k \leqq n} E\left[1 - e^{-\left[p\left(\frac{k}{n}\right) - p\left(\frac{k-1}{n}\right)\right]}\right]$$

$$= \max_{k \leqq n} E\left[e^{-p\left(\frac{k-1}{n}\right)} - e^{-p\left(\frac{k}{n}\right)}\right] E\left[e^{-p\left(\frac{k-1}{n}\right)}\right]^{-1}$$

$$\leqq E[e^{-p(1)}]^{-1} \max_{k \leqq n} \left[E\left(e^{-p\left(\frac{k-1}{n}\right)}\right) - E\left(e^{-p\left(\frac{k}{n}\right)}\right)\right]$$

tends to 0 as $n \uparrow +\infty$, it follows that if $\beta < \alpha < \gamma$, then

$$7)\quad e^{-\gamma \liminf_{n\uparrow+\infty} \sum_{k\leq n} E(l_{nk})} = \limsup_{n\uparrow+\infty} \prod_{k\leq n} e^{-\gamma E(l_{nk})}$$
$$\leq \limsup_{n\uparrow+\infty} \prod_{k\leq n} [1 - \alpha E(l_{nk})]$$
$$\leq \limsup_{n\uparrow+\infty} \prod_{k\leq n} E(e^{-\alpha l_{nk}})$$
$$= \limsup_{n\uparrow+\infty} E\left(e^{-\alpha \sum_{k\leq n} l_{nk}}\right)$$
$$= E\left[e^{-\alpha c(1)}\right]$$
$$= \liminf_{n\uparrow+\infty} E\left(e^{-\alpha \sum_{k\leq n} l_{nk}}\right)$$
$$= \liminf_{n\uparrow+\infty} \prod_{k\leq n} E(e^{-\alpha l_{nk}})$$
$$\leq \liminf_{n\uparrow+\infty} \prod_{k\leq n} [1 - \beta E(l_{nk})]$$
$$= e^{-\beta \limsup_{n\uparrow+\infty} \sum_{k\leq n} E(l_{nk})},$$

and, letting $\beta \uparrow \alpha$ and $\gamma \downarrow \alpha$, one finds

$$8)\quad E[e^{-\alpha c(1)}] = e^{-\alpha \lim_{n\uparrow+\infty} \sum_{k\leq n} E(l_{nk})} \qquad \alpha \geq 0,$$

which is impossible unless $c(1)$ is independent of the sample path.

Coming to the proof that $\mathfrak{p}([t_1, t_2) \times [l_1, l_2))$ $(t_1 < t_2,\ 0 < l_1 < l_2)$ is Poisson distributed, if

9 $l_{nk} = 0$ *or* 1 *according as* $\mathfrak{p}\left(\left[\frac{k-1}{n}, \frac{k}{n}\right) \times [l_1, l_2)\right) = 0$ *or* ≥ 1,

then, much as in 6),

$$10)\quad [1 - e^{-\alpha}] \max_{k\leq n} P(l_{nk} = 1)$$
$$\leq \max_{k\leq n} E[1 - e^{-\alpha l_{nk}}]$$
$$\leq E[e^{-\alpha p(1)}]^{-1} \max_{k\leq n} \left(E\left[e^{-\alpha p\left(\frac{k-1}{n}\right)}\right] - E\left[e^{-\alpha p\left(\frac{k}{n}\right)}\right]\right)$$

tends to 0 as $n \uparrow +\infty$, and as in 7), one finds

$$11)\quad E[e^{-\alpha \mathfrak{p}([0,1)\times[l_1, l_2))}] = \lim_{n\uparrow+\infty} E\left[e^{-\alpha \sum_{k\leq n} l_{nk}}\right]$$
$$= \lim_{n\uparrow+\infty} \prod_{k\leq n} E(e^{-\alpha l_{nk}})$$
$$= \lim_{n\uparrow+\infty} \prod_{k\leq n} [1 - (1 - e^{-\alpha}) P(l_{nk} = 1)]$$
$$= \exp\left[-(1 - e^{-\alpha}) \lim_{n\uparrow+\infty} \sum_{k\leq n} P(l_{nk} = 1)\right] \quad \alpha > 0,$$

completing the proof in case $t_1 = 0$ and $t_2 = 1$.

Because of the meaning of $\mathfrak{p}$, it is clear that it is differential in t, and, to complete the proof that $\mathfrak{p}$ is POISSON measure, it suffices to check the independence of $\mathfrak{D}^+ \equiv \mathfrak{p}([0,1) \times (l_1, l_2])$ and

$$\mathfrak{D}_- \equiv \int_{(0,l_1]} l\,\mathfrak{p}([0,1) \times dl) \qquad (0 < l_1 < l_2).$$

Consider for this purpose,

12a) $$l^-_{nk} = \int_{(0,\,l_1]} l\,\mathfrak{p}\left(\left[\frac{k-1}{n}, \frac{k}{n}\right) \times dl\right)$$

12b) $$l^+_{nk} = \mathfrak{p}\left(\left[\frac{k-1}{n}, \frac{k}{n}\right) \times (l_1, l_2]\right),$$

let $\mathfrak{f}$ denote a (non-stochastic) subset of $1, 2, \ldots, n$, and let $\mathfrak{k}$ denote the (stochastic) set of integers $k \leqq n$ such that $l^+_{nk} > 0$.

Because the event $\mathfrak{D}_+ = m$ can be expressed as the sum

$$\bigcup_{\mathfrak{f}} \bigcup_{\mathfrak{t}\in\mathfrak{T}} (l^+_{nk} = t_k,\, k \leqq n),$$

where $\mathfrak{T}$ is the class of all sets $\mathfrak{t} = (t_1, t_2, \ldots, t_n)$ of non-negative integers such that

13a) $$t_k > 0 \quad \textit{or} \quad = 0 \quad \textit{according as} \quad k \in \mathfrak{f} \quad \textit{or not}$$

13b) $$t_1 + t_2 + \cdots + t_n = m,$$

it follows that

14) $$\begin{aligned} &E\left[\mathfrak{D}_+ = m,\, e^{-\alpha \sum_{k\notin\mathfrak{k}} l^-_{nk}}\right] \\ &= \sum_{\mathfrak{f}} \sum_{\mathfrak{t}\in\mathfrak{T}} E\left[l^+_{nk} = t_k,\, k \leqq n,\, e^{-\alpha \sum_{k\notin\mathfrak{f}} l^-_{nk}}\right] \\ &= \sum_{\mathfrak{f}} \sum_{\mathfrak{t}\in\mathfrak{T}} \prod_{k\in\mathfrak{f}} P(l^+_{nk} = t_k) \prod_{k\notin\mathfrak{f}} E\left[l^+_{nk} = 0,\, e^{-\alpha l^-_{nk}}\right] \\ &= \sum_{\mathfrak{f}} \sum_{\mathfrak{t}\in\mathfrak{T}} \prod_{k\in\mathfrak{f}} P(l^+_{nk} = t_k) \prod_{k\notin\mathfrak{f}} P(l^+_{nk} = 0) \\ &\quad \times \frac{\prod_{k\notin\mathfrak{f}} E\left[l^+_{nk} = 0,\, e^{-\alpha l^-_{nk}}\right] \prod_{k\in\mathfrak{f}} E\left[l^+_{nk} = 0,\, e^{-\alpha l^-_{nk}}\right] \prod_{k\in\mathfrak{f}} P(l^+_{nk} = 0)}{\prod_{k\in\mathfrak{f}} P(l^+_{nk} = 0) \prod_{k\notin\mathfrak{f}} P(l^+_{nk} = 0) \prod_{k\in\mathfrak{f}} E\left[l^+_{nk} = 0,\, e^{-\alpha l^-_{nk}}\right]} \\ &= \sum_{\mathfrak{f}} \sum_{\mathfrak{t}\in\mathfrak{T}} P(l^+_{nk} = t_k,\, k \leqq n) \times \frac{E[e^{-\alpha\mathfrak{D}_-} \mid \mathfrak{D}_+ = 0]}{\prod_{k\in\mathfrak{f}} E\left[e^{-\alpha l^-_{nk}} \,\middle|\, l^+_{nk} = 0\right]} \end{aligned}$$

But the sets $\mathfrak{f}$ and $\mathfrak{k}$ figuring in 14) contain at most m members; thus,

15) $$\lim_{n\uparrow+\infty} \prod_{k\in\mathfrak{f}} E\left[e^{-\alpha l^-_{nk}} \,\middle|\, l^+_{nk} = 0\right] = 1,$$

and as $n \uparrow +\infty$, 14) goes over into

16) $$E[\mathfrak{D}_+ = m, e^{-\alpha \mathfrak{D}_-}] = \lim_{n\uparrow+\infty} \sum_{\mathfrak{f}} \sum_{t\in\mathfrak{T}} P(l^+_{nk} = t_k, k \leqq n)\, E[e^{-\alpha\mathfrak{D}_-} \mid \mathfrak{D}_+ = 0] = P[\mathfrak{D}_+ = m]\, E[e^{-\alpha\mathfrak{D}_-} \mid \mathfrak{D}_+ = 0],$$

completing the proof.

4.11. Eigen-differential expansions for Green functions and transition densities

Given a non-singular diffusion modified as in 4.5, let $Q^\bullet$ be the unit interval closed at 0 if $E_0(e^{-\mathfrak{m}_{0+}}) = E_{0+}(e^{-\mathfrak{m}_0}) = 1$, closed at 1 if $E_1(e^{-\mathfrak{m}_{1-}}) = E_{1-}(e^{-\mathfrak{m}_1}) = 1$, and open otherwise, introduce the stopping time $\mathfrak{m} = \min(t : x(t) \notin Q^\bullet)$, and let us use the idea of eigen-differential expansion to construct transition densities $p(t, a, b)$ such that

1) $$P_a[x(t) \in db, t < \mathfrak{m}] = p(t, a, b)\, m(db) \qquad (t, a, b) \in (0, +\infty) \times Q^\bullet \times Q^\bullet,$$

where

2) $0 \leqq p(t, a, b) = p(t, b, a)$ *is continuous on* $(0, +\infty) \times Q^\bullet \times Q^\bullet$

3) $p(t, a, b) = \int_{Q^\bullet} p(s, a, c)\, p(t-s, c, b)\, m(dc) \quad t > s > 0, \quad a, b \in Q^\bullet$

4a) $$\frac{\partial}{\partial t} p(t, a, b) = \mathfrak{G}^\bullet\, p(t, a, b)^*$$

4b) $$p(t, 0+, b) = 0 \quad \textit{if} \quad E_0(e^{-\mathfrak{m}_{0+}}) = 0, \quad b \in Q^\bullet$$
$$p(t, 1-, b) = 0 \quad \textit{if} \quad E_1(e^{-\mathfrak{m}_{1-}}) = 0, \quad b \in Q^\bullet$$

4c) $$p_2^+(t, 0+, b) = 0 \quad \textit{if} \quad E_0(e^{-\mathfrak{m}_{0+}}) = 1,\ E_{0+}(e^{-\mathfrak{m}_0}) = 0,\ b \in Q^{\bullet\,**}$$
$$p_2^-(t, 1-, b) = 0 \quad \textit{if} \quad E_1(e^{-\mathfrak{m}_{1-}}) = 1,\ E_{1-}(e^{-\mathfrak{m}_1}) = 0,\ b \in Q^\bullet,$$

$\mathfrak{G}^\bullet$ being the local differential operator described at the end of 4.7.

The method is outlined only; see H. P. McKean, jr. [2] for the proof in a special case, S. Karlin and J. McGregor [1] for a beautiful method which can be adapted to the present needs, and H. Weyl [1] for the classical eigen-differential expansion for Sturm-Liouville operators. As a general source of information about Hilbert space, B. Sz-Nagy [1] is suggested.

---

* $\mathfrak{G}^\bullet$ is applied to a or to b.

** $$p_2^+(t, a, b) = \lim_{\xi\downarrow a} \frac{p(t, \xi, b) - p(t, a, b)}{s(\xi) - s(a)},$$
$$p_2^-(t, a, b) = \lim_{\xi\uparrow b} \frac{p(t, a, b) - p(t, a, \xi)}{s(b) - s(\xi)} \quad \textit{etc.}$$

Given $\alpha > 0$, let g_1 and g_2 be the solutions of

$$\text{5a)} \qquad (\mathfrak{G}^\bullet g_1)(b) = \alpha g_1(b) \qquad 0 \leqq b < 1$$

$$\text{5b)} \qquad (\mathfrak{G}^\bullet g_2)(b) = \alpha g_2(b) \qquad 0 < b \leqq 1$$

described in 4.6, let B be the WRONSKIAN $g_1^+ g_2 - g_1 g_2^+$ (= constant > 0), introduce the GREEN function

$$\text{6)} \qquad G(a, b) = G(b, a) = B^{-1} g_1(a)\, g_2(b) \quad a \leqq b, \;\; a, b \in Q^\bullet,$$

and let us prove that, for $f \in D$,

$$\text{7)} \qquad \int_{Q^\bullet} G(a, b)\, f(b)\, m(db) = E_a\left[\int_0^{\mathfrak{m}} e^{-\alpha t} f(x_t)\, dt\right] \qquad a \in Q^\bullet$$

is the solution of

$$\text{8a)} \qquad u \in C(Q^\bullet)$$

$$\text{8b)} \qquad (\alpha - \mathfrak{G}^\bullet)\, u(b) = f(b) \qquad b \in Q^\bullet$$

subject to

$$\text{9a)} \quad u(0+) = 0 \quad \textit{if} \quad E_0(e^{-\mathfrak{m}_{0+}}) = 0, \quad E_{0+}(e^{-\mathfrak{m}_0}) = 1$$

$$\text{9b)} \quad u(1-) = 0 \quad \textit{if} \quad E_1(e^{-\mathfrak{m}_{1-}}) = 0, \quad E_{1-}(e^{-\mathfrak{m}_1}) = 1$$

$$\text{10a)} \quad u^+(0+) = 0 \quad \textit{if} \quad E_0(e^{-\mathfrak{m}_{0+}}) = 1, \quad E_{0+}(e^{-\mathfrak{m}_0}) = 0$$

$$\text{10b)} \quad u^-(1-) = 0 \quad \textit{if} \quad E_1(e^{-\mathfrak{m}_{1-}}) = 1, \quad E_{1-}(e^{-\mathfrak{m}_1}) = 0$$

(see 2.6 for a special case).

As in 2.6.16),

$$\text{11)} \qquad \int_{Q^\bullet} G(a, b)\, m(db) \leqq \alpha^{-1};$$

for example, if $0 \in Q^\bullet$, if $E_{1-}(e^{-\mathfrak{m}_1}) = 0$, and if $0 < a < 1$, then

$$\begin{aligned}
\text{12)} \quad \alpha \int_{Q^\bullet} & G(a, b)\, m(db) \\
&= B^{-1}\left[g_2(a) \int_{0 \leqq b < a} \mathfrak{G}^\bullet g_1\, dm + g_1(a) \int_{a \leqq b < 1} \mathfrak{G}^\bullet g_2\, dm\right] \\
&\leqq B^{-1}\left[g_2(a) \int_{0 < b < a} g_1^+(db) + g_1(a) \int_{a \leqq b < 1} g_2^+(db)\right] \\
&\qquad\qquad + B^{-1} g_2(a)\, \alpha\, g_1(0)\, m(0) \\
&= B^{-1}[B - g_2(a)\, g_1^+(0) + g_1(a)\, g_2^-(1)] + B^{-1} g_2(a)\, \alpha\, g_1(0)\, m(0) \\
&= 1 + B^{-1} g_2(a)\, [\alpha\, g_1(0)\, m(0) - g_1^+(0)] + B^{-1} g_1(a)\, g_2^-(1) \\
&\leqq 1.
\end{aligned}$$

Both

$$\text{13a)} \qquad u \equiv \int_{Q^\bullet} G f\, dm$$

and

13b) $$G_\alpha^\bullet f = E_\bullet\left[\int_0^{\mathfrak{m}} e^{-\alpha t} f(x_t)\,dt\right]$$

are solutions of 8), 9), and 10) as is clear in the case of 13a) on differentiating and using table 4.6.1, and in the case of 13b), on noting

14) $$\begin{aligned} G_\alpha f &= G_\alpha^\bullet f + E_\bullet[e^{-\alpha\mathfrak{m}}(G_\alpha f)(x_{\mathfrak{m}})] \\ &= G_\alpha^\bullet f + \frac{g_2}{g_2(0)}\frac{P_0[\mathfrak{m}_{0+} = +\infty]}{\varkappa(0)+\alpha} f(0) + \\ &\quad + \frac{g_1}{g_1(1)}\frac{P_1[\mathfrak{m}_{1-} = +\infty]}{\varkappa(1)+\alpha} f(1). \end{aligned}$$

But then $u^\bullet = u - G_\alpha^\bullet f \in C(Q^\bullet)$ is a solution of $(\mathfrak{G}^\bullet u^\bullet)(\xi) = \alpha u^\bullet(\xi)$ $(\xi \in Q^\bullet)$ subject to 9) and 10), and, as such, $u^\bullet \equiv 0$ on $Q^\bullet$ in agreement with 7); for example, if $0 \in Q^\bullet$, if $E_1(e^{-\mathfrak{m}_{1-}}) = 0$, and if $E_{1-}(e^{-\mathfrak{m}_1}) = 1$ then, as in the identification of $\mathfrak{G}$ as a global differential operator in 4.7, $u^\bullet(b) = \text{constant} \times g_2$ $(0 \leqq b < 1)$ because $g_1(1-) > 0$, and the constant has to be 0 because

$$0 < k(0)\, g_2(0) - g_2^+(0) + m(0)\,(\mathfrak{G}^\bullet g_2)(0+).$$

Now think of $G_\alpha^\bullet$ as acting on the (real) HILBERT space H of (speed) measurable functions h defined on $Q^\bullet$ with $\|h\|_2 = \sqrt{\int_{Q^\bullet} h^2\,dm} < +\infty$, modulo null functions.

Because $\int G\,dm \leqq \alpha^{-1}$ and $G(a, b) = G(b, a)$,

15) $$\begin{aligned} &\|G_\alpha^\bullet h\|_2^2 \\ &= \int dm\left(\int G^{\frac12} G^{\frac12} h\,dm\right)^2 \\ &\leqq \int m(da)\int G(a,b)\,m(db)\int G(a,b)\,h(b)^2\,m(db) \\ &\leqq \alpha^{-1}\int\int G(a,b)\,m(da)\,h(b)^2\,m(db) \\ &\leqq \alpha^{-2}\int h(b)^2\,m(db) \\ &= \alpha^{-2}\|h\|_2^2, \end{aligned}$$

i.e., $G_\alpha^\bullet$ *is bounded* $(\|G_\alpha^\bullet\|_2 \leqq \alpha^{-1})$ *and symmetric* $(G_\alpha^{\bullet *} = G_\alpha^\bullet)$.

$G_\alpha^\bullet$ is also *positive-definite*; indeed, using

16) $$G_\alpha^\bullet - G_\beta^\bullet + (\alpha - \beta)\, G_\alpha^\bullet G_\beta^\bullet = 0 \qquad \alpha, \beta > 0$$

to check that

$$\lim_{\varepsilon \downarrow 0} \varepsilon^{-1}[G_{\alpha+\varepsilon}^\bullet - G_\alpha^\bullet] = -G_\alpha^{\bullet 2}$$

in the operator norm and noting $\|G_\alpha^\bullet\|_2 \downarrow 0$ $(\alpha \uparrow +\infty)$, it is found that $G_\alpha^\bullet = \int_\alpha^{+\infty} G_\beta^{\bullet 2}\,d\beta$ is *non-negative definite*. A second application of

16) now implies that the null space $H \cap (h: G^\bullet_\alpha h = 0)$ is independent of α, and it follows that if $(G^\bullet_\alpha h, h) = 0$, then $0 = \lim_{\beta \uparrow +\infty} \beta G^\bullet_\beta h = h$ as desired.

$G^\bullet_\alpha$ is now *invertible*, and using 16) once more, one finds that the range $G^\bullet_\alpha H \equiv H(\mathfrak{D})$ is independent of α and that

17a) $$G^{\bullet -1}_\alpha = \alpha - \mathfrak{D}$$

17b) $$\mathfrak{D} \equiv 1 - G^{\bullet -1}_1 : H(\mathfrak{D}) \to H.$$

$\mathfrak{D}$ is *symmetric* ($\mathfrak{D}^* = \mathfrak{D}$) and *non-positive definite* ($\mathfrak{D} = \lim_{\varepsilon \downarrow 0} (\varepsilon - G^{-1}_\varepsilon) \leqq 0$).

$H(\mathfrak{D})$ is the class of functions $h \in H$ such that, for some $u \in D$ and $h^\bullet \in H$,

18a) $$u \equiv h \quad a.e.\ (m)$$

18b) $$h^\bullet(\xi)\, m(d\xi) = u^+(d\xi) - u(\xi)\, k(d\xi) \qquad 0 < \xi < 1.$$

subject to

19a) $$h^\bullet(0)\, m(0) = u^+(0) - u(0)\, k(0) \qquad \text{if } \ 0 \in Q^\bullet$$

19b) $$h^\bullet(1)\, m(1) = -u^-(1) - u(1)\, k(1) \qquad \text{if } \ 1 \in Q^\bullet$$

20a) $$u(0+) = 0 \quad \text{if } \ E_0(e^{-\mathfrak{m}_{0+}}) = 0, \quad E_{0+}(e^{-\mathfrak{m}_0}) = 1$$

20b) $$u(1-) = 0 \quad \text{if } \ E_1(e^{-\mathfrak{m}_{1-}}) = 0, \quad E_{1-}(e^{-\mathfrak{m}_1}) = 1$$

21a) $$u^+(0+) = 0 \quad \text{if } \ E_0(e^{-\mathfrak{m}_{0+}}) = 1, \quad E_{0+}(e^{-\mathfrak{m}_0}) = 0$$

21b) $$u^-(1-) = 0 \quad \text{if } \ E_1(e^{-\mathfrak{m}_{1-}}) = 1, \quad E_{1-}(e^{-\mathfrak{m}_1}) = 0$$

22a) $$u(0+) = 0 \quad \text{if } \ E_0(e^{-\mathfrak{m}_{0+}}) = E_{0+}(e^{-\mathfrak{m}_0}) = 0,$$
$$+\infty > \int_{0+}^{1/2} [s(\tfrac{1}{2}) - s(b)]^2\, m(db)$$

22b) $$u(1-) = 0 \quad \text{if } \ E_1(e^{-\mathfrak{m}_{1-}}) = E_{1-}(e^{-\mathfrak{m}_1}) = 0$$
$$+\infty > \int_{1/2}^{1-} [s(b) - s(\tfrac{1}{2})]^2\, m(db),*$$

and $\mathfrak{D}$ is the map $h \in H(\mathfrak{D}) \to h^\bullet$.

Now introduce the spectral representations

23a) $$\mathfrak{D} = \int_{-\infty}^{0+} \gamma\, \mathfrak{p}(d\gamma)$$

23b) $$G^\bullet_\alpha = \int_{-\infty}^{0+} \frac{\mathfrak{p}(d\gamma)}{\alpha - \gamma},$$

where $\mathfrak{p}(d\gamma)$ is the appropriate spectral measure from $(-\infty, 0]$ to projections ($\mathfrak{p}(-\infty, 0] = 1$).

---

* 22) is an instance of H. Weyl's limit circle case.

Given $\Gamma = (\gamma_1, \gamma_2]$ $(-\infty < \gamma_1 < \gamma_2 \leq 0)$, $\mathfrak{p}(\Gamma)$ is a CARLEMAN operator; in detail, if $\mathfrak{e} = \mathfrak{e}(\gamma, \cdot) = (e_1(\gamma, \cdot), e_2(\gamma, \cdot))$ is the solution of

24a) $$(\mathfrak{G}^\bullet\, \mathfrak{e})(\xi) = \gamma\, \mathfrak{e}(\xi) \qquad 0 < \xi < 1$$

24b) $$\mathfrak{e}(\gamma, \tfrac{1}{2}) = (1, 0), \quad \mathfrak{e}^+(\gamma, \tfrac{1}{2}) = (0, 1)$$

and if

25a) $$\mathfrak{e}(\gamma, 0) = \mathfrak{e}(\gamma, 0+) \qquad 0 \in Q^\bullet$$

25b) $$\mathfrak{e}(\gamma, 1) = \mathfrak{e}(\gamma, 1-) \qquad 1 \in Q^\bullet,$$

then

26a) $$\mathfrak{p}(\Gamma)\, h = \int_{Q^\bullet} e(\Gamma, a, b)\, h(b)\, m(db)$$

26b) $$e(\Gamma, a, b) = \int_{\Gamma} \mathfrak{e}(\gamma, a)\, \mathfrak{f}(d\gamma)\, \mathfrak{e}(\gamma, b)$$

26c) $$\int_{Q^\bullet} |e(\Gamma, a, b)|^2\, m(db) < +\infty \quad a.e. \quad (m),$$

where $\mathfrak{f}(d\gamma)$ is a BOREL measure from $(-\infty, 0]$ to 2×2 symmetric non-negative definite matrices

27a) $$\mathfrak{f}(d\gamma) = \begin{Vmatrix} f_{11}(d\gamma) & f_{12}(d\gamma) \\ f_{21}(d\gamma) & f_{22}(d\gamma) \end{Vmatrix}$$

27b) $$f_{11}(d\gamma) \geq 0, \quad f_{22}(d\gamma) \geq 0, \quad f_{12}(d\gamma) = f_{21}(d\gamma),$$
$$f_{12}(d\gamma)^2 \leq f_{11}(d\gamma)\, f_{22}(d\gamma)$$

and $\mathfrak{e}\mathfrak{f}\mathfrak{e}$ is the inner product of $\mathfrak{e}$ and $\mathfrak{f}\mathfrak{e}$ in 2-space.

$\mathfrak{h} = \int_{Q^\bullet} \mathfrak{e}\, h\, dm$ is the so-called FOURIER transform; it satisfies the PLANCHEREL theorem

28) $$\int_{-\infty}^{0+} \mathfrak{h}\, \mathfrak{f}(d\gamma)\, \mathfrak{h} = \|h\|_2^2$$

23b) implies the eigen-differential expansion

29a) $$G(a, b) = \int_{-\infty}^{0+} (\alpha - \gamma)^{-1}\, \mathfrak{e}(\gamma, a)\, \mathfrak{f}(d\gamma)\, \mathfrak{e}(\gamma, b)$$
$$\alpha > 0, \quad (a, b) \in Q^\bullet \times Q^\bullet,$$

and inverting 29a) as a LAPLACE transform, 1), 2), 3), and 4) are found to hold for the kernel

29b) $$p(t, a, b) = \int_{-\infty}^{0+} e^{\gamma t}\, \mathfrak{e}(\gamma, a)\, \mathfrak{f}(d\gamma)\, \mathfrak{e}(\gamma, b)$$
$$(t, a, b) \in (0, +\infty) \times Q^\bullet \times Q^\bullet;$$

the proof of 3) is based upon the estimates

30a) $$\int_{-\infty}^{0+} (\alpha-\gamma)^{-1} f_{11}(d\gamma) = B^{-1} g_1(\tfrac{1}{2})\, g_2(\tfrac{1}{2}) < +\infty$$

30b) $$\int_{-\infty}^{0+} (\alpha-\gamma)^{-2} f_{22}(d\gamma) \leqq (\alpha B)^{-1} g_1^+(\tfrac{1}{2})\, g_2^+(\tfrac{1}{2}) < +\infty$$

and

31) $|e_1(\gamma,\xi)| + |e_2(\gamma,\xi)| < e^{\text{constant}\sqrt{|\gamma|}}$ $(\gamma \downarrow -\infty)$ *on closed subintervals of* $Q^\bullet$,

permitting the differentiation of 29b) under the integration sign.

Given $0 < b < 1$, if $p(t,\xi,\eta)$ $(\xi < \eta < b)$ is the kernel 29b) for the stopped diffusion

$$[x(t \wedge \mathfrak{e}) : t \geqq 0, P_a : a \leqq b] \qquad \mathfrak{e} = \mathfrak{m}_b,$$

then $p(t, a, b-) = 0$ $(a < b)$ and

32) $$\frac{\partial}{\partial t} P_a[\mathfrak{m}_b < t] = -p_3^-(t, a, b) \qquad t > 0, \quad a < b,$$

leading to the identification of $u(t, a) \equiv P_a[\mathfrak{m}_b < t]$ as the solution of

33a) $$\frac{\partial}{\partial t} u(t,a) = \mathfrak{G}^\bullet u(t,a) \qquad t > 0, \quad a \in [0, b) \cap Q^\bullet$$

33b) $$u(0+, a) = 0 \qquad a < b$$

34a) $$u(t, 0+) = 0 \qquad t > 0, \quad E_0(e^{-\mathfrak{m}_{0+}}) = 0$$

34b) $$u_2^+(t, 0+) = 0 \qquad t > 0, \quad E_0(e^{-\mathfrak{m}_{0+}}) = 1, \quad E_{0+}(e^{-\mathfrak{m}_0}) = 0$$

34c) $$\lim_{a \uparrow b} u(t,a) = 1 \qquad t > 0.$$

If 0 and 1 are exit or entrance, then (see table 4.6.1), the trace $tr(G_\alpha^\bullet) \equiv \int_{Q^\bullet} G(\xi,\xi)\, m(d\xi)$ is $< +\infty$, with the result that $G_\alpha^\bullet$ is compact and $\mathfrak{f}(d\gamma)$ is concentrated on a series of simple eigenvalues

35) $$0 \geqq \gamma_1 \geqq \gamma_2 > \text{ etc. } \downarrow -\infty, \quad \sum_{n \geqq 2} \frac{1}{\gamma_n} > -\infty.$$

If $s(1) - s(0)$ and $m(0, 1)$ are both $< +\infty$, then

36) $$\lim_{n \uparrow +\infty} \frac{-\gamma_n}{\pi^2 n^2} = \left[\int_0^1 \sqrt{s(db)\, m(db)}\right]^{-2},$$

$\int_0^1 \sqrt{s(db)\,m(db)}$ being the HELLINGER integral $\inf \sum_{n\geq 1} \sqrt{s(Q_n)\,m(Q_n)}$ $\left(\bigcup_{n\geq 1} Q_n \supset Q\right)$ (see H. P. MCKEAN, JR. and D. B. RAY [1], and, for the classical case, COURANT and HILBERT [1]).

S. KARLIN and J. MCGREGOR [1, 2, 4] have developed similar eigendifferential expansions in connection with the birth and death processes described in problem 4.2.2.

Problem 1. Compute the FOURIER transform for the standard BROWNian motion.

$$\left[e_1(\gamma,\xi) = \cos\sqrt{2|\gamma|}\,\xi, \quad e_2(\gamma,\xi) = \frac{\sin\sqrt{2|\gamma|}\,\xi}{\sqrt{2|\gamma|}},\right.$$

$$\left. f_{11}(d\gamma) = \frac{d\gamma}{2\pi\sqrt{2|\gamma|}}, \quad f_{12} = f_{21} = 0, \quad f_{22}(d\gamma) = \frac{\sqrt{2|\gamma|}\,d\gamma}{2\pi}.\right]$$

Problem 2. Express the FOURIER transform for the BESSEL motion attached to $\mathfrak{G}^\bullet = \frac{1}{2}\left(\frac{d^2}{db^2} + \frac{d-1}{b}\frac{d}{db}\right)$ $(d \geq 2)$ in the special form

$$h = \int_{-\infty}^{0+} e_1(\gamma,\cdot)\, f_{11}(d\gamma) \int_0^{+\infty} e_1(\gamma,b)\, h(b)\, 2b^{d-1}\,db$$

choosing for e_1 the solution of

$$(\mathfrak{G}^\bullet e_1)(b) = \gamma\, e_1(b) \quad (0 < b < +\infty), \quad e_1(0) = 1, \quad e_1^+(0) = 0.$$

Evaluate $p(t,a,b)$ using 29b) and A. ERDÉLYI [1: 186 (39)], and check it against 2.7.4).

$$\left[e_1(\gamma,b) = 2^{\frac{d}{2}-1}\,\Gamma\!\left(\frac{d}{2}\right)\left(\sqrt{2|\gamma|}\,b\right)^{1-\frac{d}{2}} J_{\frac{d}{2}-1}\!\left(\sqrt{2|\gamma|}\,b\right),\right.$$

$$f_{11}(d\gamma) = 2^{-\frac{d}{2}}\,\Gamma\!\left(\frac{d}{2}\right)^{-2} |\gamma|^{\frac{d}{2}-1}\,d\gamma, \quad \text{and}$$

$$\left. p(t,a,b) = \frac{e^{-\frac{b^2+a^2}{2t}}}{2t\,(ab)^{\frac{d}{2}-1}}\, I_{\frac{d}{2}-1}\!\left(\frac{ab}{t}\right) \qquad a, b > 0.\right]$$

Problem 3. Given a *conservative* non-singular diffusion, define

$$p_0 \equiv 1, \quad p_1 \equiv s, \quad p_n(b) = \int_{\frac{1}{2}}^{b\cdot} s(d\xi) \int_{\frac{1}{2}+}^{\xi} p_{n-2}(\eta)\, m(d\eta) \quad n \geq 2;$$

the problem is to show that $p(t,\xi,\eta)$ is an *analytic* function of η $(0 < \eta < 1)$ in the sense that for each $t > 0$ and each $0 < \xi < 1$, $p(t,\xi,\eta)$ can be expanded as a *power series* $\sum_{n\geq 0} c_n p_n(\eta)$ with coefficients

$$c_{2n} = \mathfrak{G}^{\bullet n}\, p(t,\xi,\tfrac{1}{2})$$

$$c_{2n+1} = (\mathfrak{G}^{\bullet n}\, p)_3^+(t,\xi,\tfrac{1}{2}).$$

[Because $e_1 = \sum_{n\geqq 0} \gamma^n p_{2n}$, $e_2 = \sum_{n\geqq 0} \gamma^n p_{2n-1}$ ($p_{-1} \equiv 0$), and

$$\sum_{n\geqq 0} \int_{-\infty}^{0} e^{\gamma t}|\gamma|^n \,|\mathfrak{e}(\gamma,\xi)\,\mathfrak{f}(d\gamma)\,(p_{2n}(\eta), p_{2n-1}(\eta))|$$
$$\leqq \sum_{n\geqq 0} \int_{-\infty}^{0+} e^{\gamma t}|\gamma|^n e^{c\sqrt{|\gamma|}}\, 2[f_{11}(d\gamma) + f_{22}(d\gamma)]\, c^n\,(n!)^{-2}$$
$$< +\infty$$

for some constant c, an application of 29b) implies

$$\begin{aligned}
p(t,\xi,\eta) &= \int_{-\infty}^{0+} e^{\gamma t}\,\mathfrak{e}(\gamma,\xi)\,\mathfrak{f}(d\gamma)\,\mathfrak{e}(\gamma,\eta) \\
&= \int_{-\infty}^{0+} e^{\gamma t}\mathfrak{e}(\gamma,\xi)\,\mathfrak{f}(d\gamma) \sum_{n\geqq 0}\gamma^n (p_{2n}(\eta), p_{2n-1}(\eta)) \\
&= \sum_{n\geqq 0} p_{2n}(\eta) \int_{-\infty}^{0+} e^{\gamma t}\gamma^n\,\mathfrak{e}(\gamma,\xi)\,\mathfrak{f}(d\gamma)\,(1,0) \\
&\quad + \sum_{n\geqq 0} p_{2n-1}(\eta) \int_{-\infty}^{0+} e^{\gamma t}\gamma^n\,\mathfrak{e}(\gamma,\xi)\,\mathfrak{f}(d\gamma)\,(0,1),
\end{aligned}$$

and to complete the proof, it suffices to note that

$$\begin{aligned}
c_{2n} &\equiv \int_{-\infty}^{0+} e^{\gamma t}\gamma^n\,\mathfrak{e}(\gamma,\xi)\,\mathfrak{f}(d\gamma)\,(1,0) \\
&= \int_{-\infty}^{0+} e^{\gamma t}\,\mathfrak{e}(\gamma,\xi)\,\mathfrak{f}(d\gamma)\,\mathfrak{G}^{\bullet n}\,\mathfrak{e}(\gamma,\tfrac{1}{2}) \\
&= \mathfrak{G}^{\bullet n} p(t,\xi,\tfrac{1}{2})
\end{aligned}$$

and

$$\begin{aligned}
c_{2n+1} &\equiv \int_{-\infty}^{0+} e^{\gamma t}\gamma^n\,\mathfrak{e}(\gamma,\xi)\,\mathfrak{f}(d\gamma)\,(0,1) \\
&= \int_{-\infty}^{0+} e^{\gamma t}\,\mathfrak{e}(\gamma,\xi)\,\mathfrak{f}(d\gamma)\,(\mathfrak{G}^{\bullet n}\,\mathfrak{e})^+(\tfrac{1}{2}) \\
&= (\mathfrak{G}^{\bullet n} p)_3^+(t,\xi,\tfrac{1}{2}).]
\end{aligned}$$

Problem 4. If $p(t,a,b) \leqq p(t,a,a)$ for each $(t,a,b) \in (0,+\infty) \times Q^\bullet \times Q^\bullet$, then, after a change of scale,

$$\begin{aligned}
x(t) &= b(t) & t &< \mathfrak{e} \\
&= \infty & t &\geqq \mathfrak{e},
\end{aligned}$$

where b is a standard BROWNian motion and $\mathfrak{e}$ is an independent exponential holding time with conditional distribution $P_{\bullet}(\mathfrak{e} > t \mid \mathsf{B}) = e^{-\varkappa t}$ $(\varkappa \geqq 0)$.

$\left[\int_{-\infty}^{0+} e^{\gamma t}\, \mathfrak{e}(\gamma, a)\, \mathfrak{f}(d\gamma)\, \mathfrak{e}^+(\gamma, a)\right.$ is continuous from the right $(0<a<1)$, and so

$$\lim_{b \downarrow a} \frac{p(t, b, b) - p(t, a, a)}{s(b) - s(a)} = 2 \int_{-\infty}^{0+} e^{\gamma t}\, \mathfrak{e}(\gamma, a)\, \mathfrak{f}(d\gamma)\, \mathfrak{e}^+(\gamma, a)$$

$$= 2 \lim_{b \downarrow a} \frac{p(t, a, b) - p(t, a, a)}{s(b) - s(a)} \equiv 0 \qquad 0 < a < 1,$$

proving that $p(t, a, a) \equiv c_1(t)$ does not depend upon $0<a<1$. Because of this $c_2(\alpha) \equiv \int_0^{+\infty} e^{-\alpha t} c_1\, dt = B^{-1} g_1(a)\, g_2(a)$ is also constant $(0 < a < 1)$, and using this to solve $B = g_1^+ g_2 - g_1 g_2^+$ for $g_1 =$ constant $\times\, e^{c_3 s}$ and $g_2 =$ constant $\times\, e^{-c_3 s}$ $[c_3 = (2c_2)^{-1}]$, an application of $\mathfrak{G}^\bullet g_1 = \alpha g_1$ implies $k(db) = c_4\, s(db)$ and $m(db) = c_5\, s(db)$, where $0 \leqq c_4$ and $0 < c_5$ do not depend upon α and $c_3(\alpha) = \sqrt{c_4 + c_5 \alpha}$. $s(0) = -\infty$ because $s(0) > -\infty$ would entail either

$$0 = \varkappa(0)\, g_1(0) + (\mathfrak{G}^\bullet g_1)(0) = [\varkappa(0) + \alpha]\, g_1(0)$$

or

$$0 = k(0)\, g_1(0) - g_1^+(0) + m(0)\, (\mathfrak{G}^\bullet g_1)(0)$$
$$= \left[k(0) - \sqrt{c_4 + c_5 \alpha} + \alpha\, m(0)\right] g_1(0),$$

contradicting $g_1(0) > 0$; the proof of $s(1) = +\infty$ is the same, and now a change of scale brings $\mathfrak{G}^\bullet$ into the form $\mathfrak{G}^\bullet = \frac{1}{2} D^2 - c_6$ $(Q = R^1,\ c_6 = c_4/c_5 \geqq 0)$.]

Problem 5. Give a proof that

$$p(t, a, b) > 0 \qquad (t, a, b) \in (0, +\infty) \times Q^\bullet \times Q^\bullet$$

using the eigen-differential expansion 29b) and the CHAPMAN-KOLMOGOROV identity 3).

$$[p(t, a, a) = \int_{-\infty}^{0+} e^{\gamma t}\, \mathfrak{e}(\gamma, a)\, \mathfrak{f}(d\gamma)\, \mathfrak{e}(\gamma, a) > 0 \qquad (t > 0, 0 < a < 1)$$

because

$$\mathfrak{e}(\gamma, a)\, \mathfrak{f}(d\gamma)\, \mathfrak{e}(\gamma, a) \geqq 0$$

and

$$\int_{-\infty}^{0+} (\alpha - \gamma)^{-1}\, \mathfrak{e}(\gamma, a)\, \mathfrak{f}(d\gamma)\, \mathfrak{e}(\gamma, a) = G(a, a) > 0.$$

Given $s > 0$ and $a, b \in Q^\bullet$ such that $p(s, a, b) = 0$, the CHAPMAN-KOLMOGOROV identity

$$p(s, a, b) = \int_{Q^\bullet} p(s-t, a, c)\, p(t, c, b)\, m(dc) \qquad (t < s)$$

coupled with $p(s-t, a, a) > 0$ implies $p(t, a, b) \equiv 0$ $(t < s)$. But $p(t) \equiv p(t, a, b)$ is a LAPLACE transform, and, as such, has to be $\equiv 0$ $(t > 0)$, contradicting $\int_0^{+\infty} e^{-\alpha t} p(t)\, dt = G(a, b) > 0$.]

Problem 6 after KARLIN and MCGREGOR [5]. Consider the space of points $c^\bullet \in (0, 1)^n$ $(n \geq 2)$ such that $c_1 < c_2 <$ *etc.* and, for $t > 0$, define $e(t, a^\bullet, b^\bullet)$ to be the determinant

$$\begin{vmatrix} p(t, a_1, b_1) & p(t, a_1, b_2) & \dots & p(t, a_1, b_n) \\ p(t, a_2, b_1) & p(t, a_2, b_2) & \dots & p(t, a_2, b_n) \\ \vdots & \vdots & \vdots & \vdots \\ p(t, a_n, b_1) & p(t, a_n, b_2) & \dots & p(t, a_n, b_n) \end{vmatrix}$$

Consider also the joint sample path $x(t) = (x_1(t), x_2(t), etc.) \in [0, 1]^n$ of n independent particles starting at $a^\bullet$, let $P_{a^\bullet}$ be the corresponding law, and let $\mathfrak{m}$ be the crossing time $\sup(t: x_1(s) < x_2(s) < etc., s < t)$. Given $B^\bullet = B_1 \times B_2 \times etc. \subset (0, 1)^n$ such that B_1 lies to the left of B_2, B_2 to the left of B_3, *etc.*,

$$P_{a^\bullet}[\mathfrak{m} > t, x(t) \in B^\bullet] = \int_{B^\bullet} e(t, a^\bullet, b^\bullet)\, m(db^\bullet),$$

where $m(db^\bullet) = m(db_1)\, m(db_2)$ *etc.* Give a direct probabilistic proof of this using

$$\begin{aligned} P_{a^\bullet}[\mathfrak{m} > t, x(t) \in B^\bullet] &= \sum_{\mathfrak{o}} (\mathfrak{o})\, P_{a^\bullet}[\mathfrak{m} > t, x(t) \in \mathfrak{o} B^\bullet] \\ &= \sum_{\mathfrak{o}} (\mathfrak{o})\, (P_{a^\bullet}[x(t) \in \mathfrak{o} B^\bullet] - P_{a^\bullet}[\mathfrak{m} < t, x(t) \in \mathfrak{o} B^\bullet]) \\ &= \int_{B^\bullet} e(t, a^\bullet, b^\bullet)\, m(db^\bullet) - \sum_{\mathfrak{o}} (\mathfrak{o})\, P_{a^\bullet}[\mathfrak{m} < t, x(t) \in \mathfrak{o} B^\bullet], \end{aligned}$$

where $\mathfrak{o}$ is a permutation of $1, 2, \dots, n$, $(\mathfrak{o}) = \pm 1$ according as $\mathfrak{o}$ is even or odd, and $\mathfrak{o} B^\bullet = B_{\mathfrak{o}1} \times B_{\mathfrak{o}2} \times$ etc., and deduce that $e(t, a^\bullet, b^\bullet) > 0$ for each $t > 0$ and each choice of $a^\bullet$ and $b^\bullet$ (see KARLIN and MCGREGOR [5] for additional information and G. PÓLYA [*1*] for the original result in this direction).

[Because $P_{a^\bullet}[\mathfrak{m} > t, x(t) \in \mathfrak{o} B^\bullet] = 0$ unless $\mathfrak{o} = 1$, $P_{a^\bullet}[\mathfrak{m} > t, x(t) \in B^\bullet] = \int_{B^\bullet} e\, m(db^\bullet) - \sum_{\mathfrak{o}} (\mathfrak{o})\, P_a^\bullet[\mathfrak{m} < t, x(t) \in \mathfrak{o} B^\bullet]$ is clear, and it is also clear that the joint motion starts afresh at time $\mathfrak{m}$. But a per-

mutation that interchanges two of the coinciding particles is odd and has no effect on the future motion, so
$\sum_{\mathfrak{o}} (\mathfrak{o})\, P_{a^\bullet}[\mathfrak{m} < t, x(t) \in \mathfrak{o} B^\bullet] = 0,$

$$\text{leaving} \quad P_{a^\bullet}[\mathfrak{m} > t, x(t) \in B^\bullet] = \int_{B^\bullet} e\, m(db^\bullet)$$

as desired. $e(t, a^\bullet, b^\bullet) \geqq 0$ is now clear, and an application of the CHAPMAN-KOLMOGOROV identity $e(t, a^\bullet, b^\bullet) = \int e(t-s, a^\bullet, c^\bullet) e(s, c^\bullet, b^\bullet) \times m(dc^\bullet)$ to the case $a^\bullet = b^\bullet$ plus the fact that $e(t, a^\bullet, b^\bullet) = e(t, b^\bullet, a^\bullet)$ implies $e(t, a^\bullet, a^\bullet) > 0$ for each $t > 0$ (if not, then $e(t/2, a^\bullet, \ldots) \equiv 0$ and $P_{a^\bullet}[\mathfrak{m} > t/2] = 0$ which is not possible). Given $(t, a^\bullet, b^\bullet)$ such that $e(t, a^\bullet, b^\bullet) = 0$, a second application of the CHAPMAN-KOLMOGOROV identity implies $e(t-s, a^\bullet, a^\bullet)\, e(s, a^\bullet, b^\bullet) = 0$ $(s < t)$, *i.e.*, $e(s, a^\bullet, b^\bullet) = 0$ $(s < t)$, and so $e(s, a^\bullet, b^\bullet) \equiv 0$ $(s > 0)$ as the LAPLACE transform of a (signed) measure. But for $t > 0$, $0 < e(t, ., .)$ near the diagonal $a^\bullet = b^\bullet$, and using the obvious HEINE-BOREL covering to select a chain $c_1^\bullet, c_2^\bullet$, *etc.* of $l < +\infty$ points leading from $a^\bullet$ to $b^\bullet$ such that $0 < e(t, c_1^\bullet, c_2^\bullet), e(t, c_2^\bullet, c_3^\bullet)$, *etc.*, a final application of the CHAPMAN-KOLMOGOROV identity implies $e(lt, a^\bullet, b^\bullet) > 0$, contradicting $e(., a^\bullet, b^\bullet) \equiv 0$ and completing the proof.]

Problem 7. Check that

$$t^{-1} G(t^{-1}, \xi, \xi) \geqq p(t, \xi, \xi) \qquad t > 0.$$

[$p(t, \xi, \xi)$ is convex in t, and so

$$\alpha \int_0^{+\infty} e^{-\alpha t} p(t, \xi, \xi)\, dt > p\,(\alpha \int_0^{+\infty} e^{-\alpha t}\, t\, dt, \xi, \xi) = p(\alpha^{-1}, \xi, \xi)].$$

Problem 8. D is transient if and only if

$$\lim_{\alpha \downarrow 0} G(a, b) = \int_0^{+\infty} p(t, a, b)\, dt < +\infty \qquad a, b \in Q^\bullet$$

(see problem 4.4.4 for the definition of transience).

[Because the scale distance between 2 points $a < b$ of $Q^\bullet$ is $< +\infty$,

$$\begin{aligned}
&\int_a^b \lim_{\alpha\downarrow 0} G(\xi, \xi)^{-1} s(d\xi) \\
&\quad = \lim_{\alpha\downarrow 0} \int_a^b \frac{B}{g_1 g_2} s(d\xi) \\
&\quad = \lim_{\alpha\downarrow 0} \int_a^b \left(\frac{g_1^+}{g_1} - \frac{g_2^+}{g_2}\right) s(d\xi) \\
&\quad = \lim_{\alpha\downarrow 0} \lg \frac{g_1(b)}{g_1(a)} \frac{g_2(a)}{g_2(b)} \\
&\quad = -\lg P_a(\mathfrak{m}_b < +\infty)\, P_b(\mathfrak{m}_a < +\infty);
\end{aligned}$$

now use the solution of problem 4.4.4 and the fact that the magnitude of the slope of G is $\leqq B^{-1} \times [g_1^+ g_2 - g_1 g_2^+] \leqq 1$.]

Problem 9. Compute $\lim_{\alpha \downarrow 0} G(a, b)$ in the transient case.

$$\Big[\lim_{\alpha\downarrow 0} G(a,b) = \lim_{\alpha\downarrow 0} G(b,a) = h_1(a)\, h_2(b) \qquad a \leqq b,$$

$$0 < h_1 \in \uparrow, \quad 0 < h_2 \in \downarrow, \quad h_1^+ h_2 - h_1 h_2^+ \equiv 1$$

$$(\mathfrak{G}^\bullet h_1)(\xi) = 0 \qquad (0 \leqq \xi < 1),$$

$$(\mathfrak{G}^\bullet h_2)(\xi) = 0 \qquad (0 < \xi \leqq 1).\Big]$$

Problem 10. Compute

$$E_\xi(\mathfrak{m}^n) \quad 0 < a < \xi < b < 1, \ \mathfrak{m} = \mathfrak{m}_a \wedge \mathfrak{m}_b$$

in the conservative case. Use the solution of problem 9.

[Consider the (transient) stopped diffusion

$$[x(t \wedge \mathfrak{m}) : t \geqq 0, P_\xi(B) : a \leqq \xi \leqq b]$$

with generator $\mathfrak{G}$ and GREEN operators $G_\alpha^\bullet f = E_\bullet\left[\int_0^{\mathfrak{m}} e^{-\alpha t} f(x)\, dt\right]$.

$u = 1 - E_\bullet(e^{-\alpha\mathfrak{m}})$ is a solution of

$$(\alpha - \mathfrak{G})\, u(\xi) = \alpha \quad (a < \xi < b), \quad u(a+) = u(b-) = 0,$$

and so $u = G_\alpha^\bullet 1$.

$G_\alpha^\bullet - G_\beta^\bullet + (\alpha - \beta)\, G_\alpha^\bullet G_\beta^\bullet = 0 \ (\alpha, \beta > 0)$, and using this, it develops that

$$\begin{aligned} E_\bullet(\mathfrak{m}^n) &= -\lim_{\alpha\downarrow 0} (-)^n \frac{\partial^n u}{\partial \alpha^n} \\ &= \lim_{\alpha\downarrow 0} n!\,[-\alpha\, G_\alpha^{\bullet n+1} 1 + G_\alpha^{\bullet n} 1] \\ &= n!\, G_{0+}^{\bullet n} 1, \end{aligned}$$

$G_{0+}^\bullet$ being the GREEN operator based on the GREEN function

$$\frac{[s(\xi) - s(a)]\,[s(b) - s(\eta)]}{s(b) - s(a)} \quad a < \xi \leqq \eta < b.]$$

Problem 11. Given a persistent non-singular diffusion, if e is a non-negative BOREL measure defined on $Q^\bullet$ such that

$$\int e(da)\, P_a\big(x(t) \in B\big) = e(B) \qquad (t > 0, B \subset Q^\bullet),$$

then e is a constant multiple of m on $Q^\bullet$ (see G. MARUYAMA and H. TANAKA [*1*] for a different solution and additional information; see also 8.2).

[Because $\int e(da)\, p(t, a, b)\, m(db) = e(db)$ $(t > 0,\ db \subset Q^{\bullet})$, $f \equiv de/dm$ satisfies $G^{\bullet}_{\alpha} f = f/\alpha$. But then $\mathfrak{G}^{\bullet} f = 0$ on $Q^{\bullet}$, showing that $f^{+} =$ *constant*, and $f =$ *constant* follows from $f \geqq 0$ in case $s(0) = -\infty$ and $s(1) = +\infty$, from $f^{+}(0) = (\mathfrak{G}^{\bullet} f)(0)\, m(0)$ in case $s(0) > -\infty$, and from $-f^{-}(1) = (\mathfrak{G}^{\bullet} f)(1)\, m(1)$ in case $s(1) < +\infty$ (see problem 4.6.6).]

Problem 12. Given a conservative non-singular diffusion with $P_0[\mathfrak{m}_{0+} = 0] = 1$, if $p(t, b)$ is the passage time density

$$\frac{\partial}{\partial t} P_0[\mathfrak{m}_b < t] \text{ [see 32), 33), 34)], if } 0 < h(t) \in \uparrow (t > 0),$$

and if $\lim_{t \downarrow 0} P_0[\mathfrak{m}_{h(t)} \leqq t] = 0$, then $\int_{0+} p(t, h)\, dt < +\infty$ implies $P_0[x(t) < h(t),\ t \downarrow 0] = 1$ (see 1.8 for a special case). $\int_{0+} p(t, h)\, dt = +\infty$ *should entail* $P_0[x(t) \geqq h(t)$ *i. o.*, $t \downarrow 0] = 1$, *but the proof eludes us.*

4.12. Kolmogorov's test

KOLMOGOROV's test for the 1-dimensional BROWNian motion (see 1.9) will now be established using the elegant method of M. MOTOO [7].

Consider a persistent non-singular diffusion on $Q = [0, +\infty)$ with scale $-\infty < s(0) < s(+\infty) = +\infty$, speed measure $m(Q) < +\infty$, sample paths $w: t \to x(t)$, and probabilities $P_a(B)$, and let $\mathfrak{e}_1 < \mathfrak{e}_2 <$ *etc.* be the successive passage times to $a > 0$ via $b > a$.

$m(Q) < +\infty$ implies

1) $$\gamma = E_a(\mathfrak{e}_1) < +\infty,$$

indeed, using 4.2.28) and problem 4.5.1

2) $$\begin{aligned} E_a(\mathfrak{e}_1) &= E_a(\mathfrak{m}_b) + E_b(\mathfrak{m}_a) \\ &= \int^{b} s[a \vee \xi, b)\, m(d\xi) + \int_{a} s(a, b \wedge \xi]\, m(d\xi) \\ &= [s(b) - s(a)]\, m(Q) < +\infty. \end{aligned}$$

Because the excursions $x(t): \mathfrak{e}_{n-1} \leqq t < \mathfrak{e}_n$ are independent and identical in law, the same is true of the differences $\mathfrak{e}_n - \mathfrak{e}_{n-1}$ $(n \geqq 2)$, and an application of the strong law of large numbers implies

3a) $$P_a\left[\lim_{n \uparrow +\infty} n^{-1} \mathfrak{e}_n = \gamma\right] = 1,$$

or, what interests us,

3b) $$P_a\left[\frac{n}{2}\gamma < \mathfrak{e}_{n-1} < \mathfrak{e}_n < 2n\gamma,\ n \uparrow +\infty\right] = 1.$$

Given $h = h(t)$ increasing to $+\infty$ with t, it follows that

$$4)\qquad (w: \mathfrak{x}_n > h(2n\gamma)\ i.o.) \quad \mathfrak{x}_n = \max(x(t): \mathfrak{e}_{n-1} \leqq t < \mathfrak{e}_n)$$
$$\subset (w: \mathfrak{x}_n > h(\mathfrak{e}_n)\ i.o.)$$
$$\subset (w: x(t) > h(t)\ i.o.,\quad t \uparrow +\infty)$$
$$\subset (w: \mathfrak{x}_n > h(\mathfrak{e}_{n-1})\quad i.o.)$$
$$\subset \left(w: \mathfrak{x}_n > h\left(\frac{n}{2}\gamma\right)\quad i.o.\right),$$

and using the probabilistic significance of the scale s to check

$$5)\qquad P_a[\mathfrak{x}_n > h(n\gamma)]$$
$$= P_b[\mathfrak{m}_{h(n\gamma)} < \mathfrak{m}_a] \qquad h(n\gamma) > b$$
$$= \frac{s(b) - s(a)}{s(h(n\gamma)) - s(a)}$$
$$\sim \text{constant} \times s[h(n\gamma)]^{-1} \qquad n \uparrow +\infty,$$

the independence of the $\mathfrak{x}_n$ coupled with the BOREL-CANTELLI lemmas implies the alternative

$$6)\qquad P_a[x(t) > h(t)\quad i.o.,\quad t \uparrow +\infty] = 0 \text{ or } 1$$
$$\textit{according as } \int^{+\infty} s(h)^{-1}\,dt \quad \textit{converges or diverges.}$$

Consider now the radial (BESSEL) part $\sqrt{x_1(t)^2 + \cdots + x_d(t)^2}: t \geqq 0$ of the standard d-dimensional BROWNian motion of 2.7, let R be its standard description with sample paths $w: t \to r(t) \in [0, +\infty)$, BOREL algebra B, and probabilities $P_\bullet(B)$, and let us show that 6) can be applied to the motion $\mathsf{R}^\bullet \equiv [r^\bullet(t) = e^{-t} r^2(e^t - 1): t \geqq 0, P_\bullet]$.

Given a MARKOV time $\mathfrak{m}^\bullet$ for $\mathsf{R}^\bullet$, $\mathfrak{m} = e^{\mathfrak{m}^\bullet} - 1$ is a MARKOV time for R, and introducing the BOREL algebra

$$7)\quad \mathsf{B}^\bullet_{\mathfrak{m}^\bullet +} = (B^\bullet: B^\bullet \cap (\mathfrak{m}^\bullet < s) \in \mathsf{B}[r^\bullet(t), t \leqq s] \textit{ for each } s \geqq 0)$$
$$= \mathsf{B}_{\mathfrak{m}+} = (B: B \cap (\mathfrak{m} < s) \in \mathsf{B}[r(t), t \leqq s] \textit{ for each } s \geqq 0)$$

and using the fact that the scaling $r(t) \to \sqrt{s}\, r(t/s)$ $(s > 0)$ maps $[r, P_a]$ into $[r, P_{\sqrt{s}a}]$, it is found that

$$8)\qquad P_\bullet(r^\bullet(t + \mathfrak{m}^\bullet) \leqq b \mid \mathsf{B}^\bullet_{\mathfrak{m}^\bullet +})$$
$$= P_\bullet(e^{-(t+\mathfrak{m}^\bullet)} r^2(e^{\mathfrak{m}^\bullet}(e^t - 1) + \mathfrak{m}) \leqq b \mid \mathsf{B}_{\mathfrak{m}+})$$
$$= P_a(e^{-t} s\, r^2((e_t - 1)/s) \leqq b) \qquad s = e^{-\mathfrak{m}^\bullet}, \quad a = r(\mathfrak{m})$$
$$= P_{\sqrt{s}\,a}(e^{-t} r^2(e^t - 1) \leqq b)$$
$$= P_{\sqrt{r^\bullet(\mathfrak{m}^\bullet)}}(r^\bullet(t) \leqq b);$$

in brief, $\mathsf{R}^\bullet$ *is a conservative non-singular diffusion.*

Computing its generator $\mathfrak{G}^\bullet$, if $p(t, a, b)$ is the BESSEL transition density relative to the speed measure $2b^{d-1}\,db$ [see 2.7.3) and 4)], then

9) $$P_a(r^\bullet(t) \in db) = p(e^t - 1, \sqrt{a}, e^{t/2}\sqrt{b})\, e^{dt/2}\, b^{(d/2)-1}\, db$$
$$= p(1 - e^{-t}, e^{-t/2}\sqrt{a}, \sqrt{b})\, b^{(d/2)-1}\, db$$
$$(t, a, b) \in (0, +\infty) \times [0, +\infty)^2,$$

and since

10) $$\left[\frac{\partial}{\partial t} - 2a\frac{\partial^2}{\partial a^2} - (d-a)\frac{\partial}{\partial a}\right] p(1 - e^{-t}, e^{-t/2}\sqrt{a}, \sqrt{b})$$
$$= e^{-t}\left[p_1 - \frac{1}{2}p_{22} - \frac{d-1}{2e^{-t/2}\sqrt{a}}p_2\right] = 0$$
$$(t, a, b) \in (0, +\infty)^3$$

and $P_0[r^\bullet(t) > 0] = 1$ for $t > 0$, it appears that $\mathfrak{G}^\bullet$ is the differential operator based on the *scale*

11a) $$s(b) - s(a) = \int_a^b \xi^{-d/2}\, e^{\xi/2}\, d\xi \qquad a < b$$

and the *speed measure*

11b) $$m[a, b) = \frac{1}{2}\int_a^b \xi^{(d/2)-1}\, e^{-\xi/2}\, d\xi \qquad 0 < a < b$$
$$m(0) = 0$$

with an *entrance* barrier at 0 and

12) $$u^+(0) = 0 \qquad u \in D(\mathfrak{G}^\bullet).$$

Because $s(+\infty) = +\infty$ and $m(Q) < +\infty$, 6) can be applied to $\mathsf{R}^\bullet$ and translating 6) into BESSEL language and noting the estimate $s(b) \sim 2b^{-(d/2)}\, e^{b/2}$ $(b \uparrow +\infty)$, it develops that if $h(t) \uparrow +\infty$ with t and if $a > 0$, then

13) $$P_a[r(t) > \sqrt{t+1}\, h(t+1) \quad i.o., \quad t \uparrow +\infty]$$
$$= P_a[r^\bullet(t) > h(e^t)^2 \quad i.o., \quad t \uparrow +\infty] = 0 \text{ or } 1$$
according as $\int^{+\infty} h(t)^d\, e^{-h(t)^2/2}\, \frac{dt}{t}$ *converges or diverges,*

or, what interests us, that

14) $$P_0[r(t) > \sqrt{t}\, h(t) \quad i.o., \quad t \uparrow +\infty]$$
$$= \int_0^{+\infty} P_0(r(1) \in d\xi)\, P_\xi[r(t) > \sqrt{t+1}\, h(t+1) \quad i.o. \quad t \uparrow +\infty]$$
$$= 0 \text{ or } 1 \text{ according as } \int^{\infty} h^d\, e^{-h^2/2}\, \frac{dt}{t} \text{ converges or diverges.}$$

When $d = 1$, this is KOLMOGOROV's test, and when $d \geqq 2$, it is the DVORETSKY-ERDÖS test (see A. DVORETSKY and P. ERDÖS [*1*]).

The same method shows that if $0 < h(t) \downarrow 0$ as $t \uparrow +\infty$, then

15) $P_0[r(t) < \sqrt{t}\, h(t)$ *i.o.,* $t \uparrow +\infty] = 0$ *or* 1 *according as*

$$\left.\begin{array}{ll} \int^{+\infty} |\lg h|^{-1} \frac{dt}{t} & (d=2) \\ \int^{+\infty} h^{d-2} \frac{dt}{t} & (d \geqq 3) \end{array}\right\} \textit{converges or diverges.}$$

Because P_0 is unchanged by the transformation $x(t) \to t x(1/t)$, it follows from 15) that if $0 < h(t) \downarrow 0$ as $t \downarrow 0$, then

16) $P_0[r(t) < \sqrt{t}\, h(t)$ *i.o.* $t \downarrow 0] = 0$ *or* 1 *according as*

$$\left.\begin{array}{ll} \int_{0+} |\lg h|^{-1} \frac{dt}{t} & (d=2) \\ \int_{0+} h^{d-2} \frac{dt}{t} & (d \geqq 3) \end{array}\right\} \textit{converges or diverges}$$

(see A. Dvoretsky and P. Erdös [*1*] for the case $d \geqq 3$ and F. Spitzer [*1*] for the case $d = 2$; another method leading to a cruder result will be found in problem 7.11.2).

Problem 1. Give a complete proof of 15).

5. Time changes and killing

5.1 Construction of sample paths: a general view

Given a 1-dimensional diffusion, singular or not, a complete description of its generator $\mathfrak{G}$ is now before us. $\mathfrak{G}$ coincides on its domain with a differential operator $\mathfrak{G}^\bullet$ of degree $\leqq 2$ expressed in terms of *invariants* (scale, speed measure, *etc.*) via the formulas 4.1.8) [or 4.1.31,32,33)] and each invariant has a simple probabilistic meaning embodied in the formulas 4.1.7) [or 4.1.22, 23b, 23c, and 26)].

But the actual situation is even better, *to wit*, each choice of invariants specifies a differential operator $\mathfrak{G}^\bullet$ via the formulas 4.1.8) [or 4.1.31, 32, and 33)], and $\mathfrak{G}^\bullet$ generates a non-singular (or singular) diffusion, the original invariants being expressible in terms of this diffusion via the formulas 4.1.7) [or 4.1.22, 23b, 23c, and 26)]; in brief, *a perfect correspondence is obtained between differential operators* $\mathfrak{G}^\bullet$ *on the one hand and non-singular or singular diffusions on the other.*

The proof occupies the rest of the present chapter, but first a rough explanation of the leading ideas (time substitution, annihilation, shunting) in series of special but typical examples.

Beginning with the simplest example, if D is a standard Brownian motion with sample paths $t \to x(t)$ and if $\sigma > 0$ is constant, then a non-standard description of the diffusion associated with $\mathfrak{G}^{\bullet} u = \frac{1}{2}\sigma^2 u''$ is the scaled standard BROWNian motion $x^{\bullet} = \sigma x$, or, what is the same in distribution, $x^{\bullet}(t) = x(\sigma^2 t) : t \geqq 0$.

Now suppose $0 < \sigma$ is continuous but not constant.

Using the first prescription, the suggestion is that $x^{\bullet}$ will be like $\sigma(\xi)\, x$ while it is near $x^{\bullet} = \xi$, *i.e.*,

1) $$x^{\bullet}(t) = a + \int_0^t \sigma[x^{\bullet}(s)]\, d\overset{\bullet}{x} \qquad a = x^{\bullet}(0);$$

this is the idea of K. ITÔ [2]. But to deal with the general non-singular diffusion, the second prescription is better; this suggests that $x^{\bullet}$ is a standard BROWNian motion run with a new clock $t^{\bullet} = t^{\bullet}(t)$ that grows like $\sigma^2(\xi)\, t$ while $x^{\bullet}$ is near ξ, *i. e.*,

2a) $$dt^{\bullet} = \sigma^2(x^{\bullet})\, dt,$$

or, what is the same,

2b) $$dt = \sigma^{-2}(x^{\bullet})\, dt^{\bullet} = \sigma^{-2}[x(t^{\bullet})]\, dt^{\bullet}.$$

2b) states that $t^{\bullet}$ is the inverse function $\mathfrak{f}^{-1}$ of the additive functional $\mathfrak{f}(t) = \int_0^t \sigma^{-2}[x(s)]\, ds$, and noting that $\int_0^t \sigma^{-2}(x)\, ds = \int \mathfrak{t}(t, \xi)\, m(d\xi)$, where $\mathfrak{t}$ is the standard BROWNian local time and $m(d\xi) = 2\sigma^{-2}(\xi)\, d\xi$ is the speed measure of $\mathfrak{G}^{\bullet}$, it is now just a small jump to conjecture that a (non-standard) description of the diffusion on $Q = R^1$ associated with

3) $$\mathfrak{G}^{\bullet} u = \frac{u^+(d\xi)}{m(d\xi)}$$

is

4a) $$x^{\bullet}(t) = x(\mathfrak{f}^{-1})$$

with

4b) $$\mathfrak{f} = \int \mathfrak{t}\, dm;$$

see 5.2 for the proof.

Because the *time substitution* $t \to \mathfrak{f}^{-1}$ depends upon the BROWNian path, it is often called a *stochastic clock*; the idea occurs in G. HUNT [2(2)] and a hint of it is contained in P. LÉVY [3: 276]. H. TROTTER suggested the possibility to us. VOLKONSKIĬ [1] obtained such time substitutions but did not find the local time integral for f; see also 8.3.

A second instructive example is that of the non-singular diffusion on $[0, +\infty)$ with generator

5a) $$\mathfrak{G}^{\bullet} u = \frac{u^+(d\xi)}{m(d\xi)}$$

5b) $$m(0)\, (\mathfrak{G}^{\bullet} u)(0) = u^+(0);$$

the method of time substitution still applies with

6) $$\mathfrak{f} = \int_{0-} \mathfrak{t}(t, \xi)\, m(d\xi),$$

the time substitution of 2.11 for the reflecting BROWNian motion being a special case.

Now let us explain how to construct the non-singular diffusion with *killing* associated with

7) $$\mathfrak{G}^{\bullet} u = \frac{u^{+}(d\xi) - u(\xi)\, k(d\xi)}{m(d\xi)}$$

on $Q = R^1$. Suppose to begin with that k has a constant density $\varkappa \geqq 0$ relative to m so that $\mathfrak{G}^{\bullet} = \mathfrak{G} - \varkappa$ with $\mathfrak{G}$ as in 3); then $x^{\bullet}$ is just the motion x associated with $\mathfrak{G}$, killed (*i.e.*, sent off to the extra state ∞) after an exponential holding time $\mathfrak{m}_{\infty}$ with distribution

8) $$P_{\bullet}[\mathfrak{m}_{\infty} > t \mid x] = e^{-\varkappa t};$$

this suggests that if $\varkappa$ is continuous but not constant, then the chance that the dot particle, moving along the undotted path should not be annihilated before time t ought to be

9) $$\bigcap_{s \leqq t} [1 - \varkappa[x(s)]\, ds] = e^{-\int_0^t \varkappa[x(s)]\, ds} \text{ *}.$$

The time substitution 4) permits us to conclude that the local times

10) $$\mathfrak{t}(t, a) = \lim_{b \downarrow a} \frac{\text{measure}(s : a \leqq x(s) < b,\, s \leqq t)}{m[a, b)}$$

exist (5.4), so 9) can be expressed by means of a local time integral:

11) $$\int_0^t \varkappa[x(s)]\, ds = \int \mathfrak{t}(t, \xi)\, k(d\xi),$$

and using this formula, the suggestion is that for a general killing measure $k(db)$, the dot motion is identical in law to the undotted motion, annihilated (sent off to ∞) at time $\mathfrak{m}_{\infty}$ with conditional law

12) $$P_{\bullet}[\mathfrak{m}_{\infty} > t \mid x] = e^{-\int \mathfrak{t}\, dk};$$

the proof occupies 5.6. As the reader will easily believe, the relation between the motion on $[0, +\infty)$ with generator 5) and the motion associated with

13 a) $$\mathfrak{G}^{\bullet} u = \frac{u^{+}(d\xi) - u(\xi)\, k(d\xi)}{m(d\xi)}$$

13 b) $$m(0)\, (\mathfrak{G}^{\bullet} u)\, (0) = u^{+}(0) - u(0)\, k(0)$$

* $\bigcap$ is meant to suggest a continuous product.

is just the same, except that the local time integral $\int_{0-} \mathfrak{t}\, dk$ is used in 12) (see 5.6; the elastic BROWNian motion of 2.3 is the simplest example of this).

As an example exhibiting all the essential features of the conservative case, let $\mathfrak{G}^\bullet$ be the differential operator

$$
\begin{aligned}
14)\qquad (\mathfrak{G}^\bullet u)(a) &= \tfrac{1}{2} u''(a) && a \in Q \\
&= \lim_{b \downarrow a} \frac{u(b) - u(a)}{s(b) - s(a)} && a \in [0,1) - Q \\
&= 0 && a = 1,
\end{aligned}
$$

$Q = \bigcup_{n \geq 1} (l_n, r_n)$ being the intervals of the complement of the standard CANTOR set K and s_+ the *shunt scale*

$$
15)\qquad s_+(b) = \sum_{l_n \leq b} (r_n \wedge b - l_n) + s_+([0, b) \cap K),
$$

with the understanding that

$$
16a)\qquad u^+(l_n) = 0 \qquad (n \geq 1)
$$

$$
16b)\qquad \mathfrak{G}^\bullet u \in C[0, 1] \qquad u \in D(\mathfrak{G}^\bullet).
$$

$x^\bullet$ can be built up out of BROWNian motions on $[l_n, r_n)$ with reflection at l_n, hitched end to end, with a lag during the passage of the CANTOR set so that for paths starting at $0 \leq a \leq 1$,

$$
17)\qquad s_+([a, b) \cap K) = \text{measure}(s : x^\bullet(s) \in K, s \leq t) \qquad b = x^\bullet(t), t \leq \mathfrak{m}_1^\bullet.
$$

The simplest method of doing this is to begin with a reflecting BROWNian motion x on $[0, +\infty)$ stopped at 1 and to perform the time substitution $t \to \mathfrak{f}^{-1}$ with

$$
\begin{aligned}
18)\qquad \mathfrak{f}(t) &= \sum_{a < r_n \leq b} \text{measure}(s : l_n < x(s) < r_n, s \leq \mathfrak{m}_{r_n}) + s_+([a, b) \cap K) \\
a &= x(0), \quad b = a \vee x(t), \quad \mathfrak{m}_{r_n} = \min(t : x = r_n);
\end{aligned}
$$

the proof will be found in 5.10 and problems 5.10.1 and 2; see also 5.11 for a discussion of annihilation on the shunts.

Creation instead of annihilation is introduced to obtain a sample path model of $\mathfrak{G}^\bullet = (1/2) D^2 + \varkappa$ $(\varkappa > 0)$; this model occupies 5.12 bis 15.

5.2. Time changes: $Q = R^1$

Given a standard BROWNian motion on $Q = R^1$ with scale $s(db) = db$ and local times $\mathfrak{t} = \mathfrak{t}(t, b)$, let $m(db)$ be a speed measure on R^1 and let us prove that if $\mathfrak{f}^{-1}$ is the inverse function of

$$
1)\qquad \mathfrak{f}(t) = \mathfrak{f}(t, w) = \int_{R^1} \mathfrak{t}(t, b)\, m(db) \qquad t \geq 0,
$$

then $[x(\mathfrak{f}^{-1}), P_\bullet]$ is a (non-standard) description of the conservative diffusion with the BROWNian scale and speed measure m, *i.e.*, with generator

2) $$\mathfrak{G}^\bullet u = \frac{u^+(db)}{m(db)}.$$

Because $\mathfrak{t}(t, b) \in C([0, +\infty) \times R^1)$ and vanishes outside

$$\min_{s \leqq t} x(s) < b < \max_{s \leqq t} x(s) \qquad (t > 0),$$

3a) $$\mathfrak{f}(t \pm) = \mathfrak{f}(t) < +\infty \qquad t \geqq 0,$$

esp.,

3b) $$\mathfrak{f}(0+) = 0.$$

Because $\mathfrak{t}(t_2, b) > \mathfrak{t}(t_1, b)$ $(t_2 > t_1)$ on a set of positive speed measure and $\mathfrak{t}(+\infty, .) \equiv +\infty$ (see problem 2.8.7),

4a) $$\mathfrak{f}(t_2) > \mathfrak{f}(t_1)$$

and

4b) $$\mathfrak{f}(+\infty) \equiv +\infty,$$

and it follows that $\mathfrak{f}^{-1}$ has the same properties, *esp.*, *$x(\mathfrak{f}^{-1})$ is continuous.*

Now let $w^\bullet : t \to x^\bullet(t)$, $\mathsf{B}^\bullet_t$, $\mathsf{B}^\bullet$, and $P^\bullet_a(B^\bullet) = P_a[x(\mathfrak{f}^{-1}) \in B^\bullet]$ be the sample paths, BOREL algebras, and probabilities for the standard description of the motion $\mathsf{D}^\bullet = [x(\mathfrak{f}^{-1}), P_\bullet]$, let $w^{-1} : t \to x^{-1}(t)$ denote the sample path $x(\mathfrak{f}^{-1})$, and let us prove that $\mathsf{D}^\bullet$ is *simple* MARKOV.

Given $B^\bullet \in \mathsf{B}^\bullet_t$ with indicator function $e^\bullet(w^\bullet)$, $e(w) \equiv e^\bullet(w^{-1})$ *is measurable* $\mathsf{B}_{\mathfrak{f}^{-1}(t)+}$.

Because of GALMARINO's lemma (see 3.2), it is enough to show that if

5a) $$\mathfrak{f}^{-1}(t, u) < s$$

5b) $$x(\theta, u) = x(\theta, v) \qquad \theta \leqq s,$$

then

6) $$e(u) = e(v).$$

But, if $\theta \leqq t$, then 5a) implies $\mathfrak{f}^{-1}(\theta, u) \leqq s$, 5b) then implies $\mathfrak{f}^{-1}(\theta, u) = \mathfrak{f}^{-1}(\theta, v)$ because $\mathfrak{f}^{-1}(\theta)$ is a MARKOV time, and the resulting

7) $$x(\theta, u^{-1}) = x(\mathfrak{f}^{-1}(\theta, u), u) = x(\mathfrak{f}^{-1}(\theta, v), s) = x(\theta, v^{-1}) \qquad \theta \leqq t,$$

coupled with the fact that $e^\bullet(w^\bullet)$ is the indicator of a member of $\mathsf{B}^\bullet_t$ implies

8) $$e(u) = e^\bullet(u^{-1}) = e^\bullet(v^{-1}) = e(v)$$

as desired.

But now it follows that, with $B^\bullet$, $e^\bullet$, and e as above,

$$\begin{aligned}
9)\quad & P^\bullet_\bullet[B^\bullet, x^\bullet(t+s) \in db] \\
&= E_\bullet[e^\bullet(w^{-1}), x^{-1}(t+s) \in db] \\
&= E_\bullet[e(w), x(\mathfrak{f}^{-1}(t+s)) \in db] \\
&= E_\bullet[e(w), x(\mathfrak{m} + \mathfrak{f}^{-1}(s, w^+_\mathfrak{m})) \in db] \qquad \mathfrak{m} = \mathfrak{f}^{-1}(t) \\
&= E_\bullet[e(w), P^\bullet(x[\mathfrak{f}^{-1}(s, w^+_\mathfrak{m}), w^+_\mathfrak{m}] \in db \mid \mathsf{B}_{\mathfrak{m}+})] \\
&= E_\bullet[e(w), P_{x(\mathfrak{m})}(x[\mathfrak{f}^{-1}(s)] \in db)] \\
&= E_\bullet[e^\bullet(w^{-1}), P^\bullet_{x^{-1}(t)}(x^\bullet(s) \in db)] \\
&= E^\bullet_\bullet[B^\bullet, P^\bullet_{x^\bullet(t)}(x^\bullet(s) \in db)],
\end{aligned}$$

and that finishes the proof of the simple MARKOV character.

As to the *strict* MARKOVian character, it is now enough (see 3.6) to show that the GREEN operators

$$10)\qquad G^\bullet_\alpha f = E^\bullet_\bullet\left[\int_0^{+\infty} e^{-\alpha t} f(x^\bullet)\, dt\right] = E_\bullet\left[\int_0^{+\infty} e^{-\alpha t} f(x^{-1})\, dt\right]$$

map $C(R^1)$ into itself (see problems 2 and 3 below for an alternative proof).

Given $a < b$, if $u = G^\bullet_\alpha f$ for $f \in C(R^1)$ and if $\mathfrak{m} = \mathfrak{m}_b$, then

$$11)\qquad \mathfrak{f}^{-1}(t + \mathfrak{f}(\mathfrak{m})) = \mathfrak{m} + \mathfrak{f}^{-1}(t, w^+_\mathfrak{m}),$$

$$\begin{aligned}
12)\qquad u(a) &= E_a\left[\int_0^{\mathfrak{f}(\mathfrak{m})} e^{-\alpha t} f(x^{-1})\, dt\right] \\
&\quad + E_a\left[e^{-\alpha \mathfrak{f}(\mathfrak{m})} \int_0^{+\infty} e^{-\alpha t} f[x(\mathfrak{f}^{-1}(t, w^+_\mathfrak{m}), w^+_\mathfrak{m})]\, dt\right] \\
&= E_a\left[\int_0^{\mathfrak{f}(\mathfrak{m})} e^{-\alpha t} f(x^{-1})\, dt\right] + E_a[e^{-\alpha \mathfrak{f}(\mathfrak{m})}]\, u(b),
\end{aligned}$$

and letting $b \downarrow a$ in 12), $P_\bullet\left[\lim_{b \downarrow a} \mathfrak{f}(\mathfrak{m}) = \mathfrak{f}(0+) = 0\right] \equiv 1$ implies $u(a+) = u(a)$; the proof of $u(a-) = u(a)$ is similar.

$\mathsf{D}^\bullet = [x(\mathfrak{f}^{-1}), P_\bullet]$ is now seen to be a *conservative diffusion*, and using

$$13\text{a})\qquad \mathfrak{m}^\bullet_b(w^{-1}) = \min(t : x(\mathfrak{f}^{-1}) = b) = \mathfrak{f}(\mathfrak{m}_b)$$

together with the evaluation

$$\begin{aligned}
13\text{b})\qquad & E_\xi[\mathfrak{t}(\mathfrak{m}_a \wedge \mathfrak{m}_b, \eta)] = G(\xi, \eta) \qquad a < \xi, \eta < b \\
& G(\xi, \eta) = G(\eta, \xi) = \frac{(\xi - a)(b - \eta)}{b - a} \qquad \xi \leqq \eta
\end{aligned}$$

(see problem 1) to compute its hitting probabilities and mean exit times

14a)
$$p^\bullet_{ba}(\xi) = P^\bullet_\xi(\mathfrak{m}^\bullet_b < \mathfrak{m}^\bullet_a) = P_\xi[\mathfrak{f}(\mathfrak{m}_b) < \mathfrak{f}(\mathfrak{m}_a)]$$
$$= P_\xi(\mathfrak{m}_b < \mathfrak{m}_a) = \frac{\xi - a}{b - a} \qquad a < \xi < b$$

14b)
$$e^\bullet_{ab}(\xi) = E^\bullet_\xi(\mathfrak{m}^\bullet_a \wedge \mathfrak{m}^\bullet_b) = E_\xi[\mathfrak{f}(\mathfrak{m}_a) \wedge \mathfrak{f}(\mathfrak{m}_b)]$$
$$= E_\xi[\mathfrak{f}(\mathfrak{m}_a \wedge \mathfrak{m}_b)] = E_\xi\left[\int_a^b \mathfrak{t}(\mathfrak{m}_a \wedge \mathfrak{m}_b, \eta)\, m(d\eta)\right]$$
$$= \int_a^b G(\xi, \eta)\, m(d\eta) \qquad a < \xi < b,$$

it appears that *its scale is a positive multiple of the Brownian scale and its speed measure the reciprocal multiple of m, as desired.*

Problem 1. Use the formula for the standard BROWNian local time $\mathfrak{t}$

$$E_\xi\left[\int_0^{+\infty} e^{-\alpha t}\, \mathfrak{t}(dt, \eta)\right] = \frac{e^{-\sqrt{2\alpha}|\xi - \eta|}}{2\sqrt{2\alpha}} \qquad \alpha > 0$$

to give a direct proof of 13b).

$$[E_\xi[\mathfrak{t}(\mathfrak{m}_a \wedge \mathfrak{m}_b, \eta)]$$
$$= \lim_{\alpha \downarrow 0} E_\xi\left[\int_0^{\mathfrak{m}} e^{-\alpha t}\, \mathfrak{t}(dt, \eta)\right] \qquad \mathfrak{m} = \mathfrak{m}_a \wedge \mathfrak{m}_b$$
$$= \lim_{\alpha \downarrow 0} E_\xi\left[\int_0^{+\infty} e^{-\alpha t}\, \mathfrak{t}(dt, \eta) - e^{-\alpha \mathfrak{m}} \int_0^{+\infty} e^{-\alpha t}\, \mathfrak{t}(dt, \eta, w^+_{\mathfrak{m}})\right]$$
$$= \lim_{\alpha \downarrow 0} E_\xi\left[\frac{e^{-\sqrt{2\alpha}|\xi - \eta|} - e^{-\alpha \mathfrak{m}}\, e^{-\sqrt{2\alpha}|x(\mathfrak{m}) - \eta|}}{2\sqrt{2\alpha}}\right]$$
$$= \frac{1}{2} E_\xi[|x(\mathfrak{m}) - \eta| - |\xi - \eta|]$$
$$= \frac{1}{2}\left[|a - \eta| \frac{b - \xi}{b - a} + |b - \eta| \frac{\xi - a}{b - a} - |\xi - \eta|\right]$$
$$= G(\xi, \eta).]$$

Problem 2. Prove that if $\mathfrak{m}^\bullet$ $(w^\bullet)$ is a MARKOV time for the standard description of $x(\mathfrak{f}^{-1})$, then $\mathfrak{m}(w) = \mathfrak{f}^{-1}(\mathfrak{m}^\bullet(w^{-1}), w)$ is a MARKOV time for the standard BROWNian motion, and that if $B^\bullet \in \mathbf{B}^\bullet_{\mathfrak{m}^\bullet +}$, *i.e.*, if $B^\bullet \cap (\mathfrak{m}^\bullet < t) \in \mathbf{B}^\bullet$ $(t \geqq 0)$, then $B \equiv (w : w^{-1} \in B^\bullet) \in \mathbf{B}_{\mathfrak{m}+}$, *i.e.*, $B \cap (\mathfrak{m} < t) \in \mathbf{B}_t$ $(t \geqq 0)$ (use GALMARINO's theorem; see 3.2).

Problem 3. Give a direct proof of the strict MARKOVian character of $x(\mathfrak{f}^{-1})$ using the standard description, the result of problem 2, and the method used in 9).

5.3. Time changes: $Q = [0, +\infty)$

Given a speed measure $m(db)$ on $Q \subset R^1$ and standard BROWNian local times $\mathfrak{t}$, the statement of 5.1 was that if $\mathfrak{f}^{-1}$ is the inverse function of $\mathfrak{f} = \int \mathfrak{t}\, dm$, then the time change $t \to \mathfrak{f}^{-1}$ sends the standard BROWNian motion into a (non-standard) description of the diffusion $\mathrm{D}^\bullet$ with the BROWNian scale and speed measure m, *i.e.*, with generator

1) $$\mathfrak{G}^\bullet = \frac{u^+(db)}{m(db)}.$$

The case $Q = R^1$ was justified; the cases

2a) $$Q = [0, +\infty), \quad \int_{0+}^{1} \xi\, dm = +\infty$$

2b) $$Q = [0, +\infty), \quad m(0, 1) = +\infty, \quad \int_{0+}^{1} \xi\, dm < +\infty$$

2c) $$Q = [0, +\infty), \quad m(0, 1) < +\infty$$

will be treated here; the case $Q = [0, 1]$ is left to the reader. In case 2a), 0 is ***neither exit nor entrance.*** In case 2b), 0 is ***exit but not entrance.*** In case 2c), 0 is ***exit and entrance***, and two possibilities have to be distinguished according as

3a) $$(\mathfrak{G}^\bullet u)(0) = 0$$

or

3b) $$m(0)\, (\mathfrak{G}^\bullet u)(0) = u^+(0) \qquad 0 \leq m(0) < +\infty.$$

The precise statement of our result is as follows: *if*

4) $$m(0) = +\infty \qquad \textit{in case } 3\text{a}),$$

if

5a) $$\mathfrak{f}(t) = \int_{0+} \mathfrak{t}(t, \xi)\, m(d\xi) \quad \textit{in case } 2\text{a}) \textit{ and } 2\text{b})$$

5b) $$\mathfrak{f}(t) = \int_{0-} \mathfrak{t}(t, \xi)\, m(d\xi) \quad \textit{in case } 2\text{c}) = 3)$$

with the convention $0 \times +\infty = 0$, *and if*

6) $$\mathfrak{f}^{-1}(t) = \max(s : \mathfrak{f}(s) = t)$$

then the motion $[x(\mathfrak{f}^{-1}), P_a : a \geq 0]$ *is a (non-standard) description of the diffusion* $\mathrm{D}^\bullet$ *with generator* $\mathfrak{G}^\bullet$.

Given a BROWNian path starting in $[0, +\infty)$, let us have a good look at $\mathfrak{f} = \int_0^{+\infty} \mathfrak{t}\, dm$.

Granting

7a) $P_l[\mathfrak{f}(\mathfrak{m}_0) < +\infty] = 0$ *or* 1 *according as*

$$\int_0^1 \xi\, dm \text{ diverges or converges} \qquad (l > 0)$$

and

7b) $P_l[\mathfrak{f}(\mathfrak{m}_1) < +\infty] = 0$ *or* 1 *according as*

$$m[0, 1) = +\infty \text{ or } m[0, 1) < +\infty \qquad (1 > l \geqq 0),$$

the principal features of $\mathfrak{f}$ and $\mathfrak{f}^{-1}(t)$ are as follows:

in case 2a), $\mathfrak{f}$ is continuous $(t < \mathfrak{m}_0)$, $\mathfrak{f}(t_1) < \mathfrak{f}(t_2)$ $(t_1 < t_2 < \mathfrak{m}_0)$, $\mathfrak{f}(\mathfrak{m}_0) = +\infty$, $\mathfrak{f}^{-1}$ is continuous $(t \geqq 0)$, $\mathfrak{f}(\mathfrak{f}^{-1}) = t$ $(t \geqq 0)$, $\mathfrak{f}^{-1} < \mathfrak{m}_0$ $(t \geqq 0)$, and $\mathfrak{f}^{-1}(+\infty) = \mathfrak{m}_0$ (see diagram 1);

in case 2b) *and* 3a) the same, except that $\mathfrak{f}(\mathfrak{m}_0) < +\infty$, $\mathfrak{f}(\mathfrak{m}_0+) = +\infty$, $\mathfrak{f}(\mathfrak{f}^{-1}) = t \wedge \mathfrak{f}(\mathfrak{m}_0)$, $\mathfrak{f}^{-1}(t) < \mathfrak{m}_0$ $(t < \mathfrak{f}(\mathfrak{m}_0))$, and $\mathfrak{f}^{-1}(t) = \mathfrak{m}_0$ $(t > \mathfrak{f}(\mathfrak{m}_0))$ (see diagram 2);

in case 3b), $\mathfrak{f}$ is continuous $(t \geqq 0)$, $\mathfrak{f}(t_1) < \mathfrak{f}(t_2)$ $(t_1 < t_2)$ on each Brownian excursion $x(t) : t \in \mathfrak{Z}_n^+$ $(\bigcup_{n \geqq 1} \mathfrak{Z}_n^+ \equiv (t : x_t > 0))$ leading into $(0, +\infty)$, $\mathfrak{f}$ is flat on each excursion $x(t) : t \in \mathfrak{Z}_n^-$ $(\bigcup_{n \geqq 1} \mathfrak{Z}_n^- \equiv (t : x_t < 0))$ leading into $(-\infty, 0)$, $\mathfrak{f}^{-1}(t) = \mathfrak{f}^{-1}(t+) < +\infty$ $(t \geqq 0)$, $\mathfrak{f}(\mathfrak{f}^{-1}) = t$ $(t \geqq 0)$, and $x(\mathfrak{f}^{-1})$ is continuous (see diagram 3).

7a) is proved as follows.

Given a Brownian path starting at $l > \xi > 0$, $\mathfrak{m}^\bullet = \mathfrak{f}(\mathfrak{m}_\varepsilon)$ is identical in law to the passage time to ε of a diffusion on R^1 with the

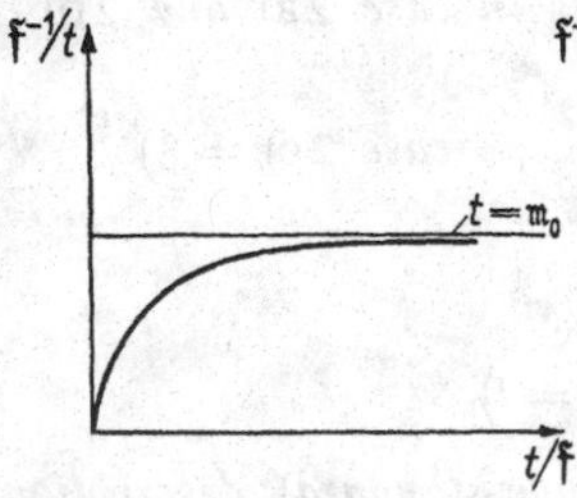

Diagram 1

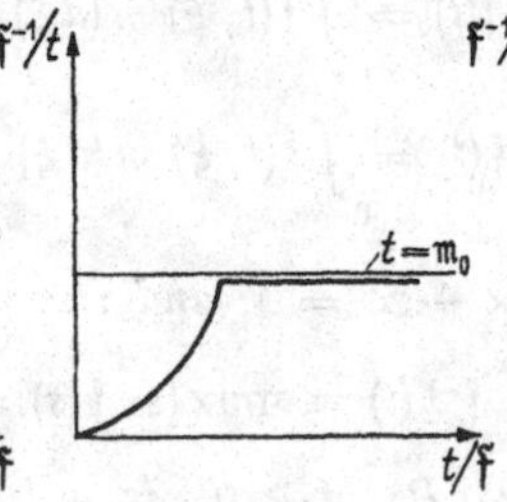

Diagram 2

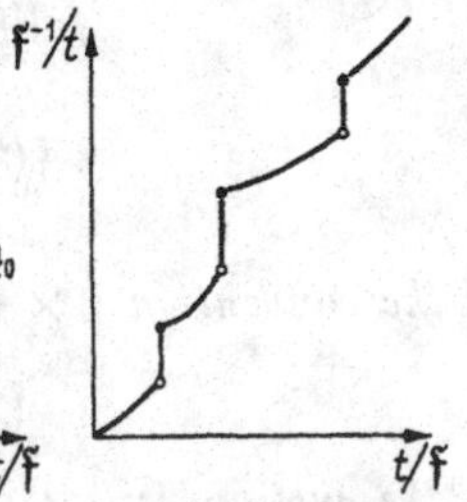

Diagram 3

Brownian scale and speed measure $m^\bullet = m$ on $[\varepsilon, +\infty)$ (see 5.2), so $E_l[e^{-\alpha \mathfrak{m}_\varepsilon^\bullet}] = \frac{g_2(l)}{g_2(\varepsilon)}$ where g_2 is the decreasing solution of $\frac{g^+(d\xi)}{m(d\xi)} = \alpha g(\xi)$

($\xi > 0$), and since

8) $$g_2(0+) \gtreqless +\infty \quad \textit{according as} \quad \int_0^1 \xi \, dm \gtreqless +\infty,$$

$P_l[\mathfrak{f}(\mathfrak{m}_0) = +\infty] = 1$ if $\int_0^1 \xi \, dm = +\infty$, while $P_l[\mathfrak{f}(\mathfrak{m}_0) < +\infty] = 1$ if $E_l[\int_0^l \mathfrak{t}(\mathfrak{m}_0, \xi) \, dm] = \int_0^l \xi \, dm < +\infty$.

7b) is immediate from the fact that $\mathfrak{t}(t, \cdot) > 0$ in the neighborhood of 0 for $t > \mathfrak{m}_0$.

Given $t \geqq 0$,

9) $$(w : \mathfrak{f}^{-1}(t) < s) = (w : t < \mathfrak{f}(s)) \qquad s \geqq 0$$

implies that $\mathfrak{m} = \mathfrak{f}^{-1}(t)$ *is a Markov time*; also,

10) $$\mathfrak{f}^{-1}(t + s) = \mathfrak{m} + \mathfrak{f}^{-1}(s, w_{\mathfrak{m}}^{+})$$

because, in cases 2b) and 3a), $t > \mathfrak{f}(\mathfrak{m}_0)$ $(< +\infty)$ implies

11a) $$\mathfrak{f}^{-1}(t + s) = \mathfrak{m}_0 = \mathfrak{f}^{-1}(t) = \mathfrak{m}$$

(see diagram 2) and

11b) $$\mathfrak{f}^{-1}(s, w_{\mathfrak{m}}^{+}) = \mathfrak{f}^{-1}(s, w_{\mathfrak{m}_0}^{'}) = 0,$$

while, in the other cases,

12a) $$\mathfrak{f}(\mathfrak{f}^{-1}(t)) = t$$

12b) $$\mathfrak{f}(s + \mathfrak{m}) = \mathfrak{f}(\mathfrak{m}) + \mathfrak{f}(s, w_{\mathfrak{m}}^{+}) = t + \mathfrak{f}(s, w_{\mathfrak{m}}^{+})$$

(see diagrams 1 and 3), and therefore

13) $$\begin{aligned} \mathfrak{f}^{-1}(t + s) &= \max(\theta : \mathfrak{f}(\theta) \leqq t + s) \\ &= \mathfrak{m} + \max(\theta : \mathfrak{f}(\theta + \mathfrak{m}) \leqq t + s) \\ &= \mathfrak{m} + \max(\theta : \mathfrak{f}(\theta, w_{\mathfrak{m}}^{+}) \leqq s) \\ &= \mathfrak{m} + \mathfrak{f}^{-1}(s, w_{\mathfrak{m}}^{+}) \end{aligned}$$

as desired.

$[x(\mathfrak{f}^{-1}), P_\bullet]$ can now be identified as a conservative non-singular diffusion on $[0, +\infty)$ and its (local) generator $\mathfrak{G}^\bullet$ can be evaluated on $(0, +\infty)$ just as in 5.2; the only novel point is the evaluation of the generator at 0, and for this, it is enough to note that $P_0[\mathfrak{f}^{-1} = x(\mathfrak{f}^{-1}) \equiv 0] = 1$ unless 3b) holds, in which case DYNKIN's formula

14) $(\mathfrak{G}^\bullet u)(0) = \lim_{\varepsilon \downarrow 0} E_0(\mathfrak{m}^\bullet)^{-1}[u(\varepsilon) - u(0)] \qquad \mathfrak{m}^\bullet = \max(t : x(\mathfrak{f}^{-1}) = \varepsilon)$

coupled with

15)
$$\begin{aligned} E_0(\mathfrak{m}^\bullet) &= E_0(\mathfrak{f}(\mathfrak{m}_\varepsilon)) \\ &= \int_{0-}^{\varepsilon} E_0[\mathfrak{t}(\mathfrak{m}_\varepsilon, b)]\, dm \\ &= \int_{0-}^{\varepsilon} (\varepsilon - b)\, dm \\ &= \int_0^{\varepsilon} m[0, b)\, db \\ &\sim \varepsilon\, m(0) \quad \textit{in case} \quad m(0) > 0 \\ &= o(\varepsilon) \quad \textit{in case} \quad m(0) = 0 \end{aligned}$$

implies the expected

16a) $$(\mathfrak{G}^\bullet u)(0) = \lim_{\varepsilon \downarrow 0} E_0(\mathfrak{m}^\bullet)^{-1}[u(\varepsilon) - u(0)] = \frac{u^+(0)}{m(0)} \qquad m(0) > 0$$

16b) $$u^+(0) = 0 \qquad m(0) = 0.$$

5.4. Local times

Consider the non-standard description

$$\mathsf{D}^\bullet = [x^\bullet = x(\mathfrak{f}^{-1}), P_\bullet] \qquad \mathfrak{f} = \int_Q \mathfrak{t}\, dm$$

of the general conservative non-singular diffusion on $Q = R^1$ or $[0, +\infty)$ or $[0, 1]$ with scale $s(b) - s(a) = b - a$, based on the standard BROWNian motion with sample paths $w : t \to x(t)$, local times $\mathfrak{t}$, and probabilities $P_a(B)$.

$\mathsf{D}^\bullet$ has a *local time*

1a) $$\mathfrak{t}^\bullet(t, b) = \frac{\text{measure}(s : x^\bullet(s) \in db,\ s \leqq t)}{m(db)}$$

1b) $$\mathfrak{t}^\bullet(t, b) = \mathfrak{t}(\mathfrak{f}^{-1}(t), b)$$

at each of the non-trap points of Q; indeed, if $f \in C(Q)$ and if t is smaller than the dot passage time to the traps, *i.e.*, if $\mathfrak{f}^{-1}(t)$ is smaller than the standard BROWNian passage time to the traps $\mathfrak{f}^{-1}(+\infty)$, then

2) $$\int_0^t f(x^\bullet(s))\,ds$$
$$= \int_0^t f(x(\mathfrak{f}^{-1}))\,ds$$
$$= \int_0^{\mathfrak{f}^{-1}(t)} f((x(s))\,\mathfrak{f}(ds)$$
$$= \lim_{n\uparrow+\infty} \sum_{k2^{-n}<\mathfrak{f}^{-1}(t)} f(x(k2^{-n}))\,\mathfrak{f}[(k-1)\,2^{-n}-, k2^{-n})$$
$$= \lim_{n\uparrow+\infty} \int_Q \sum_{k2^{-n}<\mathfrak{f}^{-1}(t)} f(x(k2^{-n}))\,\mathfrak{t}([(k-1)\,2^{-n}, k2^{-n}), b)\,m(db)$$
$$= \int_Q \int_0^{\mathfrak{f}^{-1}(t)} f(x(s))\,\mathfrak{t}(ds, b)\,m(db)$$
$$= \int_Q \mathfrak{t}(\mathfrak{f}^{-1}(t), b)\,f(b)\,m(db)$$

because $\mathfrak{t}(ds, b)$ is flat off the visiting set $(s\colon x(s) = b)$, and thinking of the original integral as

$$\int_Q \text{measure}\,(s : x^\bullet(s) \in db, s \leqq t)\,f(b),$$

1) follows.

$\mathfrak{t}^\bullet$ can be used to express the transition densities $p^\bullet(t, a, b)$ of 4.11:

3) $$\frac{\partial}{\partial t} E_a[\mathfrak{t}^\bullet(t, b)] = p^\bullet(t, a, b) \qquad (t, a, b) \in (0, +\infty) \times Q^\bullet \times Q^\bullet,$$

as will now be proved.

Given points $a < b$ of $Q^\bullet$,

4a) $$\gamma \equiv E_a\left[\int_0^{+\infty} e^{-\alpha t}\,\mathfrak{t}^\bullet(dt, a)\right]$$
$$= E_a\left[\int_0^{\mathfrak{m}_b^\bullet} e^{-\alpha t}\,\mathfrak{t}^\bullet(dt, a)\right] + E_a(e^{-\alpha\mathfrak{m}_b^\bullet})\,E_b(e^{-\alpha\mathfrak{m}_a^\bullet})\,\gamma,$$

i.e.,

4b) $$\gamma = \frac{E_a\left[\int_0^{\mathfrak{m}_b^\bullet} e^{-\alpha t}\,\mathfrak{t}^\bullet(dt, a)\right]}{1 - E_a\left(e^{-\alpha\mathfrak{m}_b^\bullet}\right) E_b\left(e^{-\alpha\mathfrak{m}_a^\bullet}\right)},$$

and introducing the GREEN function

$$G^\bullet = \int_0^{+\infty} e^{-\alpha t}\,p^\bullet dt = B^{-1} g_1 g_2, \quad B = g_1^+ g_2 - g_1 g_2^+$$

and letting $b \downarrow a$, it follows that

5) $$\gamma = \lim_{b \downarrow a} \frac{g_1(b)\, g_2(a)\, (b-a)^{-1} E_a\left[\int_0^{\mathfrak{m}_b^\bullet} e^{-\alpha t}\, \mathfrak{t}^\bullet(dt, a)\right]}{g_2(b)\dfrac{g_1(b)-g_1(a)}{b-a} - g_1(b)\dfrac{g_2(b)-g_2(a)}{b-a}}$$

$$= B^{-1} g_1(a)\, g_2(a) \lim_{b \downarrow a} (b-a)^{-1} E_a\left[\int_0^{\mathfrak{m}_b^\bullet} e^{-\alpha t}\, \mathfrak{t}^\bullet(dt, a)\right]$$

$$= B^{-1} g_1\, g_2$$

$$= G^\bullet(a, a)$$

because (see problem 2.8.3)

6a) $$E_a\left[\int_0^{\mathfrak{m}_b^\bullet} (1 - e^{-\alpha t})\, \mathfrak{t}^\bullet(dt, a)\right]$$

$$\leqq E_a[(1 - e^{-\alpha \mathfrak{m}_b^\bullet})\, \mathfrak{t}^\bullet(\mathfrak{m}_b^\bullet, a)]$$

$$\leqq \sqrt{E_a[(1 - e^{-\alpha \mathfrak{m}_b^\bullet})^2]}\, \sqrt{E_a[\mathfrak{t}^\bullet(\mathfrak{m}_b^\bullet, a)^2]}$$

$$= o(E_a[\mathfrak{t}(\mathfrak{m}_b, a)^2]^{1/2}) = o(b-a)$$

and

6b) $$E_a[\mathfrak{t}^\bullet(\mathfrak{m}_b^\bullet, a)] = E_a[\mathfrak{t}(\mathfrak{m}_b, a)] = b - a.$$

But now

7) $$E_a\left[\int_0^{+\infty} e^{-\alpha t}\, \mathfrak{t}^\bullet(dt, b)\right]$$

$$= E_a(e^{-\alpha \mathfrak{m}_b^\bullet})\, E_b\left[\int_0^{+\infty} e^{-\alpha t}\, \mathfrak{t}^\bullet(dt, b)\right]$$

$$= E_a(e^{-\alpha \mathfrak{m}_b^\bullet})\, G^\bullet(b, b)$$

$$= G^\bullet(a, b),$$

and, 3) is immediate on inverting the LAPLACE transform and differentiating.

5.5. Subordination and chain rule

Given a conservative non-singular diffusion D with state interval Q, sample paths $w: t \to x(t)$, and probabilities $P_\bullet(B)$, a second conservative non-singular diffusion $D^\bullet$ is said to be *subordinate to* D provided

1a) *its state interval* $Q^\bullet$ *is part of* Q,

1b) *it has the same (natural) scale,*

2a) *if* $l^\bullet = \inf Q^\bullet$ *is a shunt of* $Q^\bullet$, *then either* $l = \inf Q$ *is a shunt of* Q *or* $-\infty = l < l^\bullet$,

2b) *if* $r^\bullet = \sup Q^\bullet$ *is a shunt of* $Q^\bullet$, *then either* $r = \sup Q$ *is a shunt of* Q *or* $+\infty = r > r^\bullet$.

$D^\bullet$ is subordinate to D if there is a time substitution $t \to \mathfrak{f}^{-1}$ mapping D upon $D^\bullet$; note that if $l > -\infty$ is neither exit nor entrance, then

$$3) \qquad 0 < P_a\left[\lim_{t\uparrow+\infty} x(t) = l\right] \qquad l < a < r$$

(see problem 4.6.7), and no time substitution $t \to \mathfrak{f}^{-1}$ can alter this fact or make $x(\mathfrak{f}^{-1})$ behave as if $l^\bullet$ $(\geqq l)$ were a shunt.

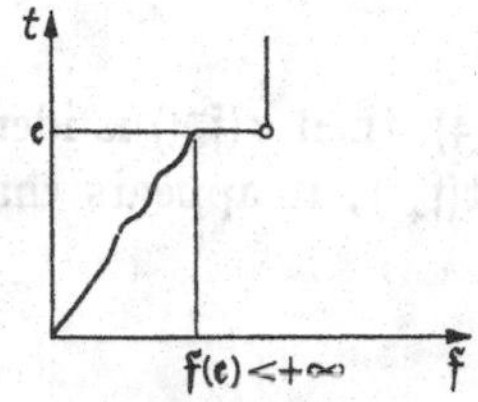

Diagram 1a

On the other hand, *if* $D^\bullet$ *is subordinate to* D, if $m^\bullet$ is the speed measure of $D^\bullet$, and if $\mathfrak{e}$ is the passage time of the sample path $w: t \to x(t)$ to the traps of $Q^\bullet$, *then the time substitution* $t \to \mathfrak{f}^{-1}$ *based upon*

$$4) \qquad \mathfrak{f}(t) = \int_{Q^\bullet} \mathfrak{t}\, dm^\bullet \qquad t \leqq \mathfrak{e}$$

$$= +\infty \qquad t > \mathfrak{e}$$

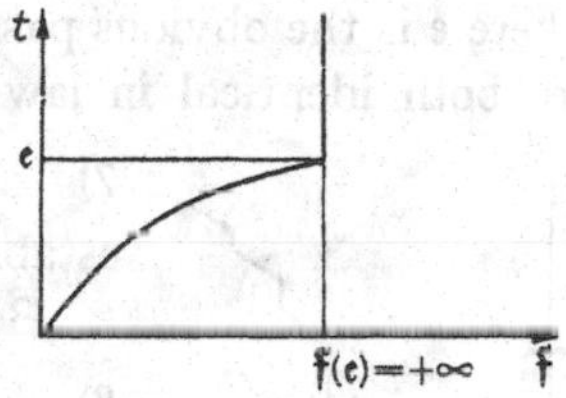

Diagram 1b

maps D *upon* $D^\bullet$, *i.e.*,

$$5) \qquad P_a^\bullet(B) = P_a[x(\mathfrak{f}^{-1}) \in B^\bullet]$$

$$a \in Q^\bullet, \quad B^\bullet \in \mathsf{B}$$

as will be proved below in the special case:

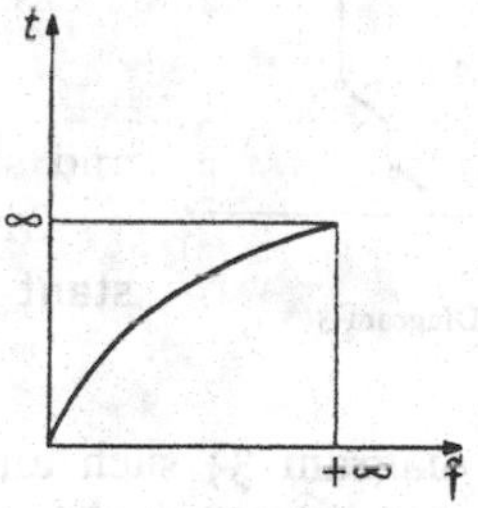

Diagram 2a

$$Q = [l, +\infty) \qquad l \geqq -\infty$$

$$Q^\bullet = [l^\bullet, +\infty) \qquad l^\bullet \geqq -\infty$$

$$s[a, b) = s^\bullet[a, b) = b - a \qquad b < a.$$

Given $l^\bullet > -\infty$ and a sample path $w: t \to x(t)$ starting at a point $x(0) \geqq l$, *if* $l^\bullet$ *is a trap of* $Q^\bullet$, then $\mathfrak{e} < +\infty$ and $\mathfrak{f}$ has the aspect of diagram 1a or of diagram 1b according as $l^\bullet$ is exit or not, while, *if* $l^\bullet$ *is a shunt of* $Q^\bullet$, then either $l = -\infty$ or $l > -\infty$ is a shunt of Q and $\mathfrak{f}$ has the aspect of diagram 2a or of diagram 2b according as $l = l^\bullet$ or $l > l^\bullet$; in both cases the identification of $x(\mathfrak{f}^{-1})$ proceeds as in 5.3.

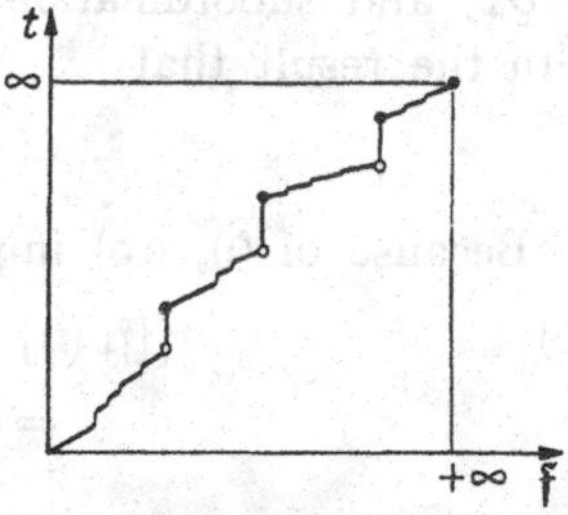

Diagram 2b

As to the case $l^\bullet = -\infty$, if $x(0) > -\infty$, then $\mathfrak{f}$ has the aspect of diagram 2a, and the same is true if $x(0) = l^\bullet = -\infty$ is a shunt of $Q^\bullet$

because then $P_{-\infty}[\mathfrak{m}_0 < +\infty] = 1$ and $E_{-\infty}[\mathfrak{f}(\mathfrak{m}_0)] = \int_{-\infty}^{0-} |\xi|\, dm < +\infty$; on the other hand, if $x(0) = l^\bullet = -\infty$ is a trap of $Q^\bullet$ then $\mathfrak{f}(0+) \equiv +\infty$, $x(\mathfrak{f}^{-1}) \equiv -\infty$, and the proof is complete.

Given conservative non-singular $\mathsf{D}_\pm$ subordinate to D, if

6a)
$$\mathfrak{f}_\pm(t) = \int_{Q_\pm} \mathfrak{t}\, dm_\pm \qquad t \leqq \mathfrak{e}_\pm$$
$$= +\infty \qquad t > \mathfrak{e}_\pm$$

as in 4), then $x(\mathfrak{f}_+^{-1})$ is identical in law to D_+, and using its local times $\mathfrak{t}_+ = \mathfrak{t}(\mathfrak{f}_+^{-1})$, it appears that if

6b)
$$\mathfrak{f}(t) = \int_{Q_-} \mathfrak{t}_+\, dm_- \qquad t \leqq \mathfrak{e}$$
$$= +\infty \qquad t > \mathfrak{e},$$

where $\mathfrak{e}$ is the obvious passage time of $x(\mathfrak{f}_+^{-1})$, then $x(\mathfrak{f}_-^{-1})$ and $x[\mathfrak{f}_+^{-1}(\mathfrak{f}^{-1})]$ are both identical in law to D_-; this suggests the *chain rule*:

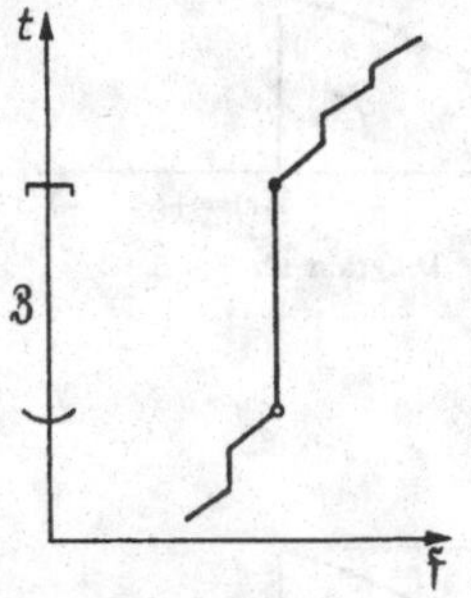

Diagram 3

7)
$$\mathfrak{f}_+^{-1}(\mathfrak{f}^{-1}) \equiv \mathfrak{f}_-^{-1} \qquad x(0) \in Q^-$$

established below.

Because $\theta \leqq \mathfrak{f}^{-1}(t)$ is the same as $\mathfrak{f}(\theta) \leqq t$,

8)
$$\mathfrak{f}_+^{-1}[\mathfrak{f}^{-1}(t)]$$
$$= \max(s : \mathfrak{f}_+(s) \leqq \mathfrak{f}^{-1}(t))$$
$$= \max(s : \mathfrak{f}[\mathfrak{f}_+(s)] \leqq t),$$

and so it is enough to prove that $\mathfrak{f}(\mathfrak{f}_+) = \mathfrak{f}_-$.

Given $t \geqq 0$, $\mathfrak{f}_+^{-1}[\mathfrak{f}_+(t)] = t$ unless $\mathfrak{f}_+$ is constant on a half-closed interval

$$\mathfrak{Z} = (t_1, t_2] \qquad (t_1 \leqq t < t_2)$$

as in diagram 3; such an interval $\mathfrak{Z}$ corresponds to an excursion of the sample path into $R^1 - Q_+$ or to its arrival at time $t = t_1$ at a trap of Q_+, and subordination implies that $\mathfrak{f}_-$ is likewise constant on $\mathfrak{Z}$ with the result that

9)
$$\mathfrak{f}_-[\mathfrak{f}_+^{-1}(\mathfrak{f}_+(t))] = \mathfrak{f}_-(t).$$

Because of 9), 6b) implies

10)
$$\mathfrak{f}[\mathfrak{f}_+(t)]$$
$$= \mathfrak{f}_-[\mathfrak{f}_+^{-1}(\mathfrak{f}_+(t))] = \mathfrak{f}_-(t) \qquad \mathfrak{f}_+(t) \leqq \mathfrak{e}$$
$$= +\infty \qquad \mathfrak{f}_+(t) > \mathfrak{e},$$

and it remains to prove

11)
$$\mathfrak{f}_-(t) = +\infty \qquad \textit{in case } \mathfrak{f}_+(t) > \mathfrak{e}.$$

But, since $\mathfrak{f}_+^{-1}(\mathfrak{e}) \geqq \mathfrak{e}_-$, it appears that $\mathfrak{e} \geqq \mathfrak{f}_+(\mathfrak{e}_-)$, and hence $\mathfrak{f}_+(t) > \mathfrak{e}$ implies $t > \mathfrak{e}_-$, which, in turn, implies $\mathfrak{f}_-(t) = +\infty$ as desired.

5.6. Killing times

Consider a non-singular differential operator

1) $$(\mathfrak{G}^\bullet u)(b) = \frac{u^+(db) - u(b)\, k(db)}{m(db)}$$

$$u^+(a) = \lim_{b \downarrow a} \frac{u(b) - u(a)}{b - a}, \qquad 0 < b < +\infty,$$

with k and m finite on closed subintervals of $[0, +\infty)$ and either

2a) $$(\mathfrak{G}^\bullet u)(0)\, m(0) = u^+(0) - u(0)\, k(0) \qquad 0 \leqq k(0), \quad m(0) < +\infty,$$

or

2b) $$(\mathfrak{G}^\bullet u)(0) = \varkappa\, u(0) \qquad 0 \leqq \varkappa < +\infty,$$

and let us show how to construct the corresponding diffusion; it will be apparent that the method used below can also be applied to the most general non-singular differential operator of 4.7.

The idea is to introduce the non-singular *conservative* diffusion $\mathfrak{D}$ with generator

3) $$(\mathfrak{G}\, u)(b) = \frac{u^+(db)}{m(db)} \qquad 0 < b < +\infty,$$

4a) $$(\mathfrak{G}\, u)(0)\, m(0) = u^+(0) \qquad \textit{in case } 2\text{a}),$$

4b) $$(\mathfrak{G}\, u)(0) = 0 \qquad \textit{in case } 2\text{b}),$$

sample paths $w: t \to x(t) \in [0, +\infty)$, Borel algebras B, probabilities $P_a(B)$, and local times

5a) $$\mathfrak{t}(t, b) = \frac{\text{measure}(s: x(s) \in db,\ s \leqq t)}{m(db)} \qquad b > 0$$

5b) $$\mathfrak{t}(t, 0) = \lim_{\varepsilon \downarrow 0} \frac{\text{measure}(s: x(s) < \varepsilon,\ s \leqq t)}{m[0, \varepsilon)} \qquad \textit{in case } 2\text{a}),$$

to adjoin to its path space an extra coordinate $0 \leqq \mathfrak{m}_\infty \leqq +\infty$, to extend $P_\bullet$ to this new coordinate according to the rule

6) $$P_\bullet(\mathfrak{m}_\infty > t \mid \mathsf{B}) = e^{-\mathfrak{k}(t)} \qquad t \geqq 0$$

with

7a) $$\mathfrak{k}(t) = \int_{0-}^{+\infty} \mathfrak{t}(t, b)\, k(db) \qquad \textit{in case } 2\text{a})$$

7b) $$\mathfrak{k}(t) = \int_{0+}^{+\infty} \mathfrak{t}(t, b)\, k(db) + \varkappa \text{ measure}(s: x(s) = 0,\ s \leqq t) \qquad \textit{in case } 2\text{b}),$$

and, introducing the new sample paths

$$8) \qquad w^\circ : t \to x^\circ(t) = x(t) \qquad t < \mathfrak{m}_\infty$$
$$= \infty \qquad t \geq \mathfrak{m}_\infty,$$

to prove that $\mathsf{D}^\bullet = [x^\circ, P_\bullet]$ is *a (non-standard) description of the diffusion with generator* $\mathfrak{G}^\bullet$.

The construction of the elastic BROWNian motion explained in 2.3 was a special case of this prescription.

Consider for the proof that $\mathsf{D}^\bullet$ is *simple Markov*, the sample paths $w^\bullet : t \to x^\bullet(t)$, killing time $\mathfrak{m}^\bullet_\infty = \min(t : x^\bullet = \infty)$ BOREL algebras B, and probabilities $P^\bullet_a(B^\bullet) \equiv P_a(w^\circ \in B^\bullet)$ of its standard description.

Given $s \geq 0$ and $db \subset [0, +\infty)$, one finds

$$\begin{aligned} 9) \quad & P^\bullet_a(B^\bullet, x^\bullet(t+s) \in db) \\ &= P^\bullet_a(B^\bullet, x^\bullet(t+s) \in db, \mathfrak{m}^\bullet_\infty > t+s) \\ &= P_a(B, x(t+s) \in db, \mathfrak{m}_\infty > t+s) \\ &= E_a[B, x(t+s) \in db, P_a(\mathfrak{m}_\infty > t+s \mid \mathsf{B})] \\ &= E_a[B, x(t+s) \in db, e^{-\mathfrak{f}(t+s)}] \\ &= E_a\left[B, e^{-\mathfrak{f}(t)}, x(s, w_t^+) \in db, e^{-\mathfrak{f}(s, w_t^+)}\right] \\ &= E_a\left[B, e^{-\mathfrak{f}(t)}, E_{x(t)}(x(s) \in db, e^{-\mathfrak{f}(s)})\right] \\ &= E_a\left[B, \mathfrak{m}_\infty > t, P_{x(t)}(x(s) \in db, \mathfrak{m}_\infty > s)\right] \\ &= E^\bullet_a\left[B^\bullet, P_{x^\bullet(t)}(x^\bullet(s) \in db)\right] \end{aligned}$$

where $P^\bullet_\infty[x^\bullet(s) \in [0, +\infty)] = 0$ was used in the final step; this finishes the proof.

Because the GREEN operators

$$\begin{aligned} 10) \qquad G^\bullet_\alpha f &= E^\bullet_\bullet\left[\int_0^{\mathfrak{m}^\bullet_\infty} e^{-\alpha t} f(x^\bullet)\, dt\right] \\ &= E_\bullet\left[\int_0^{\mathfrak{m}_\infty} e^{-\alpha t} f(x)\, dt\right] \\ &= E_\bullet\left[\int_0^{+\infty} e^{-\mathfrak{f}}\, d\mathfrak{f} \int_0^{s} e^{-\alpha t} f(x)\, dt\right] \\ &= E_\bullet\left[\int_0^{+\infty} e^{-\alpha t} f(x)\, dt \int_t^{+\infty} e^{-\mathfrak{f}}\, d\mathfrak{f}\right] \\ &= E_\bullet\left[\int_0^{+\infty} e^{-\alpha t} e^{-\mathfrak{f}} f(x)\, dt\right] \end{aligned}$$

map $C(R^1)$ into itself (use the method of 5.2.14)), $\mathsf{D}^\bullet$ is *a (non-singular) diffusion*, and for the rest, it is enough to compute its scale $s^\bullet$, killing measure $k^\bullet$, speed measure $m^\bullet$, and killing rate $\varkappa$, using the probabilistic formulas of 4.4.

Given $0 \leqq a < b < +\infty$, the hitting probability

11) $$\begin{aligned} p^\bullet_{ba}(\xi) &= P^\bullet_\xi(\mathfrak{m}^\bullet_b < \mathfrak{m}^\bullet_a) \\ &= P_\xi(\mathfrak{m}_b < \mathfrak{m}_a \wedge \mathfrak{m}_\infty) \\ &= E_\xi[\mathfrak{m}_b < \mathfrak{m}_a, e^{-\mathfrak{k}(\mathfrak{m}_b)}] \qquad a < \xi < b \end{aligned}$$

is continuous, and noting that

$$\mathfrak{t}(\mathfrak{m}, \eta) \qquad \mathfrak{m} = \mathfrak{m}_a \wedge \mathfrak{m}_b, \quad a < \eta < b$$

is invariant under time substitutions, it follows from the evaluation

$$E_\xi[\mathfrak{t}(\mathfrak{m}, \eta)] = G(\xi, \eta) \qquad a < \xi, \quad \eta < b$$

$$G(\xi, \eta) = G(\eta, \xi) = \frac{(\xi - a)(b - \eta)}{b - a} \qquad \xi \leqq \eta$$

(see problem 5.2.1) that

12) $$\begin{aligned} &\int_a^b G(\xi, \eta)\, p^\bullet_{ba}(\eta)\, k(d\eta) \\ &= E_\xi\left[\int_a^b \left(\int_0^{\mathfrak{m}} p^\bullet_{ba}(x_t)\, \mathfrak{t}(dt, \eta)\right) k(d\eta)\right] \\ &= E_\xi\left[\int_0^{\mathfrak{m}} p^\bullet_{ba}(x_t)\, \mathfrak{k}(dt)\right] \\ &= \lim_{n\uparrow+\infty} E_\xi\Big[\sum_{l2^{-n} \leqq \mathfrak{m}} p^\bullet_{ba}(x_{l2^{-n}})\, \mathfrak{k}[(l-1)\,2^{-n}, l\,2^{-n})\Big] \\ &= \lim_{n\uparrow+\infty} \sum_{l\geqq 1} E_\xi[l\,2^{-n} < \mathfrak{m}, \mathfrak{k}[(l-1)\,2^{-n}, l\,2^{-n}), \\ &\qquad\qquad E_{x(l2^{-n})}[\mathfrak{m}_b < \mathfrak{m}_a, e^{-k(\mathfrak{m})}]] \\ &= \lim_{n\uparrow+\infty} \sum_{l\geqq 1} E_\xi[l\,2^{-n} < \mathfrak{m}, \mathfrak{k}[(l-1)\,2^{-n}, l\,2^{-n}), \\ &\qquad\qquad \mathfrak{m}_b(w^+_{l2^{-n}}) < \mathfrak{m}_a(w^+_{l2^{-n}}), e^{-\mathfrak{k}(\mathfrak{m}(w^+_{l2^{-n}}), w^+_{l2^{-n}})}] \\ &= \lim_{n\uparrow+\infty} \sum_{l\geqq 1} E_\xi[\mathfrak{m}_b < \mathfrak{m}_a, e^{-\mathfrak{k}(\mathfrak{m})}, l\,2^{-n} < \mathfrak{m}, \\ &\qquad\qquad e^{\mathfrak{k}(l2^{-n})}\, \mathfrak{k}[(l-1)\,2^{-n}, l\,2^{-n})] \\ &= E_\xi\left[\mathfrak{m}_b < \mathfrak{m}_a, e^{-\mathfrak{k}(\mathfrak{m})} \int_0^{\mathfrak{m}} e^{\mathfrak{k}(t)}\, \mathfrak{k}(dt)\right] \\ &= E_\xi[\mathfrak{m}_b < \mathfrak{m}_a, 1 - e^{-\mathfrak{k}(\mathfrak{m})}] \\ &= \frac{\xi - a}{b - a} - p^\bullet_{ba}(\xi) \end{aligned}$$

i.e.,

13a) $$p^{\bullet +}_{ba}(d\xi) = p^\bullet_{ba}(\xi)\, k(d\xi) \qquad a < \xi < b.$$

$p^\bullet_{ab}(\xi) = P^\bullet_\xi(\mathfrak{m}^\bullet_a < \mathfrak{m}^\bullet_b)$ satisfies

$$\text{13 b)} \qquad p^{\bullet +}_{ab}(d\xi) = p^\bullet_{ab}(\xi)\, k(d\xi) \qquad a < \xi < b$$

for similar reasons; *esp.*,

$$\text{14)} \qquad 0 < p^{\bullet +}_{ba}(\xi)\, p^\bullet_{ab}(\xi) - p^{\bullet +}_{ab}(\xi)\, p^\bullet_{ba}(\xi) \textit{ is a constant} \qquad a < \xi < b.$$

As to the mean exit time,

$$\begin{aligned}
\text{15)} \qquad e^\bullet_{ab}(\xi) &= E^\bullet_\xi(\mathfrak{m}^\bullet_a \wedge \mathfrak{m}^\bullet_b \wedge \mathfrak{m}^\bullet_\infty) \qquad\qquad a < \xi < b \\
&= E_\xi(\mathfrak{m}_a \wedge \mathfrak{m}_b \wedge \mathfrak{m}_\infty) \\
&= E_\xi(\mathfrak{m}, \mathfrak{m} < \mathfrak{m}_\infty) + E_\xi(\mathfrak{m}_\infty, \mathfrak{m}_\infty < \mathfrak{m}) \qquad \mathfrak{m} = \mathfrak{m}_a \wedge \mathfrak{m}_b \\
&= E_\xi\left[\mathfrak{m}\, e^{-\mathfrak{k}(\mathfrak{m})} + \int_0^{\mathfrak{m}} t\, e^{-\mathfrak{k}}\, d\mathfrak{k}\right] \\
&= E_\xi\left[\mathfrak{m}\, e^{-\mathfrak{k}(\mathfrak{m})} - t\, e^{-\mathfrak{k}}\Big|_0^{\mathfrak{m}} + \int_0^{\mathfrak{m}} e^{-\mathfrak{k}}\, dt\right] \\
&= E_\xi\left[\int_0^{\mathfrak{m}} e^{-\mathfrak{k}}\, dt\right]
\end{aligned}$$

is continuous, and computing as in 12), it is found that

$$\begin{aligned}
\text{16)} \qquad & \int_a^b G(\xi, \eta)\, e^\bullet_{ab}(\eta)\, k(d\eta) \\
&= E_\xi\left[\int_0^{\mathfrak{m}} e^\bullet_{ab}(x_t)\, \mathfrak{k}(dt)\right] \\
&= E_\xi\left[\int_0^{\mathfrak{m}} \mathfrak{k}(dt) \int_0^{\mathfrak{m}(w_t^+)} e^{-\mathfrak{k}(s,\, w_t^+)}\, ds\right] \\
&= E_\xi\left[\int_0^{\mathfrak{m}} e^{-\mathfrak{k}}\, ds \int_0^{s} e^{\mathfrak{k}}\, d\mathfrak{k}\right] \\
&= E_\xi\left[\int_0^{\mathfrak{m}} (1 - e^{-\mathfrak{k}})\, ds\right] \\
&= e_{ab}(\xi) - e^\bullet_{ab}(\xi) \qquad\qquad e_{ab}(\xi) = E_\xi(\mathfrak{m}_a \wedge \mathfrak{m}_b),
\end{aligned}$$

i.e.,

$$\text{17)} \qquad -\left[e^{\bullet +}_{ab}(d\xi) - e^\bullet_{ab}(\xi)\, k(d\xi)\right] = -e^+_{ab}(d\xi) = m(d\xi) \qquad a < \xi < b.$$

Because of 14),

$$\begin{aligned}
\text{18)} \qquad s^\bullet(d\xi) &= \text{constant} \times \left[p^\bullet_{ba}(d\xi)\, p^\bullet_{ab}(\xi) - p^\bullet_{ab}(d\xi)\, p^\bullet_{ba}(\xi)\right] \\
&= \text{constant} \times s(d\xi),
\end{aligned}$$

and putting the constant $= 1$, 13) implies $k^\bullet = k$ and 17) implies $m^\bullet = m$ on $(0, +\infty)$.

It remains to compute $k(0)$ and $m(0)$ in case 2a) and $\varkappa$ in case 2b). But, *in case* 2b),

19) $$P_0(\mathfrak{m}^\bullet_\infty > t) = P_0(\mathfrak{m}_\infty > t) = E_0[e^{-\varkappa \mathfrak{t}(t, 0)}] = e^{-\varkappa t},$$

as it should be, and *in case* 2a), the hitting probability

20a) $$p^\bullet_b(\xi) = P^\bullet_\xi(\mathfrak{m}^\bullet_b < +\infty) \qquad 0 \leqq \xi < b$$

and the mean exit time

20b) $$e^\bullet_b(\xi) = E^\bullet_\xi(\mathfrak{m}^\bullet_b \wedge \mathfrak{m}^\bullet_\infty) \qquad 0 < \xi < b$$

can be computed in terms of the Green function

$$E_\xi[\mathfrak{t}(\mathfrak{m}_b, \eta)] = G(\xi, \eta) = G(\eta, \xi) = 1 - \frac{\eta}{b} \quad 0 \leqq \xi \leqq \eta < b$$

much as in 14) and 16), with the result that

21a) $$\int_{[0, b)} G(\xi, \eta)\, p^\bullet_b(\eta)\, k(d\eta) = 1 - p^\bullet_b(\xi)$$

21b) $$\int_{[0, b)} G(\xi, \eta)\, e^\bullet_b(\eta)\, k(d\eta) = e_b(\xi) - e^\bullet_b(\xi),$$

i.e.,

22a) $$k^\bullet(0) = \frac{p^{\bullet +}_b(0)}{p^\bullet_b(0)} = k(0)$$

22b) $$\begin{aligned} m^\bullet(0) &= -[e^{\bullet +}_b(0) - e^\bullet_b(0)\, k(0)] \\ &= -e^+_b(0) = m(0), \end{aligned}$$

as desired.

It is apparent that the dot motion has *local times*

23a) $$\mathfrak{t}^\bullet(t, b, w^\bullet) = \frac{\text{measure}(s : x^\bullet(s) \in db,\ s \leqq t)}{m(db)}$$

23b) $$\mathfrak{t}^\bullet(t, b, w^\bullet) = \mathfrak{t}(t \wedge \mathfrak{m}_\infty, b, w).$$

Problem 1. Check the formula

$$\begin{aligned} p^\bullet(t, a, b) &= \frac{\partial}{\partial t} E^\bullet_a[\mathfrak{t}^\bullet(t, b)] \qquad (t, a, b) \in (0, +\infty) \times Q^\bullet \times Q^\bullet, \\ Q^\bullet &= [0, +\infty) \qquad \textit{in case } 2a) \\ &= (0, +\infty) \qquad \textit{in case } 2b) \end{aligned}$$

(see 5.4 for the conservative case and 4.11 for $p^\bullet(t, a, b)$).

[Given $\alpha > 0$ and $(a, b) \in Q^\bullet \times Q^\bullet$,

$$E^\bullet_a\left[\int_0^{+\infty} e^{-\alpha t}\, \mathfrak{t}^\bullet(dt, b)\right]$$

$$= E_a\left[\int_0^{+\infty} e^{-\mathfrak{f}(t)}\, \mathfrak{t}(dt, b)\right] \qquad \mathfrak{f}(t) = \alpha \int \mathfrak{t}\, dm + k$$

$$= E_a\left[\int_0^{+\infty} e^{-t}\, \mathfrak{t}(\mathfrak{f}^{-1}(dt), b)\right],$$

and using problem 5.3.1 and the chain rule for time substitutions of 5.5, this can be identified as the GREEN function for the dot motion:

$$G^\bullet(a, b) = B^{-1} g_1(a)\, g_2(b) \qquad (a \leq b)$$

$$g^+(d\xi) = g(\xi)\,[\alpha\, m(d\xi) + k(d\xi)] \qquad \xi > 0,\ g = g_1, g_2$$

$$g_1^+(0) = g_1(0)\,[\alpha\, m(0) + k(0)] \qquad \textit{in case } 2\text{a})$$

$$g_1(0) = 0 \qquad \textit{in case } 2\text{b});$$

now use

$$G^\bullet = \int_0^{+\infty} e^{-\alpha t}\, p^\bullet\, dt.]$$

Problem 2. Evaluate $P^\bullet_\bullet[\mathfrak{m}^\bullet_\infty \in dt,\ x^\bullet(\mathfrak{m}^\bullet_\infty -) \in db]$ on $(0, +\infty) \times Q^\bullet \times \times Q^\bullet$ as $p^\bullet(t, a, b)\, dt\, k(db)$ using the formula of problem 1.

[Given $\alpha > 0$, $a \in Q^\bullet$, and $f \in C[0, +\infty)$ vanishing at 0,

$$\int_0^{+\infty} e^{-\alpha t}\, dt \int_0^{+\infty} p^\bullet(t, a, b)\, f\, dk$$

$$= \int_0^{+\infty} E_a\left[\int_0^{+\infty} e^{-\alpha t}\, \mathfrak{t}(dt, b)\right] f\, dk$$

$$= E_a\left[\int_0^{+\infty} e^{-\alpha t}\, e^{-\mathfrak{f}}\, \mathfrak{f}(dt)\, f(x_t)\right]$$

$$= E^\bullet_a\Big]e^{-\alpha\, \mathfrak{m}^\bullet_\infty} f(x^\bullet(\mathfrak{m}^\bullet_\infty -))\Big];$$

now invert the LAPLACE transform.]

Problem 3. Give a precise meaning to the statement:

$$\lim_{t \downarrow 0} t^{-1}\, m(db)\, P^\bullet_b(\mathfrak{m}^\bullet_\infty \leq t) = k(db) \qquad b > 0.$$

[Given $0 < a < b < c$, if $I = (a, b]$, then

$$\lim_{t\downarrow 0} t^{-1} \int_I m(d\xi)\, P^\bullet_\xi(\mathfrak{m}^\bullet_\infty \leqq t,\, x^\bullet(\mathfrak{m}^\bullet_\infty -) = 0)$$
$$\leqq \lim_{t\downarrow 0} t^{-1} P_a(\mathfrak{m}_0 \leqq t)\, m(I) = 0 \qquad \text{in case 2b)},$$
$$\lim_{t\downarrow 0} t^{-1} \int_I m(d\xi)\, P^\bullet_\xi(\mathfrak{m}^\bullet_\infty \leqq t,\, x^\bullet(\mathfrak{m}_\infty -) \geqq c)$$
$$\leqq \lim_{t\downarrow 0} t^{-1} P_b(\mathfrak{m}_c \leqq t)\, m(I) = 0,$$

and so

$$\lim_{t\downarrow 0} t^{-1} \int_I m(d\xi)\, P^\bullet_\xi(\mathfrak{m}^\bullet_\infty \leqq t)$$
$$= \lim_{t\downarrow 0} t^{-1} \int_I m(d\xi)\, P^\bullet_\xi(\mathfrak{m}^\bullet_\infty \leqq t,\, 0 < x^\bullet(\mathfrak{m}^\bullet_\infty -) < c)$$
$$= \lim_{t\downarrow 0} t^{-1} \int_I m(d\xi) \int_0^t ds \int_{0+}^c p^\bullet(s, \xi, \eta)\, k(d\eta)$$
$$= \lim_{t\downarrow 0} t^{-1} \int_0^t ds\, k(d\eta)\, P^\bullet_\eta[x^\bullet(s) \in I]$$
$$= k(I),$$

if neither a nor b is a jump of k.]

Problem 4. Consider a non-singular diffusion on R^1 with generator

$$\mathfrak{G}^\bullet u = \frac{u^+(db) - u(b)\, k(db)}{m(db)} \qquad u^+(a) = \lim_{b\downarrow a} \frac{u(b) - u(a)}{b - a},$$

sample paths $w^\bullet$, killing time $\mathfrak{m}^\bullet_\infty$, and probabilities $P^\bullet_a(B^\bullet)$, and introduce the composite motion

$$x(t) \equiv x^\bullet_n(t - \mathfrak{m}^\bullet_{n-1} - \cdots - \mathfrak{m}^\bullet_1)$$
$$\mathfrak{m}^\bullet_1 + \cdots + \mathfrak{m}^\bullet_{n-1} \leqq t < \mathfrak{m}^\bullet_1 + \cdots + \mathfrak{m}^\bullet_n$$

where $\mathfrak{m}^\bullet_1 + \cdots + \mathfrak{m}^\bullet_{n-1} = 0$ if $n = 0$ and, conditional on $x(t) : t < \mathfrak{m}^\bullet_1 + \cdots + \mathfrak{m}^\bullet_n$, the motion $x^\bullet_{n+1}(t) : t < \mathfrak{m}^\bullet_{n+1}$ is identical in law to $x^\bullet(t) : t < \mathfrak{m}^\bullet_\infty$ starting at the place $l_n = x^\bullet_n(\mathfrak{m}_n -)$ (see the diagram).

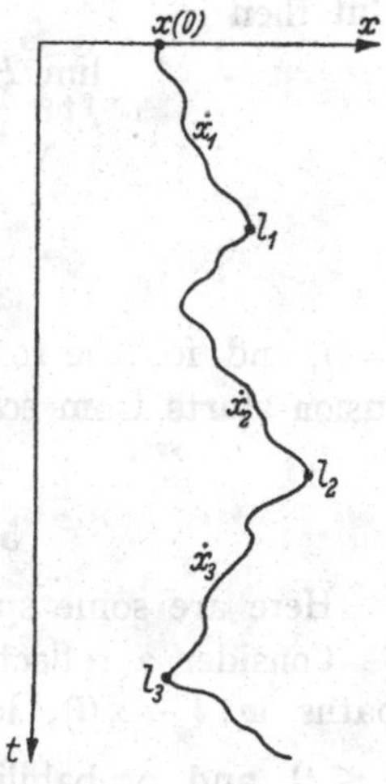

Diagram 1

The problem is to show that

$$P^\bullet_\bullet[\mathfrak{m}^\bullet_1 + \mathfrak{m}^\bullet_2 + \cdots + \mathfrak{m}^\bullet_n \uparrow +\infty] = 1$$

and to identify the composite motion as the conservative diffusion with generator $\mathfrak{G}u = u^+(db)/m(db)$.

[Consider the conservative motion with generator $\mathfrak{G}$, sample paths $w: t \to x(t)$, local times $\mathfrak{t}$, and probabilities $P_a(B)$, and kill it at time $\mathfrak{m}_\infty$ with conditional law $P_\bullet(\mathfrak{m}_\infty > t \mid \mathsf{B}) = e^{-\mathfrak{k}(t)}$ ($\mathfrak{k} = \int \mathfrak{t}\, dk$) to obtain a (non-standard) description of the dot motion. With this model, it is clear that

$$P_\bullet^\bullet[\mathfrak{m}_1^\bullet + \mathfrak{m}_2^\bullet + \cdots + \mathfrak{m}_n^\bullet \in dt, l_n \in dl]$$
$$= E_\bullet\left[e^{-\mathfrak{k}(t)} \frac{\mathfrak{k}(t)^{n-1}}{(n-1)!} \mathfrak{k}(dt), x(t) \in dl\right] \qquad n \geqq 1;$$

indeed, the formula is evident for $n = 1$, and assuming it for $n - 1$, it is found that

$$E_\bullet^\bullet[\mathfrak{m}_1^\bullet + \mathfrak{m}_2^\bullet + \cdots + \mathfrak{m}_n^\bullet \leqq t, f(l_n)]$$
$$= E_\bullet\left[\int_0^t e^{-\mathfrak{k}(s)} \frac{\mathfrak{k}(s)^{n-2}}{(n-2)!} \mathfrak{k}(ds)\, E_{x(s)}\left[\int_0^{t-s} e^{-\mathfrak{k}(\theta)} f(x_\theta)\, \mathfrak{k}(d\theta)\right]\right]$$
$$= E_\bullet\left[\int_0^t e^{-\mathfrak{k}(s)} \frac{\mathfrak{k}(s)^{n-2}}{(n-2)!} \mathfrak{k}(ds) \int_s^t e^{-\mathfrak{k}(\theta - s, w_s^+)} f(x_\theta)\, \mathfrak{k}(d\theta)\right]$$
$$= E_\bullet\left[\int_0^t \frac{\mathfrak{k}(s)^{n-2}}{(n-2)!} \mathfrak{k}(ds) \int_s^t e^{-\mathfrak{k}(\theta)} f(x_\theta)\, \mathfrak{k}(d\theta)\right]$$
$$= E_\bullet\left[\int_0^t e^{-\mathfrak{k}(\theta)} \frac{\mathfrak{k}(\theta)^{n-1}}{(n-1)!} f(x_\theta)\, \mathfrak{k}(d\theta)\right].$$

But then

$$\lim_{n \uparrow +\infty} P_\bullet^\bullet[\mathfrak{m}_1^\bullet + \mathfrak{m}_2^\bullet + \cdots + \mathfrak{m}_n^\bullet \leqq t]$$
$$= \lim_{n \uparrow +\infty} E_\bullet\left[\int_0^t e^{-\mathfrak{k}} \frac{\mathfrak{k}^{n-1}}{(n-1)!} d\mathfrak{k}\right]$$

$= 0$, and for the rest, it is enough to note that the conservative diffusion starts from scratch at time $\mathfrak{m}_\infty$.]

5.7. Feller's Brownian motions

Here are some special cases of the results of 5.3 and 5.6.

Consider a reflecting Brownian motion on $[0, +\infty)$ with sample paths $w: t \to x(t)$, local time $\mathfrak{t}(t) = \lim_{\varepsilon \downarrow 0} (2\varepsilon)^{-1}$ measure $(s: x(s) \leqq \varepsilon, s \leqq t)$, and probabilities $P_a(B)$, and let $\mathfrak{G}^\bullet$ be $1/2 \times$ *the second deri-*

vative applied to $u \in C^2[0, +\infty)$, subject to the conditions:

$$1) \qquad p_1 u(0) - p_2 u^+(0) + p_3(\mathfrak{G}^\bullet u)(0) = 0$$
$$0 \leqq p_1, p_2, p_3$$
$$p_1 + p_2 + p_3 = 1.$$

If $p_2 = 0$, then

$$2) \quad \begin{aligned} x^\bullet(t) &= x(t) && t < \mathfrak{m}_0 = \min(t : x(t) = 0) \\ &= 0 && \mathfrak{m}_0 \leqq t < \mathfrak{m}_\infty \\ &= \infty && t \geqq \mathfrak{m}_\infty \end{aligned}$$

$$P_\bullet(\mathfrak{m}_\infty > \mathfrak{m}_0 + t \mid \mathsf{B}) = e^{-\frac{p_1}{p_3}t}$$

is a non-standard description of the motion associated with $\mathfrak{G}^\bullet$, while, if $p_2 > 0$, then a possible (non-standard) description is

$$3) \quad \begin{aligned} x^\bullet(t) &= x(\mathfrak{f}^{-1}(t)) && t < \mathfrak{m}_\infty \\ &= \infty && t \geqq \mathfrak{m}_\infty \end{aligned}$$

$$\mathfrak{f} = t + \frac{p_3}{p_2}\mathfrak{t}(t)$$

$$P_\bullet(\mathfrak{m}_\infty > t \mid \mathsf{B}) = e^{-\frac{p_1}{p_2}\mathfrak{t}(\mathfrak{f}^{-1}(t))}.$$

If $p_3 = 0 < p_2$, then $x^\bullet$ is the elastic BROWNian motion as described in 2.3.

Consider the case $p_2 = 0 < p_3$, and let $\mathfrak{Z} = (t : x(t) = 0)$. Because $\mathfrak{f}(t_2) - \mathfrak{f}(t_1) > t_2 - t_1$ if $(t_1, t_2) \cap \mathfrak{Z} \neq \varnothing$, the clock $\mathfrak{f}^{-1}$ *counts actual time off* $\mathfrak{Z}^\bullet = \mathfrak{f}(\mathfrak{Z})$ *but runs too slow on* $\mathfrak{Z}^\bullet$, *i.e.*, $x^\bullet = x(\mathfrak{f}^{-1})$ *looks like a reflecting* BROWNian *motion off* $\mathfrak{Z}^\bullet$ *but lingers too long at the barrier as if the going were a little sticky at that point.*

W. FELLER [4, 9] discovered that 1) is the most general (local) boundary condition that can be imposed upon the restriction of $D^2/2$ to $[0, +\infty)$; it had not been noticed before him that conditions involving $(\mathfrak{G}^\bullet u)(0)$ were possible.

FELLER studied the case in which the sample path is permitted to *jump* from the barrier back into $(0, +\infty)$. $\mathfrak{G}^\bullet$ is then $D^2/2$ applied to $u \in C^2[0, +\infty)$ subject to the conditions:

$$4) \qquad p_1 u(0) - p_2 u^+(0) + p_3(\mathfrak{G}^\bullet u)(0) = \int_{0+}^{+\infty} [u(l) - u(0)]\, p(dl)$$
$$0 \leqq p_1, p_2, p_3, p(dl)$$
$$p_1 + p_2 + p_3 + \int_{0+} l \wedge 1\, p(dl) = 1,$$

and *the entering sample path jumps out from* 0 *like the germ of the one-sided differential process with generator*

$$p_2 u^+(0) + \int_{0+}^{+\infty} [u(l) - u(0)]\, p(dl); \tag{5}$$

see 7.20 for an explanation of 5), and K. ITÔ and H. P. MCKEAN, JR. [2] for a complete picture of the sample paths associated with $\mathfrak{G}^\bullet$. S. WATANABE [1] obtained similar results for stable processes.

5.8. Ikeda's example

N. IKEDA pointed out to us an interesting example of a simple MARKOVian motion with continuous sample paths which is not a diffusion (see 4.1 for the simplest such example).

Consider a reflecting BROWNian motion with sample paths $w: t \to$

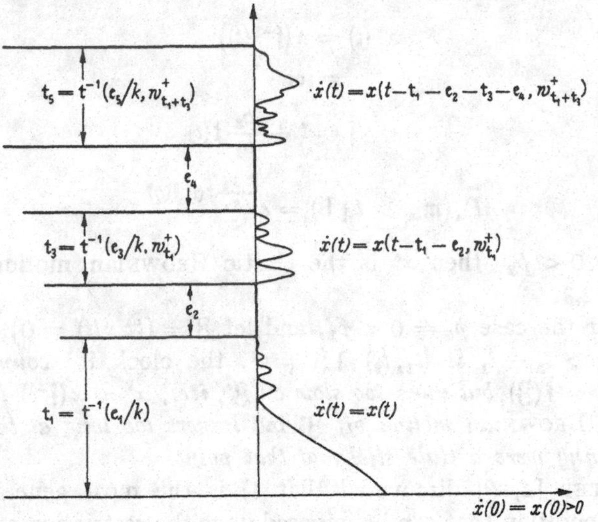

Diagram 1 a

$x(t)$, local time $\mathrm{t} = \lim_{\varepsilon \downarrow 0} (2\varepsilon)^{-1}$ measure $(s: x(s) \leqq \varepsilon,\ s \leqq t)$ and probabilities $P_a(B)$, let e_n $(n \geqq 0)$ be independent exponential holding times with conditional law

$$P_\bullet(\mathrm{e}_n > t \mid \mathrm{B}) = e^{-t} \qquad n \geqq 1, \tag{1}$$

let $t^{-1}(t)$ be the inverse function $\mathrm{t}^{-1}(t) = \min(s: \mathrm{t}(s) = t)$, and let k be a positive number. Define new sample paths $x^\bullet$ as in diagram 1, and note that since

$$P_\bullet(\mathrm{t}_1 > t \mid \mathrm{B}) = P_\bullet(\mathrm{e}_1 > k\, \mathrm{t}(t) \mid \mathrm{B}) = e^{-k\,\mathrm{t}(t)}, \tag{2}$$

the excursions alternating with the black intervals on which $x^\bullet \equiv 0$ are independent copies of the *elastic Brownian motion* corresponding to $u^+(0) = ku(0)$.

$\mathsf{D}^\bullet = [x^\bullet, P_\bullet]$ is *simple* MARKOV because $P_\bullet[x(t) = 0] \equiv 0$ $(t > 0)$; it cannot be *strict* MARKOV because $E_0(e^{-\mathfrak{m}^\bullet_{0+}})$ is neither 0 nor 1 [see 3.3.3 a)].

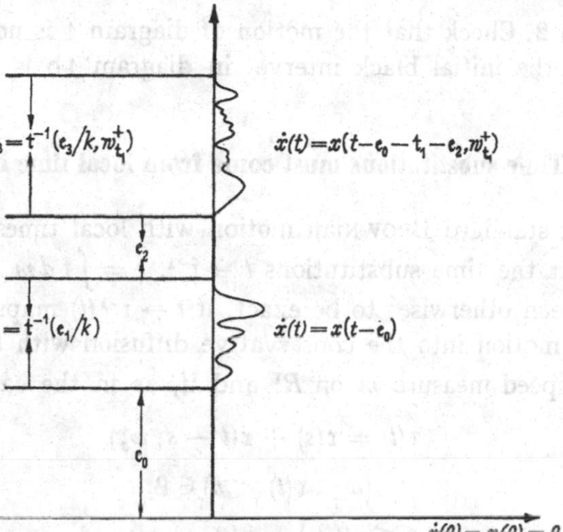

Diagram 1b

Given $f \in B[0, +\infty)$, $G^\bullet_\alpha f \equiv E_\bullet\left[\int_0^{+\infty} e^{-\alpha t} f(x^\bullet_t)\, dt\right]$ belongs to $D \equiv B[0, +\infty) \cap C(0, +\infty) \cap (u : u(0+)$ *exists*$)$*, and using the obvious $G^\bullet_\alpha - G^\bullet_\beta + (\alpha - \beta) G^\bullet_\alpha G^\bullet_\beta = 0$ $(\alpha, \beta > 0)$, it is clear that $G^\bullet_\alpha$ applied to D is $1:1$ and that a generator $\mathfrak{G}^\bullet : D(\mathfrak{G}^\bullet) \equiv G_1 D \to D$ can be introduced as in 3.7. $\mathfrak{G}^\bullet$ is similar to the elastic BROWNian generator; in fact, $D(\mathfrak{G}^\bullet)$ is the class of functions $u \in D \cap C^2(0, +\infty)$ such that

3 a) $\quad u^+(0+) = [u(0+) - u(0)]\, k$

3 b) $\quad u''(0+)$ *exists*,

and

4 a) $\quad \mathfrak{G}^\bullet = \frac{1}{2} D^2$ *on* $(0, +\infty)$

4 b) $\quad (\mathfrak{G}^\bullet u)(0) = u(0+) - u(0)$.

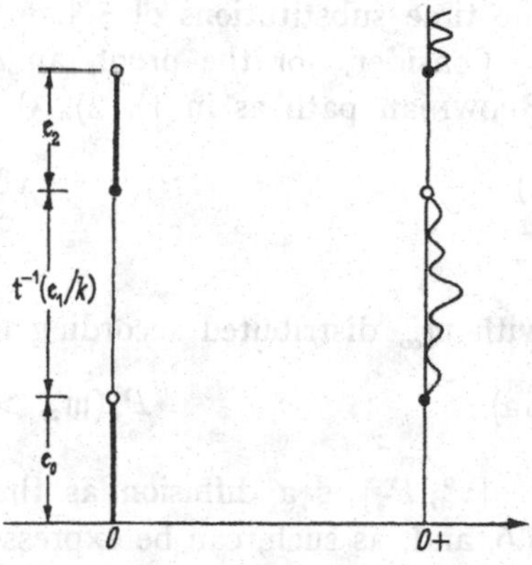

Diagram 2

* $B[0, +\infty)$ is the class of all bounded BOREL measurable functions defined on $[0, +\infty)$.

D. B. RAY [3] has found that the state space of a *simple Markovian* motion can be ramified so as to make the sample path start afresh at *each* MARKOV time; the ramification for the present case is shown in diagram 2.

Problem 1. Give the detailed evaluation of $\mathfrak{G}^{\bullet}$.

Problem 2. Check that the motion of diagram 1 is not even *simple* MARKOV if the initial black interval in diagram 1b is suppressed.

5.9. Time substitutions must come from local time integrals

Given a standard BROWNian motion with local times $\mathfrak{t}$, it is to be proved that the time substitutions $t \to \mathfrak{f}^{-1}$, $\mathfrak{f} = \int \mathfrak{t}\, dm$ of 5.2 could not have been otherwise; to be exact, if $t \to \mathfrak{r}^{-1}(t)$ maps the standard BROWNian motion into the conservative diffusion with the BROWNian scale and speed measure m on R^1 and if, as in the case $\mathfrak{r} = \int \mathfrak{t}\, dm$,

1) $$\mathfrak{r}(t) = \mathfrak{r}(s) + \mathfrak{r}(t-s, w_s^+) \qquad t \geqq s$$

2) $$(a \leqq \mathfrak{r}(t) < b) \in \mathsf{B}_t \qquad t \geqq 0$$

3a) $$0 \leqq \mathfrak{r}(t\pm) = \mathfrak{r}(t) < +\infty$$

3b) $$\mathfrak{r}(0+) = 0,$$

then

4) $$\mathfrak{r}(t) = \mathfrak{f}(t) = \int \mathfrak{t}\, dm \qquad t \geqq 0$$

up to a negligible set of BROWNian paths; a similar result holds for the time substitutions of 5.3.

Consider, for the proof, an *additive functional* $\mathfrak{r}$ of the standard BROWNian path as in 1), 2), 3) and introduce the sample paths

5) $$x^{\circ} = x(t) \qquad t < \mathfrak{m}_{\infty}$$
$$= \infty \qquad t \geqq \mathfrak{m}_{\infty},$$

with $\mathfrak{m}_{\infty}$ distributed according to the conditional law

6a) $$P_{\bullet}(\mathfrak{m}_{\infty} > t \mid \mathsf{B}) = e^{-\mathfrak{r}(t)} \qquad t \geqq 0.$$

$[x^{\circ}, P_{\bullet}]$ is a diffusion as the reader will prove using method of 5.6, and, as such, can be expressed as a conservative diffusion $x^{\bullet}$ with local times $\mathfrak{t}^{\bullet}$, killed at time $\mathfrak{m}_{\infty}^{\bullet}$ with conditional law.

6b) $$P_{\bullet}^{\bullet}(\mathfrak{m}_{\infty}^{\bullet} > t \mid \mathsf{B}^{\bullet}) = e^{-\mathfrak{r}^{\bullet}(t)} \quad \mathfrak{r}^{\bullet} = \int \mathfrak{t}^{\bullet}\, dk, \ dk \geqq 0.$$

Because $[x(t): t < \mathfrak{m}_\infty, \mathfrak{m}_\infty]$ is identical in law to $[x^\bullet(t): t < \mathfrak{m}_\infty^\bullet, \mathfrak{m}_\infty^\bullet]$, x and $x^\bullet$ are identical in law, *i.e.*, $x^\bullet$ is a standard BROWNian motion (see problem 5.6.4).

But now x and $x^\bullet$ can be identified, and choosing $t \geqq 0$, $A \in \mathsf{B}_t$, and using 2), it follows that

10a) $$E_\bullet[A, e^{-\mathfrak{r}(t)}] = P_\bullet(A, \mathfrak{m}_\infty > t) = P_\bullet(A, \mathfrak{m}_\infty^\bullet > t) = E_\bullet[A, e^{-\mathfrak{r}^\bullet(t)}],$$

which implies

10b) $$P_\bullet\left[\mathfrak{r}(t) = \mathfrak{r}^\bullet(t) = \int \mathfrak{t}\, dk, t \geqq 0\right] = 1,$$

thanks to 3a).

Now suppose that the time substitution $t \to \mathfrak{r}^{-1}$ maps the standard BROWNian motion into the conservative diffusion on R^1 with speed measure m.

Because $x(\mathfrak{r}^{-1})$ has to be continuous, $\mathfrak{r}(t_1) < \mathfrak{r}(t_2)$ $(t_1 < t_2)$, and computing the mean exit time

11) $$\begin{aligned} e_{ab}(\xi) &= E_\xi[\min(t : x(\mathfrak{r}^{-1}) \notin (a, b))] \\ &= E_\xi[\mathfrak{r}(\mathfrak{m}_a \wedge \mathfrak{m}_b)] \\ &= \int_a^b E_\xi[\mathfrak{t}(\mathfrak{m}_a \wedge \mathfrak{m}_b, \eta)]\, k(d\eta) \\ &= \int_a^b G(\xi, \eta)\, k(d\eta) \qquad a < \xi < b, \end{aligned}$$

$$G(\xi, \eta) = G(\eta, \xi) = \frac{(\xi - a)(b - \eta)}{b - a} \qquad \xi \leqq \eta$$

as in 5.2, it follows that $k = m$; *in brief*, $\mathfrak{r} = \int \mathfrak{t}\, dm$ *as stated.*

5.10. Shunts

Consider non-overlapping subintervals $Q_n = [l_n, r_n)$ of $Q = [0, 1]$ with scales $s = s_n$ and speed measures $m = m_n$ attached, let $K_+ = Q - \bigcup_{n \geqq 1} (l_n, r_n)$, let s_+ be a (shunt) scale with

1a) $$0 < s_+[a, b) = s_+(b) - s_+(a) \qquad a < b$$

1b) $$s_+[l_n, b) = \int_{l_n}^b m[l_n, \xi)\, s(d\xi) \qquad l_n \leqq b < r_n$$

1c) $$s_+[0, 1) < +\infty.$$

let $D(\mathfrak{G}^\bullet)$ be the class of functions $u \in C(Q)$ such that, for some $u^\bullet \in C(Q)$,

2a) $$\int_{[a,b)} u^\bullet\, m(d\xi) = u^+(b) - u^+(a) \qquad l_n < a < b < r_n$$

2b) $$u^\bullet(l_n)\, m(l_n) = u^+(l_n)$$

3) $$\int_{[a,b)\cap K_+} u^\bullet\, s_+(d\xi) = \int_{[a,b)\cap K_+} u\, d\xi \qquad a < b$$

4) $$u^\bullet(1) \equiv 0,$$

and let $\mathfrak{G}^\bullet$ be the differential operator $u \to u^\bullet$.

Given a conservative singular diffusion on Q with generator $\mathfrak{G}$ and

5a) $$E_0(\mathfrak{m}_1) < +\infty$$

5b) $$P_1(\mathfrak{m}_{1-} \wedge \mathfrak{m}_\infty = +\infty) = 1,$$

if l_n $(n \geqq 1)$ are the right isolated shunts of Q, if r_n is the smallest shunt $> l_n$, and if p_{ba} and e_1 are the hitting probabilities and mean exit time $P_\bullet(\mathfrak{m}_b < \mathfrak{m}_a)$ and $E_\bullet(\mathfrak{m}_1)$, then $\mathfrak{G}$ can be identified with the differential operator $\mathfrak{G}^\bullet$ on putting

6) $$s(d\xi) = \text{constant} \times p_{ba}(d\xi) \quad l_n < a < \xi < b < r_n$$

7a) $$m(d\xi) = -e_1^+(d\xi) \qquad l_n < \xi < r_n$$

7b) $$m(l_n) = -e_1^+(l_n)$$

8) $$s_+(d\xi) = -e_1(d\xi) \qquad 0 \leqq \xi \leqq 1$$

(see 4.8 and 4.9).

On the other hand, *each such differential operator* $\mathfrak{G}^\bullet$ *generates a conservative singular diffusion with* $E_0(\mathfrak{m}_1) < +\infty$ *and* $P_1(\mathfrak{m}_1 \wedge \mathfrak{m}_\infty = +\infty) = 1$ as will now be proved; this will show that *the differential operators* $\mathfrak{G}^\bullet$ *are in* 1 : 1 *correspondence with the conservative singular diffusions with* $E_0(\mathfrak{m}_1) < +\infty$ *and* $P_1(\mathfrak{m}_1 \wedge \mathfrak{m}_\infty = +\infty) = 1$ as suggested in 5.1.

On Q_n, $\mathfrak{G}^\bullet$ agrees with the generator of a conservative non-singular diffusion stopped at r_n with sample paths $t \to x_n(t)$, probabilities $P_a^n(B)$ $(l_n \leqq a < r_n)$, and

9) $$\begin{aligned} e_{r_n}(l_n) &= \int_{Q_n} [s(r_n) - s(\xi)]\, m(d\xi) \\ &= \int_{Q_n} m[l_n, \xi)\, s(d\xi) \\ &= s_+(Q_n) < +\infty \end{aligned} \qquad e_{r_n} = E_\bullet^n(\mathfrak{m}_{r_n})$$

(see 5.2, 5.3); on a shunt interval $[a, b) \subset K_+$, $\mathfrak{G}^\bullet$ generates the motion of translation at speed $+1$ in the shunt scale:

10) $$x(s_+[a, \xi)) = \xi \qquad a \leqq \xi < b;$$

and the idea is to hitch these motions end to end, arranging little hesitations as the particle crosses ∂K_+ so as to have

$$11)\qquad \text{measure}\,\big(s : x(s) \in K_+ \cap [0, 1),\, s \leqq t\big) = s_+\big([a, b) \cap K_+\big) \qquad a = x(0), \quad b = x(t).$$

Consider, for this purpose, a point $0 \leqq a \leqq 1$, choose sample paths

$$w_n : t \to x_n(t) \qquad n \geqq 1$$

according to the law

$$P = \underset{n\geqq 1}{\times} P^n_{l_n} \quad (\times \text{ stands for direct product}),$$

let

$$12)\quad t_a(b) = \sum_{r_n \in (a, b]} [\mathfrak{m}_{r_n}(w_n) - \mathfrak{m}_{l_n \vee a}(w_n)] + s_+\big([a, b) \cap K_+\big) \quad a < b \leqq 1,$$

and, noting

$$13)\qquad E\big(t_a(b)\big) \leqq E\big(t_0(1)\big)^* = \sum_{n \geqq 1} E^n_{l_n}(\mathfrak{m}_{r_n}) + s_+(K_+) = s_+[0, 1) < +\infty \qquad a < b,$$

introduce the *continuous* sample path

$$14)\quad \begin{array}{llll} x_a(t) & = x_n(t) & t < t_a(r_n) & l_n \leqq a = x_n(0) < r_n \\ & = x_n(t - t_a(l_n)) & t_a(l_n) \leqq t < t_a(r_n), & a \leqq l_n \\ & = b & t = t_a(b) & a \leqq b < 1 \\ & = 1 & t \geqq t_a(1) & \end{array}$$

as in diagram 1 below and let us prove that the motion $D^\bullet$ with probabilities $P^\bullet_a(B^\bullet) \equiv P(x_a \in B^\bullet)$ $(0 \leqq a \leqq 1)$ is *simple Markov, i.e.*,

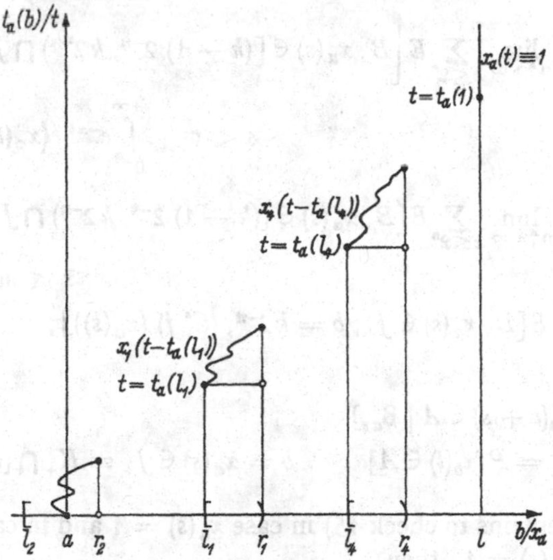

Diagram 1

* E is the expectation based upon P.

that

15) $$P[x_a(t+s)\in A \mid x_a(\theta):\theta\leqq s] = P[x_b(t)\in A] \qquad b = x_a(s).$$

Given $a<b$, x_a *starts afresh at the passage time* $\mathfrak{m}_{ab}=\min(t: x_a(t)=b)$, *i.e.*, $x_a(t+\mathfrak{m}_{ab}) = x_b(t)$ $(t\geqq 0)$, and using the obvious

16) $$\lim_{b\downarrow a} E(e^{-\mathfrak{m}_{ab}}) = \lim_{a\uparrow b} E(e^{-\mathfrak{m}_{ab}}) = 1,$$

it follows as in 3.6 that the GREEN operators

$$(G_\alpha^\bullet f)(a) = E\left[\int_0^{+\infty} e^{-\alpha t} f(x_a)\, dt\right]$$

map $C(Q)$ into itself.

Choosing $s>0$ and $B\in \mathsf{B}[x_a(\theta):\theta\leqq s]\equiv \mathsf{B}_{as}$, if

17) $$\mathfrak{m}=\mathfrak{m}_{ab} \qquad b = ([2^n x_a(s)]+1)\,2^{-n},$$

then $x_a(s)\in J_+\equiv K_+\cap[0,1)-\bigcup_{n\geqq 1} l_n$ implies

18a) $$\mathfrak{m}>s$$

18b) $$\lim_{n\uparrow+\infty}\mathfrak{m}=s,$$

and it follows that

19) $$E\left[B, x_a(s)\in J_+, \int_0^{+\infty} e^{-\alpha t} f(x_a(t+s))\,dt\right]$$

$$= \lim_{n\uparrow+\infty}\sum_{k\leqq 2^n} E\left[B, x_a(s)\in[(k-1)\,2^{-n}, k2^{-n})\cap J_+, b=k2^{-n}, \right.$$
$$\left. s<\mathfrak{m}_{ab}, \int_0^{+\infty} e^{-\alpha t} f(x_a(t+\mathfrak{m}_{ab}))\,dt\right]$$

$$= \lim_{n\uparrow+\infty}\sum_{k\leqq 2^n} E[B, x_a(s)\in[(k-1)\,2^{-n}, k2^{-n})\cap J_+, b=k2^{-n},$$
$$s<\mathfrak{m}_{ab}, (G_\alpha^\bullet f)(b)]$$

$$= E[B, x_a(s)\in J_+, b=k2^{-n}, (G_\alpha^\bullet f)(x_a(s))];$$

in brief,

20) $$P[x_a(t+s)\in A\mid \mathsf{B}_{as}]$$
$$= P[x_b(t)\in A] \qquad b = x_a(s)\in J_+\equiv K_+\cap[0,1)-\bigcup_{n\geqq 1} l_n,$$

and now it remains to check 15) in case $x_a(s)=1$ and in case $x_a(s)\in Q_n$.

But, if $x_a(s)=1$, then

21) $$P[x_a(t+s)=1\mid \mathsf{B}_{as}] = 1 = P[x_1(t)=1],$$

as desired, while, if $b = x_a(s) \in Q_n$, then, *conditional on* B_{as},

22) $$x_a(t+s) = x_n(t+s-t_a(l_n \vee a)) \qquad t < t_a(r_n) - s$$
$$= x_{r_n}(t - [t_a(r_n) - s]) \qquad t \geqq t_a(r_n) - s$$

is identical in law to x_b as is evident from the diagram on noting that $t_a(l_n \vee a)$ and $x_a(t): t \leqq t_a(l_n \vee a)$ are independent of $x_{r_n}(t)$ $(t \geqq 0)$ and that $x_n(t)$ $(t \geqq 0)$ starts afresh at times $s - t_a(l_n \vee a)$.

Because the GREEN operators of $\mathsf{D}^\bullet$ map $C(Q)$ into itself, $\mathsf{D}^\bullet$ *is a diffusion*, and for the identification of its generator with the differential operator $\mathfrak{G}^\bullet$, it suffices to note

23a) $$P^\bullet_\xi(\mathfrak{m}_b < \mathfrak{m}_a) = P(\mathfrak{m}_{\xi b} < \mathfrak{m}_{\xi a}) = \frac{s(\xi) - s(a)}{s(b) - s(a)} \qquad l_n < a < \xi < b < r_n$$

23b) $$E^\bullet_\xi(\mathfrak{m}_1)$$
$$= E(t_\xi(1))$$
$$= \sum_{\xi < r_n} E^n_{\xi \vee l_n}(\mathfrak{m}_{r_n}) + s_+([\xi, 1) \cap K_+)$$
$$= s_+[\xi, 1) \qquad 0 \leqq \xi \leqq 1,$$

and to use the formulas 6), 7), and 8).

If K_+ contains no interval and $\sum_{n \geqq 1} s(Q_n) < +\infty$, then a second method is available: changing scales so as to have

24) $$\sum_{n \geqq 1} s_+(Q_n \cap d\xi) = d\xi \qquad 0 \leqq \xi \leqq 1,$$

if x is a BROWNian motion with reflection at 0, a conservative trap at 1, and local times $\mathfrak{t}$, and if $\mathfrak{f}^{-1}$ is the inverse function of

25) $$\mathfrak{f}(t) = \sum_{a < r_n} \int_{Q_n} \mathfrak{t}(\mathfrak{m}_{r_n}, \xi)\, m(d\xi) + s_+([a, b) \cap K_+)$$
$$a = x(0), \qquad b = a \vee x(t), \qquad \mathfrak{m}_{r_n} = \min(t : x(t) = r_n)$$

then $x^\bullet = x(\mathfrak{f}^{-1})$ is a (non-standard) description of the desired motion; in the special case

26a) $$m(d\xi) = 2\,d\xi$$

26b) $$s_+(K_+) = 0$$

26c) $$\mathfrak{f}(t) = \sum_{x(0) < r_n} \text{measure}$$
$$(s : x(s) \in Q_n,\ s \leqq t \wedge \mathfrak{m}_{r_n}),$$

the effect of the time substitution is to shift the black arcs of diagram 2 down on the time scale until a continuous path is achieved (see 2.11 for a special case).

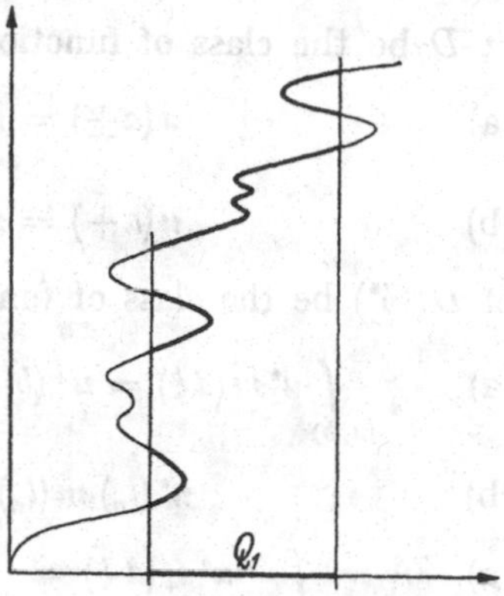

Diagram 2

Problem 1. Give a complete justification of the time substitution $t \to \mathfrak{f}^{-1}$ in the manner of 5.2 and 5.3.

Problem 2. Consider $x^{\bullet} = x(\mathfrak{f}^{-1})$ in the special case 26), taking for K_+ the standard CANTOR set $[0, 1] - \left(\frac{1}{3}, \frac{2}{3}\right) - \left(\frac{1}{9}, \frac{2}{9}\right) - \left(\frac{7}{9}, \frac{8}{9}\right) -$ *etc.*; the problem is to evaluate the shunt scale and $E_0[e^{-\alpha \min(t: x^{\bullet}(t) = 1)}]$.

$$\left[s_+[0, \xi) = \sum_{l_n \leqq \xi} (r_n \wedge \xi - l_n)^2 \qquad 0 \leqq \xi \leqq 1, \right.$$

$$\left. E_0[e^{-\alpha \min(t: x^{\bullet}(t) = 1)}] = \prod_{n \geqq 1} (\cosh 3^{-n} \sqrt{2\alpha})^{-2^{n-1}}. \right]$$

5.11. Shunts with killing

Consider non-overlapping $Q_n = [l_n, r_n) \subset Q = [0, 1]$ as in 5.10 with scales $s = s_n$, killing measures $k = k_n$, and speed measures $m = m_n$ attached, let s_+ be a (shunt) scale with

1a) $$0 < s_+[a, b) \equiv s_+(b) - s_+(a) \qquad a < b$$

1b) $$s_+[l_n, b) = \int_{l_n}^{b} p(\xi)^{-1} s(d\xi) \int_{l_n-}^{\xi} p(\eta)\, m(d\eta) \qquad l_n < b < r_n$$

$$p^+(d\xi) = p(\xi)\, k(d\xi) \quad l_n < \xi < r_n, \quad p^+(l_n) = p(l_n)\, k(l_n), \quad p(r_n-) = 1$$

1c) $$s_+[0, 1) < +\infty,$$

let k_+ be a (shunt) killing measure with

2a) *jumps of magnitude* $\leqq 1$

2b) $$k_+[l_n, b) = \int_{l_n}^{b} p(\xi)^{-1} s(d\xi) \int_{l_n-}^{\xi} p(\eta)\, k(d\eta) \qquad l_n < b < r_n$$

2c) $$k_+[0, 1) < +\infty,$$

let D be the class of functions $u \in B(Q)$ such that

3a) $$u(a\pm) = u(a) \qquad a \in \bigcup_n (l_n, r_n)$$

3b) $$u(a+) = u(a) \qquad a \in K_+ \cap [0, 1),$$

let $D(\mathfrak{G}^{\bullet})$ be the class of functions $u \in D$ such that, for some $u^{\bullet} \in D$,

4a) $$\int_{[a,b)} u^{\bullet}\, m(d\xi) = u^+(b) - u^+(a) - \int_{[a,b)} u\, k(d\xi) \qquad l_n < a < b < r_n$$

4b) $$u^{\bullet}(l_n)\, m(l_n) = u^+(l_n) - u(l_n)\, k(l_n)$$

5a) $$\int_{[a,b) \cap K_+} u^{\bullet} s_+(d\xi) = \int_{[a,b) \cap K_+} u(d\xi) - \int_{[a,b) \cap K_+} u\, k_+(d\xi) \qquad a < b$$

5b) $$u(1-) = [1 - k_+(1)]\, u(1)$$

6) $$u^\bullet(1) = 0,$$

and let $\mathfrak{G}^\bullet$ be the differential operator $u \to u^\bullet$.

Given a singular non-conservative diffusion on $Q = [0, 1]$ with

7a) $$E_0(\mathfrak{m}_1 \wedge \mathfrak{m}_\infty) < +\infty$$

7b) $$P_1(\mathfrak{m}_{1-} \wedge \mathfrak{m}_\infty = +\infty) = 1,$$

its generator $\mathfrak{G}$ can be identified with $\mathfrak{G}^\bullet$ on introducing the hitting probabilities and mean exit time

8a) $$p_{ab}(\xi) = P_\xi(\mathfrak{m}_a < \mathfrak{m}_b) \qquad a < \xi < b$$

8b) $$p_b(\xi) = P_\xi(\mathfrak{m}_b < +\infty) \qquad \xi < b$$

8c) $$e_1(\xi) = E_\xi(\mathfrak{m}_1 \wedge \mathfrak{m}_\infty)$$

and putting

9a) $$s(d\xi) = \text{constant} \times [p_{ba}(d\xi)\, p_{ab}(\xi) - p_{ab}(d\xi)\, p_{ba}(\xi)] \qquad l_n < \xi < r_n$$

9b) $$k(d\xi) = \frac{p^+_{ab}(d\xi)}{p_{ab}(\xi)} = \frac{p^+_{ba}(d\xi)}{p_{ba}(\xi)} \qquad l_n < \xi < r_n$$

$$k(l_n) = \frac{p^+_{ab}(l_n)}{p_{ab}(l_n)} = \frac{p^+_{ba}(l_n)}{p_{ba}(l_n)}$$

9c) $$m(d\xi) = -[e^+_1(d\xi) - e_1(\xi)\, k(d\xi)] \qquad l_n < \xi < r_n$$

$$m(l_n) = -[e^+_1(l_n) - e_1(l_n)\, k(l_n)]$$

10a) $$k_+(d\xi) = \frac{p_1(d\xi)}{p_1(\xi)} \qquad 0 < \xi \leqq 1$$

10b) $$s_+(d\xi) = -[e_1(d\xi) - e_1(\xi)\, k_+(d\xi)] \qquad 0 < \xi \leqq 1$$

(see 4.8 and 4.9).

As in 5.10, our purpose is to prove that *each differential operator* $\mathfrak{G}^\bullet$ *generates a diffusion with* $E_0(\mathfrak{m}_1 \wedge \mathfrak{m}_\infty) < +\infty$ *and* $P_1(\mathfrak{m}_1 \wedge \mathfrak{m}_\infty = +\infty) = 1$; *this will complete the program suggested in* 5.1.

Consider, for this purpose, the modified differential operator $\mathfrak{G}$ arising from $\mathfrak{G}^\bullet$ upon the suppression of both killing measures, and appealing to 5.10, introduce the sample paths $w: t \to x(t)$, Borel algebras B, probabilities $P_a(B)$, and local times

11) $$\mathfrak{t}(t, b) = \frac{\text{measure}(s : x(s) \in db,\ s \leqq t)}{m(db)} \qquad t \geqq 0,\ l_n \leqq b < r_n,\ n \geqq 1$$

of the standard description of the corresponding (conservative) motion.

$P_\bullet(B)$ is now extended to $\mathsf{B} \times \mathsf{B}[0, +\infty]$ according to the rule

12a) $$P_\bullet(\mathfrak{m}_\infty > t \mid \mathsf{B}) = \mathsf{e}(t) \qquad t \geqq 0$$

12b) $$\mathsf{e}(t) = e^{-\sum_{a < r_n} \int_{[l_n \vee a,\, r_n)} \mathfrak{t}(t,\xi)\, k(d\xi)} \times \bigcap_{(a, b] \cap K_+} [1 - k_+(d\xi)]$$

$$a = x(0), \quad b = x(t),$$

where, as in 4.8, $\bigcap_{(a,b]\cap K_+} [1 - k_+(d\xi)]$ is meant as a suggestive notation for $e^{-j_+((a,b]\cap K_+)}(1 - \varkappa_1)(1 - \varkappa_2)$ *etc.*, j_+ is the continuous part of k_+, and the numbers $0 < \varkappa \leqq 1$ are the jumps of $k_+(d\xi)$ $(a < \xi \leqq b)$; introducing the new sample paths

13)
$$x^\bullet(t) = x(t) \qquad t < \mathfrak{m}_\infty$$
$$= \infty \qquad t \geqq \mathfrak{m}_\infty,$$

it is claimed that $\mathsf{D}^\bullet = [x^\bullet, P_\bullet]$ *is a (non-standard) description of the desired motion.*

$\mathsf{D}^\bullet$ is *simple Markov* as the reader can check using the method of 5.6 and the obvious multiplication rule

14)
$$\mathfrak{e}(t) = \mathfrak{e}(s)\,\mathfrak{e}(t - s, w_s^+) \qquad t \geqq s \geqq 0.$$

Besides, 14) combined with the method of 5.2.14) implies that the GREEN operators

15)
$$G_\alpha^\bullet f = E_\bullet\left[\int_0^{\mathfrak{m}_\infty} e^{-\alpha t} f(x)\,dt\right]$$
$$= E_\bullet\left[\int_0^{+\infty} e^{-\alpha t} E_\bullet(\mathfrak{m}_\infty > t \mid \mathsf{B})\, f(x)\,dt\right]$$
$$= E_\bullet\left[\int_0^{+\infty} e^{-\alpha t}\,\mathfrak{e}(t)\, f(x)\,dt\right]$$

map $B(Q)$ into D, showing that $\mathsf{D}^\bullet$ *is a diffusion* and for the rest, it suffices to compute its invariants $s^\bullet$, $k^\bullet$, $m^\bullet$, $s_+^\bullet$, $k_+^\bullet$.

But, on Q_n, $s^\bullet = s$, $k^\bullet = k$ and $m^\bullet = m$ as is clear from 5.6 and 12); thus, $s_+^\bullet = s_+$ and $k_+^\bullet = k_+$ inside Q_n, and since

16)
$$p_1^\bullet(a) = P_a(\mathfrak{m}_1^\bullet < +\infty) \qquad \mathfrak{m}_1^\bullet = \min(t : x^\bullet(t) = 1)$$
$$= P_a(\mathfrak{m}_1 < \mathfrak{m}_\infty)$$
$$= E_a(\mathfrak{e}(\mathfrak{m}_1))$$
$$= E_a\left[e^{-\sum_{a < r_n} \int_{[l_n \vee a,\, r_n)} \mathfrak{t}(\mathfrak{m}_{r_n}, \xi)\, k(d\xi)}\right] \bigcap_{(a,1]\cap K_+} [1 - k_+(db)]$$
$$= \prod_{a < r_n} E_a\left[e^{-\int_{[l_n \vee a,\, r_n)} \mathfrak{t}(\mathfrak{m}_{r_n}, \xi)\, k(d\xi)}\right] \bigcap_{(a,1]\cap K_+} [1 - k_+(db)]$$
$$= \prod_{a < r_n} P_a[\mathfrak{m}_{l_n \vee a} < +\infty]^{*}\, E_{l_n \vee a}\left[e^{-\int_{[l_n \vee a,\, r)} \mathfrak{t}(\mathfrak{m}_{r_n}, \xi)\, k(d\xi)}\right]^{**}$$
$$\times \bigcap_{(a,1]\cap K_+} [1 - k_+(db)]$$

---

* $P_a[\mathfrak{m}_{l_n \vee a} < +\infty] = 1.$

** $E_{l_n \vee a}\left[e^{-\int_{(l_n \vee a,\, r_n)} \mathfrak{t}(\mathfrak{m}_{r_n}, \xi)\, k(d\xi)}\right] = P_{l_n \vee a}(\mathfrak{m}_{r_n} < \mathfrak{m}_\infty) = p_{r_n}^\bullet(l_n \vee a).$

$$= \prod_{a<r_n} p^{\bullet}_{r_n}(l_n \vee a) \bigcap_{(a,1]\cap K_+} [1 - k_+(db)]$$
$$= \prod_{a<r_n} e^{-k_+[l_n\vee a,\, r_n)*} \bigcap_{(a,1]\cap K_+} [1 - k_+(db)]$$
$$= \bigcap_{(a,1]} [1 - k_+(db)],$$

one has

17) $$k^{\bullet}_+(da) = \frac{p^{\bullet}_1(da)}{p^{\bullet}_1(a)} = k_+(da) \qquad 0 < a \leqq 1,$$

while since

18) $$e^{\bullet}_1(a) = E_a(\mathfrak{m}^{\bullet}_1 \wedge \mathfrak{m}^{\bullet}_\infty)$$
$$= E_a(\mathfrak{m}_1 \wedge \mathfrak{m}_\infty)$$
$$= E_a[E_a(\mathfrak{m}_\infty, \mathfrak{m}_\infty < \mathfrak{m}_1 \mid \mathsf{B})] + E_a[\mathfrak{m}_1 P_a(\mathfrak{m}_\infty > \mathfrak{m}_1 \mid \mathsf{B})]$$
$$= E_a\left[-\int_0^{\mathfrak{m}_1} t\, d\mathfrak{e} + \mathfrak{m}_1\, \mathfrak{e}(\mathfrak{m}_1)\right]$$
$$= E_a\left[\int_0^{\mathfrak{m}_1} \mathfrak{e}\, dt\right]$$
$$= \sum_{a<r_n} E_a\left[\int_{\mathfrak{m}_{l_n\vee a}}^{\mathfrak{m}_{r_n}} \mathfrak{e}\, dt\right] + E_a\left[\int_{[a,1)\cap(K_+ - \bigcup_{n\geqq 1} l_n)} \mathfrak{e}(\mathfrak{m}_b)\, d\mathfrak{m}_b\right]$$
$$= \sum_{a<r_n} E_a\left[\mathfrak{e}(\mathfrak{m}_{l_n\vee a})\, E_{l_n\vee a}\left(\int_0^{\mathfrak{m}_{r_n}} \mathfrak{e}\, dt\right)\right]$$
$$+ E_a\left[\int_{[a,1)\cap K_+} \mathfrak{e}(\mathfrak{m}_b)\, s_+(db)\right]^{**}$$
$$= \sum_{a<r_n} P_a[\mathfrak{m}^{\bullet}_{l_n\vee a} < +\infty] E_{l_n\vee a}(\mathfrak{m}^{\bullet}_{r_n}) + \int_{[a,1)\cap K_+} E_a[\mathfrak{e}(\mathfrak{m}_b)]\, s_+(db)$$
$$= \sum_{a<r_n} p^{\bullet}_{l_n\vee a}(a)^{***} \int_{l_n\vee a}^{r_n} p^{\bullet}_b(l_n \vee a)\dagger\, s_+(db) + \int_{[a,1)\cap K_+} p^{\bullet}_b(a)\, s_+(db)$$
$$= \sum_{a<r_n} \int_{l_n\vee a}^{r_n} p^{\bullet}_b(a)\, s_+(db) + \int_{[a,1)\cap K_+} p^{\bullet}_b(a)\, s_+(db)$$
$$= \int_a^1 p^{\bullet}_b(a)\, s_+(db),$$

---

* $e^{-k_+[a,b)} = P^{\bullet}_a(\mathfrak{m}^{\bullet}_b < +\infty)$ on each non-singular interval; see 4.8.14) and the footnote attached.

** $d\mathfrak{m}_b = s_+(db)$ on $[a,1) \cap K_+$; see 5.10.12) and 5.10.14).

*** $p^{\bullet}_b(a) = P_a(\mathfrak{m}^{\bullet}_b < +\infty)$.

† See 4.8.18).

one has

19) $$s^{\bullet}_{+}(da) = -[e^{\bullet}_{1}(da) - k^{\bullet}_{+}(da)\, e^{\bullet}_{1}(a)] = s_{+}(da) \qquad 0 \leqq a < 1,$$

completing the proof.

5.12. Creation of mass

A special model for the creation of mass will be described and used to discuss a non-linear parabolic equation (see G. HUNT [2 (3)] for a different approach to creation).

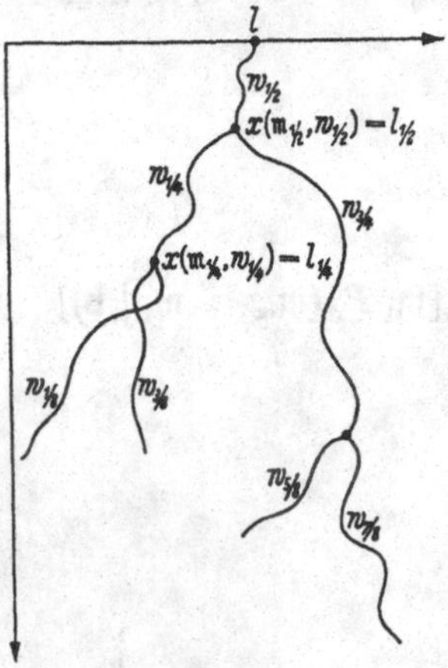

Diagram 1

Given an integral $\mathfrak{k}(t) = \int \mathfrak{t}(t, \xi)\, k(d\xi)$ of standard BROWNian local times relative to a non-negative BOREL measure $k(d\xi)$ on R^1 $(k(R^1) > 0)$, consider the shower pictured in diagram 1, where, for each $k = 1, 3, \dots, 2^n - 1$ and each $n \geqq 1$, $w_{k/2^n}$ is a standard BROWNian path starting from scratch at the indicated ramification point, $\mathfrak{m}_{k/2^n} \leqq \leqq +\infty$ is subject to the conditional law

$$P_{\bullet}(\mathfrak{m}_{k/2^n} > t \mid w_{k/2^n}) = e^{-\mathfrak{k}(t+,\, w_{k/2^n})} \qquad (t \geqq 0),$$

and, conditional on the ramification point, $w_{k/2^n}$ and $\mathfrak{m}_{k/2^n}$ are independent of the path up to the ramification and also independent of all the subshowers splitting off before the ramification time.

Given $e_n = 0$ or 1 $(n \geqq 1)$, the path

1) $$\mathfrak{w}: t \to x(t - \mathfrak{t}_{\cdot e_1 \dots e_{n-2} + 2^{-n+1}},\, w_{\cdot e_1 \dots e_{n-1} + 2^{-n}})$$
$$\mathfrak{t}_{\cdot e_1 \dots e_{n-2} + 2^{-n+1}} \leqq t < \mathfrak{t}_{\cdot e_1 \dots e_{n-1} + 2^{-n}}$$

with

2) $$\mathfrak{t}_{\cdot e_1 \dots e_{n-1} + 2^{-n}} \equiv \sum_{m \leqq n} \mathfrak{m}_{\cdot e_1 \dots e_{m-1} + 2^{-m}} \qquad (n \geqq 1)$$
$$\equiv 0 \qquad (n = 0)$$

is standard BROWNian and the ramification times $\mathfrak{t}_{\cdot e_1 \dots e_{n-1} - 2^{-n}}: n \geqq 1$ are the jump times of a POISSON process p subject to the conditional law

3) $$P_{\bullet}[p(t) = j \mid \mathfrak{w}] = e^{-\mathfrak{k}}\, \mathfrak{k}^j / j! \qquad j \geqq 0, \qquad \mathfrak{k} = \mathfrak{k}(t+, w);$$

the proofs are left to the reader (see problem 5.6.4).

In addition, the shower is strict MARKOV in the sense that if $\mathfrak{m}$ is a MARKOV time (*i.e.*, if the indicator of $(\mathfrak{m} < t)$ is a function of the shower for times $s \leqq t$), and if, up to time $\mathfrak{m}$, $n < +\infty$ particles have been created, then the future shower is identical in law to the superposition of n independent showers springing from the points that the n particles occupied at time $t = \mathfrak{m}$.

Here is a grander model combining creation and annihilation. Given $0 \leqq k_n(dl)$ $(0 \leqq n \neq 1)$ such that $k = \sum_{0 \leqq n \neq 1} k_n$ is finite on compacts, select a standard BROWNian path w_1, a ramification time $\mathfrak{m}_1$, and an integer n_1 subject to the conditional law

4) $$P_{\bullet}[\mathfrak{m} \in dt, n_1 = n \mid w] = e^{-\mathfrak{k}(t+)} \mathfrak{k}_n(dt) \quad t \geqq 0, \; 0 \leqq n \neq 1,$$

and, in case $n_1 = 0$, stop; otherwise, select n_1 independent standard BROWNian paths w_2 starting at $l_1 = x(\mathfrak{m}_1, w_1)$, to each of these attach independent ramification times $\mathfrak{m}_2$ and integers n_2 subject to the law $P_{\bullet}[\mathfrak{m} \in dt, n_2 = n \mid w = w_2]$, and so on.

Below, *simple creation* refers to the first model, *creation and annihilation* to the second.

Problem 1. Compute the generator $\mathfrak{G}$ of the model of creation and annihilation, viewing it as a MARKOVian motion on the space Q of points

$$l^n : l = (l_1, l_2, \ldots, l_d), \; l_1 < l_2 < \cdots < l_d, \; n = (n_1, n_2, \ldots, n_d),$$
$$1 \leqq n_1, n_2, \ldots, n_d$$

in the special case $k_m(d\xi) = 2d\xi$ $(0 \leqq m \neq 1)$. Here l stands for the *positions* occupied by the particles and n for the *number* of particles at those points.

[Given $u : Q \to R^1$ such that $u(l^n) \in C^2(R^d)$ as a function of $(l_1, l_2, \ldots, l_d)$ for each d and n, one finds

$$(\mathfrak{G}\, u)\,(l^n) = \tfrac{1}{2} \Delta\, u(l^n) + \sum_{\substack{i \leqq d \\ 0 \leqq m \neq 1}} n_i [u(l^{n+(m-1)e_i}) - u(l^n)],$$

where Δ is the d-dimensional LAPLACE operator, and in the sum, $e_1 = (1, 0, \ldots, 0)$, $e_2 = (0, 1, \ldots, 0)$, *etc.*]

5.13. A parabolic equation

Consider the model of diagram 5.12.1 (simple creation) with $k(dl) = 2\,|l|^{\gamma}\, dl$ $(\gamma > 0)$.

Given $(t, a, b) \in (0, +\infty) \times R^2$, if $\#(t, db)$ is the number of particles $\in db$ at the instant t, then

1) $$\begin{aligned} E_a[\#(t, db)] &= \sum_{n \geqq 0} \sum_{e_1, e_2, \ldots, e_n} P_a[\mathfrak{t}_{\bullet e_1 \ldots e_{n-1}+2^{-n}} \leqq t < \mathfrak{t}_{\bullet e_1 \ldots e_n + 2^{-n-1}}, \\ &\qquad x(t - \mathfrak{t}_{\bullet e_1 \ldots e_{n-1}+2^{-n}}, w_{\bullet e_1 \ldots e_n + 2^{-n-1}}) \in db] \\ &= \sum_{n \geqq 0} 2^n E_a \left[e^{-\mathfrak{f}} \frac{\mathfrak{f}^n}{n!}, x(t) \in db \right] \\ &= E_a[e^{\mathfrak{f}}, x(t) \in db] \\ &= e(t, a, b)\, db. \end{aligned}$$

$e(t, a, b)$ is the fundamental solution of

$$\frac{\partial u}{\partial t} = \mathfrak{G}^{\bullet}\, u, \qquad \mathfrak{G}^{\bullet} = D^2/2 + |l|^{\gamma}$$

in the sense that, if $0 \leqq f \in C(R^1)$ and *if* the problem

2a) $$\frac{\partial u}{\partial t} = \mathfrak{G}^\bullet u$$

2b) $$u(0+,.) = f$$

has a non-negative solution u for $t < t_1$, then $u \equiv \int e f$ $(t < t_1)$; the proof occupies the rest of this section.

Given $n \geqq 1$, the eigendifferential expansion of 4.11 can be applied to the kernel $e^{-tn^\gamma} \times e^n(t, a, b)$:

3) $$e^n(t, a, b)\, db = E_a[e^{\int_0^t |x(s)|^\gamma ds}, \mathfrak{m}_{-n} \wedge \mathfrak{m}_{+n} > t, x(t) \in db] \qquad t > 0, \quad |a|, |b| < n;$$

this justifies

4a) $$\frac{\partial e^n}{\partial t} = \mathfrak{G}^\bullet e^n *$$

4b) $$\lim_{a \downarrow -n} e^n(t, a, b) = \lim_{a \uparrow +n} e^n(t, a, b) = 0 \qquad t > 0, \ |b| < n$$

4c) $$\lim_{t \downarrow 0} \int_{|b-a| < \delta} e^n(t, a, b)\, db = 1 \qquad \delta > 0$$

4d) $$\lim_{t \downarrow 0} \int_{|b-a| > \delta} e^n(t, a, b)\, db = 0 \qquad \delta > 0;$$

in brief, e^n is the fundamental solution of the problem

5a) $$\frac{\partial u}{\partial t} = \mathfrak{G}^\bullet u \qquad |b| < n$$

5b) $$u(t, \pm n) = 0 \qquad t > 0.$$

Consider now a solution $v < +\infty$ of 2) for $t < t_1$; if $|a| < n$, then

6) $$\frac{\partial}{\partial s} \int_{|b|<n} e^n(t-s, a, b)\, v(s, b)\, db = -\frac{1}{2} e^n_3 v \Big|_{-n}^{n} \geqq 0,$$

and since $e^n \uparrow e$ $(n \uparrow +\infty)$, it follows that

7) $$v \geqq \int_{|b|<n} e^n f\, db \quad \uparrow u \equiv \int e f\, db \quad n \uparrow +\infty, \quad t < t_1.$$

Because u can be expressed as

8) $$u(t, b) = E_a\left[e^{\int_0^t |x(s)|^\gamma ds} f(x_t)\right]$$

$$= \int_{-\infty}^{+\infty} g(t, a, b) f\, db + E_a\left[\int_0^t e^{\int_{t-s}^t |x(\theta)|^\gamma d\theta} |x(t-s)|^\gamma ds\, f(x_t)\right]$$

$$= \int_{-\infty}^{+\infty} g(t, a, b) f(b)\, db + \int_0^t ds \int_{-\infty}^{+\infty} g(t-s, a, b)\, |b|^\gamma u(b)\, db,$$

* $\mathfrak{G}^\bullet$ is applied to a or to b.

it is a solution of 2) as long as it is $< +\infty$, as straight-forward differentiation shows, and combining this with 7b), u is found to be *the smallest solution of* 2).

Given $\gamma > 2$, *then* $e(t, a, b) \equiv +\infty$ $(t > 0)$ because the *tied* BROWNian motions

9a) $[x(s): s \leqq t, P_{ab}(B) \equiv P_a(B \mid x(t) = b)]$

9b) $\left[a + x(s) + (b-a)\frac{s}{t} : s \leqq t, P_{00}\right]$

are identical in law, and so

10) $e(2t, a, b)$

$$= \int_{-\infty}^{+\infty} e(t, a, \xi)\, e(t, \xi, b)\, d\xi$$

$$= \int_{-\infty}^{+\infty} E_0\left[\exp\left(\int_0^t \left|a + x(s) + (\xi - a)\frac{s}{t}\right|^\gamma ds\right) \,\Big|\, x(t) = 0\right] \times$$
$$\times g(t, a, \xi)\, e(t, \xi, b)\, d\xi$$

$$\geqq \text{constant} \times E_0\left[\int_{-\infty}^{+\infty} \exp\left(\int_0^t \left|a + x(s) + (\xi - a)\frac{s}{t}\right|^\gamma ds - \text{constant} \times \right.\right.$$
$$\left.\left. \times\, \xi^2\right) d\xi \,\Big|\, x(t) = 0\right]$$

$$\equiv +\infty$$

because the inner integral explodes; *esp.*, $t_1 = 0$ unless $f = u \equiv 0$.

On the other hand, *if* $\gamma < 2$, *then* $\int_{-\infty}^{+\infty} e(t, a, b)\, db < +\infty$ $(t > 0)$ because

11) $\int_{-\infty}^{+\infty} e(t, a, b)\, db$

$$= E_a\left[e^{\int_0^t |x(s)|^\gamma ds}\right]$$

$$= E_0\left[e^{\int_0^t |a + x(s)|^\gamma ds}\right]$$

$$\leqq 2E_0\left[e^{t(|a| + \max_{s \leqq t} x(s))^\gamma}\right]$$

$$= 4\int_0^{+\infty} \frac{e^{-b^2/2t}}{\sqrt{2\pi t}}\, e^{(|a| + b)^\gamma}\, db$$

$$< +\infty,$$

while, *if* $\gamma = 2$, *then*

12a) $e(t, a, b) < +\infty$ *or* $= +\infty$ *according as* $t < \pi/\sqrt{2}$ *or not*

12b) $\int_{-\infty}^{+\infty} e(t, a, b)\, db < +\infty$ *or* $= +\infty$ *according as* $t < \pi/2\sqrt{2}$ *or not*

(see problem 1 below), *so it remains to prove that if* $\gamma \leqq 2$ *and if* v *solves* 2) *for* $t < t_1$, *then* $v \equiv u\ (t < t_1)$; the method used below goes back to A. Tihonov [1].

Given such a solution v, $w \equiv v - u$ is a non-negative solution of

$$\text{13a)}\qquad \frac{\partial w}{\partial t} = \mathfrak{G}^{\bullet} w$$

$$\text{13b)}\qquad w(0+, .) = 0,$$

and

$$\text{14)}\qquad z(t, a) \equiv \int_0^t \left[w(s, a) + 2 \int_0^a db \int_0^b |c|^{\gamma} w(s, c)\, dc \right] ds + \int_0^t \left[w(s, -a) + 2 \int_0^{-a} db \int_0^b |c|^{\gamma} w(s, c)\, dc \right] ds$$

is an *even* function of a and has the properties

$$\text{15a)}\qquad \frac{\partial z}{\partial t} \geqq w(t, -a) + w(t, a) = \int_0^t [\mathfrak{G}^{\bullet} w(t, -a) + \mathfrak{G}^{\bullet} w(t, a)]\, ds = \tfrac{1}{2} z_{22} \geqq 0 \qquad t < t_1$$

$$\text{15b)}\qquad z \geqq 0$$

$$\text{15c)}\qquad z(0+, .) = 0.$$

According to 6) applied to w,

$$\text{16)}\qquad w(t, a) = \int_0^t -\tfrac{1}{2} e_3^n w \,\big|_{-n}^{n}\, ds \qquad |a| < n, \quad t < t_1,$$

and using the bound

$$\text{17)}\qquad \mp e_3^n(t, a, \pm n) \leqq e^{t n^{\gamma}} \frac{4n}{t} g(t, a, \pm n) < \text{constant} \times e^{t n^{\gamma}} e^{-n^2/3t} \qquad |a| < n,$$

which the reader can easily check from 3), it follows that

$$\text{18)}\qquad w(t, a) < \text{constant} \times e^{t n^{\gamma}} e^{-n^2/3t} \int_0^t [w(s, -n) + w(s, +n)]\, ds < \text{constant} \times e^{t n^{\gamma}} e^{-n^2/3t} z(t, n).$$

Also,

19)
$$\frac{\partial}{\partial s}\int_{-n}^{n} g(t-s,a,b)\,z(s,b)\,db$$
$$=-\tfrac{1}{2}g_3 z+\tfrac{1}{2}g z_2\Big|_{-n}^{+n}+\int_{-n}^{+n}\left(-\tfrac{1}{2}z_{22}+z_1\right)g\,db$$
$$\geqq 0,$$

i.e.,

20)
$$z(t,a)\geqq\int_{-\infty}^{+\infty} g(t-s,a,b)\,z(s,b)\,db,$$

and since $0\leqq z_2$ [see 15a)]

21)
$$z\left(\frac{t_1}{2},0\right)\geqq\int_{n}^{2n} g\left(\frac{t_1}{6},0,b\right)z\left(\frac{t_1}{3},b\right)db$$
$$>\text{constant}\times e^{-12n^2/t_1}\,z\left(\frac{t_1}{3},n\right).$$

Returning to 18) with this bound in hand, it develops that for $|a|<n$ and $t<t_1/3$,

22)
$$w(t,a)<\text{constant}\times e^{tn^\gamma}\,e^{-n^2/3t}\,e^{12n^2/t_1},$$

which tends to 0 as $n\uparrow+\infty$ provided $t+12/t_1<1/3t$; in brief, $w\equiv 0$ for small t, and this can be propagated, proving that $w\equiv 0$ $(t<t_1)$.

Problem 1 after H. Trotter (see also R. C. Cameron and W. T. Martin [*1*]). Prove that, if $\gamma=2$ and $\theta=\sqrt{2}t$, then

$$e(t,a,b)=\frac{e^{-\cot\theta\,[a^2-2ab\sec\theta+b^2]/\sqrt{2}}}{\sqrt{\pi\sqrt{2}\sin\theta}}\qquad t<\frac{\pi}{\sqrt{2}}$$
$$\equiv+\infty\qquad t\geqq\frac{\pi}{\sqrt{2}}$$

and

$$\int_{-\infty}^{+\infty} e(t,a,b)\,db=\sqrt{\sec\theta}\,e^{-a^2\tan\theta/\sqrt{2}}\qquad t<\frac{\pi}{2\sqrt{2}}$$
$$=+\infty\qquad t\geqq\frac{\pi}{2\sqrt{2}}$$

[$e(t,a,b)\equiv e^{-\cot\theta}$ *etc.* satisfies 2), so it must be the fundamental solution; another method is to express the fundamental solution

$$e_\alpha(t,a,b)\,db=E_a\left[e^{-\alpha\int_0^t x(s)^2\,ds},\ x(t)\in db\right]\qquad\alpha>0$$

in terms of eigenvalues $\gamma_n = -\sqrt{2\alpha}\,(n+\frac{1}{2})$ and eigenfunctions

$$f_n = e^{-\sqrt{2\alpha}\,b^2/2} H_n\left(\sqrt[4]{2\alpha}\,b\right) \quad \text{of} \quad \mathfrak{G}^{\bullet} = \frac{1}{2}D^2 - \alpha\, b^2:$$

$$e_\alpha(t,a,b) = \sum_{n \geqq 0} e^{-\sqrt{2\alpha}(n+\frac{1}{2})t}\, e^{-\sqrt{2\alpha}\left(\frac{a^2+b^2}{2}\right)} \frac{H_n\left(\sqrt[4]{2\alpha}\,a\right) H_n\left(\sqrt[4]{2\alpha}\,b\right)}{\sqrt{\pi}\, 2^n\, n!/\sqrt[4]{2\alpha}}$$

$$= \frac{e^{-\frac{\sqrt{2\alpha}}{2}\coth\sqrt{2\alpha}\,t\,[a^2 - 2ab\,\mathrm{sech}\sqrt{2\alpha}\,t + b^2]}}{\sqrt{2\pi\sinh\sqrt{2\alpha}\,t/\sqrt{2\alpha}}}$$

(see A. Erdélyi [*1* (2): 194 (22)]), and then to continue both sides from $\alpha > 0$ to $\alpha = -1$ in the complex plane.]

5.14. Explosions

Consider the model of diagram 5.12.1 (simple creation), denote by $\mathfrak{e}$ the explosion time $\sup(t : \#(t, R^1) < +\infty)$, and let us check that $P_\bullet(\mathfrak{e} = +\infty)$ *is constant*: $P_\bullet(\mathfrak{e} = +\infty) \equiv 0$ *or* $\equiv 1$.

Given $a < b$, if w is a (standard Brownian) branch of the full shower starting at a and if $\mathfrak{m}$ is its passage time to b, then the subshower beginning with $w^+_{\mathfrak{m}}$ is a replica (in law) of the full shower starting at b; this implies $P_a(\mathfrak{e} = +\infty) \leqq P_b(\mathfrak{e} = +\infty)$, and now the 01 law $P_\bullet(\mathfrak{e} = +\infty) \equiv 0$ or 1 is immediate from

1) $$P_\bullet(\mathfrak{e} = +\infty) = E_\bullet[P_{l_{\frac{1}{2}}}(\mathfrak{e} = +\infty)^2] \qquad l_{\frac{1}{2}} = x(\mathfrak{m}_{\frac{1}{2}}, w_{\frac{1}{2}}).$$

$P_\bullet(\mathfrak{e} = +\infty)$ is to be computed in some special cases.

Given $n \geqq 1$, if

$$\mathfrak{t}_{\bullet e_1 e_2 \cdots e_{n-1} + 2^{-n}} \equiv \sum_{m \leqq n} \mathfrak{m}_{\bullet e_1 e_2 \cdots e_{m-1} + 2^{-m}} > t$$

for each $(e_1, e_2, \ldots, e_{n-1})$, then $\#(t, R^1) \leqq 2^{n-1}$ so that

2) $$P_\bullet[\#(t, R^1) > 2^{n-1}] \leqq \sum_{e_1, e_2, \ldots, e_{n-1} = 0,1} P_\bullet[\mathfrak{t}_{\bullet e_1 e_2 \cdots e_{n-1} + 2^{-n}} \leqq t]$$

$$= 2^{n-1} P_\bullet[\mathfrak{t}_{2^{-n}} \leqq t] = 2^{n-1} E_\bullet\left[e^{-\mathfrak{f}} \sum_{m \geqq n} \frac{\mathfrak{f}^m}{m!}\right] \qquad \mathfrak{f} = \mathfrak{f}(t+)$$

$$= 2^{n-1} E_\bullet\left[\frac{\mathfrak{f}^n}{(n-1)!} \int_0^1 s^{n-1} e^{-s\mathfrak{f}}\, ds\right]$$

$$= \frac{1}{2} E_\bullet\left[e^{\mathfrak{f}} \frac{(2\mathfrak{f})^n}{(n-1)!} \int_0^1 s^{n-1} e^{-(s+1)\mathfrak{f}}\, ds\right]$$

$$\leqq \frac{1}{2} E_\bullet(e^{\mathfrak{f}}) \frac{(2n/e)^n}{(n-1)!} \int_0^1 \frac{s^{n-1}}{(s+1)^n}\, ds\, *$$

* Use $\mathfrak{f}^n e^{-(1+s)\mathfrak{f}} \leqq [n/e(1+s)]^n$.

$$= \frac{1}{2} E_{\bullet}(e^{\mathfrak{f}}) \frac{(2n/e)^n}{(n-1)!} \int_0^{1/2} \frac{t^{n-1}}{1-t} dt$$

$$\leq E_{\bullet}(e^{\mathfrak{f}}) \frac{(n/e)^n}{n!} \sim \frac{E_{\bullet}(e^{\mathfrak{f}})}{\sqrt{2\pi n}},$$

and now the MARKOVian nature of the shower shows that if $E_{\bullet}[e^{\mathfrak{f}(t+)}] < +\infty$ for some t, then

$$1 = P_{\bullet}[\#(t, R^1) < +\infty] = P_{\bullet}[\#(2t, R^1) < +\infty]$$
$$= P_{\bullet}[\#(3t, R^1) < +\infty] = etc.,$$

or, what is the same, $P_{\bullet}(\mathfrak{e} = +\infty) = 1$.

Suppose now that

$$\mathfrak{f}(t) = 2 \int_{R^1} \mathfrak{t}(t, l) \, |l|^{\gamma} \, dl = \int_0^t |x(s)|^{\gamma} ds$$

and let us prove that

3) $\quad P_{\bullet}(\mathfrak{e} = +\infty) = 1$ *or* 0 *according as* $\gamma \leq 2$ *or* $\gamma > 2$.

If $\gamma \leq 2$, then

4) $$E_a[e^{\mathfrak{f}(t)}] < E_a\left[e^{t(1+\max_{s \leq t} x(s))^2}\right]$$

$$\leq 2\sqrt{2} \int_0^{+\infty} e^{t[1+(|a|+b)^2]} \frac{e^{-b^2/2t}}{\sqrt{\pi t}} db$$

$$< +\infty \qquad\qquad t < 1/\sqrt{2},$$

which implies that $P_{\bullet}(\mathfrak{e} = +\infty) = 1$.

Consider for the proof that $P_{\bullet}(\mathfrak{e} < +\infty) = 1$ for $\gamma > 2$, the *envelope* $e_+(t) =$ *the smallest l such that the shower lies to the left of l up to time t*, take $\frac{1}{2} < \beta \leq 1$ such that $(1-\beta)\gamma > 2\beta > 1$ and then take $1 < \alpha < 2\beta$, define

$$l_n = 0 \quad (n = 0) \qquad = \sum_{1 \leq m \leq n} m^{-\beta} \qquad (n \geq 1),$$

and note that the event

$$B_+: \quad e_+\Big(2 \sum_{2 \leq m < n} m^{-\alpha}\Big) \geq l_n \qquad (n \geq 1)$$

and the resulting $e_+\left(2 \sum_{n \geq 2} n^{-\alpha}\right) = +\infty$ cannot occur unless $\#\left(\sum_{n \geq 2} n^{-\alpha}, R^1\right) = +\infty$, *i.e.*, unless $\mathfrak{e} \leq 2 \sum_{n \geq 2} n^{-\alpha} < +\infty$.*

---

* $\mathfrak{e} < +\infty$ if and only if the shower is unbounded at $t = \mathfrak{e}$; the picture is that of a line storm moving out toward $\pm\infty$ and raining down harder and harder as it travels.

Our task is now to find a manageable event $B_- \subset B_+$ and using the 01 law for $P_\bullet(\mathfrak{e} < +\infty)$, to conclude from $P_1(B_-) > 0$ that $P_\bullet(\mathfrak{e} < +\infty) \equiv 1$.

Consider for this purpose a single BROWNian path w_n starting at $l_{n-1} = \sum_{1 \leq m < n} m^{-\beta}$ $(n \geq 2)$, let new BROWNian paths split off it at the jump times of the POISSON process p with (conditional) law

$$P_{l_{n-1}}[p(t) = j \mid \mathsf{B}] = e^{-\mathfrak{t}(t)}\, \mathfrak{t}(t)^j / j! \quad j \geq 0,$$

let j_n be the number of such paths splitting off before time $n^{-\alpha} \wedge \mathfrak{m}(l_{n-2}, w_n)$*, notice that the chance that none of these hit l_n before time $2n^{-\gamma}$ is $< E_{l_{n-1}}[P_0[\mathfrak{m}((n-1)^{-\beta} + n^{-\beta}) > n^{-\alpha}]^{j_n}]$, and let B_- be the event that some of the j_2 paths splitting off w_2 before time $2^{-\alpha} \wedge \mathfrak{m}(l_0, w_2)$ hit l_2 before time $2 \cdot 2^{-\alpha}$; that, if w_2^* is the winner in point of time of this race to l_2, then some of the j_3 paths splitting off $w_3 = (w_2^*)^+_{\mathfrak{m}(l_2, w_2^*)}$ before time $3^{-\alpha} \wedge \mathfrak{m}(l_1, w_3)$ hit l_3 before time $2 \cdot 3^{-\alpha}$; that if w_3^* is the winner in point of time of this race to l_3, then some of the j_4 paths

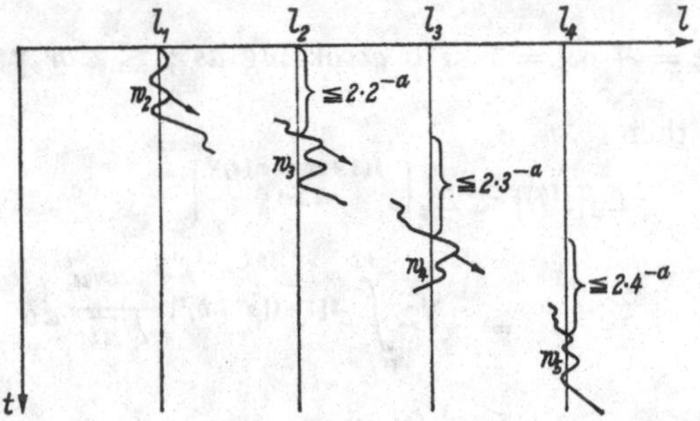

Diagram 1

splitting off $w_4 = (w_3^*)^+_{\mathfrak{m}(l_3, w_3^*)}$ before time $4^{-\alpha} \wedge \mathfrak{m}(l_2, w_4)$ hit l_4 before time $2 \cdot 4^{-\alpha}$; *etc.* as in diagram 1, depicting part of the original shower starting at $l_1 = 1$.

$B_+ \supset B_-$, and

5)
$$\begin{aligned} P_1(B_+) &\geq P_1(B_-) \\ &\geq (1 - E_{l_1}[P_0(\mathfrak{m}(1^{-\beta} + 2^{-\beta}) > 2^{-\alpha})^{j_2}]) \\ &\quad \times (1 - E_{l_2}[P_0(\mathfrak{m}(2^{-\beta} + 3^{-\beta}) > 3^{-\alpha})^{j_3}]) \\ &\quad \times \textit{etc.} \end{aligned}$$

is *positive* if

6)
$$\begin{aligned} &E_{l_1}[P_0(\mathfrak{m}(1^{-\beta} + 2^{-\beta}) > 2^{-\alpha})] \\ &\quad + E_{l_2}[P_0(\mathfrak{m}(2^{-\beta} + 3^{-\beta}) > 3^{-\alpha})^{j_3}] \\ &\quad + \textit{etc.} \\ &\quad < +\infty. \end{aligned}$$

* $\mathfrak{m}(l) = \mathfrak{m}(l, w)$ is the passage time $\min(t : x(t) = l)$.

But since $\alpha < 2\beta$,

7) $$P_0\big(\mathfrak{m}((n-1)^{-\beta}+n^{-\beta}) > n^{-\alpha}\big) = \int\limits_{n^{-\alpha}[(n-1)^{-\beta}+n^{-\beta}]^{-2}}^{+\infty} \frac{e^{-1/2t}}{\sqrt{2\pi t^3}}\,dt < \tfrac{1}{2}$$

for $n \uparrow +\infty$, and since $(1-\beta)\gamma > 2\beta > \alpha$,

8) $$\begin{aligned}\sum_{n\geq 3} & E_{l_{n-1}}[2^{-j_n}] \\ &= \sum_{n\geq 3} E_{l_{n-1}}\Big[\sum_{m\geq 0} 2^{-m} e^{-\mathfrak{k}(n^{-\alpha}\wedge\mathfrak{m}(l_{n-2}))} \frac{[\mathfrak{k}(n^{-\alpha}\wedge\mathfrak{m}(l_{n-2}))]^m}{m!}\Big] \\ &= \sum_{n\geq 3} E_{l_{n-1}}[e^{-\frac{1}{2}\mathfrak{k}(n^{-\alpha}\wedge\mathfrak{m}(l_{n-2}))}] \\ &< \sum_{n\geq 3} E_{l_{n-1}}[e^{-\frac{1}{2}(n^{-\alpha}\wedge\mathfrak{m}(l_{n-2}))l_{n-2}^{\gamma}}] \\ &< \sum_{n\geq 3} e^{-\frac{1}{2}n^{-\alpha}l_{n-2}^{\gamma}} + \sum_{n\geq 3} E_{l_{n-1}}[e^{-\frac{1}{2}\mathfrak{m}(l_{n-2})l_{n-2}^{\gamma}}] \\ &< \sum_{n\geq 3} e^{-\frac{1}{2}n^{-\alpha}l_{n-2}^{\gamma}} + \sum_{n\geq 3} e^{-l_{n-2}^{\gamma/2}(n-1)^{-\beta}} \\ &= \sum_{n\geq 3} e^{-\frac{1}{2}n^{-\alpha}\left(\sum_{i\leq n-2} i^{-\beta}\right)^{\gamma}} + \sum_{n\geq 3} e^{-(n-1)^{-\beta}\left(\sum_{i\leq n-2} i^{-\beta}\right)^{\gamma/2}} \\ &\leq \sum_{n\geq 1} e^{-\frac{1}{2}n^{(1-\beta)\gamma-\alpha}} + \sum_{n\geq 1} e^{-n^{\frac{1}{2}(1-\beta)\gamma-\beta}} \\ &< +\infty,\end{aligned}$$

which, in conjunction with 6), implies $P_1(B_-) > 0$ and completes the proof.

5.15. A non-linear parabolic equation

Given a non-negative number γ, let $u = u(t, l)$ be the solution of

1a) $$\frac{\partial u}{\partial t} = \frac{1}{2}\frac{\partial^2 u}{\partial l^2} + u(1-u)\,|l|^{\gamma} \qquad t > 0,\ l \in R^1$$

1b) $$0 \leq u \leq 1$$

1c) $$u(0+, .) = 0.$$

$u = 0$ *if* $\gamma \leq 2$, *while if* $\gamma > 2$ *an additional solution is* $u = P_\bullet(\mathfrak{e} < t)$, $\mathfrak{e}$ being the explosion time of the model of 5.14 with $\mathfrak{k} = \int_0^t |x|^{\gamma}\,ds$; *for this second solution,*

2a) $$0 < u < 1 \qquad (t > 0)$$

2b) $$u \uparrow 1 \qquad (t \uparrow +\infty);$$

the proof is indicated below. A. KOLMOGOROV, I. PETROVSKII, and N. PISCOUNOV [1] discussed a similar problem.

N. LEVINSON made us the following simple proof that $u \equiv 0$ if $\gamma \leqq 2$. Define $v = u\, e^{-2t(1+l^2)}$; then

3) $$\frac{\partial v}{\partial t} = \frac{1}{2}\frac{\partial^2 v}{\partial l^2} + \frac{\partial}{\partial l}[4t\,l\,v] + [-2t + 8t^2 l^2 + (1-u)\,|l|^\gamma - 2(1+l^2)],$$

and since the last bracket is $\leqq 0$ for $t \leqq 1/3$,

4) $$\begin{aligned} \int_{-n}^{n} v\,dl &\leqq \left(\int_{n}^{n+1} db + \int_{-n-1}^{-n} da\right)\int_a^b v\,dl \\ &\leqq \int_0^t ds \left(\int_{-n-1}^{-n} + \int_n^{n+1}\right)[\tfrac{1}{2}v_2(s,l) + 4s\,l\,v(s,l)]\,dl \\ &\leqq \int_0^t ds\, 2e^{-2n^2 s} \\ &< n^{-2} \downarrow 0 \quad (n \uparrow +\infty). \end{aligned}$$

But then $u \equiv 0\,(t \leqq 1/3)$ and this propagates to all later times.

Suppose now that $\gamma > 2$, put $u = P_\bullet(\mathfrak{e} < t)$ and $v = (1-u)^2 |l|^\gamma$, and use

5) $$\begin{aligned} 1 - u &= P_\bullet(\mathfrak{e} \geqq t) \\ &= P_\bullet(\mathfrak{m}_{1/2} \geqq t) + E_\bullet\left[\int_0^t P_\bullet(\mathfrak{m}_{1/2} \in ds \mid w_{1/2})\, P_{x(s)}(\mathfrak{e} \geqq t - s)^2\right] \\ &= E_\bullet[e^{-\mathfrak{k}(t)}] + E_\bullet\left[\int_0^t e^{-\mathfrak{k}(s)}\,[1 - u(t-s, x(s))]^2\, d\mathfrak{k}\right] \\ &= E_\bullet[e^{-\mathfrak{k}(t)}] + \int_0^t ds\, E_\bullet[e^{-\mathfrak{k}(s)}\, v(t-s, x(s))] \\ &= e^{t\mathfrak{G}^\bullet} 1 + \int_0^t ds\, e^{s\mathfrak{G}^\bullet} v(t-s, \bullet) \qquad \mathfrak{G}^\bullet = \tfrac{1}{2}D^2 - |l|^\gamma, \end{aligned}$$

to conclude that u solves 1a):

6) $$\begin{aligned} -\frac{\partial u}{\partial t} &= \mathfrak{G}^\bullet e^{t\mathfrak{G}^\bullet} 1 + \int_0^t ds\, e^{s\mathfrak{G}^\bullet}\frac{\partial}{\partial t} v(t-s, \bullet) \\ &= \mathfrak{G}^\bullet e^{t\mathfrak{G}^\bullet} 1 - \int_0^t ds\, e^{s\mathfrak{G}^\bullet}\frac{\partial}{\partial s} v(t-s, \bullet) \\ &= \mathfrak{G}^\bullet (1-u) - \int_0^t ds\,\frac{\partial}{\partial s}[e^{s\mathfrak{G}^\bullet} v(t-s, \bullet)] \\ &= \mathfrak{G}^\bullet (1-u) + v. \end{aligned}$$

Because $u = P_\bullet(\mathfrak{e} < t) \leq P_\bullet(\mathfrak{m}_{1/2} < t)$, $u(0+, \cdot) = 0$, and the same bound proves that $u < 1\,(t \geq 0)$; also, $u > 0\,(t > 0)$ because if $u(t, l) = 0$ at some point $(t, l) \in (0, +\infty) \times R^1$, then 6) would make $u \equiv 0$ up to time t and $u \equiv 0$ would follow, contradicting $\lim_{t \uparrow +\infty} u = P_\bullet(\mathfrak{e} < +\infty) \equiv 1$ as proved in 5.14; this completes the proof of 1) and 2).

Problem 1. Consider the explosion time $\mathfrak{e}$ for the second model of 5.12 (creation and annihilation) with $k_n(dl) = 2c_n\,dl$, $0 \leq c_n\,(0 \leq n \neq 1)$, and $\sum_{n \neq 1} c_n = 1$, and let $c(\theta) = \sum_{n \neq 1} c_n\,\theta^n$. Check that $u = P_\bullet(\mathfrak{e} < t)$ is a solution of

$$\frac{\partial u}{\partial t} = \frac{1}{2}\frac{\partial^2 u}{\partial l^2} + [1 - u - c(1 - u)]$$

$$u(0+, \cdot) = 0, \quad 0 \leq u < 1.$$

Problem 2. Give also a proof that if $\mathfrak{f}$ is the time of extinction, then $u = P_\bullet(\mathfrak{f} < t)$ is a solution of

$$\frac{\partial u}{\partial t} = \frac{1}{2}\frac{\partial^2 u}{\partial l^2} + [c(u) - u]$$

$$u(0+, \cdot) = 0, \quad 0 \leq u < 1.$$

Problem 3. Deduce or give a direct proof that the chance $u = P_\bullet(\mathfrak{f} < +\infty)$ of ultimate extinction is a solution of

$$\tfrac{1}{2}u'' + [c(u) - u] = 0.$$

Problem 4. Prove that if $0 < c_0 \sum_{n \geq 2} c_n$, then the solution of problem 4 is $u \equiv \theta$ where $0 < \theta \leq 1$ is a root of $c(\theta) = \theta$. (1 is a root; a second root occurs if $\sum_{n \geq 2} n\,c_n > 1$.)

$\Big[\sum_{n \geq 2} n\,c_n \leq 1$ implies $c(\theta) = c_0 + \sum_{n \geq 2} c_n\,\theta^n < 1$ and

$$c'(\theta) = \sum_{n \geq 2} n\,c_n\,\theta^{n-1} < 1\,(\theta < 1),$$

so $\theta = 1$ is the sole root of $c(\theta) = \theta$ and $(1/2)\,u'' = u - c(u) \leq 0$, proving that $u(\leq 1)$ is constant $(= 1)$. Contrariwise, if $\sum_{n \geq 2} n\,c_n > 1$, then $c(1) = 1$ and $c'(\theta) > 1$ near $\theta = 1$ so that $\theta = c(\theta)$ has a second root $\theta < 1$. But u is still constant $(= 1$ or $\theta)$, for if $1 > u(l) > \theta$ and $u'(l) \geq 0$ at some point $l \in R^1$, then $u''(l) > 0$, so

$$\theta < u(\xi),\ 0 < u'(\xi),\ 0 < u''(\xi) \qquad (\xi > l)$$

contradicting $u \leq 1$; a similar contradiction is obtained if $u'(l) \leq 0$. $u < \theta$ is likewise impossible.]

6. Local and inverse local times

6.1. Local and inverse local times

We now take up the fine structure of the local time $\mathfrak{t}(t) = \mathfrak{t}(t, 0)$ and its inverse function $\mathfrak{t}^{-1}(t)$ for a persistent non-singular diffusion D^* on an interval Q containing 0 as an inside point or as a left end point, with $-u^+(0) + m(0)\,(\mathfrak{G}\,u)(0) = 0$ in the second case. A number of the statements made below hold for transient diffusions also (see esp. 6.3, 6.5, 6.6); the necessary modifications of the proofs are left to the reader.

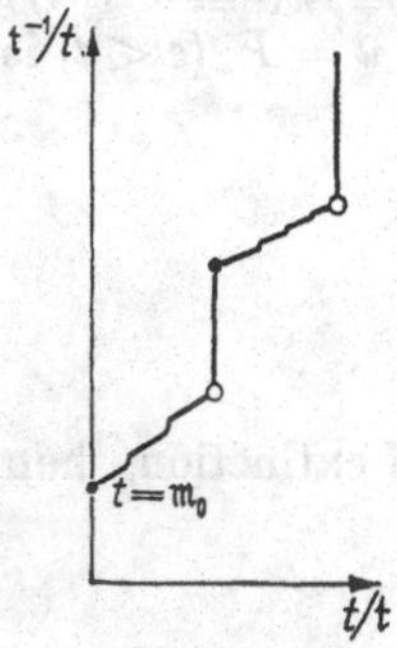

Diagram 1

$\mathfrak{t}^{-1}$ turns out to be *differential* and *homogeneous in time* just as in the standard BROWNian case (see 6.2), and the connection between it and the intervals of the complement of $\mathfrak{Z} = (t : x(t) = 0)$ comes through in part (see 6.3). $\mathfrak{t}$ can also be interpreted as a HAUSDORFF measure (see 6.5 b), and the down-crossing estimate of 2.4 is unaltered (see 6.5 a). $\mathfrak{t}^{-1}$ is useful for computing the HAUSDORFF-BESICOVITCH dimension of $\mathfrak{Z}$ (see 6.6 and 6.7).

Remove from the path space a negligible set of sample paths so that the following hold with no exceptions:

1) $\mathfrak{t} \in C[0, +\infty)$

2) $\mathfrak{m}_0 = \min(t : x(t) = 0) < +\infty$

3) $\mathfrak{t}(+\infty) = \lim_{t \uparrow +\infty} \mathfrak{t}(t) = +\infty$

4) $\mathfrak{t}(t_2) > \mathfrak{t}(t_1) \qquad (t_1, t_2) \cap \mathfrak{Z} \neq \emptyset$

5) $\mathfrak{t}(t_2) = \mathfrak{t}(t_1) + \mathfrak{t}(t_2 - t_1, w_{t_1}^+) \qquad t_2 > t_1 \geqq 0,$

define the *inverse local time*

6) $\mathfrak{t}^{-1}(t) = \max(s : \mathfrak{t}(s) = t) \qquad t \geqq 0,$

and note for future use the following simple facts:

7a) $\mathfrak{t}^{-1} \in \uparrow,$

7b) $\mathfrak{t}^{-1}(t+) = \mathfrak{t}^{-1}(t),$

7c) $\mathfrak{t}^{-1}(0) = \mathfrak{m}_0$

7d) $\mathfrak{t}^{-1}(+\infty) = \lim_{n \uparrow +\infty} \mathfrak{t}^{-1}(t) = +\infty$

7e) *the range of* $\mathfrak{t}^{-1}$ *is the set of points* $t \in \mathfrak{Z}$ *not isolated above.*

$\mathfrak{t}^{-1}$ *is to be split into 4 parts:*

8) $\mathfrak{t}^{-1}(t) = \mathfrak{m}_0 + t\,m(0) + \mathfrak{t}_-^{-1}(t) + \mathfrak{t}_+^{-1}(t),$

* See problem 4.6.6 for a discussion of persistent diffusion.

where

9a) $$\mathfrak{t}_-^{-1}(t) = \text{measure}\,(s : x(s) < 0,\ \mathfrak{m}_0 \leqq s \leqq \mathfrak{t}^{-1}(t))$$

and

9b) $$\mathfrak{t}_+^{-1}(t) = \text{measure}\,(s : x(s) > 0,\ \mathfrak{m}_0 \leqq s \leqq \mathfrak{t}^{-1}(t))$$

are differential, independent, and independent of $\mathfrak{m}_0$.

Given $t > 0$, it is clear from 5) that

10) $$\begin{aligned}\mathfrak{t}^{-1}(t) &= \max\,(s_1 : \mathfrak{t}(s_1) = t)\\ &= \max\,(s_1 : s_1 = \mathfrak{m}_0 + s_2,\ \mathfrak{t}(\mathfrak{m}_0) + \mathfrak{t}(s_2, w^+_{\mathfrak{m}_0}) = t)\\ &= \mathfrak{m}_0 + \max\,(s_2 : \mathfrak{t}(s_2, w^+_{\mathfrak{m}_0}) = t)\\ &= \mathfrak{m}_0 + \mathfrak{t}^{-1}(t, w^+_{\mathfrak{m}_0}).\end{aligned}$$

$\mathfrak{m}_0$ is therefore independent of

11) $$\mathfrak{t}_\pm^{-1}(t) = \text{measure}\,(s : x(s, w^+_{\mathfrak{m}_0}) \lessgtr 0,\ s \leqq \mathfrak{t}^{-1}(t, w^+_{\mathfrak{m}_0}))$$

and

12) $$\begin{aligned}&\mathfrak{t}^{-1}(t) - \mathfrak{m}_0 - \mathfrak{t}_-^{-1}(t) - \mathfrak{t}_+^{-1}(t)\\ &= \text{measure}\,(s : x(s, w'_{\mathfrak{m}_0}) = 0,\ s \leqq \mathfrak{t}^{-1}(t, w^+_{\mathfrak{m}_0}))\\ &= \mathfrak{t}(\mathfrak{t}^{-1}(t, w^+_{\mathfrak{m}_0}))\, m(0)\\ &= t\, m(0),\end{aligned}$$

establishing the decomposition 8).

Consider for the proof that $\mathfrak{t}_\pm^{-1}$ is *differential*, two times $t_2 \geqq t_1 > 0$, let $\mathfrak{m} = \mathfrak{t}^{-1}(t_1)$, and use 5) and

13) $$\mathfrak{t}(\mathfrak{m}) = t_1$$

to check

14) $$\begin{aligned}\mathfrak{t}^{-1}(t_2) &= \max\,(s_1 : \mathfrak{t}(s_1) = t_2)\\ &= \max\,(s_1 : s_1 = \mathfrak{m} + s_2,\ \mathfrak{t}(\mathfrak{m}) + \mathfrak{t}(s_2, w^+_{\mathfrak{m}}) = t_2)\\ &= \mathfrak{m} + \max\,(s_2 : \mathfrak{t}(s_2, w^+_{\mathfrak{m}}) = t_2 - t_1)\\ &= \mathfrak{m} + \mathfrak{t}^{-1}(t_2 - t_1, w^+_{\mathfrak{m}}).\end{aligned}$$

Using 14) and $\mathfrak{m}_0(w^+_{\mathfrak{m}}) = 0$, it results that

15) $$\begin{aligned}&\mathfrak{t}_\pm^{-1}(t_2)\\ &= \text{measure}\,[s : x(s, w^+_{\mathfrak{m}}) \gtrless 0,\ \mathfrak{m}_0 \leqq s \leqq \mathfrak{m}]\\ &+ \text{measure}\,[s : x(s, w^+_{\mathfrak{m}}) \gtrless 0,\ \mathfrak{m} \leqq s \leqq \mathfrak{m} + \mathfrak{t}^{-1}(t_2 - t_1, w^+_{\mathfrak{m}})]\\ &= \mathfrak{t}_\pm^{-1}(t_1) + \text{measure}\,[s : x(s, w^+_{\mathfrak{m}}) \gtrless 0,\ s < \mathfrak{t}^{-1}(t_2 - t_1, w^+_{\mathfrak{m}})]\\ &= \mathfrak{t}_\pm^{-1}(t_1) + \mathfrak{t}_\pm^{-1}(t_2 - t_1, w^+_{\mathfrak{m}}).\end{aligned}$$

But

16) $$(w : \mathfrak{m} < s) = (w : t_1 < \mathfrak{t}(s)) \in \mathsf{B}_s \qquad s \geqq 0,$$

so that $\mathfrak{m}$ is MARKOV. Because of

17) $\qquad \mathsf{B}(\mathfrak{f}(t): t \leq t_1) \subset \mathsf{B}_{\mathfrak{m}+} \qquad \mathfrak{f} = \mathfrak{t}_-^{-1}, \mathfrak{t}_+^{-1}, \mathfrak{t}_+^{-1} - \mathfrak{t}_-^{-1},$

it follows that

18)
$$\begin{aligned} P_\bullet(\mathfrak{f}(t_2) - \mathfrak{f}(t_1) \leq s \mid \mathsf{B}_{\mathfrak{m}+}) & \qquad t_2 \geq t_1 \\ = P_\bullet(\mathfrak{f}(t, w_{\mathfrak{m}}^+) \leq s \mid \mathsf{B}_{\mathfrak{m}+}) & \qquad t = t_2 - t_1 \\ = P_{x(\mathfrak{m})}(\mathfrak{f}(t) \leq s) = P_0(\mathfrak{f}(t) \leq s), & \end{aligned}$$

proving that $\mathfrak{t}_-^{-1}$, $\mathfrak{t}_+^{-1}$, and $\mathfrak{t}_+^{-1} - \mathfrak{t}_-^{-1}$ are differential, and, to complete the proof, it is enough to note from 9) that $\mathfrak{t}_-^{-1}$ and $\mathfrak{t}_+^{-1}$ are sums of jumps and to conclude from the differential character of $\mathfrak{t}_+^{-1} - \mathfrak{t}_-^{-1}$ that the sum ($\mathfrak{t}_-^{-1}$) of its negative jumps is independent of the sum ($\mathfrak{t}_+^{-1}$) of its positive jumps.

Problem 1. Give new proofs of the results of the present section using time changes.

[Given a standard BROWNian motion in the natural scale with local times $\mathfrak{t}$, if $\mathfrak{t}^{-1}$ is its inverse local time at 0, then

$$\int \mathfrak{t}(\mathfrak{t}^{-1}(t), \xi)\, m(d\xi), \quad \int^{0-} \mathfrak{t}(\mathfrak{t}^{-1}(t), \xi)\, m(d\xi), \quad \text{and} \quad \int_{0+} \mathfrak{t}(\mathfrak{t}^{-1}(t), \xi)\, m(d\xi)$$

are identical in law to the inverse local times $\mathfrak{t}^{\bullet -1}, \mathfrak{t}^{\bullet -1}_{\pm}$ for the diffusion $\mathsf{D}^\bullet$ with speed measure m; now check the independence and differential property using problem 2.8.4.]

6.2. Lévy measures

We will now check the LÉVY formulas:

1a)
$$\begin{aligned} E_0\left(e^{-\alpha \mathfrak{t}_-^{-1}(t)}\right) &= e^{-t g_1^-(0)/g_1(0)} \\ &= e^{-t \int_{0+}^{+\infty} (1 - e^{-\alpha l})\, n_-(dl)} \\ n_-(dl) &= \lim_{\varepsilon \downarrow 0} s(-\varepsilon, 0]^{-1} P_{-\varepsilon}(\mathfrak{m}_0 \in dl), \end{aligned}$$

1b)
$$\begin{aligned} E_0\left(e^{-\alpha \mathfrak{t}_+^{-1}(t)}\right) &= e^{+t g_2^+(0)/g_2(0)} \\ &= e^{-t \int_{0+}^{+\infty} (1 - e^{-\alpha l})\, n_+(dl)} \\ n_+(dl) &= \lim_{\varepsilon \downarrow 0} s[0, \varepsilon)^{-1} P_\varepsilon(\mathfrak{m}_0 \in dl), \end{aligned}$$

and

2)
$$\begin{aligned} E_0\left(e^{-\alpha \mathfrak{t}^{-1}(t)}\right) &= e^{-t/G(0,0)} \\ &= e^{-t\left[m(0)\alpha + \int_{0+}^{+\infty} (1 - e^{-\alpha l})\, n(dl)\right]} \\ n(dl) &= n_-(dl) + n_+(dl), \end{aligned}$$

where $G = g_1 g_2$ is the Green function for D and $m(0)$ is the jump of the speed measure at 0. A part of the above can be got from 4.10, but we prefer to derive it all anew. Because of

3) $$G(0,0)^{-1} = \frac{g_1^+(0)}{g_1(0)} - \frac{g_2^+(0)}{g_2(0)} = \frac{g_1^-(0)}{g_1(0)} + m(0)\,\alpha - \frac{g_2^+(0)}{g_2(0)}$$

and the fact that $\mathfrak{t}_-^{-1}$ and $\mathfrak{t}_+^{-1}$ are independent, it is enough to prove 1b).

Given $l > 0$, $P_\varepsilon(\mathfrak{m}_0 \leqq l)$ is convex in the scale $s(\varepsilon)$ for $\varepsilon > 0$ (see problem 4.4.2), so

4) $$\begin{aligned} &s[0,\varepsilon)^{-1}\,[P_\varepsilon(\mathfrak{m}_0 \leqq l) - 1] \\ &\quad = -s[0,\varepsilon)^{-1} P_\varepsilon(\mathfrak{m}_0 > l) \\ &\quad \downarrow n_+(l) < 0 \qquad\qquad \varepsilon \downarrow 0, \end{aligned}$$

and it results from 4) that

5) $$\begin{aligned} g_2^+(0)/g_2(0) &= \lim_{\varepsilon\downarrow 0} s[0,\varepsilon)^{-1}[g_2(\varepsilon) - g_2(0)]\,g_2(0)^{-1} \\ &= \lim_{\varepsilon\downarrow 0} s[0,\varepsilon)^{-1}\left[\int_0^{+\infty} e^{-\alpha l} P_\varepsilon(\mathfrak{m}_0 \in dl) - 1\right] \\ &= \lim_{\varepsilon\downarrow 0} s[0,\varepsilon)^{-1}\left[\int_0^{+\infty} P_\varepsilon(\mathfrak{m}_0 > l)\, d(e^{-\alpha l} - 1)\right] \\ &= -\int_0^{+\infty} n_+(l)\, d(e^{-\alpha l} - 1) \\ &= -\int_0^{+\infty} (1 - e^{-\alpha l})\, n_+(dl), \end{aligned}$$

where, for the justification of the final step, we use

6) $$\begin{aligned} 0 &\geqq \lim_{l\uparrow+\infty} n_+(l) \geqq \int_0^{+\infty} n_+(l)\, d(1 - e^{-\alpha l}) \\ &= g_2^+(0)/g_2(0) \\ &= \frac{g_2^+(\varepsilon)}{g_2(0)} - \alpha \int_{(0,\varepsilon]} \frac{g_2(\xi)}{g_2(0)}\, m(d\xi) \\ &> s[0,\varepsilon)^{-1}\,\frac{g_2(\varepsilon) - g_2(0)}{g_2(0)} - \alpha\, m(0,\varepsilon] \\ &= s[0,\varepsilon)^{-1}\,[E_\varepsilon(e^{-\alpha\mathfrak{m}_0}) - 1] - \alpha\, m(0,\varepsilon] \\ &\to 0 \qquad \alpha \downarrow 0 \end{aligned}$$

and

7) $$0 \geqq \varepsilon\, n_+(\varepsilon) \geqq e^{\varepsilon} \int_0^{\varepsilon} n_+(l)\, e^{-l}\, dl > -\infty.$$

Coming to $\mathfrak{t}_+^{-1}$, the time change $t \to \mathfrak{f}^{-1}(t)$ based on the local time integral $\mathfrak{f}(t) = \int_{0+} \mathfrak{t}(t, \xi)\, m(d\xi)$ takes D into the diffusion D_+ on $[0, +\infty)$ with invariants

$$8) \qquad s_+ = s, \quad m_+ = m \quad \textit{on} \quad (0, +\infty), \qquad m_+(0) = 0$$

and since, for $t > 0$, $\mathfrak{f}$ grows just to the left of $\mathfrak{t}^{-1}(t)$, *i.e.*, $\mathfrak{f}(\theta) < \mathfrak{f}(s)$ for $\theta < s = \mathfrak{t}^{-1}(t)$, one finds

$$\begin{aligned} 9) \qquad \mathfrak{t}_+^{-1}(t) &= \mathfrak{f}(\mathfrak{t}^{-1}(t)) \\ &= \min(s : s \geqq \mathfrak{f}(\mathfrak{t}^{-1}(t))) \\ &= \min(s : \mathfrak{f}^{-1}(s) \geqq \mathfrak{t}^{-1}(t)) \\ &= \min(s : \mathfrak{t}(\mathfrak{f}^{-1}(s)) \geqq t), \end{aligned}$$

proving that $\mathfrak{t}_+^{-1}(t)$ is identical (in law) to the inverse of the local time $\mathfrak{t}_+(t) = \mathfrak{t}_+(t, 0)$ for D_+.

But now, taking $\gamma > 0$ and using the GREEN function expansion of problem 5.6.1, one finds

$$\begin{aligned} 10) \qquad \int_0^{+\infty} e^{-\gamma t}\, dt\, E_0\left(e^{-\alpha \mathfrak{t}_+^{-1}(t)}\right) &= E_0\left[\int_0^{+\infty} e^{-\gamma t}\, e^{-\alpha \mathfrak{t}_+^{-1}(t)}\, dt\right] \\ &= E_0\left[\int_0^{+\infty} e^{-\alpha t}\, e^{-\gamma \mathfrak{t}_+(t)}\, \mathfrak{t}_+(dt)\right] \\ &= G_-(0, 0), \end{aligned}$$

where G_- is the GREEN function for the diffusion D_- on $[0, +\infty)$ with invariants

$$11) \qquad s_- = s_+, \quad m_- = m_+, \qquad k_-(0) = \gamma, \quad k_-(0, +\infty) = 0,$$

and since $G_- = g_3\, g_2$ with

$$\begin{aligned} 12) \qquad \frac{g_3^+(d\xi)}{m(d\xi)} &= \alpha\, g_3(\xi) \qquad (\xi > 0) \\ g_3^+(0) &= \gamma\, g_3(0) \\ g_3^+(0)\, g_2(0) - g_3(0)\, g_2^+(0) &= 1, \end{aligned}$$

it follows that

$$\begin{aligned} 13) \qquad G_-(0, 0) = g_3(0)\, g_2(0) &= \frac{g_2(0)}{\gamma\, g_2(0) - g_2^+(0)} \\ &= \left(\gamma - \frac{g_2^+(0)}{g_2(0)}\right)^{-1}, \end{aligned}$$

completing the proof of 1b) apart from undoing a LAPLACE transform.

Not all non-negative mass distributions $n_+(dl)$ with $\int l \wedge 1\, dn_+ < +\infty$ can arise as Lévy measures of inverse local times; in fact, it is necessary that

14) $$\frac{n_+(dl)}{dl} = \int_{-\infty}^{0+} e^{l\gamma} f_{22}(d\gamma) \qquad 0 \leqq f_{22}(d\gamma),$$

as will now be proved.

Consider the generator $\mathfrak{G}^\bullet$ for the original diffusion starting from $[0, +\infty)$ and stopped at 0, write $G = g_1 g_2$ for its Green function, and

15) $$\mathfrak{e}(\gamma, a)\, \mathfrak{f}(d\gamma)\, \mathfrak{e}(\gamma, b) \qquad \gamma \leqq 0, \quad a, b \geqq 0,$$

for the eigen-differentials of $\mathfrak{G}^\bullet$ where $\mathfrak{e} = (e_1, e_2)$ is the solution of

16) $$\dot{\mathfrak{G}}\, \mathfrak{e} = \gamma\, \mathfrak{e} \qquad \mathfrak{e}(0) = (1, 0), \qquad \mathfrak{e}^+(0) = (0, 1)$$

and $\mathfrak{f}$ is a Borel measure on $(-\infty, 0)$ to 2×2 matrices with entries $f_{11}, f_{12}, f_{21}, f_{22}$ such that

17) $$0 \leqq f_{11}, f_{22}, \quad f_{12} = f_{21}, \quad f_{12}^2 = f_{21}^2 \leqq f_{11} f_{22},$$

as in 4.11. Owing to $g_1(0) = 0$,

18) $$g_1^+(0)\, g_2(0) = g_1^+(0)\, g_2(0) - g_1(0)\, g_2^+(0) = 1,$$

and so

19) $$G_2(\alpha, 0, \varepsilon) = g_1^+(0)\, g_2(\varepsilon) = g_2(\varepsilon)/g_2(0)$$
$$= \int_0^{+\infty} e^{-\alpha l} P_\varepsilon(\mathfrak{m}_0 \in dl) \qquad \varepsilon > 0.$$

Inverting the transform, 19) goes over into

20) $$P_\varepsilon(\mathfrak{m}_0 \in dl) = \int_{-\infty}^{0+} e^{l\gamma}\, \mathfrak{e}^+(\gamma, 0)\, \mathfrak{f}(d\gamma)\, \mathfrak{e}(\gamma, \varepsilon)\, dl \qquad \varepsilon > 0,$$

and, computing the gradient at $\varepsilon = 0$, it follows from the definition of n_+ [see 1b)] that

21) $$\frac{n_+(dl)}{dl} = \int_{-\infty}^{0+} e^{l\gamma}\, \mathfrak{e}^+(\gamma, 0)\, \mathfrak{f}(d\gamma)\, \mathfrak{e}^+(\gamma, 0) = \int_{-\infty}^{0+} e^{l\gamma} f_{22}(d\gamma) \qquad l > 0.$$

It is an open problem to describe the class of non-negative mass distributions $f(d\gamma)$ on $(-\infty, 0]$ such that

$$n(dl) = dl \times \int_{-\infty}^{0+} e^{l\gamma} f(d\gamma)$$

is the Lévy measure of the inverse local time of a diffusion.

Problem 1. Check that

$$n_+[l, +\infty) < \frac{-g_2^+(l^{-1}, 0)}{g_2(l^{-1}, 0)}$$

and that

$$n[l, +\infty) < G(l^{-1}, 0, 0)^{-1}.$$

$$[n_+[l, +\infty) = \int_{-\infty}^{0} |\gamma|^{-1} e^{l\gamma} f(d\gamma) \quad \text{with} \quad 0 \leqq f(d\gamma), \quad \textit{i.e.,} \quad n_+[l, +\infty)$$

is convex, so

$$\frac{-g_2^+(\alpha, 0)}{g_2(\alpha, 0)} = \alpha \int_0^{+\infty} l e^{-\alpha l} n_+[l, +\infty)\, dl$$

$$> n_+[\alpha \int_0^{+\infty} l e^{-\alpha l}\, dl, +\infty) = n_+[\alpha^{-1}, +\infty);$$

the proof of the second part is similar.]

Problem 2. Check that $n_+(0, +\infty) = n(0, +\infty) = +\infty$.

$$[n(0, +\infty) = \lim_{\alpha \uparrow +\infty} \int_0^{+\infty} [1 - e^{-\alpha l}]\, n(dl) = \lim_{\alpha \uparrow +\infty} \left(\int_0^{+\infty} e^{-\alpha t} p(t, 0, 0)\, dt \right)^{-1}$$
$$= +\infty.]$$

6.3. t and the intervals of $[0, +\infty) - \mathfrak{Z}$

We saw in 2.2.4) that in the case of the standard BROWNian motion, if $\mathfrak{Z} = (t : x(t) = 0)$, then

1) $\lim_{\varepsilon \downarrow 0} \sqrt{\frac{\pi \varepsilon}{2}} \times$ *the number of intervals of* $[0, t] - \mathfrak{Z}$ *of length* $\geqq \varepsilon = \mathfrak{t}(t)$ $\quad t \geqq 0,$

in which $\mathfrak{t}$ is BROWNian local time at 0 and $\sqrt{2/\pi\varepsilon} = \int_{\varepsilon}^{+\infty} \frac{dl}{\sqrt{2\pi l^3}}$ is the expected number of jumps of $\mathfrak{t}^{-1}$ of length $\geqq \varepsilon$ per unit time. $n(dl) = (2\pi l^3)^{-1/2}\, dl$ is the LÉVY measure for $\mathfrak{t}^{-1}$.

With the general $n[\varepsilon, +\infty)$ *in place of* $\sqrt{2/\pi\varepsilon}$, 1) *is still true:* in fact, letting $\mathfrak{Z}^-$ designate an (open) interval of $(t : x(t) < 0)$, $\mathfrak{Z}^+$ an interval of $(t : x(t) > 0)$, and n_-, n_+, n the LÉVY measures for $\mathfrak{t}_-^{-1}$, $\mathfrak{t}_+^{-1}$, and $\mathfrak{t}^{-1}$,

2a) $$P_0\left[\lim_{\varepsilon \downarrow 0} \frac{\#(\mathfrak{Z}^- : \mathfrak{Z}^- \subset [0, t), |\mathfrak{Z}^-| \geqq \varepsilon)}{n_-[\varepsilon, +\infty)} = \mathfrak{t}(t),\ t \geqq 0\right] = 1$$

2b) $$P_0\left[\lim_{\varepsilon \downarrow 0} \frac{\#(\mathfrak{Z}^+ : \mathfrak{Z}^+ \subset [0, t), |\mathfrak{Z}^+| \geqq \varepsilon)}{n_+[\varepsilon, +\infty)} = \mathfrak{t}(t),\ t \geqq 0\right] = 1$$

and

3) $$P_0\left[\lim_{\varepsilon \downarrow 0} \frac{\#(\mathfrak{Z}^\pm : \mathfrak{Z}^\pm \subset [0, t), |\mathfrak{Z}^\pm| \geqq \varepsilon)}{n[\varepsilon, +\infty)} = \mathfrak{t}(t),\ t \geqq 0\right] = 1.$$

Consider, for the proof of 2b), the counting function

4) $\#(t, \varepsilon) =$ *the number of jumps of magnitude* $\geqq \varepsilon$ *of* $\mathfrak{t}_+^{-1}$ *up to time* t,

and keeping in mind that $n_+[\varepsilon, +\infty)$ is continuous ($\varepsilon > 0$) and that $n_+(0, +\infty) = +\infty$, let

5) $$n_+^{-1}(s) = \min(t : n_+[t, +\infty) \leqq s).$$

$[\#(t, n_+^{-1}(s)) : s \geqq 0, P_0]$ is differential and Poisson distributed. Because

6) $$E_0[\#(t, n_+^{-1}(s))] = t\, n_+[n_+^{-1}(s), +\infty) = t\, s,$$

it is homogeneous in its temporal parameter s, and thanks to the strong law of large numbers,

7) $$P_0\left[\lim_{\varepsilon\downarrow 0} \frac{\#(t,\varepsilon)}{n_+[\varepsilon, +\infty)} = \lim_{s\uparrow+\infty} \frac{\#(t, n_+^{-1}(s))}{s} = t, t \geqq 0\right] = 1.$$

But

8) $$\#(\mathfrak{t}(t), \varepsilon) = \#(\mathfrak{Z}^+ : \mathfrak{Z}^+ \subset [0, t], |\mathfrak{Z}^+| \geqq \varepsilon)$$

if $\mathfrak{t}^{-1}(\mathfrak{t}(t), 0) = t$, *i.e.*, if $t \in \mathfrak{Z}$ is not isolated from above (in $\mathfrak{Z}$), and since $\#(\mathfrak{Z}^+ : \mathfrak{Z}^+ \subset [0, t), |\mathfrak{Z}^+| \geqq \varepsilon)$ and $\mathfrak{t}(t)$ are increasing in t, since $\mathfrak{Z}$ is closed and perfect, and since $\mathfrak{t}$ is continuous and flat outside $\mathfrak{Z}$, the strong law

9) $$P_0\left[\lim_{\varepsilon\downarrow 0} \frac{\#(\mathfrak{Z}^+ : \mathfrak{Z}^+ \subset [0, t), |\mathfrak{Z}^+| \geqq \varepsilon)}{n[\varepsilon, +\infty)} = \mathfrak{t}(t), \mathfrak{t}^{-1}(\mathfrak{t}) = t, t \geqq 0\right] = 1,$$

which is clear from 7), implies the desired 2b).

The proofs of 2a) and 3) are similar.

We also saw in 2.2.6) that for the standard Brownian motion,

10) $$\lim_{\varepsilon\downarrow 0} \sqrt{\frac{\pi}{2\varepsilon}} \times \textit{the total length of the intervals of } [0, t] - \mathfrak{Z} \textit{ of length} < \varepsilon = \mathfrak{t}(t) \qquad t > 0,$$

in which $\sqrt{2\varepsilon/\pi}$ is the expected sum of the jumps of $\mathfrak{t}^{-1}$ of magnitude $< \varepsilon$ per unit time.

A natural conjecture is

11) $$\lim_{\varepsilon\downarrow 0} \left(\int_0^\varepsilon l\, n_+(dl)\right)^{-1} \times \textit{the total length of the intervals } \mathfrak{Z}^+ \subset [0, t] \textit{ of length} < \varepsilon = \mathfrak{t}(t) \qquad t \geqq 0,$$

i.e., substituting $\mathfrak{t}^{-1}(t)$ in place of t and $n_+^{-1}(s)$ in place of ε,

12) $$\lim_{s\uparrow+\infty} \left(\int_0^{n_+^{-1}(s)} l\, n_+(dl)\right)^{-1} \int_0^{n_+^{-1}(s)} l\, \#(t, dl) = \lim_{s\uparrow+\infty} \frac{\sum_{s_n > s} n_+^{-1}(s_n)}{\int_s^{+\infty} n_+^{-1}(\theta)\, d\theta} = t \qquad t \geqq 0,$$

in which the $s_n (n \geqq 1)$ are the jump times of the (homogeneous) POISSON process $\# (t, n_+^{-1}(s)) : s \geqq 0$.

Because of strong law of large numbers,

$$13) \qquad \lim_{n\uparrow\infty} n^{-1} s_n = t^{-1},$$

and it is a simple matter to check 12) from 13) *if $n_+^{-1}(s)$ is a power of s times a function $f(s)$ such that $f(t\,s) \sim f(s)$ as $s \uparrow +\infty$ for each $t > 0$* [see 6.7.4) for the special case: $ds = d\xi$, $dm = 2\,|\xi|^\beta\, d\xi$ $(\beta > -1)$, $dn = \text{constant} \times l^{-(1+\alpha)}\, dl$, $\alpha = (\beta + 2)^{-1}$]. But 11) is *false* in general as the counter example of the next section proves.

6.4. A counter example: $\mathfrak{t}$ and the intervals of $[0, +\infty) - \mathfrak{Z}$.

6.3.11) is false in general: *for example, it is possible to have $0 < c_1 < e^s n_+^{-1}(s) < c_2 < +\infty$ as $s \uparrow +\infty$ and then 6.3.11) would force $\limsup_{n\uparrow+\infty} e^{s_n} \sum_{k\geqq n} e^{-s_k} \leqq c_2/c_1 < +\infty$, violating the fact that the law of $e^{s_n} \sum_{k\geqq n} e^{-s_k}$ is independent of n and has as its support the whole of $[1, +\infty)$.*

Consider for the construction of such n_+, the function $\psi(t) = (1/2) \int_0^1 t^s \Gamma(s+1)^{-1}\, ds$ and its inverse function θ and let $\mathfrak{G}$ be the generator with the natural scale and the speed measure $dm = (1/4)\,(\theta')^2 d\xi$ on $[0, +\infty)$.

Consider the increasing and decreasing solutions g_1 and g_2 of $\mathfrak{G}\, g = \alpha\, g$ with $g_1^+ g_2 - g_1 g_2^+ = 1$, $g_1(0) = 1$, and $g_1^+(0) = 0$.

$g_3 = g_1(\xi) \int_\xi^{+\infty} g_1^{-2}\, d\eta$ satisfies $g_3^+ < 0$, $\mathfrak{G}\, g_3 = \alpha\, g_3$, and $g_3^+(0) = g_2^+(0) = -1$, so that $g_3 = g_2$, or what interests us,

$$1) \qquad g_2(0) = \int_0^{+\infty} g_1^{-2}\, d\xi.$$

Given $t > 0$, $0 < \psi'(t) \in \downarrow$ so that $0 < \theta'(t) \in \uparrow$ and $\theta''(t) > 0$. Now $\psi(e^t)$ is convex so that $\lg\theta(t)$ is concave and $\theta''(\theta')^{-2} < \theta^{-1}$. Besides $\psi'(0+) = +\infty$ so that $\theta'(0+) = 0$, and putting $g_4(\theta) = g_1(\xi)$, it follows from $\mathfrak{G}\, g_1 = \alpha\, g_1$ and the facts just cited that

$$2) \qquad g_4''(\theta) + \theta''(\theta')^{-2} g_4'(\theta) = \frac{\alpha}{4}\, g_4(\theta)$$

$$g_4(0+) = 1$$

$$g_4'(0+) = \lim_{\varepsilon\downarrow 0} \frac{g_1'(\varepsilon)}{\theta'(\varepsilon)} \leqq \lim_{\varepsilon\downarrow 0} 4\alpha\, g_1(\varepsilon)\, \theta'(\varepsilon) = 0$$

and

3) $$g_4'' < \frac{\alpha}{4} g_4 < g_4'' + \theta^{-1} g_4' \qquad \theta > 0.$$

But then

4) $$I_0(\sqrt{\alpha}\,\theta/2) < g_4(\theta) < e^{\frac{1}{2}\sqrt{\alpha}\,\theta},$$

and 1) implies

5) $$\int_0^{+\infty} e^{-\sqrt{\alpha}\,\theta} d\xi < g_2(0) < \int_0^{+\infty} I_0\left(\frac{1}{2}\sqrt{\alpha}\,\theta\right)^{-2} d\xi$$

$$< \frac{\lg\alpha}{\sqrt{\alpha}} + \int_{\lg\alpha/\sqrt{\alpha}}^{+\infty} I_0\left(\frac{1}{2}\sqrt{\alpha}\,\theta\right)^{-2} d\xi$$

$$< \frac{\lg\alpha}{\sqrt{\alpha}} + \int_0^{+\infty} e^{-\sqrt{\gamma\alpha}\,\theta} d\xi,$$

where $\gamma \uparrow 1$ as $\alpha \uparrow +\infty$ and is independent of ξ (use the fact that $I_0(\xi) \sim \frac{e^\xi}{\sqrt{2\pi\xi}}$ as $\xi \uparrow +\infty$).

Working out the estimates with the aid of

6) $$\int_0^{+\infty} e^{-\sqrt{\alpha}\,\theta} d\xi = \int_0^{+\infty} e^{-\sqrt{\alpha}\,t} \psi'(t)\, dt = \frac{1}{2}\int_0^1 \alpha^{-s/2}\, ds$$

$$= \frac{1-\alpha^{-1/2}}{\lg\alpha},$$

one finds

7) $$\frac{1-\alpha^{-1/2}}{\lg\alpha} < g_2(0) < \frac{1-(\gamma\alpha)^{-1/2}}{\lg\gamma\alpha} \qquad \alpha \uparrow +\infty,$$

and thanks to

8) $$g_2(0)^{-1} = [g_1(0)\, g_2(0)]^{-1} = \int_0^{+\infty} (1 - e^{-\alpha l})\, n_+(dl)$$

$$= \alpha \int_0^{+\infty} e^{-\alpha l}\, n_+[l, +\infty)\, dl,$$

it follows that

9) $$\frac{\lg\gamma\alpha}{1-(\gamma\alpha)^{-1/2}} < \alpha \int_0^{+\infty} e^{-\alpha l}\, n_+[l, +\infty)\, dl < \frac{\lg\alpha}{1-\alpha^{-1/2}} \qquad \alpha \uparrow +\infty.$$

But $n_+(dl)/dl$ is the LAPLACE transform of a (positive) spectral measure $f(d\gamma)$. $n_+[l, +\infty) = \int_{-\infty}^{0+} |\gamma|^{-1} e^{l\gamma} f(d\gamma)$ is therefore convex, and the conclusion is that

$$10)\quad \frac{\lg\alpha}{1-\alpha^{-1/2}} > \alpha\int_0^{+\infty} e^{-\alpha l}\, n_+[l,+\infty)\, dl > n_+\left[\alpha\int_0^{+\infty} e^{-\alpha l}\, l\, dl, +\infty\right)$$

$$= n_+\left[\frac{1}{\alpha}, +\infty\right) \qquad \alpha\uparrow+\infty.$$

On the other hand, since now $\frac{|\lg l|}{1-\sqrt{l}} - n_+[l,+\infty)$ is positive and since $n_+[l,+\infty)$ and $\frac{|\lg l|}{1-\sqrt{l}}$ are both decreasing for small l,

$$11)\quad 2e^{-1}\left[\frac{\lg\alpha}{1-\alpha^{-1/2}} - n_+\left[\frac{1}{2\alpha}, +\infty\right)\right]$$

$$< \frac{\alpha}{e}\int_{1/2\alpha}^{1/\alpha}\left[\frac{|\lg l|}{1-\sqrt{l}} - n_+[l,+\infty)\right] dl$$

$$< e^{-\alpha/8} - \alpha\int_0^{+\infty} e^{-\alpha l}\left[(\lg l)(1+2\sqrt{l}) + n_+[l,+\infty)\right] dl$$

$$< e^{-\alpha/8} + \lg\alpha - \int_0^{+\infty} e^{-l}\lg l\, dl$$

$$+ \frac{\lg\alpha}{\alpha}\int_0^{+\infty} e^{-l}\sqrt{l}\, dl - \frac{1}{\sqrt{\alpha}}\int_0^{+\infty} e^{-l}\sqrt{l}\lg l\, dl - \frac{\lg\gamma\alpha}{1-(\gamma\alpha)^{-1/2}}$$

$$< \text{constant}\times\frac{\lg\alpha}{\sqrt{\alpha}} - \int_0^{+\infty} e^{-l}\lg l\, dl$$

$$< \tfrac{3}{2} \qquad \alpha\uparrow+\infty,$$

and therefore

$$12)\quad |\lg l| - 4 < \frac{\left|\lg\frac{l}{2}\right|}{1-\sqrt{l/2}} - \frac{3e}{4} < n_+[l,+\infty) < |\lg l| + 1 \qquad l\downarrow 0,$$

proving the desired estimate

$$13)\quad c_1 e^{-s} < n_+^{-1}(s) < c_2 e^{-s} \qquad s\uparrow+\infty$$

with $c_1 = e^{-4}$ and $c_2 = e$

6.5a. t and downcrossings

Because the statement

1) $\lim_{\varepsilon\downarrow 0} s[0,\varepsilon)\times$ *the number of times* $x(s): s<t$ *crosses down from* ε *to* 0 $= \mathfrak{t}(t)$

is correct for the standard Brownian motion (see 2.4) and invariant under the time substitution $t\to \mathfrak{f}^{-1}$ [see 5.4.1 b)], it applies to the present case also.

6.5b. t as Hausdorff measure

Given $t_2 > t_1$, if $\mathfrak{z}^-$ and $\mathfrak{z}^+$ are the first and last roots of

$$x(s) = 0 : t_1 \leqq s \leqq t_2 \quad (\mathfrak{z}^- = \mathfrak{z}^+ = 0 \text{ if no such roots occur}),$$

and if $\mathfrak{t}(t)$ is local time at 0, then much as in 2.5,

1) $$E_0[\mathfrak{t}(t_2) - \mathfrak{t}(t_1) \mid x(s) : s \leqq t_1, \mathfrak{z}^-, \mathfrak{z}^+, x(s) : s \geqq t_2]$$
$$= E_0[\mathfrak{t}(t) \mid x(t) = 0]$$
$$= \frac{\int_0^t p(\theta, 0, 0)\, p(t-\theta, 0, 0)\, d\theta}{p(t, 0, 0)}$$
$$\equiv h(t) \qquad t = \mathfrak{z}_+ - \mathfrak{z}_-,$$

where $p(t, \xi, \eta)$ is the transition kernel of 4.11, and this implies (compare 2.5) that

2) $$\lim_{n\uparrow\infty} \sum_{i\leqq 2^n} h(\operatorname{diam} \mathfrak{Z} \cap [(i-1)\, 2^{-n} t, i 2^{-n} t]) = \mathfrak{t}(t) \qquad t \geqq 0,$$

where $\mathfrak{Z} = (t : x(t) = 0)$; the proof is left to the reader as a problem.

$\mathfrak{t}$ is therefore something like HAUSDORFF *h-measure cut down to* $\mathfrak{Z}$*; it is plausible that $h(0+) = 0$ and that $h(t) \in \uparrow$ for small t but the proof eludes us.

6.5c. t as diffusion

D. B. RAY [4] proved that for a class of MARKOV times $\mathfrak{e}$, including independent exponential holding times and passage times, the conditional local times $[\mathfrak{t}(\mathfrak{e}, \xi) : \xi \in Q,\ P_{ab}(B) = P_a(B \mid x(\mathfrak{e}) = b)]$ can be described in terms of the 2 and 4-dimensional BESSEL processes as in 2.8. $\mathfrak{t}(\mathfrak{m}_1, \xi)$ is invariant under time substitutions, so the passage time case is as before, and the description of $\mathfrak{t}(\mathfrak{e}, \xi)$ for an exponential holding time $\mathfrak{e}$ with conditional law $P_\bullet(\mathfrak{e} > t \mid B) = e^{-t}$ is not changed except that the end points of the state interval come in, and for the scaling of the BESSEL processes, the solutions of $\mathfrak{G}\, g = g$ take the place of the $e^{\pm\xi}$ used in 2.8.

6.5d. Excursions

H. P. MCKEAN [5] described the excursions $[x(t) : t \in \mathfrak{Z}_n, P_0]$ $(n \geqq 1)$ corresponding to the open intervals of the complement of $\mathfrak{Z} = (t : x = 0)$ in case $Q = [0, +\infty)$ and $u^+(0) = 0$, $u \in D(\mathfrak{G})$; the standard BROWNian scaling is not available, but otherwise the description of 2.9 and 2.10 is unchanged, except that the 3-dimensional BESSEL process is replaced by the diffusion associated with $\mathfrak{G}^\bullet = h^{-1}\, \mathfrak{G}\, h$, $h = h(\gamma, \bullet)$ being the

* See note 2.5.2 for the definition of HAUSDORFF measure.

solution of

$$1)\qquad \mathfrak{G} h = \gamma h,\ h(0) = 0,\ h^+(0) = 1$$

with

$$2a)\qquad \gamma = \gamma_1 = \textit{the greatest negative root of } h^-(\gamma, +\infty) = 0$$

or

$$2b)\qquad \gamma = 0$$

according as $+\infty$ is entrance or not. $\mathfrak{G}^\bullet$ can also be described as the generator of the original motion *conditioned* so as not to land at 0 at a positive time.

6.6. Dimension numbers[1]

We use the inverse local time $\mathfrak{t}^{-1}$ to prove that *the dimension number* $\gamma = \dim \mathfrak{Z}$ *of* $\mathfrak{Z} = (t : x = 0)$ *is constant*; for example, $\dim \mathfrak{Z} = 1/2$ for the standard BROWNian motion as we saw in 2.5.

Consider, for the proof, a partition $\mathfrak{Q}$ of $[0, 1]$ into a finite number of non-overlapping intervals $Q_i : i \leqq n$ with rational endpoints, let $|\mathfrak{Q}| = \max_{i \leqq n} |Q_i|$, and let $e_i = 0$ or 1 according as $\mathfrak{Z} \cap Q_i = \varnothing$ or not $(i \leqq n)$.

Given $0 \leqq \alpha \leqq 1$, it is clear that $\sum_{i \leqq n} e_i |Q_i|^\alpha$ is a BOREL function of the sample path $x(t) : t \leqq 1$. Because the class of possible partitions $\mathfrak{Q}$ is countable,

$$1)\qquad \Lambda^\alpha(\mathfrak{Z} \cap [0, 1]) = \liminf_{\varepsilon \downarrow 0\, |\mathfrak{Q}| < \varepsilon} \sum_{i \leqq n} e_i |Q_i|^\alpha$$

is also BOREL, and recalling the definition of the dimension number, it results that $\dim(\mathfrak{Z} \cap [0, 1])$, and with it

$$\gamma = \dim(\mathfrak{Z}) = \sup_{n \geqq 1} \dim(\mathfrak{Z} \cap [n-1, n)),$$

is a BOREL function of the path.

Given $m(0) > 0$,

$$2)\qquad \text{measure}\,[\mathfrak{Z} \cap [0, \mathfrak{t}^{-1}(t))] = \text{measure}\,(s : x(s) = 0,\ s \leqq \mathfrak{t}^{-1}(t))$$
$$= \mathfrak{t}(\mathfrak{t}^{-1}(t))\, m(0) = t\, m(0) > 0$$

for each $t > 0$, so that $\gamma = 1$.

Consider now the case $m(0) = 0$ and let $\mathfrak{W}$ stand for the range $(\mathfrak{t}^{-1}(t) : t > 0)$. $\mathfrak{Z} \supset \mathfrak{W}$ and $\mathfrak{Z} - \mathfrak{W}$ is countable, so that $\dim(\mathfrak{W}) = \gamma$

[1] See note 2.5.2 for the definition of HAUSDORFF-BESICOVITCH dimension numbers.

is BOREL, and it follows from the properties of the POISSON integral $\int_0^{+\infty} l\, \mathfrak{p}([0,t]\times dl)$ for $\mathfrak{t}^{-1}$ that $\gamma = \dim(\mathfrak{W})$ is measurable over $\bigcap_{n\geq 1} \mathsf{B}[\mathfrak{p}(dt\times dl): t\geqq 0,\ l\leqq n^{-1}]$ and so is constant, thanks to the differential character of $\mathfrak{p}$ and the KOLMOGOROV 01 law.

Problem 1. $\gamma = \dim \mathfrak{Z} = \dim \mathfrak{Z}\cap[0,\varepsilon)$ for each $\varepsilon > 0$ for paths starting at 0.

6.7. Comparison tests

Consider $\gamma = \dim \mathfrak{Z}$ as a function of the speed measure m. Because the time substitution $t\to\varepsilon t$ maps $\mathfrak{t}^{-1}(t)\to\varepsilon^{-1}\mathfrak{t}^{-1}(\varepsilon t)$ and $\mathfrak{Z}\to\varepsilon^{-1}\mathfrak{Z}$, it is clear that

1)
$$\begin{aligned}\dim\mathfrak{Z} &= \dim(\mathfrak{t}^{-1}(t): t\leqq 1)\\ &= \dim \varepsilon^{-1}\mathfrak{Z} = \dim(\mathfrak{t}^{-1}(\varepsilon t): t\leqq 1)\\ &= \lim_{\varepsilon\downarrow 0}\dim \mathfrak{Z}\cap[0,\mathfrak{t}^{-1}(\varepsilon)],\end{aligned}$$

and since $\mathfrak{t}^{-1}(\varepsilon)\downarrow \mathfrak{m}_0$ as $\varepsilon\downarrow 0$, *γ is a local function of m, i.e.*, it depends only on the behavior of m in the immediate neighborhood of 0.

Given a pair of diffusions D and $\mathsf{D}^\bullet$ with speed measures m and $m^\bullet \leqq m$, it is to be proved that $\gamma^\bullet \leqq \gamma$.

The proof is not hard: in fact, using the time substitution $t\to \mathfrak{f}^{-1}(t)$ $(\mathfrak{f} = \int \mathfrak{t}\, dm^\bullet)$ mapping x into $x^\bullet = x(\mathfrak{f}^{-1})$, it is clear that $\mathfrak{f}^{-1}$ maps $\mathfrak{Z}^\bullet = (t: x^\bullet = 0)$ into $\mathfrak{Z} = (t: x = 0)$, and since $\mathfrak{f}(\mathfrak{f}^{-1}) = t$, it follows that $\mathfrak{Z}^\bullet \subset \mathfrak{f}(\mathfrak{Z})$. Using

2)
$$\mathfrak{f}(dt) = \int \mathfrak{t}(dt,\xi)\, m^\bullet(d\xi) \leqq \int \mathfrak{t}(dt,\xi)\, m(d\xi) = dt$$

and the result of O. FROSTMAN cited in the note placed at the end of this section, it follows that

3)
$$\begin{aligned}\dim \mathfrak{Z}^\bullet &\leqq \dim \mathfrak{f}(\mathfrak{Z})\\ &= \sup\left(\alpha: \inf_{\substack{e\geq 0\\ e(\mathfrak{f}(\mathfrak{Z}))=1}} \int_{\mathfrak{f}(\mathfrak{Z})\times\mathfrak{f}(\mathfrak{Z})} \frac{e(da)\,e(db)}{|a-b|^\alpha} < +\infty\right)\\ &\leqq \sup\left(\alpha: \inf_{\substack{e\geq 0\\ e(\mathfrak{Z})=1}} \int_{\mathfrak{Z}\times\mathfrak{Z}} \frac{e(da)\,e(db)}{|\mathfrak{f}(a)-\mathfrak{f}(b)|^\alpha} < +\infty\right)\\ &\leqq \sup\left(\alpha: \inf_{\substack{e\geq 0\\ e(\mathfrak{Z})=1}} \int_{\mathfrak{Z}\times\mathfrak{Z}} \frac{e(da)\,e(db)}{|a-b|^\alpha} < +\infty\right)\\ &= \dim(\mathfrak{Z}),\end{aligned}$$

completing the proof.

At no extra cost, it is found that $\gamma^\bullet \leqq \gamma$ *if* $m^\bullet \leqq m$ *near* 0.

t^{-1} and γ will be computed for the special case:

4) $\quad s(d\xi) = d\xi, \qquad m(d\xi) = 2|\xi|^\beta d\xi, \qquad Q = R^1, \qquad \beta > -1:$

the result is that t^{-1} is stable with characteristic exponent $\gamma = \dim \mathfrak{Z} = (\beta + 2)^{-1}$, and this stock of dimension numbers will lead to a comparison test, permitting us to read off the dimension number of the zeros of most diffusions encountered in practice.

Given $\beta > -1$ and $\alpha > 0$, the solution $g_2 = g_2(\alpha, \xi)$ of

5) $$\alpha g_2(\xi) = \tfrac{1}{2}|\xi|^{-\beta} g_2''(\xi) \quad g_2 \in \downarrow, g_2(+\infty) = 0, g_2(0) = 1,$$

is invariant under the substitution

6) $$\alpha \to c\alpha, \quad \xi \to \xi/c^\gamma \qquad \gamma = (\beta + 2)^{-1}, c > 0;$$

i.e.,

7) $$g_2(\alpha, \xi) = g_2(1, \alpha^\gamma \xi),$$

and therefore

8) $$\int_0^{+\infty} (1 - e^{-\alpha l})\, n(dl) = 2\int_0^{+\infty} (1 - e^{-\alpha l})\, n_+(dl)$$

$$= -2\frac{g_2^+(\alpha, 0)}{g_2(\alpha, 0)} = -2\alpha^\gamma \frac{g_2^+(1, 0)}{g_2(1, 0)} \qquad \alpha \geqq 0,$$

proving that t^{-1} is stable with characteristic exponent $\gamma = (\beta + 2)^{-1}$.

Computing $p(t) = p(t, 0, 0) = \text{constant} \times t^{\gamma-1}$ from 8), we find that the HAUSDORFF function $h = p^{-1}\, p \otimes p$ of 6.5b is a constant multiple of t^γ. $\dim \mathfrak{Z} \leqq \gamma$ is now clear from 6.5b.2), and since

9) $$E_0\left(\int_{\mathfrak{Z}\cap[0, t^{-1}(1))\times \mathfrak{Z}\cap[0, t^{-1}(1))} |s - t|^{-\alpha}\, t(ds)\, t(dt)\right)$$

$$= E_0\left(\int_0^1\int_0^1 |t^{-1}(s) - t^{-1}(t)|^{-\alpha}\, ds\, dt\right)$$

$$= \text{constant} \int_0^1 ds \int_0^1 dt \int_0^{+\infty} \theta^{\alpha-1}\, d\theta\, E_0(e^{-\theta t^{-1}(|t-s|)})$$

$$= \text{constant} \int_0^1 ds \int_0^1 dt \int_0^{+\infty} \theta^{\alpha-1}\, d\theta\, e^{-\text{constant}\,|t-s|\,\theta^\gamma}$$

$$= \text{constant} \int_0^1\int_0^1 |s - t|^{\alpha/\gamma}\, ds\, dt$$

$$< +\infty \qquad \alpha < \gamma,$$

FROSTMAN's lemma implies

10) $$\dim \mathfrak{Z} = \sup\left(\alpha : \inf_{\substack{e \geqq 0 \\ e(\mathfrak{Z}) = 1}} \int_{\mathfrak{Z}\times\mathfrak{Z}} \frac{e(da)\, e(db)}{|a - b|^\alpha} < +\infty\right) \geqq \gamma,$$

and the proof is complete.

Our comparison test now reads as follows: *if* $s(d\xi) = d\xi$, *if* $\beta > -1$, *and if* $m(d\xi) \geqq (\leqq)$ constant $\times$ $|\xi|^\beta \, d\xi$ *near* $\xi = 0$, *then* $\gamma = \dim \mathfrak{Z} \geqq (\leqq) (\beta + 2)^{-1}$; the proof is clear.

Additional information on the computation of dimension numbers can be found in BLUMENTHAL and GETOOR [1].

Problem 1. Can $\gamma = 0$ serve as the dimension number of the visiting set $\mathfrak{Z}$ of a diffusion?

[Given $s(d\xi) = d\xi$ and $m(d\xi) = e^{-|\xi|^{-1}} d\xi$, $m(d\xi) \leqq |\xi|^\beta \, d\xi$ near $\xi = 0$ for each $\beta > -1$, proving $\gamma \leqq \inf_{\beta > -1} (\beta + 2)^{-1} = 0$.]

Note 1. Dimension numbers and fractional dimensional capacities. We need the result of O. FROSTMAN [1], *that for compact* $B \subset [0, +\infty)$,

1) $$\dim(B) = \sup\left(\beta : \inf_e \int \frac{e(da)\, e(db)}{|a - b|^\beta} < +\infty\right),$$

where e *runs over the class of non-negative Borel measures of total mass* $+1$ *loaded up on* B.

Here is the proof.

Given $\beta > 0$ such that $+\infty > \int |a - b|^{-\beta} \, e(da)\, e(db)$, select compact $Q \subset B$ such that $e(Q) > 0$ and $\int_Q |a - b|^{-\beta} \, e(db) \leqq \varkappa < +\infty$ $(a \in Q)$, and cover Q with closed intervals Q_n $(n \geqq 1)$. Because

2) $$\begin{aligned} 0 < e(Q) &\leqq \sum_{n \geqq 1} e(Q \cap Q_n) \\ &\leqq \sum_{n \geqq 1} |Q_n|^\beta \inf_{a \in Q \cap Q_n} \int_{Q \cap Q_n} |a - b|^{-\beta} \, e(db) \\ &\leqq \sum_{n \geqq 1} |Q_n|^\beta \varkappa, \end{aligned}$$

it is immediate that

3) $$\Lambda^\beta(B) \geqq \Lambda^\beta(Q) \geqq e(Q)/\varkappa > 0,$$

proving

4) $$\dim(B) \geqq \beta,$$

and to complete the proof, it is enough to check that

$$+\infty > \inf \int |a - b|^{-\beta} \, e(da)\, e(db) \quad \textit{for} \quad \beta < \dim(B).$$

Given $\beta < \alpha < \dim(B)$ so that $\Lambda^\alpha(B) = +\infty$, A. S. BESICOVITCH [2] showed that it is possible to select compact $Q_1 \subset B$ with

5) $$0 < \Lambda^\alpha(Q_1) < +\infty.$$

Next, according to BESICOVITCH [1], it is possible to select compact $Q_2 \subset Q_1$ with

6) $$0 < \Lambda^\alpha(Q_2) < +\infty$$

and

7) $$\lim_{\varepsilon\downarrow 0}(2\varepsilon)^{-\alpha}\,\Lambda^{\alpha}(Q_2\cap(a-\varepsilon,a+\varepsilon))\leqq 1 \qquad a\in Q_2.$$

Now select compact $Q_3\subset Q_2$ such that

8) $$0<\Lambda^{\alpha}(Q_3)<+\infty$$

and

9) $$(2\varepsilon)^{-\alpha}\,\Lambda^{\alpha}(Q_3\cap(a-\varepsilon,a+\varepsilon))<2 \qquad \varepsilon<\delta,\ a\in Q_3$$

for suitable $\delta>0$ and define

10) $$e(db)=\Lambda^{\alpha}(Q_3)^{-1}\,\Lambda^{\alpha}(Q_3\cap db);$$

then

11) $$\begin{aligned}\int_{B\times B}\frac{e(da)\,e(db)}{|a-b|^{\beta}} &=\int_{Q_3}e(da)\left[\int_{|a-b|\geqq\delta}\frac{e(db)}{|a-b|^{\beta}}+\sum_{n\geqq 1}\int_{2^{-n}\delta\leqq|a-b|<2^{-n+1}\delta}\frac{e(db)}{|a-b|^{\beta}}\right]\\ &\leqq\int_{Q_3}e(da)\left[\delta^{-\beta}+\sum_{n\geqq 1}(2^{-n}\delta)^{-\beta}\,2(2.2^{-n+1}\delta)^{\alpha}\right]\Lambda^{\alpha}(Q_3)^{-1}\\ &=\delta^{-\beta}+2^{2+\alpha}\delta^{\alpha-\beta}\sum_{n\geqq 1}2^{-n(\alpha-\beta)}\,\Lambda^{\alpha}(Q_3)^{-1}\\ &<+\infty,\end{aligned}$$

completing the proof.

6.8. An individual ergodic theorem[1]

Consider a persistent diffusion D with state interval Q, scale s, speed measure m, and local times t, and define the local time integral (additive functional)

1) $$e(t)=\int \mathrm{t}(t,\xi)\,e(d\xi)$$

for any non-negative measure e, and let us prove that

2) $$P_{\bullet}\left[\lim_{t\uparrow+\infty}\frac{e_1(t)}{e_2(t)}=\frac{e_1(Q)}{e_2(Q)}\right]=1 \quad \textit{in case} \quad 0<e_2(Q)<+\infty.$$

C. Derman [1] proved a special case of 2):

3) $$P_{\bullet}\left[\lim_{t\uparrow+\infty}\frac{\int_0^t f[x(s)]\,ds}{\int_0^t g[x(s)]\,ds}=\frac{\int_Q f\,dm}{\int_Q g\,dm}\right]=1$$

[1] 7.17 contains similar results for the 2-dimensional Brownian motion.

for the standard BROWNian motion by a method of W. DOBLIN [1] used below; see also M. MOTOO and H. WATANABE [1].

Given $a < b$, let $\mathfrak{m}^1 < \mathfrak{m}^2 <$ *etc.* be the successive passage times to a via b, *i.e.*,

4) $$\mathfrak{m}^n = \mathfrak{m}^{n-1} + \mathfrak{m}(w^+_{\mathfrak{m}^{n-1}}) \qquad (n \geqq 1),$$
$$\mathfrak{m}^0 = \mathfrak{m}_a, \; \mathfrak{m} = \mathfrak{m}_b + \mathfrak{m}_a(w^+_{\mathfrak{m}_b}).$$

Because the excursions $x(t) : \mathfrak{m}^{n-1} \leqq t \leqq \mathfrak{m}^n$ $(n \geqq 1)$ are independent and identical in law, so are the increments of $\mathfrak{e}$:

5) $$e_n = \mathfrak{e}(\mathfrak{m}^n) - \mathfrak{e}(\mathfrak{m}^{n-1}) \qquad (n \geqq 1),$$

Using the strong law of large numbers and the evaluation of problem 1:

6) $$E_a[\mathfrak{t}(\mathfrak{m}, \xi)] = s(b) - s(a),$$

one obtains

7) $$P_a\left[\lim_{n\uparrow+\infty} \frac{\mathfrak{e}(\mathfrak{m}^n)}{n} = [s(b) - s(a)]\, e(Q)\right] = 1,$$

i.e.,

8) $$P_a\left[\lim_{t\uparrow+\infty} \frac{\mathfrak{e}(t)}{\min(n : \mathfrak{m}^n \geqq t)} = [s(b) - s(a)]\, e(Q)\right] = 1,$$

and 2) follows.

Besides 3), there are several interesting special cases of 2):

9) $$\lim_{t\uparrow+\infty} \frac{\text{measure}(s : x(s) \in A, s \leqq t)}{\text{measure}(s : x(s) \in B, s \leqq t)} = \frac{m(A)}{m(B)}$$
$$0 < m(B) < +\infty, \quad A, B \in \mathbf{B}(Q).$$

10) $$\lim_{t\uparrow+\infty} \frac{\mathfrak{e}(t)}{\mathfrak{t}(t, a)} = e(Q) \qquad a \in Q,$$

11) $$\lim_{t\uparrow+\infty} \frac{\mathfrak{t}(t, a)}{\mathfrak{t}(t, b)} = 1 \qquad a, b \in Q,$$

12) $$\lim_{t\uparrow+\infty} \frac{\mathfrak{e}(t)}{t} = \frac{e(Q)}{m(Q)} \qquad e(Q) < +\infty,$$

13) $$\lim_{t\uparrow+\infty} \frac{\mathfrak{t}(t, a)}{t} = m(Q)^{-1} \qquad a \in Q$$

(in case $m(Q) = +\infty$, 12) and 13) are to be understood with the convention $1/+\infty = 0$).

If $\mathbf{D}$ *is a standard Brownian motion, then*

14) $$\lim_{t\uparrow+\infty} P_\bullet\left[\frac{2\mathfrak{e}(t)}{e(R^1)\sqrt{t}} \leqq b\right] = \sqrt{\frac{2}{\pi}} \int_0^b e^{-c^2/2}\, dc$$
$$b \geqq 0, \; 0 < e(R^1) < +\infty;$$

in the special case $e(d\xi) = f\,d\xi$, this becomes the KALLIANPUR-ROBBINS law [*1*]:

$$15)\qquad \lim_{t\uparrow+\infty} P_\bullet\left[\frac{\int_0^t f[x(s)]\,ds}{\sqrt{t}\int f\,d\xi} \leqq b\right] = \sqrt{\frac{2}{\pi}}\int_0^b e^{-c^2/2}\,dc$$

$$b \geqq 0, \quad 0 < \int f\,d\xi < +\infty.$$

14) follows from 9) and the properties of the BROWNian local times as follows:

$$\lim_{t\uparrow+\infty} P_a\left[\frac{2e(t)}{e(R^1)\sqrt{t}} \leqq b\right]$$

$$= \lim_{t\uparrow+\infty} P_a\left[\frac{e(t)}{\mathfrak{t}(t,a)\,e(R^1)}\,\frac{2\mathfrak{t}(t,a)}{\sqrt{t}} \leqq b\right]$$

$$= \lim_{t\uparrow+\infty} P_a\left[\frac{2\mathfrak{t}(t,a)}{\sqrt{t}} \leqq b\right]$$

$$= \lim_{t\uparrow+\infty} P_0\left[\frac{2\mathfrak{t}(t,0)}{\sqrt{t}} \leqq b\right] = P_0[2\mathfrak{t}(1,0) \leqq b]$$

$$= P_0[\mathfrak{t}^+(1) \leqq b] = P_0[\mathfrak{t}^-(1) \leqq b] = \sqrt{\frac{2}{\pi}}\int_0^b e^{-c^2/2}\,dc$$

(see 2.1 for $\mathfrak{t}^+$ and $\mathfrak{t}^-$).

Given $e(Q) < +\infty$, $\lim_{t\uparrow+\infty} t^{-1}e(t) = e(Q)/m(Q)$ by 12), and it is an interesting problem to describe the behavior of $e^\bullet = e - t\,e(Q)/m(Q)$ as $t\uparrow+\infty$. H. TANAKA [*1*] solved this problem in the special case $Q = R^1$, $s(b) - s(a) = b - a$, $m(Q) < +\infty$ by showing that if $\sigma^2 m(Q) \equiv \int_Q m(0, b)^2\,db < +\infty$, then

$$\lim_{t\uparrow+\infty} P_\bullet(a \leqq e^\bullet(t)/\sigma\sqrt{2t} < b) = \frac{1}{\sqrt{2\pi}}\int_a^b e^{-c^2/2}\,dc$$

and

$$P_\bullet\left(\overline{\lim_{t\uparrow+\infty}}\; e^\bullet(t)/\sigma\sqrt{2t\lg_2 t} = 1\right) = 1.$$

Problem 1. Derive 6) from the result of problem 2.8.3.

[Because $t(\mathfrak{m}, \xi)$ is invariant under the time change, it is enough to prove

$$17)\qquad E_a[\mathfrak{t}(\mathfrak{m}, \xi)] = b - a, \qquad \mathfrak{m} = \mathfrak{m}_b + \mathfrak{m}_a(w^+_{\mathfrak{m}_b}), \qquad a < b$$

for the standard BROWNian motion. Using problem 2.8.3, one obtains

$$E_a[\mathfrak{t}(\mathfrak{m}, \xi)] = E_\xi[\mathfrak{t}(\mathfrak{m}_b, \xi)] + E_\xi[\mathfrak{t}(\mathfrak{m}_a, \xi)] = (b - \xi) + (\xi - a) = b - a$$

in case $a < \xi < b$, and the other cases can be treated similarly.]

Problem 2. Give a new proof of 15) using the scaling properties of the standard BROWNian motion.

[Given a standard BROWNian motion starting at 0, the scaling $x(s) \to t\,x(s/t^2)$ shows that $\mathfrak{t}(t, \xi)$ is identical in law to $\sqrt{t}\,\mathfrak{t}(1, \xi/t)$; thus $2\int \mathfrak{t}\, de/e(R^1)\sqrt{t}$ is identical in law to

$$\frac{2}{e(R^1)}\int \mathfrak{t}(1, \xi/t)\, e(d\xi) \sim 2\mathfrak{t}(1, 0).]$$

Problem 3. Show how to compute the invariants of a persistent diffusion from the sample path.

[9) shows how to compute the speed measure from the path up to a constant factor, and fixing it, one can obtain the scale from the special case of 7a):

$$\lim_{n\uparrow+\infty} \frac{\mathfrak{t}(\mathfrak{m}^n, \xi)}{n} = s(b) - s(a).]$$

Problem 4. Consider the BESSEL process with $\mathfrak{G} = \frac{1}{2}\left(\frac{d^2}{dr^2} + \frac{1}{r}\frac{d}{dr}\right)$. Use the law

18) $$\lim_{n\uparrow+\infty} P_\bullet\left[\sqrt[n]{\mathfrak{m}^n} < t\right] = 4^{-1/\lg t}\ (t > 1), \quad = 0\ (t \leq 1)$$

for the successive passage times $\mathfrak{m}^n$ to 1 via 2 to check that

19) $$\lim_{t\uparrow+\infty} P_\bullet\left[\frac{2e(t)}{\lg t} < u\, e(0, +\infty)\right] = 1 - e^{-u};$$

this formula is an analogue of the KALLIANPUR-ROBBINS law 14) and will be used in 7.17 (see problem 4.6.4 for a related result).

[Use the passage time formulas

$$E_1(e^{-\alpha\mathfrak{m}_2}) = \frac{I_0(\sqrt{2\alpha})}{I_0(2\sqrt{2\alpha})}, \quad E_2(e^{-\alpha\mathfrak{m}_1}) = \frac{K_0(2\sqrt{2\alpha})}{K_0(\sqrt{2\alpha})}$$

and the estimates $K_0(x) \sim -\lg x$ and $I_0(x) - 1 \sim x^2/4$ for $x \downarrow 0$ to check

$$E_\bullet(e^{-\alpha t^{-n}\mathfrak{m}^n}) \sim \left(1 - \frac{\lg 4}{n \lg t}\right)^n \sim 4^{-1/\lg t},$$

which implies 18). To prove 19), it is enough to check

$$\begin{aligned}
&\lim_{t\uparrow+\infty} P_\bullet\,[2e(t)/\lg t < u] \\
&= \lim_{n\uparrow+\infty} P_\bullet\,[2e(\mathfrak{m}^n)/\lg \mathfrak{m}^n < u] \\
&= \lim_{n\uparrow+\infty} P_\bullet\left[2e(\mathfrak{m}^n)/n \lg \sqrt[n]{\mathfrak{m}^n} < u\right] \\
&= \lim_{n\uparrow+\infty} P_\bullet\left[2\lg 2\, e(0, +\infty)/\lg \sqrt[n]{\mathfrak{m}^n} < u\right],
\end{aligned}$$

noting that $s(r) = \lg r$; now replace u with $u\,e(0, +\infty)$ and use 18) to complete the proof of 19).]

7. Brownian motion in several dimensions

7.1. Diffusion in several dimensions

Given a nice topological space Q (for example, a several-dimensional differentiable manifold), let ∞ be an extra point which is either *isolated* or *the one-point compactification of* Q according as Q is compact or not, let $C_\infty(Q)$ be the space of bounded continuous functions $f: Q \cup \infty \to R^1$ with $f(\infty) \equiv 0$, introduce the (continuous) *sample paths* $w: t \to x(t) \in Q \cup \infty$ with $x(t) \in Q$ $(t < +\infty)$ and $x(+\infty) \equiv \infty$, define *Markov times* $\mathfrak{m}$, *shifted paths* $w^+_{\mathfrak{m}}$, and *Borel algebras* B and $\mathsf{B}_{\mathfrak{m}+}$ as usual, take *probabilities* $P_a(B)$ $(a \in Q \cup \infty,\ B \in \mathsf{B})$ with the usual properties including $P_\infty[x(t) \equiv \infty\ (t \geqq 0)] = 1$, and call the associated motion *a diffusion if it starts afresh at each Markov time, i.e., if*

1) $$P_a(w^+_{\mathfrak{m}} \in B \mid \mathsf{B}_{\mathfrak{m}+}) = P_b(B) \quad a \in Q,\ B \in \mathsf{B},\ b = x(\mathfrak{m}),\ \mathfrak{m} < +\infty$$

for each Markov time $\mathfrak{m}$ *and if, in addition, it is smooth, i.e., if the Green operators* $G_\alpha: f \to E_\bullet\left[\int_0^{+\infty} e^{-\alpha t} f\, dt\right]$ *map* $C_\infty(Q)$ *into itself:*

2) $$G_\alpha C_\infty \subset C_\infty \qquad \alpha > 0.$$

1) and 2) are not unrelated.

R. BLUMENTHAL [*1*] proved that if the motion is *simple Markov, i.e.*, if it starts afresh at each *constant* time $t \geqq 0$, and if it is also *smooth* as in 2), then 1) is automatic (the proof can be modelled on the standard BROWNian case 1.6).

Bearing in mind the result of 3.6, it is natural to conjecture that the GREEN operators map into itself a space similar to $C_\infty(Q)$ under the sole condition 1) (see E. B. DYNKIN [*7*], H. P. MCKEAN, JR. and H. TANAKA [*1*], and especially D. B. RAY [*3*] for information on this point).

Given a diffusion as described above,

3) $$G_\alpha - G_\beta + (\alpha - \beta) G_\alpha G_\beta = 0$$

as in 3.7.1), permitting the usual definition of the *generator:*

4a) $$\mathfrak{G} = 1 - G_1^{-1}$$

4b) $$D(\mathfrak{G}) = G_1 C_\infty(Q)$$

and the derivation of *Dynkin's formula*:

5) $$E_\bullet[u(x_{\mathfrak{m}})] - u = E_\bullet\left[\int_0^{\mathfrak{m}} (\mathfrak{G} u)(x_t)\, dt\right] \qquad u \in D(\mathfrak{G}),$$

where $\mathfrak{m} \equiv \min(t: x(t) \notin B)$ is the (MARKOVian) exit time from a sufficiently small open region $B \subset Q$, and it follows that if B is a neighborhood of $a \in Q$, then *either* $P_a[\mathfrak{m} = +\infty] = 1$ for all B *or* $E_a(\mathfrak{m}) < +\infty$ for all small B; in both cases

6) $$(\mathfrak{G} u)(a) = \lim_{B \downarrow a} \frac{E_a[u(x_{\mathfrak{m}})] - u(a)}{E_a(\mathfrak{m})} \qquad u \in D(\mathfrak{G}),$$

just as in 3.7.

$\mathfrak{G}$ is a *local operator*, and $(\mathfrak{G} u)(a) \geq 0$ if $u \in D(\mathfrak{G})$ has a local minimum at $a \in Q$. $\mathfrak{G}$ shares both these features with the classical elliptic differential operators (see chapter 8 for additional information about elliptic operators).

BLUMENTHAL's useful 01 law

7) $$P_\bullet(B) = 0 \text{ or } 1 \qquad B \in \mathsf{B}_{0+} \equiv \bigcap_{\varepsilon > 0} \mathsf{B}_\varepsilon$$

follows from 1).

Problem 1. Give a complete proof that if $B \subset Q$ is open and $A \subset Q$ is closed, then $\mathfrak{m}_B = \inf(t: x(t) \in B)$ and $\mathfrak{m}_A = \inf(t: t > 0, x(t) \in A)$ are both MARKOV times.

$[(w: \mathfrak{m}_B < t) = \bigcup_{j n^{-1} < t} (w: x(j n^{-1}) \in B)$, so $\mathfrak{m}_B$ is MARKOV, and, choosing open $B_1 \supset B_2$ *etc.* $\downarrow A$, the MARKOVian nature of $\mathfrak{m}_A$ is clear from

$$\mathfrak{m}_A = \inf_{j \geq 1} \sup_{n \geq 1} [j^{-1} + \mathfrak{m}_{B_n}(w_{j^{-1}}^+)]$$

and

$$\begin{aligned}(w: \mathfrak{m}_A < t) &= \bigcup_{j \geq 1} \bigcup_{i \geq 1} \bigcap_{n \geq 1} (w: \mathfrak{m}_{B_n}(w_{j^{-1}}^+) < t - j^{-1} - i^{-1}) \\ &= \bigcup_{j \geq 1} \bigcup_{i \geq 1} \bigcap_{n \geq 1} \bigcup_{k l^{-1} < t - j^{-1} - i^{-1}} (w: x(k l^{-1}, w_{j^{-1}}^+) \in B_n).]\end{aligned}$$

7.2. The standard Brownian motion in several dimensions

Consider the standard $d (\geq 2)$ dimensional BROWNian motion of section 2.10: it has continuous sample paths, its GREEN operators

1) $(G_\alpha f)(a)$

$$= 2(2\pi)^{-\frac{d}{2}} \int_{R^d} \left(\frac{\sqrt{2\alpha}}{|b - a|}\right)^{\frac{d}{2}-1} K_{\frac{d}{2}-1}(\sqrt{2\alpha}\,|b - a|)\, f(b)\, db^*$$

map $C_\infty(R^d)$ into itself, and so it is a diffusion as defined in 7.1.

DYNKIN's formula will now be used to compute its generator $\mathfrak{G}$.

---

* $2(2\pi)^{-\frac{d}{2}} \left(\frac{\sqrt{2\alpha}}{|b - a|}\right)^{\frac{d}{2}-1} K_{\frac{d}{2}-1}(\sqrt{2\alpha}\,|b - a|)$ is the LAPLACE transform of the GAUSS kernel $(2\pi t)^{-\frac{d}{2}} e^{-|b-a|^2/2t}$; see A. ERDÉLYI [*1* (1): 146 (26)]. $K_{\frac{d}{2}-1}$ is the usual modified BESSEL function.

Because the GAUSS kernel $(2\pi t)^{-d/2} e^{-|b-a|^2/2t}$ is invariant under the group of euclidean motions, it is clear that, for $a \in R^d$, the distribution $P_a[x(\mathfrak{m}_{\partial B}) \in db]$* of BROWNian hits on the spherical surface $\partial B: |b-a| = \varepsilon$ is invariant under the group of rotations about its center; as such it has to be the uniform distribution do on ∂B, an observation that was already used in 2.7.

$E_a(\mathfrak{m}_{\partial B})$ can also be computed with ease: the radial motion $r(t) = |x(t) - a|$ $(t \geqq 0)$ is the BESSEL process with scale

$$\lg r \quad (d=2) \qquad -r^{2-d}/(d-2) \qquad (d \geqq 3)$$

and speed measure $2r^{d-1}\,dr$. $\mathfrak{m}_{\partial B}$ is the BESSEL passage time to $r = \varepsilon$, and so

2) $$E_a(\mathfrak{m}_{\partial B}) = \int_0^\varepsilon m[0, r]\, s(dr) = \int_0^\varepsilon \frac{2r}{d}\, dr = \frac{\varepsilon^2}{d}$$

(see 2.7).

DYNKIN's formula now reads

3) $$(\mathfrak{G} u)(a) = \lim_{\varepsilon \downarrow 0} \frac{\int_{\partial B} u(b)\, do - u(a)}{\varepsilon^2/d} \qquad u \in D(\mathfrak{G});$$

esp., if $u \in C^2(R^d)$, then $\mathfrak{G} u = (1/2)\Delta u$, Δ being LAPLACE's operator, as the reader can easily check using the MACLAURIN expansion of u about a.

$\mathfrak{G}$ can be described as the operator

4) $$\mathfrak{G}^\bullet u = \lim_{\varepsilon \downarrow 0} d\varepsilon^{-2} \left(\int_{\partial B} u\, do - u \right)$$

applied to the class of functions $u \in C_\infty(R^d)$ such that $\mathfrak{G}^\bullet u \in C_\infty(R^d)$.

$\mathfrak{G}^\bullet \supset \mathfrak{G}$, as has just been proved, and now to obtain $\mathfrak{G}^\bullet = \mathfrak{G}$ it is enough to check that $D(\mathfrak{G}^\bullet) \subset D(\mathfrak{G})$. Given $u_1 \in D(\mathfrak{G}^\bullet)$, let $f = (1 - \mathfrak{G}^\bullet) u_1$; then $f \in C_\infty(R^d)$, $u_2 = G_1 f \in D(G)$, and $u_3 = u_2 - u_1 \in D(\mathfrak{G}^\bullet)$ is a solution of $\mathfrak{G}^\bullet u_3 = u_3$. But then $u_3 \in C_\infty(R^d)$ attains its greatest value at some point $a \in R^d$, and it follows from the definition of $\mathfrak{G}^\bullet$ that $u_3 \leqq u_3(a) = (\mathfrak{G}^\bullet u_3)(a) \leqq 0$. $u_3 \geqq 0$ for a similar reason and so $u_3 \equiv 0$, *i.e.*, $u_1 = u_2 \in D(\mathfrak{G})$ as desired.

Problem 1. $\mathfrak{G}$ is the uniform closure of $\Delta/2$ acting on the class $C_\infty(\Delta)$ of functions $u \in C^2_\infty(R^d)$ with $\Delta u \in C_\infty(R^d)$, *i.e.*, if $u \in D(\mathfrak{G})$, then it is possible to select $u_n \in C_\infty(\Delta)$ $(n \geqq 1)$ such that $\|u_n - u\|_\infty$ and $\|\Delta u_n/2 - \mathfrak{G} u\|_\infty$ tend to 0 as $n \uparrow +\infty$.

Problem 2. $\mathfrak{G}$ is a *proper* extension of $\Delta/2$ acting on $C_\infty(\Delta)$.

[Consider the class $C_\bullet(R^d)$ of continuous functions f with $\|f\|_\bullet = \|(1+r) f\|_\infty < +\infty$ and let $G_{0+} f = c(d) \int_{R^d} |a-b|^{2-d} f(b)\, db$

* ∂B means the boundary of B.

with $4c(d) = \pi^{-d/2}\,\Gamma(d/2-1)$. $\mathfrak{G}$ is closed and $(\Delta/2)\,G_{0+} = -1$ on $C_\bullet(R^d) \cap C^1(R^d)$, so $\mathfrak{G}\,G_{0+} = -1$ on $C_\bullet(R^d)$, and $D(\mathfrak{G}) \subset C_\infty(\Delta)$ would mean that $\partial^2 G_{0+}$ could be viewed as a (closed) map from $C_\bullet(R^d)$ to $C(|a|<1)$ for each second partial $\partial^2 = \partial^2/\partial x_i\,\partial x_j$ $(i, j \leq d)$. As such $\partial^2 G_{0+}$ would be bounded, and it would follow that $(\partial^2 u)(0) = \int_{R^d} (\Delta u)(\xi)\, e(d\xi)$ for $u \in C^2(R^d)$ vanishing far out with a signed measure e such that $\int_{R^d} d\,|e| < +\infty$. But then, changing u into $\delta^{-2} u(\delta \xi)$ and letting $\delta \downarrow 0$, it would follow that $(\partial^2 u)(0) = e(0)\,(\Delta u)(0)$ which is ridiculous. The above proof is adapted from K. DE LEEUW and H. MIRKIL [*1*].]

Problem 3 after P. LÉVY [*3*]. Use the scaling and differential properties of the 2-dimensional BROWNian motion to show that its sample path has 2-dimensional LEBESGUE measure 0.

[Consider the arcs

$$\begin{aligned} B_1 &= (x(t): 0 \leq t \leq 1) \\ B_2 &= (x(t): 0 \leq t \leq 2) \\ B_3 &= (x(1-t) - x(1): 0 \leq t \leq 1) \\ B_4 &= (x(1+t) - x(1): 0 \leq t \leq 1) \end{aligned}$$

and their indicators e_1, e_2, e_3, and e_4 for 2-dimensional BROWNian paths starting at 0. B_3 and B_4 are independent, and B_1, B_3, and B_4 are identical in law, as is clear by the differential property and the translation and reflection invariance of the motion. B_2 and $\sqrt{2}\,B_1$ are identical in law, as is also clear on using the measure-preserving scaling $x(t) \to x(2t)/\sqrt{2}$. Using $|B|$ to denote the area of $B \subset R^2$, $|B_3 \cup B_4| = |B_2|$ by the translation invariance of the LEBESGUE measure, and so

$$\begin{aligned} 2E_0(|B_1|) &= E_0(|B_2|) \\ &= E_0(|B_3 \cup B_4|) \\ &= E_0(|B_3| + |B_4| - |B_3 \cap B_4|) \\ &= 2E_0(|B_1|) - E_0\Big(\int e_3\, e_4\, db\Big) \\ &= 2E_0(|B_1|) - \int E_0(e_3\, e_4)\, db \\ &= 2E_0(|B_1|) - \int E_0(e_1)^2\, db \end{aligned}$$

which implies

$$E_0(|B_1|) = \int E_\bullet(e_1)\, db = 0$$

as desired. Another solution due to H. SATO is as follows. 2.7.9) states that the 2-dimensional BROWNian path never hits a point singled out in advance at a positive time. Because of this $P_a[\sup_{t>0} e_b(x(t)) = 0] = 1$ ($a \in R^2$, e_b = the indicator function of $b \in R^2$), and using FUBINI's theorem, it follows that

$$E_0[\text{measure}(x(t) : t \geq 0)] = E_0\left[\int_{R^2} \sup_{t>0} e_b(x(t))\, db\right] = 0.]$$

7.3. Wandering out to ∞

One feature of the BROWNian motion that will be used below is that in $d \geq 3$ dimensions $P_\bullet[\lim_{t\uparrow+\infty} |x(t)| = +\infty] = 1$, while in the 2-dimensional case, the BROWNian sample path $x(t)$ meets each little circle at an infinite series of times increasing to $+\infty$.

Here is the proof (see problem 4.6.3 for a different method).

Given a standard BROWNian path starting at $0 \in R^d$, let $A \subset B$ be two spheres with centers at 0, let $\mathfrak{m}_1 < \mathfrak{m}_2 <$ *etc.* be the successive passage times to ∂A via ∂B, *i.e.*, define

$$\mathfrak{m}_1 = \mathfrak{m}_{\partial A},\ \mathfrak{m}_n = \mathfrak{m}_{n-1} + \mathfrak{m}(w^+_{\mathfrak{m}_{n-1}})\ (n \geq 2),\ \mathfrak{m} = \mathfrak{m}_{\partial B} + \mathfrak{m}_{\partial A}(w^+_{\mathfrak{m}_{\partial B}}),$$

let $\mathfrak{e}_n$ be the time the path spends in A for $\mathfrak{m}_{n-1} \leq t < \mathfrak{m}_n$ ($n \geq 1$, $\mathfrak{m}_0 = 0$), and using the isotropic nature of the BROWNian motion, let γ be the (constant) value of $P_\bullet(\mathfrak{m}_{\partial A} < +\infty)$ on ∂B and e_2 the (constant) value of $E_\bullet(\mathfrak{e}_2)$ on ∂A.

Because the $\mathfrak{m}_1 < \mathfrak{m}_2 <$ *etc.* are MARKOV times and the motion is isotropic, $\mathfrak{e}_n$ is identical in law to $\mathfrak{e}_2$ if $\mathfrak{m}_{n-1} < +\infty$ ($n \geq 2$) so

$$\begin{aligned}
1)\qquad +\infty \gtreqless \int_0^{+\infty} dt \int_A \frac{e^{-|b|^2/2t}}{(2\pi t)^{d/2}}\, db \qquad & d \lesseqgtr 2 \\
= E_0[\text{measure}(t : x(t) \in A)] & \\
= E_0(\mathfrak{m}_{\partial A}) + \sum_{n\geq 2} E_0[\mathfrak{e}_n, \mathfrak{m}_{n-1} < +\infty] & \\
= E_0(\mathfrak{m}_{\partial A}) + \sum_{n\geq 2} e_2 \gamma^{n-2}, &
\end{aligned}$$

and using

$$2)\qquad E_0(\mathfrak{m}_{\partial A}) + e_2 \leq E_0(\mathfrak{m}_{\partial B}) < +\infty$$

[see 7.2.2)], it develops that

$$3)\qquad \sum_{n\geq 0} \gamma^n \gtreqless +\infty \quad (i.e., \gamma \gtreqless 1) \textit{ according as } d \lesseqgtr 2.$$

But then, if $d \geq 3$, the path cannot meet A for large times and $P_0\left[\lim_{t\uparrow+\infty} |x(t)| = +\infty\right] = 1$ as is clear on letting $A \uparrow R^d$, while if $d = 2$, then $\mathfrak{m}_n < +\infty$ for each $n \geq 1$, $\lim_{n\uparrow+\infty} \mathfrak{m}_n = +\infty$, and selecting a disc $D \subset R^2$, the obvious estimate $\sup_{\partial A} P_\bullet(\mathfrak{m}_D = +\infty) = \gamma < 1$ coupled with

4) $$\gamma = \sup_{\partial A} \lim_{n\uparrow+\infty} E_\bullet[\mathfrak{m}_D > \mathfrak{m}_n, P_{x(\mathfrak{m}_n)}(\mathfrak{m}_D = +\infty)]$$
$$\leq \gamma \sup_{\partial A} \lim_{n\uparrow+\infty} P_\bullet(\mathfrak{m}_D > \mathfrak{m}_n)$$
$$= \gamma^2$$

justifies

5) $$P_0(\mathfrak{m}_D = +\infty) = E_0[P_{x(\mathfrak{m}_{\partial A})}(\mathfrak{m}_D = +\infty)] \leq \gamma = 0,$$

proving that the path meets each disc an infinite number of times as desired.

7.4. Greenian domains and Green functions

$D \subset R^d$ ($d \geq 2$) is said to be a domain if it is open and connected. D is said to be a GREENian domain if, in addition, there is a function $G = G(a, b) = K + H$ defined on $D \times D$ and $< +\infty$ off the diagonal such that

1) $$G \geq 0,$$

2) $$G(a, b) = G(b, a),$$

3) $$K = -(2\pi)^{-1} \lg|b - a| \qquad d = 2$$
$$= \frac{\Gamma\left(\frac{d}{2} - 1\right)}{4\pi^{d/2}} |b - a|^{2-d} \qquad d > 2,$$

4) $$\Delta_a H = \Delta_b H = 0.$$

Given a GREENian domain D, there is a smallest such function $G = G^\bullet$. $G^\bullet$ is the so-called GREEN function of D. $G^\bullet(a, b)$ should be considered as the potential at the point a due to a unit positive electrical charge placed at b inside the grounded shell ∂D.

$G^\bullet$ has a probabilistic as well as an electrical significance: $2G^\bullet(a, b)\,db$ is the expected time $\int_0^{+\infty} dt\, P_a[\mathfrak{m}_{\partial D} > t, x(t) \in db]$ that the standard BROWNian path starting at a spends in db before hitting ∂D, and this aspect of $G^\bullet$ is used below for the proof of the basic formula

5) $$\int_0^{+\infty} g^\bullet(t, a, b)\, dt = G^\bullet(a, b) \qquad \textit{if } D \textit{ is Greenian}$$
$$\equiv +\infty \qquad \textit{if } D \textit{ is not Greenian}$$

where $(a, b) \in D \times D$, and $g^\bullet$ is the absorbing BROWNian kernel:

6) $$2g^\bullet(t, a, b)\, db = P_a[\mathfrak{m}_{\partial D} > t, x(t) \in db] \qquad t > 0, (a, b) \in D \times D$$

(that $g^{\bullet}$ exists as a BOREL function is clear from the estimate $P_a[\mathfrak{m}_{\partial D} > t, x(t) \in db] \leqq 2g\,db$*; it will be found below that $g^{\bullet}$ can be modified so as to be smooth and unambiguous).

Because $D = R^d$ $(d > 2)$ is GREENian (its GREEN function is $\frac{1}{4}\pi^{-d/2}\Gamma\left(\frac{d}{2}-1\right)|b-a|^{2-d}$), each $d > 2$ dimensional domain is GREENian.

$D = R^2$ is not GREENian; in fact, it will appear that the 2-dimensional GREENian domains are those for which $P_\bullet[\mathfrak{m}_{\partial D} < +\infty] \equiv 1$ (see problem 7.8.3 where this is seen to be the same as $R^2 - D$ having positive logarithmic capacity).

Here is the proof for $d = 2$ (see G. HUNT for a similar proof); the $d > 2$ dimensional case is much easier and can be left to the reader. $g^{\bullet}$ is the first item of business.

Given a subdomain B of D such that $\overline{B} \subset D$, if $(t, a, b) \in (0, +\infty) \times B \times B$, then

$$7)\quad P_a[x(i2^{-n}t) \in \overline{B}, i < 2^n, x(t) \in db]$$
$$= P_a[x(t) \in db] - \sum_{i<2^n} P_a[x(j2^{-n}t) \in \overline{B}, j < i, x(i2^{-n}t) \notin \overline{B}, x(t) \in db]$$

decreases as $n \uparrow +\infty$ to $P_a[x(s) \in \overline{B}, s \leqq t, x(t) \in db]$, which, in turn, increases as $B \uparrow D$ to $P_a[\mathfrak{m}_{\partial D} > t, x(t) \in db]$, and using the fact that the GAUSS kernel g is continuous on $(0, +\infty) \times R^2 \times R^2$ and $\downarrow 0$ as $t \downarrow 0$ off the diagonal of $R^2 \times R^2$, it is found that the corresponding (unambiguous) densities

$$8)\quad \int_{\overline{B}^{2^n-1}} g(2^{-n}t, a, b_1)\,2db_1\, g(2^{-n}t, b_1, b_2)\,2db_2 \ldots g(2^{-n}t, b_{2^n-1}, b)$$
$$= g(t, a, b) - \sum_{i<2^n} E_a[x(j2^{-n}t) \in \overline{B}, j < i, x(i2^{-n}t) \notin \overline{B},$$
$$g((1 - i2^{-n})t, x(i2^{-n}t), b)]$$

decrease as $n \uparrow +\infty$ to some limiting function, which, in turn, increases as $B \uparrow D$ to

$$9)\quad g(t, a, b) - E_a[t > \mathfrak{m}_{\partial D}, g(\mathfrak{m}_{\partial D}, x(\mathfrak{m}_{\partial D}), b)]$$
$$\equiv g^{\bullet}(t, a, b) \qquad t > 0, (a, b) \in D \times D.$$

Because $\int_{\overline{B}^{2^n-1}} g(2^{-n}t, a, b_1)$ *etc.* $\leqq g$ is bounded for each $t > 0$, 6) is immediate. Because $g(t, a, b) = g(t, b, a)$, a glance at 8) shows that

* g is the GAUSS kernel $\frac{1}{2}(2\pi t)^{-d/2}\, e^{-|b-a|^2/2t}$ for the standard BROWNian motion (note the nuisance factors 1/2 in g and 2 in $2db$).

$g^{\bullet}(t, a, b) = g^{\bullet}(t, b, a)$, and using this and the smoothness of g described above, it follows from 9) that $g^{\bullet}(t, a, b)$ is continuous on $(0, +\infty) \times D \times D$. $g^{\bullet}(t, a, b)$ also $\downarrow 0$ as $t \downarrow 0$ off the diagonal of $D \times D$ as is clear from $g^{\bullet} \leqq g$.

Consider $G^{\bullet} = \int_0^{+\infty} g^{\bullet}\, dt$. 1) and 2) are automatic for $G^{\bullet}$, and the next item is the proof that $G^{\bullet} \equiv +\infty$ in case $P_{\bullet}[\mathfrak{m}_{\partial D} < +\infty] < 1$ at a single point of D.

Given $a \in D$ and a closed disc B: $|b - a| \leqq \varepsilon$ $(B \subset D)$, $P_a[\mathfrak{m}_{\partial D} > \mathfrak{m}_{\partial B}] = 1$, and using the notation $\mathfrak{m} = \mathfrak{m}_{\partial B}$,

$$10) \qquad P_a[\mathfrak{m}_{\partial D} = +\infty] = P_a[\mathfrak{m}_{\partial D} > \mathfrak{m},\ \mathfrak{m}_{\partial D}(w_{\mathfrak{m}}^+) = +\infty] = \int_{\partial B} P_{\bullet}[\mathfrak{m}_{\partial D} = +\infty]\, d o^*$$

implies that $u \equiv P_{\bullet}[\mathfrak{m}_{\partial D} = +\infty]$ is continuous on D. $P_{\bullet}[\mathfrak{m}_{\partial B} < +\infty] = 1$ (see 7.3) now implies

$$11) \qquad \begin{aligned} u(b) &= \lim_{\varepsilon \downarrow 0} P_b[\mathfrak{m}_{\partial D} > \mathfrak{m},\ \mathfrak{m}_{\partial D}(w_{\mathfrak{m}}^+) = +\infty] \\ &= \lim_{\varepsilon \downarrow 0} E_b[\mathfrak{m}_{\partial D} > \mathfrak{m},\ u(x(\mathfrak{m}))] \\ &\leqq u(a), \end{aligned}$$

showing that u is constant on D, and using

$$12) \qquad \begin{aligned} u &= P_{\bullet}[\mathfrak{m}_{\partial D} > n,\ \mathfrak{m}_{\partial D}(w_n^+) = +\infty] = E_{\bullet}[\mathfrak{m}_{\partial D} > n,\ u(x_n)] \\ &= u\, P_{\bullet}[\mathfrak{m}_{\partial D} > n] \downarrow u^2 \qquad n \uparrow +\infty, \end{aligned}$$

it is seen that $u \equiv 1$ in case $P_{\bullet}[\mathfrak{m}_{\partial D} < +\infty] = 1 - u < 1$ at a single point of D. But then $g^{\bullet}(t, a, b) = g(t, a, b)$ at each $(t, a, b) \in (0, +\infty) \times D \times D$, and

$$13) \qquad G^{\bullet} = \int_0^{+\infty} \frac{e^{-|b-a|^2/2t}}{4\pi t}\, dt \equiv +\infty$$

as desired.

Coming to the case $P_{\bullet}[\mathfrak{m}_{\partial D} < +\infty] \equiv 1$, the next step is the proof that $\int_B G^{\bullet}(a, b)\, db < +\infty$ for each $a \in D$ and each closed disc $B \subset D$.

Given concentric closed discs $B_1 \subset B_2 \subset D$, if $\mathfrak{m}_0 = 0 < \mathfrak{m}_1 <$ *etc.* are the successive passage times to ∂B_1 via ∂B_2 for the standard BROWNian path starting on ∂B_1 and if $e_n =$ measure $(t: x(t) \in B_1$,

* do is the uniform distribution on ∂B.

$\mathfrak{m}_{n-1} \leqq t < \mathfrak{m}_n$) ($n \geqq 1$), then

$$14)\quad 2\int_{B_1} G^\bullet(a,b)\,db$$
$$= E_a[\text{measure}(t: x(t) \in B_1, t < \mathfrak{m}_{\partial D})]$$
$$\leqq \sup_{B_1} E_\bullet(\mathfrak{m}_{\partial B_1}) + \sup_{\partial B_1} E_\bullet[\text{measure}(t: x(t) \in B_1, t < \mathfrak{m}_{\partial D})]$$
$$\leqq \frac{l_1^2}{d} + \sup_{\partial B_1} \sum_{l \geqq 1} E_\bullet(e_l, \mathfrak{m}_{l-1} < \mathfrak{m}_{\partial D}) \qquad l_1 = \operatorname{diam} B_1$$
$$\leqq \frac{l_1^2}{d} + \sup_{\partial B_1} E_\bullet(e_1) \sum_{l \geqq 1} \gamma^{l-1} \qquad \gamma = \sup_{\partial B_2} P_\bullet[\mathfrak{m}_{\partial B_1} < \mathfrak{m}_{\partial D}]$$
$$\leqq \frac{l_1^2}{d} + \sup_{\partial B_1} E_\bullet(\mathfrak{m}_{\partial B_2})(1-\gamma)^{-1}$$
$$\leqq \frac{l_1^2}{d} + \frac{l_2^2}{d}(1-\gamma)^{-1} \qquad l_2 = \operatorname{diam} B_2,$$

much as in 7.3, and to complete the proof, it is enough to check that $\gamma < 1$. But, if $u = P_\bullet[\mathfrak{m}_{\partial B_1} < \mathfrak{m}_{\partial D}]$, then for each $a \in D - B_1$ and each closed disc $B: |b-a| \leqq \varepsilon$ ($B \subset D - B_1$),

$$15)\quad u(a) = P_a[\mathfrak{m}_{\partial B_1}(w_\mathfrak{m}^+) < \mathfrak{m}_{\partial D}(w_\mathfrak{m}^+)] = \int_{\partial B} u\,do \qquad \mathfrak{m} = \mathfrak{m}_{\partial B},$$

and either $\gamma < 1$ or $u = 1$ at some point of ∂B_2, in which case $u \equiv 1$ on $D - B_1$, contradicting $P_\bullet[\mathfrak{m}_{\partial D} < +\infty] \equiv 1$.

Continuing with the case $P_\bullet[\mathfrak{m}_{\partial D} < +\infty] \equiv 1$, it is now possible to show that $G^\bullet + (2\pi)^{-1} \lg |b-a|$ is harmonic on $D \times D$ as stated in 4).

Given $a \in D$, a closed disc $B: |b-a| \leqq \varepsilon$ ($B \subset D$), and a point $b \in D - B$ and using the smoothness of $g^\bullet$ as described below 9), it is not difficult to see that

$$16)\quad g^\bullet(t,a,b) = E_a[t > \mathfrak{m}_{\partial B}, g^\bullet(t - \mathfrak{m}_{\partial B}, x(\mathfrak{m}_{\partial B}), b)],$$

and integrating this over t to get

$$17)\quad G^\bullet(a,b) = E_a[G^\bullet(x(\mathfrak{m}_{\partial B}), b)] = \int_{\partial B} G^\bullet(\cdot, b)\,do,$$

it is found, using $\int_B G^\bullet\,db < +\infty$ and 2), that

$$18)\quad G^\bullet < +\infty$$

and

$$19)\quad \Delta_a G^\bullet = \Delta_b G^\bullet = 0$$

off the diagonal of $D \times D$. $G^\bullet + (2\pi)^{-1} \lg |b-a|$ has now to be discussed on a neighborhood of the diagonal.

Given an open disc $D_1 \subset D$ with closure $\bar{D}_1 \subset D$, the corresponding absorbing BROWNian kernel $g_1^\bullet$ satisfies

20) $$g_1^\bullet(t, a, b) = g^\bullet(t, a, b) - E_a[t > \mathfrak{m}_{\partial D_1}, g^\bullet(t - \mathfrak{m}_{\partial D_1}, x(\mathfrak{m}_{\partial D_1}), b)] \qquad t > 0, (a, b) \in D_1 \times D_1$$

[see 9)], and integrating over t, it is found that

21) $$G_1^\bullet(a, b) = \int_0^{+\infty} g_1^\bullet \, dt = G^\bullet(a, b) - E_a[G^\bullet(x(\mathfrak{m}_{\partial D_1}), b)] \qquad (a, b) \in D_1 \times D_1.$$

But if a is the center of $D_1 : |b - a| < \varepsilon$, then $g_1^\bullet(t, a, b)$ depends upon t and $|b - a| = r < \varepsilon$ alone; in fact, $g_1^\bullet(t, a, b) = (2\pi)^{-1} \times$ the transition kernel $p(t, 0, r)$ for the BESSEL motion $|x(t) - a|$ $(t \geq 0)$ with scale $\lg r$ and speed measure $2r\,dr$ starting at $r = 0$ and killed at the BESSEL passage time $\mathfrak{m}_\varepsilon = \mathfrak{m}_{\partial D_1}$, and using the BESSEL local time $\mathfrak{t}$ to compute

22) $$G_1^\bullet(a, b) = (2\pi)^{-1} \int_0^{+\infty} p(t, 0, r)\, dt = (2\pi)^{-1} E_0[\mathfrak{t}(\mathfrak{m}_\varepsilon, r)]^* = -(2\pi)^{-1} \lg r/\varepsilon$$

[see problem 5.2.1], 21) goes over into

23) $$G^\bullet(a, b) + (2\pi)^{-1} \lg|b - a| = E_a[G^\bullet(x(\mathfrak{m}_{\partial D_1}), b)] + (2\pi)^{-1} \lg \varepsilon \qquad |b - a| < \varepsilon.$$

$G^\bullet + (2\pi)^{-1} \lg|b - a|$ is now seen to be a solution of 4) *near* as well as *off* the diagonal, as desired.

$G^\bullet$ has now to be identified as the *smallest* solution of 1, 2, 3, 4).

Given a bounded domain D such that $\partial D \in C^2$, $P_\bullet[\mathfrak{m}_{\partial D} < +\infty] \equiv 1$, $G^\bullet(a, b) < +\infty$ off the diagonal, and for each $b \in D$, $G^\bullet(a, b)$ tends to 0 as $a \in D$ tends to a point c of ∂D as is clear from the estimate

24) $$G^\bullet(a, b) = E_a[\mathfrak{m}_{\partial B} < \mathfrak{m}_{\partial D}, G^\bullet(x(\mathfrak{m}_{\partial B}), b)] \leq \sup_{\partial B} G^\bullet(\cdot, b)\, P_a[\mathfrak{m}_{\partial B} < \mathfrak{m}_{\partial D}]$$

and the diagram; this depicts the positions of the points a, b, c, and the circle ∂B enclosing b and shows, in addition, a closed disc B_1 touching ∂D at the single point c and a concentric closed disc B_2 excluding B; indeed, $P_a[\mathfrak{m}_{\partial B} < \mathfrak{m}_{\partial D}]$ is smaller than $P_a[\mathfrak{m}_{\partial B_2} < \mathfrak{m}_{\partial B_1}]$ on the annulus

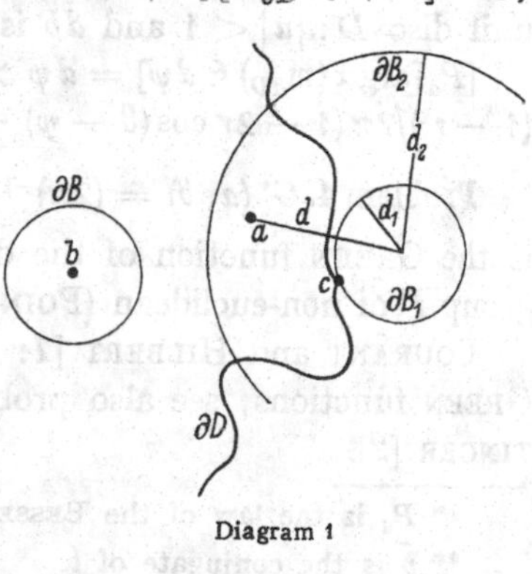

Diagram 1

* E_0 is the BESSEL expectation.

between ∂B_1 and ∂B_2, and thanks to the isotropic nature of the BROWNian motion,

$$25)\qquad P_a[\mathfrak{m}_{\partial B_2} < \mathfrak{m}_{\partial B_1}] = P_d[\mathfrak{m}_{d_2} < \mathfrak{m}_{d_1}]^* \downarrow 0 \qquad d \downarrow d_1$$

as noted in 2.7 already.

Consider now the general domain D and a subdomain D_1 such that $\overline{D}_1 \subset D$ is bounded and $\partial D_1 \in C^2$.

Given a solution G of 1, 2, 3, 4), if $g_1^\bullet$ is the absorbing BROWNian kernel for D_1 and if

$$26)\qquad G_1^\bullet = \int_0^{+\infty} g_1^\bullet \, dt \quad on \quad D_1 \times D_1$$
$$= 0 \qquad on \quad D_1 \times \partial D_1,$$

then $G - G_1^\bullet \in C(D_1 \times \overline{D}_1)$, $\Delta[G - G_1^\bullet] = 0$ on $D_1 \times D_1$, and $G - G_1^\bullet \geqq 0$ on $D_1 \times \partial D_1$. $G \geqq G_1^\bullet$ on $D_1 \times D_1$ follows, and using the fact that $g_1 \uparrow g^\bullet$ as $D_1 \uparrow D$ [see 9)], it is found that $G \geqq G^\bullet$ on $D \times D$; *in brief, D is Greenian or not according as $P_\bullet[\mathfrak{m}_{\partial D} < +\infty] \equiv 1$ or not, and in the Greenian case, $G^\bullet = \int g^\bullet \, dt$ is the Green function of D.*

Problem 1 (CHAPMAN-KOLMOGOROV identity).

$$g^\bullet(t_1 + t_2, a, b) = 2 \int_D g^\bullet(t_1, a, c)\, g^\bullet(t_2, c, b)\, dc$$
$$t_1, t_2 > 0,\ (a, b) \in D \times D.$$

Problem 2. $g^\bullet(t, a, b)$ is of class $C^\infty[(0, +\infty) \times D \times D]$ and a solution of $\partial u/\partial t = \Delta u/2$ $(t > 0)$.

[$g^\bullet \in C^\infty(D \times D)$ as is clear from 9) and $g^\bullet(t, a, b) = g^\bullet(t, b, a)$; now compute $\partial g^\bullet/\partial t$ from 9).]

Problem 3. Compute $P_a[x(\mathfrak{m}_{\partial D}) \in db]$, where a is a point of the open unit disc $D: |a| < 1$ and db is a little arc on ∂D.

[$P_a[\arg x(\mathfrak{m}_{\partial D}) \in d\psi] = d\psi \times$ the classical POISSON kernel $(1 - r^2)/2\pi(1 - 2r\cos(\theta - \psi) + r^2)$, where $r = |a|$ and $\theta = \arg a$.]

Problem 4. $G^\bullet(a, b) = (2\pi)^{-1} \lg \left| \dfrac{1 - a\bar{b}}{a - b} \right|^{**} = (2\pi)^{-1} \lg \coth [a, b]^{***}$ is the GREEN function of the disc $|a| < 1$ and is invariant under the group Γ of non-euclidean (POINCARÉ) motions† of the disc onto itself.

COURANT and HILBERT [*1*: 326—337] compute a number of other GREEN functions; see also problem 7.6.2 and MAGNUS and OBERHETTINGER [2].

* P_d is the law of the BESSEL motion starting at $d_1 < d < d_2$.

** $\bar{b}$ is the conjugate of b.

*** $[a, b]$ is the non-euclidean distance from a to b (see C. CARATHÉODORY [*1*: 69]).

† Γ is the group of all 1:1 conformal maps $a \to e^{i\psi} \dfrac{a - b}{1 - a\bar{b}}$ $(0 \leqq \psi < 2\pi,$ $|b| < 1)$ of the disc onto itself and their conjugates (see C. CARATHÉODORY [*1*: 81]).

7.5. Excessive functions

Consider a non-negative function $u \leqq +\infty$ defined on a GREENian domain D and excessive in the sense of G. HUNT [2 (1)], *i.e.*,

1a) $$u < +\infty \text{ at some point of } D$$

1b) $$E_\bullet[u(x_t), t < \mathfrak{m}_{\partial D}] \uparrow u \qquad t \downarrow 0.$$

Because $v = 2\int g^\bullet(t, a, b)\,(u \wedge n)(b)\,db$ is *excessive* for each $n \geqq 1$ and $t > 0$, u is a limit of bounded excessive functions $v \in C^\infty(D)$; for such v,

2) $$\mathfrak{G}\, v = \lim_{t \downarrow 0} t^{-1}(E_\bullet[v(x_t), t < \mathfrak{m}_{\partial D}] - v) \leqq 0$$

and

3) $$\int_{\partial B} v\, do - v(a) = E_a[v(x(\mathfrak{m}_{\partial B}))] - v(a)$$

$$= E_a\left[\int_0^{\mathfrak{m}_{\partial B}} (\mathfrak{G}\, v)(x_t)\, dt\right] \leqq 0, \qquad D \supset B,\ B: |b - a| \leqq \varepsilon,$$

and it follows that u is *superharmonic* in the sense of F. RIESZ [1 (1)], *i.e.*,

4a) $$u < +\infty \text{ at some point of } D$$

4b) $$\varliminf_{b \to a} u(b) = u(a) \qquad a \in D$$

4c) $$\int_{\partial B} u\, do \leqq u(a) \qquad a \in D, D \supset B: |b - a| \leqq \varepsilon.$$

On the other hand, if u is superharmonic on D, then so is $v = u \wedge n$ for each $n \geqq 1$ and likewise $v(a) = \int_{|b-a| \leqq \varepsilon} u(b)\, db$ on the smaller domain $D^\bullet$ of points of D at distance $> \varepsilon$ from ∂D, and using this, u is seen to be an increasing limit of bounded superharmonic functions $v \in C^2(D)$ on each domain $D^\bullet$ with closure $\subset D$. But for such v,

5) $$\mathfrak{G}\, v = \lim_{\varepsilon \downarrow 0} \frac{\int_{\partial B} v\, do - v}{\varepsilon^2/d} \leqq 0$$

and

6) $$E_\bullet[v(x_t), t < \mathfrak{m}_{\partial D^\bullet}] - v$$

$$\leqq E_\bullet[v(x(t \wedge \mathfrak{m}_{\partial D^\bullet}))] - v$$

$$= E_\bullet\left[\int_0^{t \wedge \mathfrak{m}_{\partial D^\bullet}} \mathfrak{G}\, v\, ds\right] \leqq 0$$

and it follows that u is also *excessive*; in brief, *excessive* is just the same as *superharmonic*.

Given a non-negative mass distribution $e(db)$ on D, its *potential* $u = \int_D G^\bullet\, de$ is excessive because

$$7)\qquad E_\bullet[u(x_t), t < \mathfrak{m}_{\partial D}] = 2\int_D g^\bullet(t, \bullet, a)\, da \int_D G^\bullet(a, b)\, e(db)$$

$$= \int_D e(db) \int_0^{+\infty} ds\, 2\int_D g^\bullet(t, \bullet, a)\, g^\bullet(s, a, b)\, da$$

$$= \int_D e(db) \int_0^{+\infty} ds\, g^\bullet(t+s, \bullet, b)$$

$$= \int_D \int_t^{+\infty} g^\bullet(s, \bullet, b)\, ds\, e(db)$$

$$\uparrow \int_D G^\bullet\, de = u \qquad t \downarrow 0.$$

As to the general excessive function u, the mass distribution

$$8)\qquad e_\varepsilon(db) = \varepsilon^{-1}[u(b) - E_b(u(x_\varepsilon), \varepsilon < \mathfrak{m}_{\partial D})] \times 2db \qquad \varepsilon > 0,$$

is non-negative, its potential

$$9)\qquad \int_D G^\bullet\, de_\varepsilon$$

$$= \varepsilon^{-1} E_\bullet\left(\int_0^{\mathfrak{m}_{\partial D}} [u(x_t) - E_{x(t)}(u(x_\varepsilon), \varepsilon < \mathfrak{m}_{\partial D})]\, dt \right)$$

$$= \varepsilon^{-1} \int_0^{+\infty} [v(t) - v(t+\varepsilon)]\, dt \qquad v(t) \equiv E_\bullet[u(x_t), t < \mathfrak{m}_{\partial D}]$$

$$= \varepsilon^{-1} \int_0^{\varepsilon} v\, dt - \lim_{n\uparrow+\infty} \varepsilon^{-1} \int_n^{n+\varepsilon} v\, dt$$

$$= \varepsilon^{-1} \int_0^{\varepsilon} v\, dt - h \qquad h \equiv v(+\infty)$$

increases to $u - h$ as $\varepsilon \downarrow 0$, h is harmonic (see problem 1 below), and, as is not difficult to prove, e_ε converges (in the sense of mass distributions) as $\varepsilon \downarrow 0$ to $e \equiv e^u \geqq 0$* such that

$$10)\qquad u = \int_D G^\bullet\, de^u + \textit{the greatest harmonic minorant of } u;$$

see 7.7 for the proof in a special case and 7.18 for an application.

* $e(B) = \lim_{\varepsilon\downarrow 0} e_\varepsilon(B)$ for compact $B \subset D$ with $e(\partial B) = 0$.

e^u is the *Riesz measure of* u, so named for F. RIESZ [*1* (2)] who discovered 10); it leads to a simple expression for the standard BROWNian generator

$$\mathfrak{G} u = \lim_{\varepsilon \downarrow 0} d\varepsilon^{-2}\left[\int_{\partial B} u\, do - u\right]$$

of 7.2.4): *if* $0 \geqq \mathfrak{G} u \in C(D)$, *then* u *is excessive on* D *and* $-e^u(db)/2db = \mathfrak{G}u$, *while if* u *is excessive on* D *and if* $u^\bullet = -e^u(db)/2db \in C(D)$, *then* $0 \geqq \mathfrak{G} u = u^\bullet$.

Problem 1. Given an excessive function u on a GREENian domain D, $h \equiv \lim_{t\uparrow+\infty} E_\bullet[u(x_t), t < \mathfrak{m}_{\partial D}]$ is harmonic.

[h is superharmonic, the isotropic nature of the BROWNian motion implies that, if $a \in D$ and if $D \supset B: |b-a| \leqq \varepsilon$, then

$$E_a[h(x_t), t < \mathfrak{m}_{\partial B}] \leqq h(a)\, P_a[t < \mathfrak{m}_{\partial D}],$$

and using $h = E_\bullet[h(x_t), t < \mathfrak{m}_{\partial D}]$, it follows that

$$h(a) = \lim_{\alpha\downarrow 0} E_\bullet\left[\alpha \int_0^{\mathfrak{m}_{\partial B}} e^{-\alpha t}\, h(x_t)\, dt + e^{-\alpha \mathfrak{m}_{\partial B}}\, h[x(\mathfrak{m}_{\partial B})]\right]$$

$$= \int_{\partial B} h\, do.]$$

Problem 2. Prove that $e = e^u$ if $e \geqq 0$ and if $u = \int G^\bullet\, de$ up to a harmonic function.

7.6. Application to the spectrum of $\Delta/2$

Given a compact surface $\partial D \subset R^d$ of class C^2 bounding a domain D with volume $|D|$, the operator $\Delta/2$, applied to the class $D(\Delta)$ of functions $u \in C^2(D) \cap C(\bar{D})$ such that $\Delta u \in C(D)$ and $u = 0$ on ∂D, has a countable spectrum

1) $$0 > \gamma_1 \geqq \gamma_2 \geqq etc. \downarrow -\infty\ *$$

and as H. WEYL [*2*: 41—45] discovered,

2) $$-\gamma_n \sim \text{constant} \times \left(\frac{n}{|D|}\right)^{2/d} \quad n\uparrow+\infty,$$

the constant being independent of D [its actual value is $2\pi\Gamma(d/2+1)^{d/2}$].

M. KAC [*1*: 205—209] derived 2) using the BROWNian motion, and it is his proof that will be explained below (see COURANT and HILBERT [*1*: 373—387] for the classical proof).

---

* γ is listed the same number of times as $\Delta f/2 = \gamma f$ has independent solutions $f \in D(\Delta)$.

Consider the operator $H_t f = 2 \int_D g^{\bullet}(t, a, b) f \, db$ acting on the (real) HILBERT space $L^2(D, 2db)$ with inner product $(f_1, f_2) = 2 \int_D f_1 f_2 db$ and norm $\|f\|_2 = \sqrt{(f, f)}$ and use the CHAPMAN-KOLMOGOROV identity (problem 7.4.1) and the fact that $g^{\bullet}(t, a, b) = g^{\bullet}(t, b, a)$ to construct, as in 4.11, a BOREL measure $\mathfrak{p}(d\gamma)$ from $(-\infty, 0]$ to projections such that $H_t = e^{t\mathfrak{Q}} = \int_{-\infty}^{0+} e^{t\gamma} \mathfrak{p}(d\gamma)$ $(t > 0, \mathfrak{p}(-\infty, 0] = 1)$, $\mathfrak{Q}$ being the operator $\int_{-\infty}^{0+} \gamma \, \mathfrak{p}(d\gamma)$ applied to $D(\mathfrak{Q}) = (u : \int_{-\infty}^{0+} \gamma^2 \|\mathfrak{p}(d\gamma) u\|_2^2 < +\infty)$.

Because $g^{\bullet} \leq g \leq (2\pi t)^{-d/2}/2$ and $|D| < +\infty$, H_t is of finite trace $2 \int_D g^{\bullet}(t, b, b) \, db \leq (2\pi t)^{-d/2} |D|$ and so compact, and this means that $\mathfrak{Q}$ has a countable spectrum

3) $0 \geq \gamma_1 \geq \gamma_2 \geq$ etc. $\downarrow -\infty.$ *

Given an eigenfunction u corresponding to $\gamma_1 = 0$,

4) $|u| = \left| 2 \int g^{\bullet} u \, db \right| \leq$ constant $\times t^{-d/2} \downarrow 0$ $(t \uparrow +\infty)$,

i.e.,

5) $\gamma_1 < 0$

as in 1).

Consider the eigenfunctions $f = f_1, f_2$, etc. corresponding to

$\gamma = \gamma_1 \geq \gamma_2 \geq$ etc. $(\mathfrak{Q}f = \gamma f, \|f\|_2 = 1, (f_n, f_m) = 0, n < m)$.

Because $e^{\gamma t} f = 2 \int g^{\bullet} f \, db$, problem 7.4.2 plus the bound $g^{\bullet} \leq (2\pi t)^{-d/2}/2$ ensures that $f \in C^{\infty}(D)$, and using the smoothness of ∂D, it is seen that

6) $e^{\gamma t} |f(a)| \leq 2 \int g^{\bullet}(t, a, b) |f| \, db$

$\leq \sqrt{2 \int g^{\bullet}(t, a, b) f^2 \, db} \sqrt{2 \int g^{\bullet}(t, a, b) \, db}$

$\leq \frac{(2\pi t)^{-d/4}}{\sqrt{2}} \|f\|_2 \sqrt{P_a(m_{\partial D} > t)}$

tends to 0 as a tends to ∂D, i.e., $f \in D(\Delta)$.

But $\mathfrak{Q} = \Delta/2$ on $D(\Delta) \subset D(\mathfrak{Q})$ (use problem 7.4.2) so $\Delta f/2 = \gamma f$ as desired, and using MERCER's theorem[1], one finds

7) $g^{\bullet}(t, a, b) = \sum_{n \geq 1} e^{\gamma_n t} f_n(a) f_n(b)$ $t > 0, (a, b) \in D \times D,$

_____

* γ is listed the same number of times as $\mathfrak{Q}f = \gamma f$ has independent solutions $f \in D(\mathfrak{Q})$.

[1] COURANT and HILBERT [1: 117—118].

which implies

8) $$2\int_D g^\bullet(t, a, a)\,da = \sum_{n\geq 1} e^{\gamma_n t} \qquad t > 0.$$

Because

9) $$g^\bullet(t, a, a) = \tfrac{1}{2}(2\pi t)^{-d/2} - E_a[t > \mathfrak{m}_{\partial D}, g(t - \mathfrak{m}_{\partial D}, x(\mathfrak{m}_{\partial D}), a)]$$
$$= \tfrac{1}{2}(2\pi t)^{-d/2} + \textit{an error of magnitude} \leq e^{-\varepsilon^2/t} \quad t \downarrow 0, a \in D,$$

where ε is $1/2$ the distance from a to ∂D, 8) implies

10) $$\sum_{n\geq 1} e^{\gamma_n t} \sim (2\pi t)^{-d/2} |D| \qquad t \downarrow 0,$$

and a standard TAUBERian theorem[1] converts this into

11) $$\sum_{\gamma_n > \gamma} 1 \sim \frac{|D|}{(2\pi)^{d/2}\,\Gamma\left(\frac{d}{2}+1\right)} |\gamma|^{d/2} \qquad \gamma \downarrow -\infty,$$

from which 2) is clear on putting $\gamma = \gamma_n$ and letting $n \uparrow +\infty$.

Problem 1. Compute the eigenvalues of $\Delta/2$ for the 3-dimensional domain D: $0 < \xi_1 < a$, $0 < \xi_2 < b$, $0 < \xi_3 < c$ and check WEYL's law 2).

[$\frac{\pi^2}{2}\left(\frac{l^2}{a^2} + \frac{m^2}{b^2} + \frac{n^2}{c^2}\right)$ $(l, m, n \geq 1)$ are the eigenvalues; the number of these that lie above γ $(\downarrow -\infty)$ is (about) the volume $\frac{1}{8} \times \frac{4}{3} \pi abc$ $(2|\gamma|/\pi^2)^{3/2}$ of the ellipsoidal octant

$$\frac{\xi_1^2}{a^2} + \frac{\xi_2^2}{b^2} + \frac{\xi_3^2}{c^2} \leq 2|\gamma|/\pi^2 \qquad \xi_1, \xi_2, \xi_3 > 0,$$

as it should be.]

Problem 2. Compute the GREEN function for the domain D of problem 1 (see COURANT and HILBERT [*1*: 328—333]).

7.7. Potentials and hitting probabilities

Given a GREENian domain D with GREEN function G, if B is a compact subset of D or an open subset with compact closure $\overline{B} \subset D$, then *the hitting probability* $p_B = P_\bullet[\mathfrak{m}_B < \mathfrak{m}_{\partial D}]$ *is the potential* $\int G\, d e_B$ *of a Borel measure* $e_B \geq 0$ *concentrated on* ∂B; indeed, p_B can be identified as the *Newtonian electrostatic potential* of B, meaning for compact B, that it is the *greatest* of the potentials $\int_B G\, de \leq 1$ $(e \geq 0)$,

[1] G. DOETSCH [*1*: 203—216].

and for open B, that it is the *smallest majorant* of the potentials $\int_A G\,de \leqq 1$ $(e \geqq 0,\ \bar{A} \subset B)$.

We give the proof in the case $d = 3$, $D = R^3$, $G = (4\pi)^{-1}|b-a|^{-1}$. Given bounded $B \subset R^3$, $P_\bullet[\mathfrak{m}_B(w_t^+) < +\infty] \downarrow 0$ as $t \uparrow +\infty$ because

$$P_\bullet\left[\lim_{t\uparrow+\infty} |x(t)| = +\infty\right] = 1$$

(see 7.3), and introducing the measures

1) $$e_\varepsilon(db) = \varepsilon^{-1}\, 2\,db \times P_b[\mathfrak{m}_B < +\infty,\ \mathfrak{m}_B(w_\varepsilon^+) = +\infty] \qquad \varepsilon > 0,$$

it is evident that

2) $$\begin{aligned} p_\varepsilon(a) &= \int_{R^3} G(a,b)\, e_\varepsilon(db) \\ &= E_a\left(\int_0^{+\infty} \varepsilon^{-1} P_{x(t)}[\mathfrak{m}_B < +\infty,\ \mathfrak{m}_B(w_\varepsilon^+) = +\infty]\, dt\right) \\ &= \varepsilon^{-1}\int_0^{+\infty} P_a[\mathfrak{m}_B(w_t^+) < +\infty,\ \mathfrak{m}_B(w_{t+\varepsilon}^+) = +\infty]\, dt \\ &= \varepsilon^{-1}\int_0^{+\infty} [P_a(\mathfrak{m}_B(w_t^+) < +\infty) - P_a(\mathfrak{m}_B(w_{t+\varepsilon}^+) < +\infty)]\, dt \\ &= \varepsilon^{-1}\int_0^{\varepsilon} P_a(\mathfrak{m}_B(w_t^+) < +\infty)\, dt \\ &\uparrow P_a(\mathfrak{m}_B < +\infty) = p_B(a) \qquad \varepsilon \downarrow 0. \end{aligned}$$

Consider a bounded open neighborhood A of ∂B:

3) $$\begin{aligned} & e_\varepsilon(R^3 - A) \\ &= \varepsilon^{-1}\, 2 \int_{R^3 - A} P_b[\mathfrak{m}_B < +\infty,\ \mathfrak{m}_B(w_\varepsilon^+) = +\infty]\, db \\ &\leqq \varepsilon^{-1}\, 2 \int_{R^3 - A} db \int_0^{\varepsilon} \frac{l}{\sqrt{2\pi t^3}}\, e^{-l^2/2t}\, dt^* \\ &\leqq \text{constant} \times \varepsilon^{-1} \int_{\frac{1}{2}\min_{R^3 - A} l}^{+\infty} dc \int_0^{\varepsilon} \frac{c^3}{\sqrt{2\pi t^3}}\, e^{-c^2/2t}\, dt \\ &\downarrow 0 \qquad \varepsilon \downarrow 0, \end{aligned}$$

where l is 1/3 of the distance from $b \in R^3 - A$ to ∂B; in addition,

4) $$\sup_{\varepsilon > 0} e_\varepsilon(A) \leqq 1/\inf_{A \times A} G < +\infty,$$

* $(w: \mathfrak{m}_B < +\infty,\ \mathfrak{m}_B(w_\varepsilon^+) = +\infty) \subset (w: \mathfrak{m}_{\partial B} \leqq \varepsilon)$ is used at this point.

and it follows that $\lim_{\varepsilon\downarrow 0} e_\varepsilon = e_B$ exists, is concentrated on ∂B, and has p_B as its potential; indeed, selecting $\varepsilon = \varepsilon_1 > \varepsilon_2$ *etc.* $\downarrow 0$ such that $\lim_{\varepsilon\downarrow 0} e_\varepsilon = e$ exists (in the sense of mass distributions) and using the fact that $\int G\,de$ is an excessive function,

$$
\begin{aligned}
5) \qquad P_a[\mathfrak{m}_B(w_t^+) < +\infty] &= 2\int_{R^3} g(t,a,b)\, p_B\, db \\
&= \lim_{\varepsilon\downarrow 0} 2\int_{R^3} g\, p_\varepsilon\, db \\
&= \lim_{\varepsilon\downarrow 0} 2\int_{R^3} g\, db \int G\, de_\varepsilon \\
&= \lim_{\varepsilon\downarrow 0} \int de_\varepsilon\, 2\int_{R^3} g\, G\, db \\
&= \int de\, 2\int_{R^3} g\, G\, db \\
&= 2\int_{R^3} g\, db \int G\, de
\end{aligned}
$$

implies $p_B = \int G\,de$, and the stated facts follow at once.

The potential p_B has now to be identified as the NEWTONian electrostatic potential of B.

Given *compact* B, a potential $p = \int_B G\,de \leq 1$, and a point $a \notin B$, if A is the open $1/n$ neighborhood of B and if the distance from a to ∂B is $> 1/n$, then

$$
\begin{aligned}
6) \qquad p(a) &= \int_B G(a,b)\, e(db) \\
&= \int_B E_a[\mathfrak{m}_{\partial A} < +\infty,\, G(x(\mathfrak{m}_{\partial A}), b)]\, e(db)^* \\
&= \int_{\partial A} P_a[\mathfrak{m}_{\partial A} < +\infty,\, x(\mathfrak{m}_{\partial A}) \in db]\, p(b) \\
&\leq P_a[\mathfrak{m}_{\partial A} < +\infty] \\
&= p_A(a) \downarrow p_B(a) \qquad n\uparrow +\infty,
\end{aligned}
$$

and since $p_B = 1 \geq p$ on $B - \partial B$, $p \leq p_B$ on $R^3 - \partial B$.

On ∂B itself, either $p_B(a) = 1 \geq p(a)$ or $p_B(a) < 1$. But in the second case, $P_a[\mathfrak{m}_B = 0] < 1$, which combined with BLUMENTHAL's

* See 7.4.17).

01 law shows that $P_a[\mathfrak{m}_B > 0] = 1$, and using this, it is found that

$$\begin{aligned} \text{7)} \qquad p_B(a) &= \lim_{\varepsilon\downarrow 0} P_a[\mathfrak{m}_B(w_\varepsilon^+) < +\infty] \\ &= \lim_{\varepsilon\downarrow 0} E_a[p_B(x_\varepsilon)] \\ &= \lim_{\varepsilon\downarrow 0} E_a[\varepsilon < \mathfrak{m}_B, p_B(x_\varepsilon)] \\ &\geqq \lim_{\varepsilon\downarrow 0} E_a[\varepsilon < \mathfrak{m}_B, p(x_\varepsilon)] \\ &= \lim_{\varepsilon\downarrow 0} E_a[p(x_\varepsilon)] \\ &= p(a), \end{aligned}$$

completing the proof.

Given *open* B, it is clear from the probabilistic picture that p_B is the smallest majorant of the potentials p_A as A runs through the compact subsets of B, so p_B is the NEWTONian electrostatic potential in this case also.

Problem 1 after F. SPITZER [3]. $e_\varepsilon(R^d) = C(B)$ for each $\varepsilon > 0$, where $C(B)$ is the NEWTONian capacity of B defined in the next section.

7.8. Newtonian capacities

Given a GREENian domain D and $B \subset D$ as in 7.7, the NEWTONian electrostatic capacity of B (relative to D) is defined to be the total mass $C(B)$ of the NEWTONian electrostatic distribution e_B; if B is compact and if ∂D is considered to be *grounded*, then $C(B)$ is the greatest positive electrical charge that can be placed on B and still have the potential of the corresponding electrostatic field $\leqq 1$.

$C(B)$ has the following properties:

1) $$C(A) \leqq C(B) \qquad A \subset B,$$

2) $$C(A \cup B) + C(A \cap B) \leqq C(A) + C(B),$$

3) $$C(B) \leqq C(\partial B) = C(\bar{B}),$$

4) $$C(B) = \inf_{\substack{A \text{ open} \\ A \supset B}} C(A) \qquad B \text{ compact},$$

5) $$C(B) = \sup_{\substack{A \text{ compact} \\ A \subset B}} C(A) \qquad B \text{ open},$$

6) $$C(mB) = |m|^{d-2} C(B)$$

if $D = R^d$ $(d > 2)$ and m is a rigid motion of R^d $(|m| = 1)$ or a magnification by the factor $|m| > 0$.

GAUSS [1] observed that if the potential $p = \int_B G\,de$ of a signed measure e is non-negative at each point of D, then the total (algebraic) charge $e(B)$ is likewise non-negative, as is clear on selecting some open $A \supset \overline{B}$ and noting that $e(B) = \int_B p_A\, de = \int_{\partial A} p\, de_A \geqq 0$ [use 7.4.2)].

GAUSS's observation, combined with $p_A \leqq p_B$ $(A \subset B)$, proves 1). 2) is likewise clear from

7) $p_{A\cup B} - p_B$

$$= P_\bullet[\mathfrak{m}_{A\cup B} < \mathfrak{m}_{\partial D}, \mathfrak{m}_B \geqq \mathfrak{m}_{\partial D}] \leqq P_\bullet[\mathfrak{m}_A < \mathfrak{m}_{\partial D}, \mathfrak{m}_{A\cap B} \geqq \mathfrak{m}_{\partial D}]$$
$$= p_A - p_{A\cap B},$$

and using $p_B \leqq p_{\partial B} = p_{\overline{B}}$, another application of GAUSS's trick proves 3).

Given *compact* B and open $A \downarrow B$ such that $\lim_{A\downarrow B} e_A = e$ exists, $p = \int G\, de \leqq 1$, $e(D - \partial B) = 0$, and so $C(A) \downarrow e(\partial B) \leqq C(B) \leqq \lim_{A\downarrow B} C(A)$, proving 4).

Given *open* B, compact $A \uparrow B$, and open $D \supset \overline{B}$, $p_A \uparrow p_B$, and so $C(A) = \int p_D\, de_A = \int p_A\, de_D \uparrow \int p_B\, de_D = \int p_D\, de_B = C(B)$, proving 5).

6) is immediate from the fact that G is a constant multiple of $|b - a|^{2-d}$ if $D = R^d$ $(d > 2)$.

G. CHOQUET [*1*: 117—153] proved an infinite series of inequalities similar to 2).

Here is G. HUNT's neat derivation [*2* (1): 53].

Given $A, B_1, B_2, \ldots, B_n \subset D$ and using $\mathfrak{l}$ for subsets of $1, 2, \ldots, n$, $|\mathfrak{l}|$ for the number of integers in $\mathfrak{l}$, and $B_\mathfrak{l}$ for $\bigcup_{l\in\mathfrak{l}} B_l$, it is found that

8)
$$\begin{aligned} 0 &\leqq P_\bullet[\mathfrak{m}_{A\cup B_l} < \mathfrak{m}_{\partial D}, l \leqq n, \mathfrak{m}_A \geqq \mathfrak{m}_{\partial D}] \\ &= P_\bullet[\mathfrak{m}_{A\cup B_l} < \mathfrak{m}_{\partial D}, l \leqq n] - P_\bullet[\mathfrak{m}_A < \mathfrak{m}_{\partial D}] \\ &= -\sum_{m\leqq n} (-)^m \sum_{|\mathfrak{l}|=m} P_\bullet[\mathfrak{m}_{A\cup B_\mathfrak{l}} < \mathfrak{m}_{\partial D}]^* - P_\bullet[\mathfrak{m}_A < \mathfrak{m}_{\partial D}] \\ &= -\sum_{m\leqq n} (-)^m \sum_{|\mathfrak{l}|=m} p_{A\cup B_\mathfrak{l}} - p_A, \end{aligned}$$

and an application of GAUSS' trick establishes CHOQUET's result:

9)
$$0 \leqq -\sum_{m\leqq n} (-)^m \sum_{|\mathfrak{l}|=m} C(A \cup B_\mathfrak{l}) - C(A);$$

this reduces to 2) for $n = 2$ on replacing A, B_1, B_2 by $A \cap B$, $A - B$, $B - A$.

Problem 1. Let $G = G_D(a, b)$ denote the GREEN function of a bounded domain $D \subset R^d$ $(G \equiv 0$ outside $D \times D)$. S. BERGMAN and M. SCHIFFER [*1*: 368] showed that

$$G_{D_1\cup D_2} + G_{D_1\cap D_2} \geqq G_{D_1} + G_{D_2}.$$

---

* $P_\bullet(\bigcap_{m\leqq n} E_m) = -\sum_{m\leqq n} (-)^m \sum_{|\mathfrak{l}|=m} P_\bullet(E_\mathfrak{l})$ is used at this point; this is the dual of the classical inclusion-exclusion formula of POINCARÉ (see W. FELLER [*3*: 61—62]).

Use the method of the proof of 2) to prove

$$G_{D_1 \cup D_2 \cdots \cup D_n} + \sum_{1 \leq k \leq n} (-)^k \sum_{l_1 < l_2 < \cdots < l_k} G_{D_{l_1} \cap D_{l_2} \cdots \cap D_{l_k}} \geqq 0.$$

[Given $A \in \mathbf{B}$,

$$\begin{aligned} P_a(A \cap (x(s) \in D_1 \cup D_2 \cup \cdots \cup D_n, s \leqq t)) \\ \geqq P_a\Big(\bigcup_{k \leqq n} A \cap (x(s) \in D_k, s \leqq t)\Big) \\ = - \sum_{1 \leqq k \leqq n} (-)^k \sum_{l_1 < l_2 < \cdots < l_k} P_a(A \cap (x(s) \in D_{l_1}, s \leqq t) \cap \cdots \\ \cap (x(s) \in D_{l_k}, s \leqq t)) \\ = - \sum_{1 \leqq k \leqq n} (-)^k \sum_{l_1 < l_2 < \cdots < l_k} P_a(A \cap (x(s) \in D_{l_1} \cap \cdots \cap D_{l_k}, s \leqq t)); \end{aligned}$$

now let A be the event $(x(t) \in db)$ and integrate over t.]

Problem 2 after G. CHOQUET [*1*]. Check that if B_l is open, if A_l is compact, and if $B_l \supset A_l$ $(l \leqq n)$, then

$$C\Big(\bigcup_{l \leqq n} B_l\Big) - C\Big(\bigcup_{l \leqq n} A_l\Big) \leqq \sum_{l \leqq n} [C(B_l) - C(A_l)].$$

$\Big[p_{\cup_{l \leqq n} B_l} - p_{\cup_{l \leqq n} A_l} = P_\bullet\big[\mathfrak{m}_{\cup_{l \leqq n} B_l} < \mathfrak{m}_{\partial D}, \mathfrak{m}_{\cup_{l \leqq n} A_l} \geqq \mathfrak{m}_{\partial D}\big] \leqq \sum_{l \leqq n} P_\bullet[\mathfrak{m}_{B_l} < \mathfrak{m}_{\partial D}, \mathfrak{m}_{A_l} \geqq \mathfrak{m}_{\partial D}] = \sum_{l \leqq n} [p_{B_l} - p_{A_l}]$; now use GAUSS's trick.$\Big]$

Problem 3 (see the discussion of 2-dimensional GREENian domains in 7.4). Use the connection between NEWTONian capacities and hitting probabilities in 2 dimensions to prove KAKUTANI's test[1]: that for compact $B \subset R^2$, $P_\bullet[\mathfrak{m}_B < +\infty] = 0$ or 1 according as the logarithmic capacity

$$l(B) = \exp\Big(\sup_{\substack{e \geqq 0 \\ e(B) = 1}} \int_{B \times B} \lg|b-a|\, de\, de\Big)$$

is 0 or positive (use problem 7.9.1).

[$l(B) \gtreqless 0$ according as $\sup_{\substack{e \geqq 0 \\ e(B)=1}} \int_{B \times B} \lg|b-a| \gtreqless -\infty$, and in this alternative, the negative of the GREEN function G of some disc $D \supset B$ can be used in place of $\lg|b-a|$. KELVIN's principle (see problem 7.9.1) now shows that $l(B) \gtreqless 0$ according as the NEWTONian capacity $C(B) \gtreqless 0$, *i.e.*, according as $P_\bullet[\mathfrak{m}_B < \mathfrak{m}_{\partial D}] \gtreqless 0$, KAKUTANI's alternative follows from the fact that $P_\bullet[\mathfrak{m}_B < +\infty]$ is constant ($\equiv 0$ or 1) on $R^2 - B$ (see the discussion of GREENian domains in 7.4).]

Problem 4 after F. SPITZER in collaboration with H. KESTEN and W. WHITMAN; see F. SPITZER [*3*] and W. WHITMAN [*1*]. Given a compact $d \geqq 3$ dimensional figure B, let $c(t)$ denote the volume swept

---

[1] S. KAKUTANI [*2*].

out by $B + x(s)$ up to time t; the problem is to prove that $P_0\left[\lim_{t\uparrow+\infty} t^{-1} c(t) = C(B)/2\right] = 1$.

[Consider $f(\varepsilon)$ = volume ($a : a \in B + x(s)$ *at some time* $s < \varepsilon$ *but at no time later than* ε) and use problem 7.7.1 to check

$$E_0[\varepsilon^{-1} f(\varepsilon)] = \varepsilon^{-1} \int P_a(\mathfrak{m}_B < \varepsilon, \mathfrak{m}_B(w_\varepsilon^+) = +\infty)\, da$$
$$= e_\varepsilon(R^d)/2 = C(B)/2.$$

Using $w_s^\bullet$ to denote the P_0-preserving shift $x(t, w_s^\bullet) = x(t+s) - x(s)$, one finds the lower bound: $\sum_{k\leq n} f(\varepsilon, w_{(k-1)\varepsilon}^\bullet) \leq c(n\varepsilon)$, and since the flow $w \to w_s^\bullet$ is ergodic (use KOLMOGOROV's 01 law), G. D. BIRKHOFF's ergodic theorem implies $\varliminf_{t\uparrow+\infty} t^{-1} c(t) \geq C(B)/2$. But also

$$c(n\varepsilon) \leq \sum_{k\leq n} c(\varepsilon, w_{(k-1)\varepsilon}^\bullet),$$

so that

$$\varlimsup_{t\uparrow+\infty} t^{-1} c(t) \leq \varepsilon^{-1} E[c(\varepsilon)] = \varepsilon^{-1} \int P_a(\mathfrak{m}_B < \varepsilon)\, da$$
$$= C(B)/2 + \varepsilon^{-1} \int P_a(\mathfrak{m}_B < \varepsilon, \mathfrak{m}_B(w_\varepsilon^+) < +\infty)\, da$$
$$= C(B)/2 + \text{constant} \times \varepsilon^{-1} \int E_a((1 + |x(\varepsilon)|)^{2-d}, \mathfrak{m}_B < \varepsilon)\, da$$
$$= C(B)/2 + \text{constant} \times \varepsilon^{-1} \int \frac{P_\bullet(\mathfrak{m}_B < \varepsilon)}{(1 + |a|)^{d-2}}\, da$$

using problem 7.7.1 and the obvious $P_a(\mathfrak{m}_B < \varepsilon \mid x(\varepsilon) = b) = P_b(\mathfrak{m}_B < \varepsilon \mid x(\varepsilon) = a)$, and this last integral tends to 0 as $\varepsilon \uparrow +\infty$.]

7.9. Gauss's quadratic form

GAUSS [1: 41–43] tried to base the definition of the NEWTONian electrostatic potential of a compact subset B of a GREENian domain D on the fact that the quadratic form

1) $$\mathsf{G}(e) = \tfrac{1}{2} \int_{B\times B} G\, de\, de - e(B) \qquad e \geq 0^*$$

has a strict minimum at the NEWTONian electrostatic distribution $e = e_B$.

Here is a modern proof.

Given $e_n \geq 0$ such that $\mathsf{G}(e_n) \downarrow \inf_{e\geq 0} \mathsf{G}(e)$ as $n \uparrow +\infty$, it is clear from

2) $$+\infty > \lim_{n\uparrow+\infty} \mathsf{G}(e_n) \geq \lim_{n\uparrow+\infty} \sup \left[\tfrac{1}{2}\left(\inf_{B\times B} G\right) e_n(B)^2 - e_n(B)\right]$$

that

3) $$\sup_{n\geq 1} e_n(B) < +\infty,$$

* G is the GREEN function of D.

permitting us to suppose that $\lim_{n\uparrow+\infty} e_n = e_\infty$ exists.

$\mathsf{G}(e)$ is then least for $e = e_\infty$ as is clear from

4) $$\mathsf{G}(e) \geqq \lim_{n\uparrow+\infty} \mathsf{G}(e_n) \geqq \lim_{n\uparrow+\infty} \tfrac{1}{2}\int_{B\times B} G\wedge m\, de_n\, de_n - e_n(B)$$

$$= \tfrac{1}{2}\int_{B\times B} G\wedge m\, de_\infty\, de_\infty - e_\infty(B) \uparrow \mathsf{G}(e_\infty) \qquad m\uparrow+\infty,$$

and it follows that for $e \geqq 0$,

5) $$0 \leqq \lim_{\varepsilon\downarrow 0} \varepsilon^{-1}[\mathsf{G}(e_\infty+\varepsilon e) - \mathsf{G}(e_\infty)]$$

$$= \int_{B\times B} G\, de\, de_\infty - e(B)$$

$$= \int_B (p_\infty - 1)\, de \qquad p_\infty = \int_B G\, de_\infty,$$

provided all the integrals converge, with the $=$ sign in case $e \leqq e_\infty$.

Given B such that ∂B is of class C^2, $p_B \equiv 1$ on B [see 7.4], and using 5), it is seen that

6) $$\int_{B\times B} G\, d(e_B - e_\infty)\, d(e_B - e_\infty)$$

$$= \int p_B\, de_B - \int p_\infty\, de_B - \int p_B\, de_\infty + \int p_\infty\, de_\infty$$

$$= -\int (p_\infty - 1)\, de_B + \int (p_\infty - 1)\, de_\infty$$

$$\leqq 0,$$

which is impossible unless $e_\infty = e_B$, as is clear from

7) $$\int_{B\times B} G\, de\, de \qquad\qquad e = e_B - e_\infty$$

$$= \int_0^{+\infty} dt \int_{B\times B} e(d\xi)\, e(d\eta)\, g^\bullet(t,\xi,\eta)^{*}$$

$$= \int_0^{+\infty} dt \int_{B\times B} e(d\xi)\, e(d\eta)\, 2\int_D g^\bullet(t/2, a, \xi)\, g^\bullet(t/2, a, \eta)\, da^{**}$$

$$= 2\int_0^{+\infty} dt \int_D da \left(\int_B g^\bullet(t/2, a, b)\, e(db)\right)^2.$$

Compute $\gamma(B) = \inf_{e\geqq 0} \mathsf{G}(e)$ for this special case: the result is $-\tfrac{1}{2}C(B)$, and coming back to the original case, it is clear from

8) $$\gamma(A) \leqq \gamma(B) \qquad\qquad A \supset B,$$

9) $$\gamma(B) = \tfrac{1}{2}\int p_\infty\, de_\infty - e_\infty(B) = -\tfrac{1}{2} e_\infty(B),$$

---

* $g^\bullet$ is the absorbing BROWNian kernel for D. ** See problem 7.4.1.

and 7.8.4) that

10) $$-\tfrac{1}{2}e_\infty(B) = \gamma(B) \geqq -\tfrac{1}{2}C(B),$$

which justifies

11) $$\begin{aligned}\int_{B\times B} G\,d(e_B - e_\infty)\,d(e_B - e_\infty) &= \int p_B\,de_B - 2\int p_\infty\,de_B + \int p_\infty\,de_\infty \\ &\leqq C(B) - 2\int (p_\infty - 1)\,de_B - 2C(B) + \int (p_\infty - 1)\,de_\infty + e_\infty(B) \\ &\leqq e_\infty(B) - C(B) \\ &\leqq 0.\end{aligned}$$

$e_\infty = e_B$ is now established as in 7), and the proof of GAUSS's statement is complete.

Problem 1. Give a proof of KELVIN's principle: that if B is compact, then

$$\inf_{\substack{e\geqq 0\\ e(B)=1}} \int_{B\times B} G\,de\,de = C(B)^{-1}*$$

and, if $C(B) > 0$, then

$$\int_{B\times B} G\,de\,de > C(B)^{-1} \qquad e \geqq 0,\ e(B) = 1$$

unless $e = C(B)^{-1}\times e_B$.

[Given $e \geqq 0$ with $e(B) = 1$ and $t > 0$, $\frac{t^2}{2}\int_{B\times B} G\,de\,de - t\,e(B) \geqq \int_{B\times B} G\,de_B\,de_B - C(B)$ with the $>$ sign unless $te = e_B$; putting $t = C(B)$ completes the proof if $C(B) > 0$, and if $C(B) = 0$, $\int_{B\times B} G\,de\,de = +\infty$ is immediate from

$$\frac{t^2}{2}\int_{B\times B} G\,de\,de \geqq t\,e(B) = t.]$$

7.10. Wiener's test

Given compact $B \subset R^d$ $(d \geqq 2)$, the event $\mathfrak{m}_B = 0$ is measurable B_{0+}, and so (use BLUMENTHAL's 01 law) $P_\bullet[\mathfrak{m}_B = 0] = 0$ or 1.

WIENER's test states that

1) $$P_a[\mathfrak{m}_B = 0] = 0 \text{ or } 1$$

---

* $C(B)^{-1} = +\infty$ if $C(B) = 0$.

according as

$$\left.\begin{array}{ll}\sum\limits_{n\geqq 1} n\, C(B_n) & (d=2)\\ \sum\limits_{n\geqq 1} 2^{n(d-2)}\, C(B_n) & (d\geqq 3)\end{array}\right\}\quad \textit{converges or diverges}$$

where C is NEWTONian capacity relative to a spherical surface $|b-a| = \gamma \geqq 1$ ($\gamma < +\infty$ in case $d=2$) and B_n is the intersection of B with the spherical shell $2^{-n-1} \leqq |b-a| < 2^{-n}$ ($n \geqq 1$).*

Consider for the proof the case $d = 3$, $a = 0$, $\gamma = +\infty$.

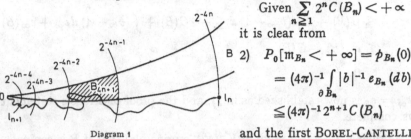

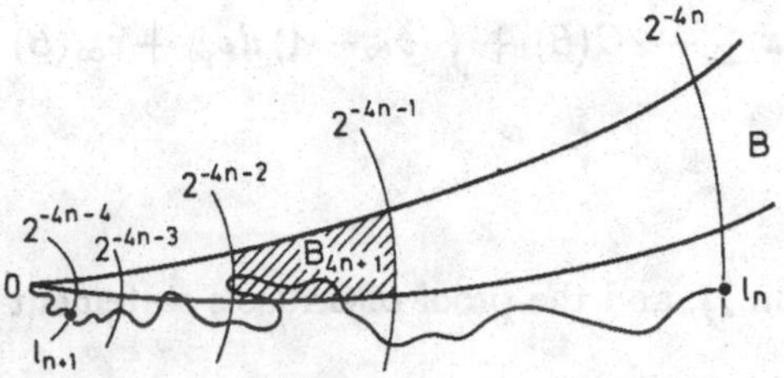

Diagram 1

Given $\sum\limits_{n\geqq 1} 2^n C(B_n) < +\infty$ it is clear from

2) $$P_0[\mathfrak{m}_{B_n} < +\infty] = p_{B_n}(0) = (4\pi)^{-1} \int\limits_{\partial B_n} |b|^{-1}\, e_{B_n}(db) \leqq (4\pi)^{-1} 2^{n+1} C(B_n)$$

and the first BOREL-CANTELLI lemma that, for all but a negligiable class of BROWNian paths, $\mathfrak{m}_{B_n} = +\infty$ from some $n = n_1 < < +\infty$ on, so that $P_0[\mathfrak{m}_B = 0] = 0$ as desired.

Contrariwise, if $\sum\limits_{n\geqq 1} 2^n C(B_n) = +\infty$, then $\sum\limits_{n\geqq 1} 2^{4n+j} C(B_{4n+j}) = +\infty$ for some $j = 0, 1, 2, 3$, and confining ourselves to the case $j = 2$, it is clear from the diagram that for $\mathfrak{m}_n = \min(t : |x(t)| = 2^{-4n})$ and $l_n = x(\mathfrak{m}_n)$ $(n \geqq 1)$,

3) $$\begin{aligned} &P_0[x(t) \in B_{4n+2} \textit{ for some } t \in [\mathfrak{m}_{n+1}, \mathfrak{m}_n) \mid \mathsf{B}_{\mathfrak{m}_{n+1}}]\\ &= P_{l_{n+1}}[x(t) \in B_{4n+2} \textit{ for some } t < \mathfrak{m}_n]\\ &\geqq P_{l_{n+1}}[x(t) \in B_{4n+2} \textit{ for some } t > 0]\\ &\quad - P_{l_{n+1}}[x(t) \in B_{4n+2} \textit{ for some } t \geqq \mathfrak{m}_n]\\ &= p_{B_{4n+2}}(l_{n+1}) - E_{l_{n+1}}[p_{B_{4n+2}}(l_n)]\\ &= (4\pi)^{-1} \int\limits_{\partial B_{4n+2}} [|b - l_{n+1}|^{-1} - E_{l_{n+1}}(|b - l_n|^{-1})]\, e_{B_{4n+2}}(db)\\ &\geqq (4\pi)^{-1} C(B_{4n+2})\, [(2^{-4n-2} + 2^{-4n-4})^{-1} - (2^{-4n} - 2^{-4n-1})^{-1}]\\ &= (4\pi)^{-1} C(B_{4n+2}) \frac{28}{15} 2^{4n}\\ &> 2^{4n-3} C(B_{4n+2}). \end{aligned}$$

---

* See COURANT and HILBERT [*2*: 284—286], O. D. KELLOGG [*1*: 330—334], and N. WIENER [*2*], for the classical proofs of WIENER's test as applied to potentials and the DIRICHLET problem. WIENER himself was interested in the DIRICHLET problem and did not give a probabilistic interpretation to his test. S. KAKUTANI [*2*] was the first to discover the connection with the BROWNian motion; see 7.12, where the DIRICHLET problem and BROWNian motion aspects of WIENER's test are brought together.

But then, using 3),

$$4)\quad d_n \equiv P_0[x(t) \notin B_{4j+2}, t \in [\mathfrak{m}_{j+1}, \mathfrak{m}_j), j \geq n]$$
$$= E_0[P_0[x(t) \notin B_{4n+2}, t \in [\mathfrak{m}_{n+1}, \mathfrak{m}_n) \mid \mathsf{B}_{\mathfrak{m}_{n+1}}], x(t) \notin B_{4j+2}, t \in [\mathfrak{m}_{j+1}, \mathfrak{m}_j), j > n]$$
$$\leq [1 - 2^{4n-3}\, C(B_{4n+2})]\, d_{n+1}$$
$$\leq [1 - 2^{4n-3}\, C(B_{4n+2})]\, [1 - 2^{4(n+1)-3}\, C(B_{4(n+1)+2})]\, d_{n+2}$$
$$\leq \textit{etc.}$$
$$\leq \prod_{j \geq n} [1 - 2^{4j-3}\, C(B_{4j+2})],$$

and since $\sum_{n\geq 1} 2^{4n+2}\, C(B_{4n+2})$ was infinite, this product is 0, and

$$5)\qquad P_0[\mathfrak{m}_B = 0] = \lim_{n\uparrow+\infty} P_0[\mathfrak{m}_B < \mathfrak{m}_n] = 1,$$

completing the proof of WIENER's test.

J. LAMPERTI [1] found another proof of WIENER's test using a nice BOREL-CANTELLI lemma.

POINCARÉ's test[1] states that $P_0[\mathfrak{m}_B = 0] = 1$ if $B \subset R^3$ contains a little cone A with summit at 0. WIENER's test can be used for the proof; indeed, $C(A_n)$ is proportional to 2^{-n} $(n \geq 1)$ (see problem 1 below), causing WIENER's sum $\sum_{n\geq 1} 2^n\, C(B_n)$ to diverge.

Problem 1 after G. HUNT [1]. Use the isotropic nature of the BROWNian motion to give a direct proof of POINCARÉ's test.

[$A \subset B$ is the cone with summit at 0. $P_0[\mathfrak{m}_B = 0] = \lim_{n\uparrow+\infty} P_0[\mathfrak{m}_B \leq \mathfrak{m}_n] \geq \limsup_{n\uparrow+\infty} P_0[l_n \in A] = \alpha > 0$, where $\alpha = (4\pi)^{-1} \times$ the solid angle of A, and an application of BLUMENTHAL's 01 law completes the proof.]

Problem 2. Prove that for the BROWNian motion in $d \geq 3$ dimensions,

$$P_\bullet(Z) = 0 \text{ or } 1 \quad \textit{according as} \quad \sum_{n\geq 1} 2^{-n(d-2)}\, C(B_n) \quad \textit{converges or diverges},$$

where B is a closed set clustering to ∞, B_n is the intersection of B with the spherical shell $2^{n-1} \leq |x| < 2^n$, C is the d-dimensional NEWTONian capacity, and Z is the event that $(t: x(t) \in B)$ clusters to $+\infty$ (see K. ITÔ and H. P. MCKEAN, JR. [1] for a similar fact for the random walk).

7.11. Applications of Wiener's test

Consider a JORDAN arc $B \subset R^2$, *i.e.*, the image of the closed unit interval $[0, 1]$ under a topological map $\mathfrak{z}: t \to R^2$, and let us use WIENER's

[1] O. D. KELLOGG [1].

test and also a direct probabilistic method to prove

1) $$P_{\bullet}[\mathfrak{m}_B = 0] = 1$$

at each point of B.

Consider the case $\mathfrak{z}(0) = 0$.

Given n so large that $2^{-n} < |\mathfrak{z}(1)|$, if $t_1 = \max(t: |\mathfrak{z}(t)| = 2^{-n-1})$ and if $t_2 = \min(t: t > t_1, |\mathfrak{z}(t)| = 2^{-n})$, then

2) $$A_1 = (\mathfrak{z}(t): t_1 \leqq t \leqq t_2) \subset B_n = B \cap (2^{-n-1} \leqq |\mathfrak{z}| < 2^{-n});$$

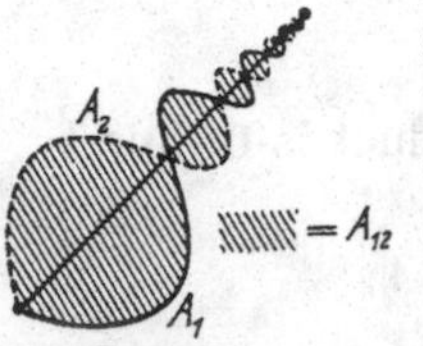

Diagram 1

in addition if A_2 is the reflection of A_1 in the straight line A connecting $\mathfrak{z}(t_1)$ and $\mathfrak{z}(t_2)$, then A lies in the (closed) complement A_{12} of the unbounded part of $R^2 - A_1 \cup A_2$ and $\partial A_{12} \subset A_1 \cup A_2$ (see diagram 1).

KELVIN's principle:

3) $$C(B)^{-1} = \inf_{\substack{e \geqq 0 \\ e(B) = 1}} \int_{B \times B} G\, de\, de^*$$

(see problem 7.9.1), coupled with the fact that $G \sim -(2\pi)^{-1} \lg|b - a|$ for small $|b - a|$, now justifies the estimates

4) $$C(A_1) \sim C(A_2)$$

and

5) $$\begin{aligned} C(A)^{-1} &< \int_{A \times A} -\lg|b - a|\, de\, de \\ &< -2\lg[\text{length of } A] \\ &< 2\lg(2^{n+1}) \end{aligned}$$

for $n \uparrow +\infty$, where de is the element of arc length divided by the total length of A, and using diagram 1, it is found that

6) $$\begin{aligned} &1/2n \\ &< C(A) \leqq C(A_{12}) = C(\partial A_{12}) \\ &\leqq C(A_1 \cup A_2) \leqq C(A_1) + C(A_2) \\ &< 3C(A_1) \leqq 3C(B_n) \qquad n \uparrow +\infty, \end{aligned}$$

causing WIENER's sum $\sum_{n \geqq 1} n\, C(B_n)$ to diverge. 1) is immediate from this.

Coming to the probabilistic proof, let Z_n be the event that, for the BROWNian motion starting at 0, $x(t): 0 < t \leqq n^{-1}$ winds about 0 and cuts itself in the sense that the (closed) complement of the unbounded part of $R^2 - (x(t): 0 \leqq t \leqq n^{-1})$ contains a disc $|\mathfrak{z}| < \varepsilon$.

* G is the GREEN function of $|\mathfrak{z}| < 1$.

$P_0(Z_1) > 0$ (see diagram 2), and using the measure-preserving scaling $x(t) \to \sqrt{n}\, x(t/n)$, it is found that

7) $0 < P_0(Z_1)$

$$= P_0[\sqrt{n}\, x(t/n) : 0 < t \leq 1 \quad \textit{winds about } 0 \textit{ and cuts itself}]$$
$$= P_0[x(t) : 0 < t \leq n^{-1} \quad \textit{winds about } 0 \textit{ and cuts itself}]$$
$$= P_0(Z_n) \qquad n \geq 1$$
$$\downarrow P_0\Big(\bigcap_{m \geq 1} Z_m\Big) \qquad n \uparrow +\infty.$$

But $\bigcap_{m\geq 1} Z_m \in \mathsf{B}_{0+}$, and using BLUMENTHAL's 01 law, 7) implies

8) $$P_0\Big(\bigcap_{n \geq 1} Z_n\Big) = 1 = P_0[\mathfrak{m}_B = 0],$$

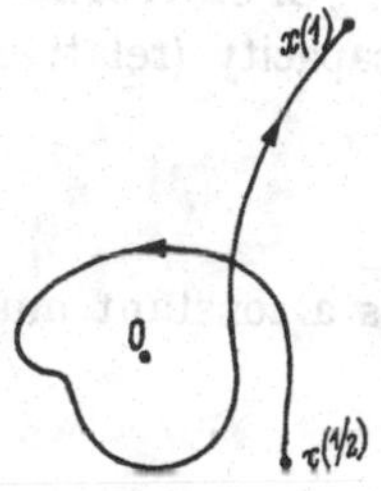

Diagram 2

thanks to the topological fact that the BROWNian path cannot wind about 0 and cut itself for small times without meeting B.

Coming to $d \geq 3$ dimensions, take a positive function $h \in C[0, +\infty)$ such that $b^{-1} h(b) \downarrow 0$ as $b \downarrow 0$, let B be the *thorn*

$$B: (b_1, b_2, \ldots, b_d) \in R^d, \sqrt{b_1^2 + \cdots + b_{d-1}^2} \leq h(b_d), b_d \geq 0,$$

and let us use WIENER's test to establish the alternative

9) $$P_0[\mathfrak{m}_B = 0] = 0 \textit{ or } 1 \textit{ according as}$$

$$\int_0^1 |\lg (h(b)/b)|^{-1} \frac{db}{b} \quad (d = 3)$$

*converges or diverges;**

$$\int_0^1 (h(b)/b)^{d-3} \frac{db}{b} \quad (d > 3)$$

for example, in the 3-dimensional case, if $h(b) = e^{-|\lg b|^{\varepsilon}}$, then

10) $$\int_0^1 |\lg (h(b)/b)|^{-1} \frac{db}{b} \textit{ converges or diverges according as } \varepsilon > 1 \textit{ or } \varepsilon \leq 1,$$

and, in the $d > 3$ dimensional case, if $h = b\,|\lg b|^{\varepsilon}$, then

11) $$\int_0^1 (h(b)/b)^{d-3} \frac{db}{b} \textit{ converges or diverges according as } \varepsilon < 0 \textit{ or } \varepsilon \geq 0.\text{**}$$

* See K. ITÔ and H. P. MCKEAN [*1*] for a similar alternative for the d-dimensional random walk.

** 11) is to be compared to POINCARÉ's test (see problem 7.10.1).

Because $b^{-1}h(b)\downarrow 0$ as $b\downarrow 0$, $B_m = B \cap (2^{-m-1} \leqq |b| \leqq 2^{-m})$ is long and thin for large m, and this makes it possible to obtain favorable estimates of the capacity of B_m in terms of the capacities of the ellipsoids:

$$E_-: \frac{b_1^2}{h(2^{-m-1})^2} + \cdots + \frac{b_{d-1}^2}{h(2^{-m-1})^2} + \frac{(b_d - 3\cdot 2^{-m-2})^2}{2^{-2m-4}} \leqq 1$$

and

$$E_+: \frac{b_1^2}{2n\,h(2^{-m})^2} + \cdots + \frac{b_{d-1}^2}{2n\,h(2^{-m})^2} + \frac{(b_d - 3\cdot 2^{-m-2})^2}{2^{-2m-2}} \leqq 1,$$

$$E_+ \supset B_m \supset E_-.$$

G. CHRYSTAL [*1*: 30] presents a neat proof that the NEWTONian capacity (relative to R^d) of the ellipsoid

$$E: \frac{b_1^2}{e_1^2} + \frac{b_2^2}{e_2^2} + \cdots + \frac{b_d^2}{e_d^2} \leqq 1$$

is a constant multiple of the reciprocal of the elliptic integral

$$\mathrm{e} = \int_0^{+\infty} \frac{dt}{\sqrt{(e_1^2+t)(e_2^2+t)\cdots(e_d^2+t)}},$$

and using

12) $$\mathrm{e} \sim 2(e'')^{2-d}\,\lg e''/e' \qquad (d=3)$$
$$\frac{d-3}{2}(e'')^{2-d}(e''/e')^{d-3} \qquad (d>3)$$

$$e_1 = e_2 = \cdots = e_{d-1} = e', \quad e_d = e'', \quad e'/e''\downarrow 0$$

to estimate the bounds $C(E_-) \leqq C(B_m) \leqq C(E_+)$ as $m\uparrow +\infty$, it appears that WIENER's sum $\sum_{m\geqq 1} 2^{-m(d-2)}\,C(B_m)$ diverges or converges according as

13) $$\sum_{m\geqq 1} |\lg[2^m h(2^{-m})]|^{-1} \quad (d=3)$$

diverges or converges.

$$\sum_{m\geqq 1} [2^m h(2^{-m})]^{d-3} \quad (d>3)$$

9) is immediate from this.

Incidentally, the proof beginning with 7) above shows that the 2-dimensional BROWNian path has an infinite number of double points; the same holds in 3 dimensions but not in 4 or more, as A. DVORETSKY, P. ERDÖS, and S. KAKUTANI [*1*] proved, using the fact that a small segment $[x(s): t \leqq s < t+h]$ of the BROWNian path has a positive capacity in 2 and 3 dimensions but not in 4. DVORETSKY, ERDÖS, and KAKUTANI [*2*] also proved that the 2-dimensional BROWNian path has an infinite number of n-fold multiple points for each $n = 2, 3, 4,$ *etc.*

and, in collaboration with S. J. TAYLOR [*1*], that the 3-dimensional BROWNian motion has no triple point; some open problems are to prove a) that the n-fold multiple points of the 2-dimensional BROWNian path have dimension 2 for each $n \geqq 2$, and b) that the double points of the 3-dimensional BROWNian path have dimension 1.

Problem 1. Check that for the 3-dimensional thorn

$$B: \sqrt{a^2 + b^2} \leqq e^{-|\lg c \lg_2 c \cdots \lg_{n-1} c| \, |\lg_n c|^\varepsilon} \qquad 0 \leqq c < e_n^{-1},*$$

$P_0[\mathfrak{m}_B = 0] = 0$ or 1 according as $\varepsilon < 1$ or $\varepsilon \geqq 1$, and use this to prove that for the 2-dimensional BROWNian motion,

$$P_0[|x(t)| > e^{-|\lg t \lg_2 t \cdots \lg_{n-1} t| \, |\lg_n t|^\varepsilon}, t \downarrow 0] = 0 \textit{ or } 1$$

according as $\varepsilon > 1$ *or not.*

This answers a question of A. DVORETSKY and P. ERDÖS [*1*: 367]; see also 4.12 for the more precise alternative of F. SPITZER [*1*: 188].

$$[\int_0^{e_n^{-1}} \left(\lg \frac{1}{c} \cdots \left(\lg_n \frac{1}{c}\right)^\varepsilon\right)^{-1} \frac{dc}{c} = \int_1^{+\infty} c^{-\varepsilon}\, dc$$

converges or diverges according as $\varepsilon > 1$ or not, proving the first statement. Consider, for the second part, the projections $x^2 (x^1)$ of the 3-dimensional BROWNian motion onto $c = 0$ $(a = b = 0)$ and let $f_n(t) = e^{-\lg \frac{1}{t} \cdots \lg_n \frac{1}{t}}$ $(t < e_n^{-1})$. $x^2(x^1)$ is a 2 (1) dimensional BROWNian motion, and using the law of the iterated logarithm for x^1, it is clear that, for $\varepsilon = 1$, $|x^2(t)| \leqq f_n[|x^1(t)|] \leqq f_n\left[\sqrt{3t \lg_2 \frac{1}{t}}\right] \leqq f_{n-1}(t)$ for an infinite number of times $t \downarrow 0$. On the other hand, $\varepsilon > 1$ implies $P_0[|x^2(t)| > f_n[|x^1(t)|], t \downarrow 0] = 1$, and using the independence of x^1 and x^2, it follows that, if $0 < P_0[|x^2(t)| < f_n(t)$ *an infinite number of times* $t \downarrow 0]$, then it will be possible to select $t_1 > t_2 >$ *etc.* $\downarrow 0$ such that $P_0[|x^1(t_n)| < t_n, \textit{i.o.}] = 1$. But $\lim_{t \downarrow 0} P_0[|x^1(t)| < t] = 0$, so this is absurd, and the proof is complete.]

7.12. Dirichlet problem

Consider the *classical* DIRICHLET problem for a bounded domain $D \subset R^d$ $(d \geqq 2)$: *given* $f \in C(\partial D)$, *to find a function* $u \in C(\overline{D})$ *such that* $\Delta u = 0$ *inside* D *and* $\lim_{\substack{a \to b \\ a \in D}} u(a) = f(b)$ *at each point* $b \in \partial D$.

H. LEBESGUE [*2*: 350—352] discovered that the DIRICHLET problem is not well-posed in $d > 2$ dimensions; his example, called *Lebesgue's thorn*, follows.

---

* $\lg_1 = \lg$, $\lg_2 = \lg(\lg)$, *etc.*; $e_1 = e$. $e_2 = e^e$, *etc*,

Consider a $d \geqq 3$ dimensional spherical surface, push a sharp thorn into its side, deforming it into the surface ∂D of diagram 1, and think of the domain D enclosed as a little chamber and the tip of the thorn as a heater, the walls (∂D) being held at temperature $f \in C(\partial D)$ $(0 \leqq f \leqq 1)$. Consider an extreme case: $f \equiv 0$ save near the tip of the thorn, at which point $f = 1$. As $t \uparrow +\infty$, the temperature u inside D should converge to the solution of the DIRICHLET problem with $u = f$ on D. But the heat radiated from the thorn is proportional to its surface area, and if its tip is sharp enough, a person sitting in the chamber will be cold no matter how close he huddles to the heater; in other words,

Diagram 1

1) $$\underline{\lim}_{\substack{a \to 0 \\ a \in D}} u(a) < 1 = f(0)$$

(see COURANT and HILBERT [2: 272—274] for a simple proof of this).

Define $b \in \partial D$ to be *singular* if $P_b[\mathfrak{m}_{R^d - D} > 0] = 1$, and imitating J. DOOB [2], let us prove that

2) $$u(a) = E_a(f[x(\mathfrak{m}_{\partial D})]) \qquad a \in D$$

is the solution of the modified Dirichlet problem: to find a bounded function u such that $\Delta u = 0$ *inside D and* $\lim_{\substack{a \to b \\ a \in D}} u(a) = f(b)$ *at each non-singular point* $b \in \partial D$.*

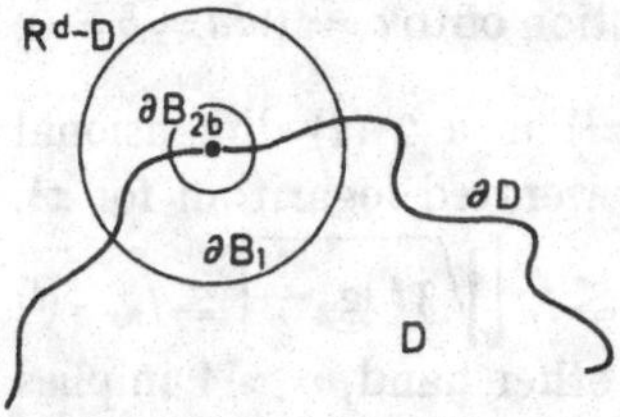

Diagram 2

Given a BROWNian path starting at $a \in D$, if $\mathfrak{m}$ is the passage time to the surface of a little closed sphere $B \subset D$ centered at a, then $\mathfrak{m}_{\partial D} = \mathfrak{m} + \mathfrak{m}_{\partial D}(w_{\mathfrak{m}}^{+})$ and $u(a) = E_a(u[x(\mathfrak{m})])$ is the arithmetic average of u over ∂B, *i.e.*, u is harmonic inside D.

Consider next a non-singular point $b \in \partial D\,(P_b[\mathfrak{m}_{R^d - D} = 0] = 1)$, let B_1 be the sphere $|b - a| < \varepsilon_1$, let G be the GREEN function of B_1, and select $B_2\colon |b - a| \leqq \varepsilon_2$ $(\varepsilon_2 < \varepsilon_1)$ as in diagram 2. Because $P_\bullet[\mathfrak{m}_{B_2 - D} < \mathfrak{m}_{\partial B_1}]$ is the potential $\int G\, de$ of a non-negative BOREL measure on $\partial(B_2 - D)$, it satisfies

3) $$\underline{\lim}_{\substack{a \to b \\ a \in D}} P_a[\mathfrak{m}_{B_2 - D} < \mathfrak{m}_{\partial B_1}] \geqq P_b[\mathfrak{m}_{B_2 - D} < \mathfrak{m}_{\partial B_1}] = 1,$$

and making ε_2 and $\varepsilon_1 \downarrow 0$ in that order, it is seen that

4) $$\lim_{\substack{a \to b \\ a \in D}} u(a) = f(b),$$

* N. WIENER [2] discovered that the modified DIRICHLET problem (with *singular point* defined in terms of WIENER's test) is well-posed. WIENER's solution is the same as the older solution of O. PERRON [1].

completing the proof that u is a solution of the modified DIRICHLET problem.

Given (if possible) a second solution v, select domains $D_n \uparrow D$ with $\overline{D}_n \subset D$ and ∂D_n of class C^2 and use POINCARÉ's test (see 7.10) to check that

5) $$u_n(a) = E_a(v[x(\mathfrak{m}_{\partial D_n})]) \qquad a \in D_n$$

converges to $v(b)$ as $a \in D_n$ approaches $b \in \partial D_n$. Because the DIRICHLET problem in its classical form is well-posed for such smooth domains, $u_n = v$ inside D_n, and granting

6) $$P_\bullet[x(\mathfrak{m}_{\partial D}) \text{ singular}] = 0,$$

the fact that v is bounded and approaches f at the non-singular points of ∂D implies

7) $$v = \lim_{n\uparrow+\infty} u_n = E_\bullet\Bigl(\lim_{n\uparrow+\infty} v[x(\mathfrak{m}_{\partial D_n})]\Bigr) = E_\bullet(f[x(\mathfrak{m}_{\partial D})]) = u.$$

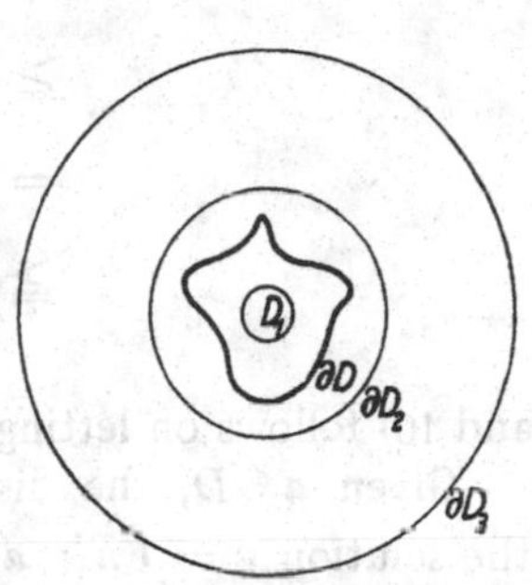

Diagram 3

Coming to the proof of 6), select open spheres $D_1 \subset D \subset D_2 \subset D_3$ as in diagram 3 ($\overline{D}_1 \subset D \subset \overline{D} \subset D_2 \subset \overline{D}_2 \subset D_3$) and, for $\varepsilon > 0$, let B_ε be the points of ∂D at which the electrostatic potential $p = P_\bullet[\mathfrak{m}_{D_2 - D} < \mathfrak{m}_{\partial D_1} \wedge \mathfrak{m}_{\partial D_3}]$ is $\leqq 1 - \varepsilon$. B_ε is compact, $\bigcup_{\varepsilon>0} B_\varepsilon$ is the set of singular points of ∂D, $p_\varepsilon = P_\bullet[\mathfrak{m}_{B_\varepsilon} < \mathfrak{m}_{\partial D_1} \wedge \mathfrak{m}_{\partial D_3}] \leqq p \leqq 1 - \varepsilon$ on B_ε, and using GAUSS' principle (see 7.9), it is found that

8) $$C(B_\varepsilon) = \int_{B_\varepsilon \times B_\varepsilon} G\, de_\varepsilon\, de_\varepsilon = \int_{B_\varepsilon} p_\varepsilon\, de_\varepsilon \leqq (1 - \varepsilon)\, C(B_\varepsilon),$$

where G is the GREEN function of $D_3 - \overline{D}_1$ and e_ε is the electrostatic distribution for B_ε. But then $C(B_\varepsilon) = 0$ $(\varepsilon > 0)$,

9) $$\begin{aligned} P_a[x(\mathfrak{m}_{\partial D}) \text{ singular}, \mathfrak{m}_{\partial D} < \mathfrak{m}_{\partial D_1}] &= \lim_{\varepsilon\downarrow 0} P_a[x(\mathfrak{m}_{\partial D}) \in B_\varepsilon, \mathfrak{m}_{\partial D} = \mathfrak{m}_{B_\varepsilon} < \mathfrak{m}_{\partial D_1}] \\ &\leqq \lim_{\varepsilon\uparrow 0} P_a[\mathfrak{m}_{B_\varepsilon} < \mathfrak{m}_{\partial D_1} \wedge \mathfrak{m}_{\partial D_3}] \\ &= 0 \qquad a \in D - D_1, \end{aligned}$$

and 6) follows on letting D_1 shrink to a point.

Coming back to LEBESGUE's thorn, its complete explanation is now before us; it is enough to prove that if $b \in \partial D$ is singular and if

$f \in C(\partial D)$ is 1 at b and <1 elsewhere, then

10) $$\lim_{\substack{a \to b \\ a \in D}} u(a) < 1 = f(b),$$

where u is the solution 2) of the modified Dirichlet problem[1]. But this is clear, for if B is the closed sphere $|b - a| \leqq \varepsilon$ with center at the singular point b, then

11) $$\begin{aligned} 1 &> E_b(f[x(\mathfrak{m}_{\partial D})]) \\ &\geqq E_b(\mathfrak{m}_{\partial B} < \mathfrak{m}_{\partial D}, f[x(\mathfrak{m}_{\partial D})]) \\ &= E_b(\mathfrak{m}_{\partial B} < \mathfrak{m}_{\partial D}, u[x(\mathfrak{m}_{\partial B})]) \\ &\geqq P_b(\mathfrak{m}_{\partial B} < \mathfrak{m}_{\partial D}) \inf_{\substack{|b-a| \leqq \varepsilon \\ a \in D}} u(a), \end{aligned}$$

and 10) follows on letting $\varepsilon \downarrow 0$.

Given $a \in D$, the distribution $h(a, db) = P_a[x(\mathfrak{m}_{\partial D}) \in db]$ used in the solution $u = \int_{\partial D} h(\cdot, db) f$ of the modified Dirichlet problem is the so-called *harmonic measure* of ∂D as viewed from a; see O. D. Kellogg [*1*] for its electrical interpretation.

7.13. Neumann problem

Consider the open ($d \geqq 2$)-dimensional ball $D: b_1^2 + \cdots + b_d^2 < 1$.

Given $f \in C(\partial D)$ such that $\int_{\partial D} f\, do = 0$*, the classical Neumann problem[2] is to find a function u harmonic on D such that $\partial u / \partial n = f$ on ∂D, where $\partial / \partial n$ denotes differentiation along the outward pointing normal.

N. Ikeda [*1*: 416—426] discovered a probabilistic method for solving Neumann problems as indicated below in the case $d = 2$.

Consider the reflecting Brownian motion D on the closed unit disc as the skew product $x = [r, \theta(\mathfrak{l})]$ of a reflecting Bessel motion and an independent standard circular Brownian motion run with the clock $\mathfrak{l}(t) = \int_0^t r(s)^{-2}\, ds$ (see 7.15), define a new diffusion $\mathsf{D}^\bullet : x^\bullet = x(\mathfrak{f}^{-1})$, where $\mathfrak{f} = t + \mathfrak{t}$ and $\mathfrak{t}$ is the Bessel local time at 1, and introduce the space $C^\bullet(D)$ of functions u defined and bounded on the closed disc, conti-

[1] 7.11.10) and 7.11.11) provide examples of singular thorns.

* do is the uniform distribution on ∂D.

[2] I. G. Petrovsky [*2*: 192—196].

nuous on the open disc, and such that $u(1, \cdot)$ is continuous on the perimeter.

Given $f \in C^\bullet(D)$ and $\alpha > 0$, it is not difficult to see that $G^\bullet_\alpha f = E^\bullet\left[\int_0^{+\infty} e^{-\alpha t} f(x^\bullet)\,dt\right]$ is continuous on the closed disc; in addition, $G^\bullet_\alpha - G^\bullet_\beta + (\alpha - \beta)\, G^\bullet_\alpha G^\bullet_\beta = 0$ $(\alpha, \beta > 0)$ as usual, so that $G^\bullet_\alpha C^\bullet(D) = D(\mathfrak{G}^\bullet)$ is independent of α, and if $G^\bullet_\alpha f = 0$ for a single $\alpha > 0$, then $0 = \lim_{\alpha \uparrow +\infty} \alpha\, G^\bullet_\alpha f = f$ on the open disc, with the result that

1)
$$
\begin{aligned}
0 &= \alpha (G^\bullet_\alpha f)(1, \theta) \\
&= E_{(1,\theta)}\left[\alpha \int_{(t:\, r(t)=1)} e^{-\alpha \mathfrak{f}} f[1, \theta(\mathfrak{l})]\, \mathfrak{f}(dt)\right] \\
&= E_{(1,\theta)}\left[\alpha \int_0^{+\infty} e^{-\alpha \mathfrak{f}} f[1, \theta(\mathfrak{l})]\, \mathfrak{t}(dt)\right] \\
&\sim f(1, \theta)\, E_{(1,\theta)}\left[\alpha \int_0^{+\infty} e^{-\alpha \mathfrak{f}}\, \mathfrak{t}(dt)\right] \qquad \alpha \uparrow +\infty \\
&\sim f(1, \theta),
\end{aligned}
$$

proving that $G^\bullet_\alpha : C^\bullet(D) \to D(\mathfrak{G}^\bullet)$ is $1:1$.

A generator $\mathfrak{G}^\bullet$ can now be defined as usual. $D(\mathfrak{G}^\bullet)$ turns out to be the class of functions $u \in C(\overline{D})$ such that

2)
$$
\begin{aligned}
\mathfrak{G}^\bullet u &= \mathfrak{G}\, u && r < 1 \\
&= -\partial u / \partial n && r = 1
\end{aligned}
$$

is of class $C^\bullet(D)$, $\mathfrak{G}$ being the (local) standard BROWNian generator as described in 7.2.3).

Given $f \in C(\partial D)$ such that $\int_{\partial D} f\, do = 0$, define $f^\bullet$ to be f on ∂D and 0 inside and introduce $u = E_\bullet[f^\bullet(x^\bullet)]$. IKEDA's solution of the NEUMANN problem for f is

3)
$$
v = G^\bullet_{0+} f^\bullet = \int_0^{+\infty} u\, dt.
$$

Consider for the proof the BESSEL expectation $E^+_\bullet$ and integrate out the circular BROWNian motion from the definition of u: the result is

4)
$$
u = E^+_r\Big[r(\mathfrak{f}^{-1}) = 1,\ \sum_{|n|>0} e^{-n^2 \mathfrak{l}(\mathfrak{f}^{-1})/2}\, c_n\, e^{in\theta}\Big],
$$

c_n being the n-th FOURIER coefficient of f, and using the estimates

$$5)\qquad \int_0^{+\infty} |u|\,dt \leqq \int_0^{+\infty} dt\, E_r^+\Big[r(\mathfrak{f}^{-1}) = 1,\ \sum_{|n|>0} e^{-n^2\mathfrak{l}(\mathfrak{f}^{-1})/2}\,|c_n|\Big]$$

$$= E_r^+\Big[\int_{(t:r(t)=1)} \sum_{|n|>0} e^{-n^2\mathfrak{l}(t)/2}\,|c_n|\,\mathfrak{f}(dt)\Big]$$

$$= E_r^+\Big[\int_0^{+\infty} \sum_{|n|>0} e^{-n^2\int_0^t r(s)^{-2}ds/2}\,|c_n|\,\mathfrak{t}(dt)\Big]$$

$$\leqq \sum_{|n|>0} |c_n|\, E_r^+\Big[\int_0^{+\infty} e^{-n^2 t/2}\,\mathfrak{t}(dt)\Big]$$

$$= \sum_{|n|>0} |c_n|\, G(n^2/2, r, 1)^*$$

$$\leqq \sum_{|n|>0} |c_n|\, G(n^2/2, 1, 1)$$

$$\leqq \sqrt{\sum_{|n|>0} |c_n|^2}\,\sqrt{\sum_{|n|>0} G(n^2/2, 1, 1)^2}$$

and

$$6)\qquad G(n^2/2, 1, 1) \sim 2/n \qquad n \uparrow +\infty,$$

$G_\alpha^\bullet f^\bullet = \int_0^{+\infty} e^{-\alpha t}\,u\,dt$ is seen to be bounded as $\alpha \downarrow 0$ and to converge to $v = G_{0+}^\bullet f^\bullet = \int_0^{+\infty} u\,dt$. But then $v = G_\alpha^\bullet(f^\bullet + \alpha v)$ as is seen on letting $\beta \downarrow 0$ in $[G_\alpha^\bullet - G_\beta^\bullet + (\alpha - \beta)\, G_\alpha^\bullet G_\beta^\bullet] f = 0$, and it is immediate that $v \in D(\mathfrak{G}^\bullet) \subset C^\bullet(D)$ and $\mathfrak{G}^\bullet v = -f^\bullet$, *i.e.*,

$$7)\qquad \mathfrak{G}^\bullet v = \mathfrak{G} u = 0 \qquad \textit{inside } D$$

$$= -\frac{\partial u}{\partial n} = -f \qquad \textit{on } \partial D$$

as desired.

7.14. Space-time Brownian motion

Consider the space-time BROWNian motion:

$$1)\qquad \mathfrak{z}(s) = [t - s, x(s)] \qquad s \geqq 0,$$

$x(s): s \geqq 0$ being the standard d-dimensional BROWNian motion and let $P_{(t,a)}$ be the measure for space-time paths starting at $(t, a) \in R^1 \times R^d$.

DOOB [4] noticed that the generator of the space-time BROWNian motion is $\mathfrak{G} = -\frac{\partial}{\partial t} + \frac{1}{2}\Delta$ and used this fact to treat heat flow

* G is the reflecting BESSEL GREEN function.

problems

2) $$-\frac{\partial u}{\partial t}+\frac{1}{2}\Delta u=0 \qquad (t,a)\in D\subset R^{d+1}$$

$$u=f\in C(\partial D) \qquad (t,a)\in\partial D$$

as DIRICHLET problems.

Confining ourselves to the case $d=1$, $D=(0,1)\times(0,1)=((t,a):0<t<1,0<a<1)$ and imitating the probabilistic solution of the DIRICHLET problem for Δ, it is plausible that the solution u of 2) is the integral of f over ∂D relative to the law of the hitting place

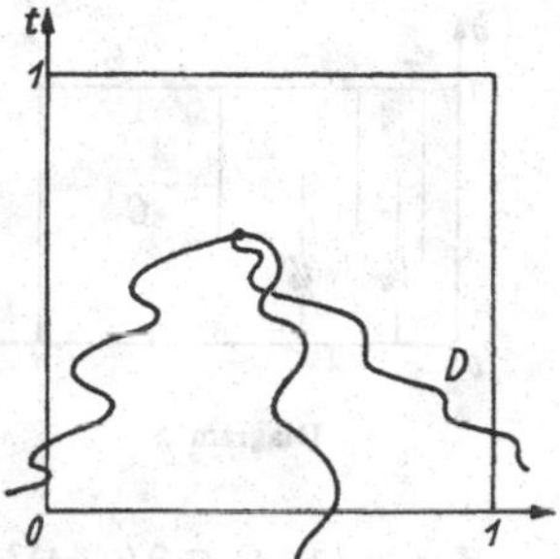

Diagram 1

3) $$\mathfrak{z}(\mathfrak{m}) \qquad \mathfrak{m}=\min(s:s>0,\mathfrak{z}(s)\in\partial D)$$

as a function of the starting point $\mathfrak{z}(0)=(t,a)\in D$ of the space-time path (see diagram 1); in fact, if $\mathfrak{m}_0$ and $\mathfrak{m}_1$ are the 1-dimensional BROWNian passage times and if $E_a:0<a<1$ is the 1-dimensional BROWNian expectation, then

4) $$\begin{aligned} u(t,a)&=E_{(t,a)}[f(\mathfrak{z}(\mathfrak{m}))]\\ &=E_a[\mathfrak{m}_0<\mathfrak{m}_1\wedge t, f(t-\mathfrak{m}_0,0)]\\ &\quad+E_a[\mathfrak{m}_1<\mathfrak{m}_0\wedge t, f(t-\mathfrak{m}_1,1)]\\ &\quad+E_a(t<\mathfrak{m}_0\wedge\mathfrak{m}_1, f(0,x(t)))]\\ &=\int_0^t f(t-s,0)\sum_{n=1}^{\infty} n\pi e^{-n^2\pi^2 s/2}\sin n\pi a\,ds\\ &\quad+\int_0^1 f(t-s,1)\sum_{n=1}^{\infty} n\pi e^{-n^2\pi^2 s/2}\sin n\pi(1-a)\,ds\\ &\quad+\int_0^1 2\sum_{n=1}^{\infty} e^{-n^2\pi^2 t/2}\sin n\pi a\sin n\pi b\, f(0,b)\,db, \end{aligned}$$

and as the reader will check, this is just the classical (DUHAMEL) solution of 2), as presented for example in G. DOETSCH [*1*: 346—66].

Of course, to be candid, the problem 2) is not well-posed since $u=f$ cannot be prescribed on $1\times(0,1)\subset\partial D$. But the integral of 4) is extended just over $\overline{D}-1\times(0,1)$ and this good fortune is due to the fact that $1\times(0,1)$ as viewed from D is of *parabolic* measure 0:

5) $$P_{(t,a)}(\mathfrak{z}(\mathfrak{m})\in 1\times(0,1))=0 \qquad (t,a)\in D.$$

DIRICHLET problems where the knowledge of f on *part* of ∂D specifies the solution look a little strange, but as the reader will recall, the same thing occurs in the classical DIRICHLET problem for plane regions possessing a prime end: for example, $0 \times [0, 1]$ is a prime end for the interior C of the mouth of LITTLEWOOD's crocodile shown in diagram 2, the BROWNian motion stopped on ∂C cannot reach it, and the values of f on this end have no influence on the DIRICHLET solution.

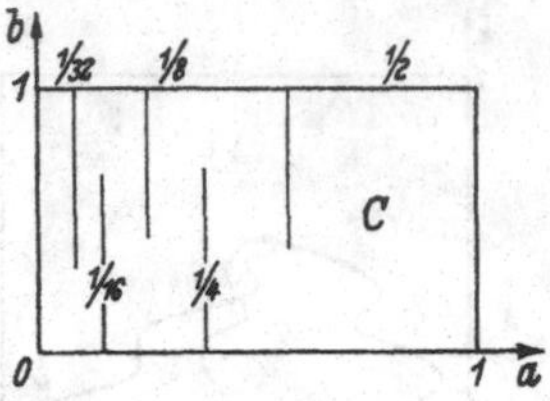

Diagram 2

Given open $D \subset R^2$, a point of D is singular for the space-time BROWNian motion if the hitting time $\mathfrak{m} = \inf\,(s : s > 0,\ \mathfrak{z}(s) \in \partial D)$ is positive except for a negligible set of paths. BLUMENTHAL's 01 law shows that the alternative is $\mathfrak{m} = 0$ except on a negligible set: for example, all the points of $1 \times (0, 1) \subset \partial (0, 1)^2$ are singular for the crocodile.

Consider, now, the region D of diagram 3 to the left of the graph of $f(-t) : t \leqq 0$.

KOLMOGOROV's test (4.12) states that if $f \in C(0, 1]$, if $f \in \uparrow$, and if $t^{-\frac{1}{2}} f \in \downarrow$, then 0 is regular or singular for D according as $\int_{0+}^{1} t^{-3/2} f e^{-f^2/2t}\, dt \gtreqless +\infty$: thus, KOLMOGOROV's test is the WIENER test for the space-time BROWNian motion (see diagram 3).

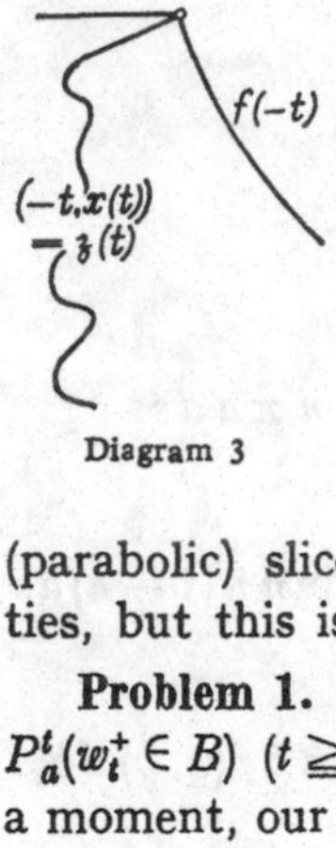

Diagram 3

ERDÖS's and FELLER's proofs of KOLMOGOROV's test are long and complicated and MOTOO's is not elementary either, so it is natural to ask if we cannot give a new proof similar to the probabilistic proof of WIENER's test, estimating the chances of meeting (parabolic) slices of the complement of D in terms of (heat) capacities, but this is an open problem.

Problem 1. Consider a simple MARKOVian motion with probabilities $P_a^t(w_t^+ \in B)$ $(t \geqq 0, a \in Q, B \in \mathsf{B})$ depending upon time (violating, for a moment, our established usage), *i.e.*, suppose $P_a^t[x(t) \in db]$ is the unit mass at $b = a$ and

$$P_a^{t_1}(w_{t_2}^+ \in B \mid x(s) : t_1 \leqq s \leqq t_2) = P_b^{t_2}(w_{t_2}^+ \in B)$$
$$t_2 > t_1,\ a \in Q,\ B \in \mathsf{B},\ b = x(t_2);$$

the problem is to prove that the *space-time* motion with state space $Q^\bullet = [0, +\infty) \times Q$ and probabilities

$$P_{a^\bullet}(B^\bullet) = P_a^s(w_s^{\bullet +} \in B^\bullet) \qquad w^\bullet : t \to (t, x(t)),\ a^\bullet = (s, a) \in Q^\bullet$$

is simple MARKOV *with probabilities no longer depending upon time, i.e.*, that at the expense of adding time as a new coordinate, the temporal dependence of the probabilities can be ignored.

Problem 2. Given a smooth positive function $\sigma^2 = \sigma^2(t, a)$ of the pair $(t, a) \in [0, +\infty) \times R^1$, if $t \to x(t)$ is the sample path of a standard 1-dimensional BROWNian motion with probabilities $P_a(B)$ and if, for $t \geq 0$, $\mathfrak{f}^{-1}$ is the inverse function of the unique solution $\mathfrak{f} = \mathfrak{f}(s)$ of

$$\mathfrak{f}(s) = \int_0^s \sigma^{-2}[t + \mathfrak{f}(r), x(r)]\, dr \qquad s \geq 0,$$

then the time-dependent motion with probabilities

$$P_a^t(w_t^+ \in B) = P_a(x(\mathfrak{f}^{-1}) \in B) \qquad (t, a) \in [0, +\infty) \times R^1$$

is simple MARKOV and $u(t, a) = P_a^t(x(t_1) < b)$ solves $\frac{\partial u}{\partial t} = \frac{1}{2}\sigma^2 \frac{\partial^2 u}{\partial a^2}$ $(t < t_1)$, while the associated space-time motion is a diffusion with generator

$$(\mathfrak{G}^\bullet u)(a^\bullet) = \frac{\partial u}{\partial t} + \frac{1}{2}\sigma^2 \frac{\partial^2 u}{\partial a^2} \qquad a^\bullet = (t, a) \in [0, +\infty) \times R^1$$

(see VOLKONSKIĬ [2] for an ingeneous alternative approach to the above time change).

7.15. Spherical Brownian motion and skew products

Define the *spherical Brownian motion* $\mathsf{BM}(S^d)$ to be the diffusion on the spherical surface $S^d: |x| = 1 \subset R^{d+1}$ with generator 1/2 the spherical LAPLACE operator $\Delta = \Delta^d$, closed up as in problem 7.2.1:

1) $$\Delta^d = (\sin\varphi)^{1-d} \frac{\partial}{\partial\varphi} (\sin\varphi)^{d-1} \frac{\partial}{\partial\varphi} + (\sin\varphi)^{-2} \overset{\bullet}{\Delta}{}^{d-1}$$

$$\varphi = \textit{colatitude}, \qquad \Delta^1 = \partial^2/\partial\theta^2.$$

$\mathsf{BM}(S^1)$ is the projection *modulo* 2π of the standard 1-dimensional BROWNian motion $\mathsf{BM}(R^1)$ onto the unit circle S^1. $\mathsf{BM}(S^2)$ is constructed as follows.

Given a circular BROWNian motion $\mathsf{BM}(S^1)$ with sample paths $t \to \theta(t)$ and an independent LEGENDRE process $\mathsf{LEG}(2)$ on $[0, \pi]$ with generator

2) $$\frac{1}{2}(\sin\varphi)^{-1} \frac{\partial}{\partial\varphi} \sin\varphi \frac{\partial}{\partial\varphi} \qquad (0 < \varphi < \pi)$$

and sample paths $t \to \varphi(t)$, the additive functional (clock)

3) $$\mathfrak{l}(t) = \int_0^t [\sin\varphi(s)]^{-2}\, ds$$

converges $(t \geq 0)$ if $0 < \varphi(0) < \pi$, and the *skew product*

4) $$x = [\varphi, \theta(\mathfrak{l})]$$

is a diffusion since θ begins afresh at time $\mathfrak{l}$.

Computing its generator $\mathfrak{G}$, if f is the product of a smooth function of colatitude $e_- = e_-(\varphi)$ vanishing at 0 and π and a smooth function of longitude $e_+ = e_+(\theta)$, then as $t \downarrow 0$,

5) $$E_\varphi \times E_\theta[f(x)]$$
$$= E_\varphi[e_-(\varphi(t))\, E_\theta[e_+(\theta(\mathfrak{l}))]]$$
$$= E_\varphi\left[e_-(\varphi(t))\left[e_+(\theta) + \frac{1}{2} e_+''(\theta)\, \mathfrak{l}\right]\right] + o(t)$$
$$= e_-(\varphi)\, e_+(\theta) + \frac{1}{2}(\sin\varphi)^{-1} \frac{\partial}{\partial\varphi} \sin\varphi \frac{\partial}{\partial\varphi} e_-(\varphi)\, e_+(\theta)\, t$$
$$+ e_-(\varphi) \frac{1}{2} e_+''(\theta)(\sin\varphi)^{-2} t + o(t),$$

i.e., $\mathfrak{G} f = \Delta f/2$ for such functions, and the desired identification of the skew product and $\mathsf{BM}(S^2)$ follows at once.

$\mathsf{BM}(S^d)$ can be factored in a similar manner as the skew product of the LEGENDRE process $\mathsf{LEG}(d)$ with generator

6) $$\frac{1}{2}(\sin\varphi)^{1-d} \frac{\partial}{\partial\varphi} (\sin\varphi)^{d-1} \frac{\partial}{\partial\varphi} \qquad (0 < \varphi < \pi)$$

and an independent spherical BROWNian motion $\mathsf{BM}(S^{d-1})$ run with the clock $\mathfrak{l} = \int_0^t (\sin\varphi)^{-2}\, ds$.

$\mathsf{BM}(R^d)$ can also be factored as the skew product of its radial (BESSEL) part $\mathsf{BES}(d)$ with generator

7) $$\frac{1}{2}\left(\frac{\partial^2}{\partial r^2} + \frac{d-1}{r} \frac{\partial}{\partial r}\right) \qquad (r > 0)$$

and an independent spherical BROWNian motion $\mathsf{BM}(S^{d-1})$ run with the clock $\mathfrak{l} = \int_0^t r(s)^{-2}\, ds$; the proof is the same.

The skew product formula $\mathsf{BES}(2) \times \mathsf{BM}(S^1) = \mathsf{BM}(R^2)$ can be used to check the result of F. SPITZER [*1*: 194] that for 2-dimensional BROWNian paths not starting at 0, the total algebraic angle $\varphi(t)$ swept out up to time t satisfies

8) $$\lim_{t \uparrow \infty} P_\bullet(2\varphi(t)/\lg t \le a) = (\pi)^{-1} \int_{-\infty}^{a} \frac{db}{1 + b^2}.$$

Because $\mathsf{BM}(S^1) = \mathsf{BM}(R^1)$ modulo 2π, φ is identical in law to a 1-dimensional BROWNian motion $\mathsf{BM}(R^1)$ run with an independent BESSEL clock $\int_0^t r_s^{-2}\, ds$; this justifies

9) $$E_\bullet[e^{i\alpha\varphi(t)}] = E_\bullet\left[e^{-\frac{\alpha^2}{2}\int_0^t r_s^{-2} ds}\right]$$

and taking a LAPLACE transform, it is seen that

10) $$f = \int_0^{+\infty} e^{-\beta t} E_\bullet[e^{i\alpha\varphi(t)}]\,dt = E_\bullet\Big[\int_0^{+\infty} e^{-\beta t} e^{-\frac{\alpha^2}{2}\int_0^t r(s)^{-2}ds}\,dt\Big] \qquad \beta > 0$$

is a radial function and satisfies

11) $$\Big[\beta - \frac{1}{2}\Big(\frac{\partial^2}{\partial r^2} + \frac{1}{r}\frac{\partial}{\partial r} - \frac{\alpha^2}{r^2}\Big)\Big] f = 1 \qquad r > 0.$$

Computing the GREEN function for 11) and solving, it develops that

12) $$f(a) = K_{|\alpha|}(\sqrt{2\beta}\,a)\int_0^a I_{|\alpha|}(\sqrt{2\beta}\,b)\,2b\,db + I_{|\alpha|}(\sqrt{2\beta}\,a)\int_a^{+\infty} K_{|\alpha|}(\sqrt{2\beta}\,b)\,2b\,db,$$

and consulting A. ERDÉLYI [*1* (1): 284 (56)] for the inverse LAPLACE transform,

13) $$E_\bullet[e^{i\alpha\varphi(t)}] = (2t)^{-1}\int_0^{+\infty} e^{-\frac{a^2+b^2}{2t}} I_{|\alpha|}(a\,b/t)\,2b\,db \qquad a = r(0) > 0$$

into which we substitute $2\alpha/\lg t$ in place of α, make $t \uparrow \infty$, and conclude as follows:

14) $$\begin{aligned} E_\bullet[e^{i\alpha 2\varphi(t)/\lg t}] &= (2t)^{-1}\int_0^{+\infty} e^{-\frac{a^2+b^2}{2t}} I_{2|\alpha|/\lg t}(a\,b/t)\,2b\,db \\ &\sim \int_0^{+\infty} e^{-b^2/2} I_{2|\alpha|/\lg t}\Big(\frac{a\,b}{\sqrt{t}}\Big)\,b\,db \\ &\to e^{-|\alpha|} \\ &= \frac{1}{\pi}\int e^{i\alpha b}(1+b^2)^{-1}\,db. \end{aligned}$$

S. BOCHNER [*2*] gives an instructive discussion of spherical differential processes; the spherical BROWNian motion is a special case.

A. R. GALMARINO [*1*] has proved that the most general $d \geqq 2$ dimensional diffusion which is *isotropic* in the sense that its law is unchanged by the action of the (full) rotation group $\mathsf{O}(d)$ can be expressed as a skew product of its radial motion and the spherical BROWNian motion $\mathsf{BM}(S^{d-1})$ run with a stochastic clock $\mathfrak{f}$ which is an additive functional of the radial motion.

Problem 1. Given a 2-dimensional BROWNian motion $\mathsf{BM}(R^2)$, use time changes and skew products to check that if we take the excursions leading into the disc $E^2: r < 1$ and rotate them clockwise as in the picture, closing up the gaps, then the resulting motion is identical in law to the reflecting BROWNian motion $\mathsf{BM}^+(E^2)$ with generator $\mathfrak{G}^+ = \frac{1}{2}\Delta$, $u^+(1,\theta) = 0$, $0 \leq \theta < 2\pi$ (u^+ = outward normal derivative).

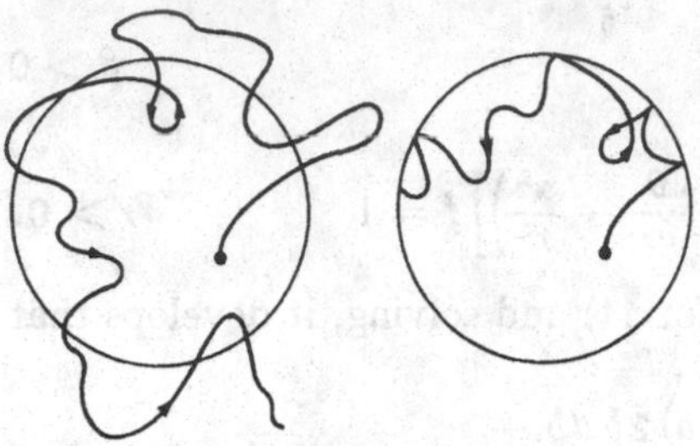

Diagram 1

[Consider the radial (BESSEL) part $\mathsf{BES}(2)$ of $\mathsf{BM}(R^2)$ and let $\mathfrak{f}(t)$ be the time it spends in $[0,1]$ up to time t. $\mathsf{BES}(2)$ *run with the clock* $\mathfrak{f}^{-1}$ is the reflecting BESSEL process and $\mathsf{BM}^+(E^2)$ is identical in law to a skew product of this sampled BESSEL process and an independent circular BROWNian motion $\mathsf{BM}(S^1)$. But the angular part of this skew product is $\theta\left[\int_0^t r(\mathfrak{f}^{-1})^{-2}\,ds\right]$ with $\theta = \mathsf{BM}(S^1)$, $r = \mathsf{BES}(2)$, and t = measure $(s: r(s) \leq 1,\ s \leq \mathfrak{f}^{-1}(t))$, and substituting $\mathfrak{f}(s)$ for s in the inner integral, the resulting $\theta\left[\int_0^{\mathfrak{f}^{-1}(t)} r^{-2}\,d\mathfrak{f}\right]$ is identical in law to the standard BROWNian angle $\theta\left[\int_0^t r(s)^{-2}\,ds\right]$ sampled on the excursions leading into E^2, *i.e.*, sampled on the support of $\mathfrak{f}$ up to time $\mathfrak{f}^{-1}(t)$.]

Problem 2. Consider the plane motion with generator $\mathfrak{G} = \Delta/2$ $(r < 1) = (r^4/2)\Delta$ $(r \geq 1)$, use the circular inversion $r \to 1/r$ to map the excursions leading out of the unit disc back into it, check that this folded motion is still a diffusion, and find its generator $\mathfrak{G}^+$.

[$\mathfrak{G}$ is invariant under circular inversion; this makes the derived motion MARKOV. $\mathfrak{G}^+ = \Delta/2$ subject to $u^+(1,\cdot) = 0$, $u \in D(\mathfrak{G}^+)$, *i.e.*, the derived motion is reflecting BROWNian.]

Problem 3. The GREEN function of the $d \geq 2$ dimensional BROWNian motion is

15)
$$G(\alpha, a, b) = \frac{1}{2}\int_0^{+\infty} e^{-\alpha t}(2\pi t)^{-d/2} e^{-|b-a|^2/2t}\,dt$$
$$= (2\pi)^{-d/2}\left(\frac{\sqrt{2\alpha}}{|b-a|}\right)^{d/2-1} K_{d/2-1}\left(\sqrt{2\alpha}\,|b-a|\right)$$

(see A. ERDÉLYI [*1* (1): 146 (29)]). Use the skew product to expand G in the form

16)
$$(2\pi)^{-1} I_0(\sqrt{2\alpha}\,|a|)\,K_0(\sqrt{2\alpha}\,|b|) + \pi^{-1}\sum_{n\geq 1} I_n(\sqrt{2\alpha}\,|a|)\,K_n(\sqrt{2\alpha}\,|b|)\cos n(\theta_2 - \theta_1)$$
$$a = (|a|, \theta_1),\ b = (|b|, \theta_2),\ |a| \leq |b|$$

in case $d = 2$ and, in case $d \geqq 3$, in the form

$$\text{17)} \quad \frac{\Gamma(d/2)}{2(d-2)\pi^{d/2}} \sum_{n \geqq 0} (2n + d - 2)\, C_n^{\frac{1}{2}(d-2)} [(\theta_1, \theta_2)] \; |a|^{1-d/2} I_{\#(n)}(\sqrt{2\alpha}\,|a|)\, |b|^{1-d/2} K_{\#(n)}(\sqrt{2\alpha}\,|b|)$$

$$a = (|a|, \theta_1),\ b = (|b|, \theta_2),\ |a| < |b|,\ \#(n) = \sqrt{n(n+d-2) + (d/2-1)^2}.$$

A. ERDÉLYI [2 (2): 44 (4)] states 16), but not 17). $C_n^{\frac{1}{2}(d-2)}$ is the GEGENBAUER polynomial figuring in the addition rule for spherical harmonics of weight n:

$$\frac{\Gamma(d/2)}{2\pi^{d/2}} \frac{2n+d-2}{d-2} C_n^{\frac{1}{2}(d-2)} [(\theta_1, \theta_2)] = \sum_{l \leqq m(n)} S_n^l(\theta_1)\, S_n^l(\theta_2),$$

where $m(n)$ is the number of spherical harmonics of weight n; see A. ERDÉLYI [2 (2): 243 (2)].

$$[(G_\alpha f)(r, \theta) = E_r\left[\int_0^{+\infty} e^{-\alpha t}\, dt \sum_{n \geqq 0} e^{-(\frac{1}{2}) n(n+d-2) \int_0^t r(s)^{-2} ds} \sum_{l \leqq m(n)} S_n^l(\theta) \int S_n^l(\psi)\, d\psi\, f(r(t), \psi)\right],$$

where $E_\cdot$ is the BESSEL expectation, and so

$$G = \sum_{n=0}^{+\infty} G^n(\alpha, |a|, |b|) \frac{\Gamma(d/2)}{2\pi^{d/2}} \frac{2n+d-2}{d-2} C_n^{\frac{1}{2}(d-2)} [(\theta_1, \theta_2)],$$

where G^n is the GREEN function for $\mathsf{BES}(d)$ killed with $\frac{1}{2} n(n+d-2) \times \times r^{d-3}\, dr$, *i.e.*, G^n is the GREEN function that goes with

$$\mathfrak{G}^n = \frac{1}{2}\left(\frac{d^2}{dr^2} + \frac{d-1}{r}\frac{d}{dr} - \frac{n(n+d-2)}{r^2}\right) \qquad (r \geqq 0):$$

$$G^n(\alpha, \xi, \eta) = \xi^{1-d/2} I_{\#(n)}(\sqrt{2\alpha}\,\xi)\, \eta^{1-d/2} K_{\#(n)}(\sqrt{2\alpha}\,\eta)$$

$$\#(n) = \sqrt{n(n+d-2) + (d/2-1)^2},\ \xi \leqq \eta.]$$

Problem 4. Construct a diffusion on the closed unit disc E^2 identical in law to $\mathsf{BM}^+(E^2)$ up to time $\mathfrak{m}_{\partial E^2}$ and such that, for smooth $u \in D(\mathfrak{G})$,

$$p_2 u_1(1-, \cdot) + p_3(\mathfrak{G} u)(1-, \cdot) = p_2^\bullet\, \partial u^\bullet/\partial\theta + \tfrac{1}{2} p_3^\bullet\, \partial^2 u^\bullet/\partial\theta^2$$

$$u^\bullet = u(1, \theta), \quad p_2, p_3, p_2^\bullet, p_3^\bullet \geqq 0, \quad p_2 + p_3 = 1$$

(use skew products and local time). A. VENTCEL' [1] has studied this motion from a different standpoint; see also N. IKEDA [1], T. UENO [2], and 8.5.

[Given a BESSEL motion on $[0, 1]$ with generator $\mathfrak{G}^+$ subject to $p_2 u^-(1) + p_3(\mathfrak{G}^+ u)(1) = 0$ for $u \in D(\mathfrak{G}^+)$, and an independent standard circular BROWNian motion θ, if $\mathfrak{t}$ is the BESSEL local time at $r = 1$ ($\mathfrak{t} = (t - \mathfrak{m}_1) \vee 0$ if $p_2 = 0$), then the desired motion is the skew product

$$\left[r(t), \theta\left[\int_0^t r(s)^{-2}\, ds + p_3^{\bullet}\, \mathfrak{t}(t)\right] + p_2^{\bullet}\, \mathfrak{t}(t)\right] \qquad t \geqq 0;$$

indeed, this motion is MARKOV, it is standard BROWNian up to time $\mathfrak{m}_{\partial E^2}$, and for smooth f, a brief computation shows that $u = G_\alpha f$ satisfies

$$p_2 u_1(1-, \bullet) + p_3(\mathfrak{G} u)(1-, \bullet) = p_2^{\bullet}\, \partial u^{\bullet}/\partial\theta + \tfrac{1}{2} p_3^{\bullet}\, \partial^2 u^{\bullet}/\partial\theta^2.]$$

7.16. Spinning

Consider the d-dimensional skew product

$$1) \qquad [r(t), \theta(\mathfrak{l}(t))] \qquad t \geqq 0,$$

of which the ***radial part*** is a non-singular conservative diffusion on $[0, +\infty)$ with scale s, speed measure m, an ***entrance non-exit*** barrier at 0 ($s(0) = -\infty$), and a ***non-exit*** barrier at $+\infty$, and the ***angular part*** is an independent $(d-1)$-dimensional spherical BROWNian motion run with the clock $\mathfrak{l}(t) = \int_0^{+\infty} \mathfrak{t}(t, r)\, l(dr)$, where $\mathfrak{t}$ is radial local time and l is a non-negative BOREL measure finite on subcompacts of $(0, +\infty)$.

A brief computation verifies that the skew product is a diffusion with (local) generator

$$2) \qquad (\mathfrak{G} u)(r, \theta) = \frac{u^+(dr, \theta) + l(dr)\, \frac{1}{2}(\Delta u)(r, \theta)}{m(dr)} \qquad (r \neq 0)$$

and GREEN operators

$$3) \qquad G_\alpha e_+ e_- = E_\bullet\left[\int_0^{+\infty} e^{-\alpha t} e_+[r(t)]\, dt \int p(\mathfrak{l}(t), \bullet, \theta)\, e_-(\theta)\, d\theta\right],$$

where Δ is the $(d-1)$-dimensional spherical LAPLACE operator, $e_+ = e_+(r)$ is a radial function, $e_- = e_-(\theta)$ is a spherical function, $E_\bullet$ is the radial expectation operator, and the kernel under the integral sign is the transition function

$$\text{4)}\qquad p(t,\theta_1,\theta_2)=\sum_{n=0} e^{\gamma_n t}\sum_{l\leq m(n)} S_n^l(\theta_1)\,S_n^l(\theta_2)$$

for the spherical BROWNian motion, γ_n being the common eigenvalue $-\frac{1}{2}n(n+d-2)$ of the spherical harmonics $S_n^l: l\leq m(n)$ of weight n $(\frac{1}{2}\Delta S_n^l=\gamma_n S_n^l)$.

Question: *can the skew product be completed so as to be a diffusion on the whole of* R^d?

The answer is *no* if

$$\text{5)}\qquad \int_0^1 s(\xi)\,\mathfrak{l}(d\xi)>-\infty,$$

for the completed GREEN operators have to map $C(R^d)$ into itself, and if $e_+\in C[0,+\infty)$, if $e_+(0)=0$, and if e_- is a spherical harmonic of weight $n>1$, then the product $e_+e_-\in C(R^d)$ and

$$\text{6)}\qquad \begin{aligned}(G_\alpha e_+e_-)(0)&=\lim_{\varepsilon\downarrow 0}(G_\alpha e_+e_-)(\varepsilon,\theta)\\ &=\lim_{\varepsilon\downarrow 0}E_\varepsilon\left[\int_0^{+\infty}e^{-\alpha t}e_+\,e^{\gamma_n\mathfrak{l}}\,dt\right]e_-(\theta)\\ &=E_0\left[\int_0^{+\infty}e^{-\alpha t}e_+\,e^{\gamma_n\mathfrak{l}}\,dt\right]e_-(\theta)\end{aligned}$$

cannot be independent of θ unless

$$E_0\left[\int_0^{+\infty}e^{-\alpha t}e^{\gamma_n\mathfrak{l}}\,dt\right]=0,$$

i.e., unless

$$\text{7)}\qquad P_0[\mathfrak{l}(0+)=+\infty]=1,$$

and, as we know from 4.6,

$$\text{8)}\qquad P_0[\mathfrak{l}(0+)=+\infty]=0 \textit{ or } 1 \textit{ according as } \int_0^1 s(\xi)\,\mathfrak{l}(d\xi) \textit{ converges or diverges (to } -\infty).$$

7) means that the sample path entering from 0 comes in spinning like a spherical BROWNian motion defined for $-\infty<t\leq 0$ as t comes in from $-\infty$.

Consider for the proof of this statement, the angular part θ^+ of the entering sample path and note that $P_0[\theta^+(t)\in d\theta\mid r(s):s\geq 0]$ must be the uniform spherical distribution.

Given $1 > t_1 = 1 - s_1 > t_2 = 1 - s_2 > \cdots > t_n = 1 - s_n > 0$ and letting $\mathfrak{l}^-(t) = \mathfrak{l}(t, w^+_{1-t})$ $(0 \leqq t \leqq 1)$, it is not difficult to see that

$$
\begin{aligned}
9) \qquad & P_0[\theta^+(1-s_1) \in d\theta_1, \theta^+(1-s_2) \in d\theta_2, \ldots, \\
& \qquad \theta^+(1-s_n) \in d\theta_n \mid r(t): 0 \leqq t \leqq 1, \theta^+(1)] \\
& = d\theta_n\, p(\mathfrak{l}(t_{n-1}-t_n), w^+_{t_n}), \theta_n, \theta_{n-1})\, d\theta_{n-1} \\
& \quad \ldots p(\mathfrak{l}(t_2 - t_3), w^+_{t_3}), \theta_3, \theta_2)\, d\theta_2 \\
& \quad p(\mathfrak{l}(t_1 - t_2), w^+_{t_2}), \theta_2, \theta_1)\, d\theta_1 \\
& \quad p(\mathfrak{l}(1 - t_1), w^+_{t_1}), \theta_1, \theta^+(1)) \\
& = p(\mathfrak{l}^-(s_1), \theta^+(1), \theta_1)\, d\theta_1 \\
& \quad p(\mathfrak{l}^-(s_2) - \mathfrak{l}^-(s_1), \theta_1, \theta_2)\, d\theta_2 \\
& \quad \ldots p(\mathfrak{l}^-(s_n) - \mathfrak{l}^-(s_{n-1}), \theta_{n-1}, \theta_n)\, d\theta_n .
\end{aligned}
$$

But this is just the statement that, conditional on $r(t): 0 \leqq t \leqq 1$ and $\theta^+(1)$, $\theta^+(1-s): 0 \leqq s \leqq 1$ is identical in law to $\theta(\mathfrak{l}^-(s)): 0 \leqq s < 1$ with $\theta(0) = \theta^+(1)$, and thanks to

$$10) \qquad \mathfrak{l}^-(1-) = \lim_{t \uparrow 1} \int_0^{+\infty} \mathfrak{t}(t, r, w^+_{1-t})\, l(dr) = \mathfrak{l}(1) = +\infty,$$

the proof of the proposed interpretation of 7) is complete.

At no extra cost, it is seen that $\int_0^1 s\, dl = -\infty$ is also sufficient for the skew product to admit a completion: in fact, if $r(0) = 0$, if θ is an independent spherical BROWNian motion defined for $-\infty < t < +\infty$, and if the law of $\theta(0)$ is the uniform spherical distribution, then it is not difficult to see that the entering sample path

$$
11) \qquad
\begin{array}{ll}
0 & t = 0 \\
[r(t), \theta(-\mathfrak{l}^-(1-t))] & 0 < t \leqq 1 \\
[r(t), \theta(\mathfrak{l}(t-1, w^+_1))] & t > 1
\end{array}
$$

is MARKOVian and identical in law to the skew product 1) on $R^d - 0$ with GREEN operators mapping $C(R^d)$ into itself.

Problem 1. Give complete proof of all the statements about 11).

Problem 2. Compute the rate of growth of $\mathfrak{l}^-(t)$ $(t \uparrow 1)$ for the generators

$$\mathfrak{G} = \frac{1}{2}\left(\frac{\partial^2}{\partial r^2} + \frac{d-1}{r}\frac{\partial}{\partial r} + \frac{1}{r^\varepsilon}\Delta\right) \qquad \varepsilon > 0,\ d \geqq 2,$$

using the ERDÖS-DVORETSKY and SPITZER tests of 4.12.

Problem 3. Give all the isotropic MARKOVian completions of the skew product 1) in case $\int_0^1 s(\xi)\, l(d\xi) > -\infty$.

7.17. An individual ergodic theorem for the standard 2-dimensional Brownian motion

Consider a standard 2-dimensional BROWNian motion. Given an additive functional $\mathfrak{e} = \mathfrak{e}(t)$ of the BROWNian path, we can determine a non-negative measure $e(db)$ such that

1) $$E_a[\mathfrak{e}(\mathfrak{m}_{\partial D})] = \int_D G_D(a, b)\, e(db) \qquad a \in D,$$

for every GREENian domain D, where G_D is the GREEN function of D. To prove this fact, it is enough to note that the left side of 1) is an excessive function of $a \in D$ and to use the results of section 7.5; see H. P. MCKEAN, JR. and H. TANAKA [1] and also 7.19 for further information on such BROWNian additive functionals.

As in the 1-dimensional case (6.8),

2a) $$P_\bullet\left[\lim_{t\uparrow+\infty} \frac{\mathfrak{e}_1(t)}{\mathfrak{e}_2(t)} = \frac{e_1(R^2)}{e_2(R^2)}\right] = 1 \quad \textit{in case} \quad 0 < e_2(R^2) < +\infty,$$

and

2b) $$\lim_{t\uparrow+\infty} P_\bullet\left[\frac{4\pi}{\lg t}\, \mathfrak{e}(t) < u\, e(R^2)\right] = 1 - e^{-u} \quad u \geq 0,\ 0 < e(R^2) < +\infty$$

as will be proved below. An elementary example of an additive functional is

3) $$\mathfrak{e}(t) = \int_0^t f[x(s)]\, ds$$

with $e(db) = 2f(b)\, db$; for this special case, 2a) was proved by [private communication] and by G. MARUYAMA and H. TANAKA [2], and 2b) was proved by G. KALLIANPUR and H. ROBBINS [1].

Coming to the proof of 2), suppose $x(0) = 0$ (this is harmless) and putting $\mathfrak{m}_l = \min(t: |x| = l)$, define $\mathfrak{m}^n$ $(n \geq 0)$ to be the successive passage times to $|a| = 1$ via $|a| = 2$, *i.e.*, let

4) $$\mathfrak{m}^0 = \mathfrak{m}_1,\ \mathfrak{m}^n = \mathfrak{m}^{n-1} + \mathfrak{m}(w^+_{\mathfrak{m}^{n-1}}) \ (n \geq 1),\ \mathfrak{m} = \mathfrak{m}_2 + \mathfrak{m}_1(w^+_{\mathfrak{m}_2}).$$

Because of the isotropic property of x, the chain $\mathfrak{e}_n = \mathfrak{e}(\mathfrak{m}^n) - \mathfrak{e}(\mathfrak{m}^{n-1}): n \geq 1$ is shift invariant and BIRKHOFF's individual ergodic theorem implies that $\gamma = \lim_{n\uparrow+\infty} \mathfrak{e}(\mathfrak{m}^n)/n$ exists. γ is measurable on the tail algebra $\bigcap_{n\geq 1} \mathsf{B}[x(t): t > n]$, and this algebra is the same as $\bigcap_{n\geq 1} \mathsf{B}[t\, x(1/t): t < 1/n]$. Because $t\, x(1/t): t > 0$ is a version of $x(t)$ with respect to P_0, γ is constant by BLUMENTHAL's 01 law, and BIRKHOFF's ergodic theorem implies that $\gamma = E_0(\mathfrak{e}_1)$, *i.e.*,

5) $$P_0\left[\lim_{n\uparrow+\infty} \frac{\mathfrak{e}(\mathfrak{m}^n)}{n} = E_0(\mathfrak{e}_1)\right] = 1.$$

Using the additive property of $\mathfrak{e}$ and the isotropic property of x, one can easily see that

$$6)\quad E_0(\mathfrak{e}_1) = E_0[\mathfrak{e}(\mathfrak{m}^1) - \mathfrak{e}(\mathfrak{m}^0)] = \frac{1}{2\pi}\int_0^{2\pi} d\theta [E_{(1,\theta)}(\mathfrak{e}(\mathfrak{m}_2)) + E_{(2,\theta)}(\mathfrak{e}(\mathfrak{m}_1))]$$

$$= \frac{1}{2\pi}\int_0^{2\pi} d\theta \left[E_{(1,\theta)}(\mathfrak{e}(\mathfrak{m}_2)) + \lim_{n\uparrow+\infty} E_{(2,\theta)}(\mathfrak{e}(\mathfrak{m}_1 \wedge \mathfrak{m}_n))\right]$$

$$= \frac{1}{2\pi}\int_0^{2\pi} d\theta \left[\int_{R^2} G_{02}((1,\theta), b)\, \mathfrak{e}(db) + \lim_{n\uparrow+\infty} \int_{R^2} G_{1n}((2,\theta), b)\, \mathfrak{e}(db)\right],$$

$$= \int_{R^2} \mathfrak{e}(db) \frac{1}{2\pi}\int_0^{2\pi} d\theta \left[G_{02}((1,\theta), b) + \lim_{n\uparrow+\infty} G_{1n}((2,\theta), b)\right] d\theta,$$

where G_{02} is the GREEN function of the domain $|a| < 2$ and G_{1n} that of $1 < |a| < n$.

The integrand $G(b)$ in the last integral is clearly a radial function $G(|b|)$ because of the isotropic property of x and is lower semi-continuous as an increasing limit of continuous functions.

Applying 6) to 5) with a non-negative radial function f, it develops that

$$7)\quad E_0\left[\int_{\mathfrak{m}^0}^{\mathfrak{m}^1} f[x(s)]\, ds\right] = 2\int_{R^2} G(|b|) f(b)\, db = 4\pi \int_0^{+\infty} G(r) f(r)\, r\, dr.$$

On the other hand we can compute this expectation in terms of the BESSEL process $\mathbf{D}^\bullet : r(t) = |x(t)|$ as

$$8)\quad E_0\left[\int_{\mathfrak{m}^0}^{\mathfrak{m}^1} f(|x(s)|)\, ds\right] = E_1^\bullet\left[\int_0^{\mathfrak{m}^\bullet} f[r(s)]\, ds\right]$$

($\mathfrak{m}^\bullet$ = the BESSEL passage time to 1 via 2)

$$= E_1^\bullet\left[\int \mathfrak{t}^\bullet(\mathfrak{m}^\bullet, r) f(r)\, 2r\, dr\right] \qquad (\mathfrak{t}^\bullet = \text{the BESSEL local time})$$

$$= 2\int_0^{+\infty} E_1^\bullet[\mathfrak{t}^\bullet(\mathfrak{m}^\bullet, r)] f(r)\, r\, dr$$

$$= 2\int_0^{+\infty} \lg 2\, f(r)\, r\, dr$$

[see 6.8.6)].

Comparing 7) with 8) we obtain $G(r) = \lg 2/2\pi$ for *almost all* r; since $G(r)$ is lower semi-continuous, this is true for *all* r, *i.e.*,

$$9)\quad P_0\left[\lim_{n\uparrow+\infty} \frac{\mathfrak{e}(\mathfrak{m}^n)}{n} = \frac{\lg 2\, e(R^2)}{2\pi}\right] = 1,$$

and 2a) follows from 9) as in section 6.8.

Because of 2a), it is enough to prove 2b) for $e(t) = \int_0^t f[|x(s)|]\, ds$ with $f \geqq 0$ and $0 < \int f(r)\, r\, dr < +\infty$; this can be done in terms of the BESSEL process $D^\bullet: r(t) = |x(t)|$ as follows:

$$P_0\left[\frac{4\pi}{\lg t} e(t) < u\, e(R^2)\right] = P_0^\bullet\left[\frac{4}{\lg t}\int_0^t f[r(s)]\, ds < 4u \int_0^{+\infty} f(r)\, r\, dr\right] = P_0^\bullet\left[\frac{2}{\lg t}\int_0^t f[r(s)]\, ds < u \int_0^{+\infty} f(r)\, 2r\, dr\right];$$

by problem 6.8.4, this tends to $1 - e^{-u}$ as $t \uparrow +\infty$.

For additional information and the ergodic theorem for general MARKOV processes, see T. HARRIS and H. ROBBINS [*1*], G. MARUYAMA and H. TANAKA [*2*] and T. UENO [*1*].

7.18. Covering Brownian motions

Given an unramified covering surface K of an open connected region $D \subset R^2$, let j denote the natural projection of K onto D, and selecting a point o of K, consider the covering path[1] starting at o that lies over the standard BROWNian path $x(t): t < \mathfrak{m}_{\partial D}$ starting at $j(o)$ as in diagram 1.

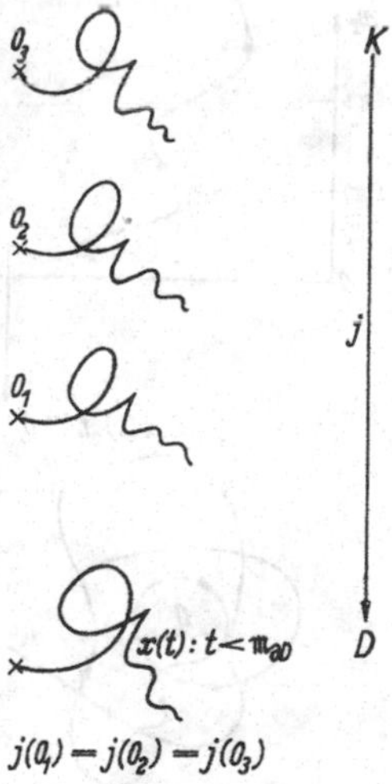

Diagram 1

Given a MARKOV time $\mathfrak{m}$ for the covering motion and projecting down via j, it is clear that $\mathfrak{m}$ is also MARKOV for the ground motion, and since the latter starts afresh at time $\mathfrak{m}$, it follows from the definition of covering path that the covering motion begins afresh at time $\mathfrak{m}$ also.

We see at once that the covering motion is a diffusion and that, in the small, its generator is $\mathfrak{G} u = \frac{1}{2}[\Delta u(j^{-1})](j)$, where j^{-1} is the local inverse of the local homeomorphism j.

With the help of coverings we can give a simple proof of P. LÉVY's result [*3*: 270] that if j is a non-constant regular function on the open connected region $D \subset R^2$ and if j' is never 0 on D, then the composition of j and the standard BROWNian motion on D is the standard BROWNian motion on $j(D)$ with a change of clock: in detail, D is an unramified covering of $j(D)$ with

[1] SEIFERT and THRELFALL [*1*: 181—185].

natural projection j, the covering motion of the standard BROWNian motion on $j(D)$ is standard BROWNian on D with a change of clock, and inverting the clock in ground and covering paths, it is seen that the projection via j of the standard BROWNian motion on D is the standard BROWNian motion on $j(D)$ run with the inverse clock.

We learn from P. LÉVY's result that, up to a change of clock, the 2-dimensional BROWNian path is a conformal invariant.

Consider, as the simplest example, the stereographic projection j of the 2-sphere S^2 onto R^2:

$$j:(x_1, x_2, x_3) \in S^2 \to \left(\frac{x_1}{1-x_3}, \frac{x_2}{1-x_3}\right) \in R^2.$$

$$\mathfrak{G} = \frac{1}{2}(1-\cos\psi)^2 \left[\frac{1}{\sin\psi}\frac{\partial}{\partial\psi}\sin\psi\frac{\partial}{\partial\psi} + \frac{1}{\sin^2\psi}\frac{\partial^2}{\partial\theta^2}\right],$$

where $0 \leq \psi \leq \pi$ is colatitude and $0 \leq \theta < 2\pi$ is longitude; thus, the covering (stereographic) BROWNian motion is the standard spherical BROWNian motion $\mathsf{BM}(S^2)$, run with the clock

$$\mathfrak{f}^{-1}(t), \quad \text{where} \quad \mathfrak{f}(t) = \int_0^t [1-\cos\psi(s)]^{-2}\,ds \quad \text{and} \quad \psi(t): t \geq 0$$

is the colatitude of $\mathsf{BM}(S^2)$.

Consider, for the next example, the projection $j: w \to e^w$ of the RIEMANN surface R^2 for $w = \lg z$ onto $R^2 - 0$ as in diagram 2.

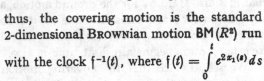

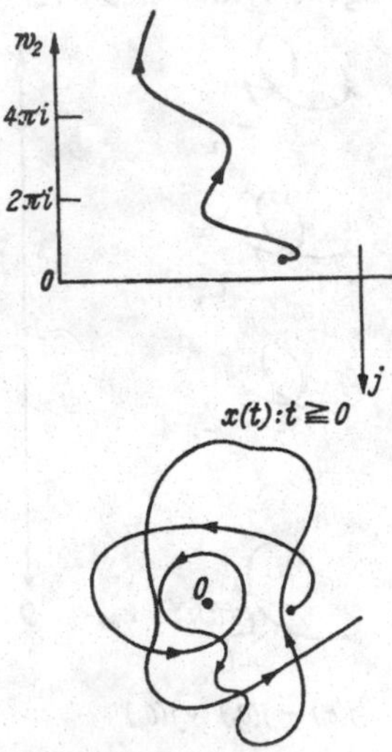

Diagram 2

$$\mathfrak{G} = \frac{1}{2}|j'(w)|^{-2}\Delta = \frac{1}{2}e^{-2w_1}\left(\frac{\partial^2}{\partial w_1^2} + \frac{\partial^2}{\partial w_2^2}\right),$$

where $\quad w = w_1 + i\,w_2;$

thus, the covering motion is the standard 2-dimensional BROWNian motion $\mathsf{BM}(R^2)$ run with the clock $\mathfrak{f}^{-1}(t)$, where $\mathfrak{f}(t) = \int_0^t e^{2x_1(s)}\,ds$ and $x_1(t): t \geq 0$ is the horizontal component of $\mathsf{BM}(R^2)$.

$\mathsf{BM}(R^2)$ hits each disk *i.o.*; thus, the covering motion does also, and since the covering path leads from strip i to strip j when the ground path winds (counterclockwise) $j - i$ times about 0, it is seen that the ground path winds counterclockwise about 0 an infinite number of times and also unwinds itself an infinite number of times (*i.e.*, the covering motion returns to the strip $0 \leq w_2 < 2\pi$).

Consider, for the third example, the projection j of the elliptic modular figure[1] of diagram 3 onto the punctured plane $R^2 - 0 \cup 1$. j is the composition of the fractional linear map $j(w) = \frac{1-i}{2}\,\frac{w+1}{w-i}$ of the open unit disc $|w| < 1$ onto $w_2 > 0$ and the modulus $j_2 = k^2 = \theta_2^4 \theta_3^{-4}$ of the JACOBI elliptic functions mapping $w_2 > 0$ onto the punctured plane. $\mathfrak{G} = \frac{1}{2}\,|j'(w)|^{-2}\left(\frac{\partial^2}{\partial w_1^2} + \frac{\partial^2}{\partial w_2^2}\right)$ for $|w| < 1$, so now the covering motion is $\mathsf{BM}(R^2)$ run with a clock $\mathfrak{f}^{-1}(t)$ that blows up as the covering path runs out to $|w| = 1$, and since the sheets of the covering surface correspond to the elements of the fundamental group of the punctured plane (free group of 2 generators), it is seen that the ground BROWNian motion loops about 0 and 1 worse and worse as $t \uparrow +\infty$ and never gets unlooped.

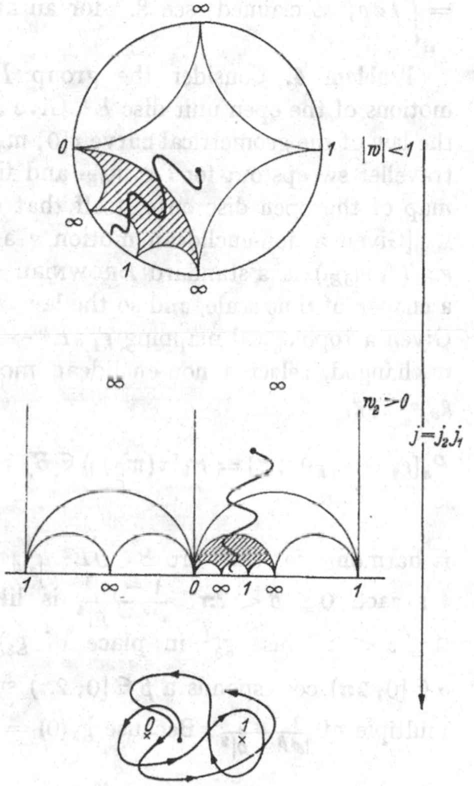

Diagram 3

S. KAKUTANI [3] is a source of additional information on covering BROWNian motions.

Problem 1. What is the reason that the ground BROWNian motion cannot unloop itself in the twice-punctured plane?

[By the ergodic theorem for $\mathsf{BM}(R^2)$ (7.17), the fraction of time that the BROWNian path spends in $|w| < 1$ up to time t is something like $(\lg t)^{-1}$. Winding about a single point depends on angle not radius; with 2 points, the path can wind about one and not the other and $(\lg t)^{-1} t$ is not enough time to unwind the loops thus produced.]

Problem 2. Prove that for the spherical BROWNian motion $\mathsf{BM}(S^2)$,

$$P.\left[\lim_{t\uparrow+\infty} t^{-1}\int_0^t f(x_s)\,ds = \int_{S^2} f\,do\right] = 1 \qquad f \in L^1(S^2)$$

using the ergodic theorem for $\mathsf{BM}(R^2)$ and the stereographic projection.

[1] COURANT and HURWITZ [1: 432].

Problem 3. Give a new proof of the result of problem 2 for $\mathsf{BM}(S^d)$ $(d \geqq 2)$ assuming that $e(f) = \lim_{t\uparrow+\infty} t^{-1} \int_0^t f(x_s)\,ds$ exists.

[Granting that $e(f)$ exists, consider the tail algebra $\mathsf{A} = \bigcap_{t>0} \mathsf{B}[x(s): s \geqq t]$. Given $A \in \mathsf{A}$, $p = P_\bullet(A)$ satisfies $\Delta p = 0$ so that p is constant, and we infer that $P_\bullet(A) = 0$ or 1. $e(f)$ is measurable A and therefore constant as regards w, and its invariance under rotations forces $e(f) = \int_{S^d} f\,do$, as claimed; see 8.7 for an amplification of this method.]

Problem 4. Consider the group Γ of non-euclidean (POINCARÉ) motions of the open unit disc E^2. Give a proof that Γ leaves unchanged the law of the geometrical curve $x[0, \mathfrak{m}_{\partial E^2})$ that the standard BROWNian traveller sweeps out for $t < \mathfrak{m}_{\partial E^2}$ and find the most general topological map of the open disc onto itself that does this.

[Given a non-euclidean motion g and a standard BROWNian path, $g\,x(t \wedge \mathfrak{m}_{\partial E^2})$ is a standard BROWNian motion stopped on ∂E^2, up to a change of time scale, and so the law of $x[0, \mathfrak{m}_{\partial E^2})$ is invariant under g. Given a topological mapping $g_1: E^2 \to E^2$ leaving the law of $x[0, \mathfrak{m}_{\partial E^2})$ unchanged, select a non-euclidean motion $g_2: g_1(0) \to 0$ and consider $g_3 = g_2 g_1$.

$$P_a[g_3\,x(\mathfrak{m}_{\partial E^2}) \in B] = P_b[x(\mathfrak{m}_{\partial E^2}) \in B] = \frac{1}{2\pi} \int_B \frac{1-|b|^2}{|e^{i\beta}-b|^2}\,d\beta \qquad (b = g_3 a)$$

is harmonic for each arc $B \subset \partial E^2$. $\frac{1-|b|^2}{|e^{i\beta}-b|^2}$ is therefore harmonic in a for each $0 \leqq \beta < 2\pi$. $\frac{1-|a|^2}{|e^{i\alpha}-a|^2}$ is likewise harmonic in b for each $0 \leqq \alpha < 2\pi$ (use g_3^{-1} in place of g_3), and it follows that to each $\alpha \in [0, 2\pi)$ corresponds a $\beta \in [0, 2\pi)$ such that $\frac{1-|a|^2}{|e^{i\alpha}-a|^2}$ is a constant multiple of $\frac{1-|b|^2}{|e^{i\beta}-b|^2}$. Because $g_3(0) = 0$, the constant is 1, and

$$\min_{0 \leqq \alpha < 2\pi} \frac{1-|a|^2}{|e^{i\alpha}-a|^2} = \frac{1-|a|^2}{1+|a|^2} \geqq \frac{1-|b|^2}{1+|b|^2}.$$

$\frac{1-|b|^2}{1+|b|^2}$ is likewise $\geqq \frac{1-|a|^2}{1+|a|^2}$, *i.e.*, $|a| = |b|$, and so g_3 is a topological map of each circle $|a| = l < 1$ onto itself. g_2 can now be adjusted so as to make g_3 orientation preserving on all the circles $|a| = l < 1$ at once, and using $\frac{1-|a|^2}{|e^{i\alpha}-a|^2} = \frac{1-|b|^2}{|e^{i\beta}-b|^2}$ to prove that $e^{-i\alpha} a$ and $e^{-i\beta} b$ have the same real part, it follows that $b = e^{i\gamma} a$ $(\gamma = \beta - \alpha)$, *i.e.*, g_3 is a rotation. g_1 can now be identified as the non-euclidean motion $g_2^{-1} g_3$, and the problem is solved.]

7.19. Diffusions with Brownian hitting probabilities

Given a bounded d-dimensional domain D, let $h(a, db) = h_{\partial D}(a, db)$ be the classical distribution of harmonic mass on ∂D as viewed from $a \in D$ and recall that for the standard BROWNian motion

1) $$h(a, db) = P_a[x(\mathfrak{m}_{\partial D}) \in db]$$

(see 7.12). A diffusion on $R^d \cup \infty$ is said to have *Brownian hitting probabilities* if 1) holds for all bounded $D \subset R^d$; in the special case $d = 1$, each diffusion in its natural scale has BROWNian hitting probabilities. Here it will be proved that the generator $\mathfrak{G}$ of a d-dimensional diffusion with BROWNian hitting probabilities can be expressed as

2) $$(\mathfrak{G} u)(a) = \frac{-e^u(da)}{m(da)},$$

m being a suitable speed measure and e^u the RIESZ measure of $u \in D(\mathfrak{G})$; see 7.5. $-e^u(a, b] = u^+(b) - u^+(a)$ for $d = 1$; see 4.2.

Given bounded $D \subset R^d$, let $f \in C_\infty(R^d)$ be $\equiv 0$ on $\bar{D}$ and positive outside; then $\mathfrak{G} G_1 f = G_1 f$ is bounded below by a positive constant c on $\bar{D}$ so that $\mathfrak{G}$ applied to $u = G_1 f/c$ is $\geqq 1$ on $\bar{D}$, and an application of DYNKIN's formula gives

3) $$e_D(a) = E_a(\mathfrak{m}_{\partial D}) \leqq E_a\left[\int_0^{\mathfrak{m}_{\partial D}} \mathfrak{G} u \, dt\right]$$
$$= E_a[u(x(\mathfrak{m}_{\partial D}))] - u(a) \leqq \|u\|_\infty < +\infty$$

inside D. Define $\mathfrak{m} = \mathfrak{m}_{\partial D_1}$ for $D_1 \subset D$; then for $a \in D_1$,

4) $$e_D(a) = E_a(\mathfrak{m} + \mathfrak{m}_{\partial D}(w_{\mathfrak{m}}^+))$$
$$= e_{D_1}(a) + \int h_{\partial D_1}(a, db)\, e_D(b)$$
$$> \int h_{\partial D_1}(a, db)\, e_D(b),$$

and using the RIESZ decomposition [see 7.5], $e_D(a)$ is found to be the potential $\int G_D(a, b)\, m(db)$* of a non-negative charge distribution m, positive on opens. Using the composition rule

5) $$G_D(a, b) = G_{D_1}(a, b) + \int h_{\partial D_1}(a, d\xi)\, G_D(\xi, b) \qquad a, b \in D_1 \subset D$$

and the resulting formula for e_{D_1}:

6) $$e_{D_1}(a) = e_D(a) - \int h_{\partial D_1}(a, db)\, e_D(b)$$
$$= \int G_D(a, b)\, m(db) - \int h_{\partial D_1}(a, d\xi) \int G_D(\xi, b)\, m(db)$$
$$= \int G_{D_1}(a, b)\, m(db),$$

one finds that m does not depend upon D.

* G_D is the GREEN function of D.

As to 2), if $\mathfrak{G}$ applied to $u_1 \in D(\mathfrak{G})$ is $\leqq -1$ on D, if $u \in D(\mathfrak{G})$, and if $u_2 = u + n\, u_1$, then $\mathfrak{G}\, u_2 \leqq 0$ on D for $n \geqq \|\mathfrak{G}\, u\|_\infty$, and for $D_1 \subset D$, DYNKIN's formula implies

$$7) \qquad v(a) - \int h_{\partial D_1}(a, db)\, v(b) = -E_a\left[\int_0^{\mathfrak{m}_{\partial D_1}} \mathfrak{G}\, v\, dt\right] \geqq 0$$

$$a \in D_1 \quad v = u_1 \text{ or } u_2.$$

A second application of the RIESZ decomposition now permits us to express $u(a) - \int h_{\partial D}(a, db)\, u(b)$ as the potential of a signed charge distribution e^u, and since non-negative potentials come from non-negative total charges, the bounds

$$8a) \qquad \alpha \int G_D(a, b)\, m(db) = \alpha\, e_D(a) \qquad\qquad \alpha \equiv \inf_D \mathfrak{G}\, u$$

$$\leqq E_a\left[\int_0^{\mathfrak{m}_{\partial D}} \mathfrak{G}\, u\, dt\right] = \int h_{\partial D}(a, db)\, u(b) - u(a) = -\int G_D(a, b)\, e^u(db)$$

$$\leqq \beta\, e_D(a) = \beta \int G_D(a, b)\, m(db) \qquad\qquad \beta \equiv \sup_D \mathfrak{G}\, u$$

show that

$$8b) \qquad \alpha\, m(D) \leqq -e^u(D) \leqq \beta\, m(D),$$

and 2) follows at once because $\mathfrak{G}\, u$ is continuous.

H. P. MCKEAN, JR. and H. TANAKA [1] described the speed measures of a wider class of motions with BROWNian hitting probabilities; in the present case, it can be proved that m is a speed measure if and only if each of its potentials $e_D = \int G_D\, dm$ is continuous on D and tends to 0 at the non-singular points of ∂D, and each of its null sets Z is *thin* in the sense that, for the standard BROWNian motion,

$$P_\bullet[Z \cap (x(t) : t > 0) = \varnothing] = 1.$$

Given such a measure, it is natural to hope that the associated diffusion will be the composition of the standard BROWNian motion with a stochastic clock $\mathfrak{f}^{-1}$, *e.g.*, the inverse function of the HELLINGER integral

$$9) \qquad \mathfrak{f}(t) = \int_{R^d} \frac{\text{measure}\,(s : x(s) \in db,\ s \leqq t)\, m(db)^*}{2\, db},$$

and something close to this is true (see H. P. MCKEAN, JR and H. TANAKA [1], R. BLUMENTHAL, R. GETOOR, and H. P. MCKEAN, JR [1], and also 8.3).

Because the 2-dimensional BROWNian motion is persistent, so is the motion attached to $\mathfrak{G}$ in case $d = 2$. In $d \geqq 3$ dimensions it drifts

* measure $(s : x(s) \in db,\ s \leqq t)$ is not a multiple of LEBESGUE measure in dimensions $d \geqq 2$, *i. e.*, local times cannot be defined.

out to ∞. $P_\bullet(\mathfrak{m}_\infty < +\infty)^* \equiv 1$ if R^d can be split into $A \cup B$ such that

10) $$\int_A |b|^{2-d} m(db) < +\infty$$

and B is thin at ∞ in the sense that, for the BROWNian motion,

11) $$P_\bullet[x(t) \in B \ i.o., \ t \uparrow +\infty] = 0.$$

$P_\bullet(\mathfrak{m}_\infty < +\infty) \equiv 0$ otherwise. WIENER's test can be used to decide if 11) holds or not.

Problem 1. Given a standard 2-dimensional BROWNian motion with components a and b, let $\mathfrak{t}$ be the local time $\lim_{\varepsilon\downarrow 0}(2\varepsilon)^{-1}$ measure $(s: 0 \leqq b(s) < \varepsilon, s \leqq t)$; the problem is to compute the speed measure $e^\bullet$ and the generator $\mathfrak{G}^\bullet$ of the motion $\mathsf{D}^\bullet: x^\bullet = x(\mathfrak{f}^{-1})$ $(\mathfrak{f} = t + 2\mathfrak{t})$ (see T. UENO [1] for a different description of $\mathsf{D}^\bullet$).

[$e^\bullet(da \times db) = 2da\, db$ on $|b| > 0$; also, $e^\bullet(da \times 0)$ is translation invariant and, as such, a constant multiple of the 1-dimensional LEBESGUE measure da. Because $\mathfrak{f} = \lim_{\varepsilon\downarrow 0}\int_0^t [1 + \varepsilon^{-1} \times \text{the indicator of } 0 \leqq b < \varepsilon](x_s)\, ds$ and

$$\int_{R^2} f\, 2da\, db + \int_{R^1\times 0} f\, 2da = \lim_{\varepsilon\downarrow 0}\int_{R^2} [1 + \varepsilon^{-1} \times \text{the indicator of } 0 \leqq b < \varepsilon]\, f\, 2da\, db,$$

$e^\bullet(da \times 0) = 2da$, and it follows that if $u \in D(\mathfrak{G}^\bullet)$ and if $A \times B$ is an open rectangle meeting $R^1 \times 0$, then, up to functions of class $C^1(A \times B)$, $-u = \int_{A\times B} G\mathfrak{G}^\bullet u\, 2da$ on $A \times B$, where G is the GREEN function of $A \times B$. Given $a \in A$, it is immediate that

$$\begin{aligned}
&\lim_{\varepsilon\downarrow 0}(2\varepsilon)^{-1}[u(a,\varepsilon) - 2u(a,0) + u(a,-\varepsilon)] \\
&\quad = -\lim_{\varepsilon\downarrow 0}(2\varepsilon)^{-1}\int_{A\times 0}\frac{1}{2\pi}\lg\frac{(a-a')^2}{(a-a')^2+\varepsilon^2}\,\mathfrak{G}^\bullet u(a',0)\,2da' \\
&\quad = \lim_{\varepsilon\downarrow 0}\frac{1}{\pi}\int_{R^1\times 0}\frac{\varepsilon}{(a-a')^2+\varepsilon^2}\,\mathfrak{G}^\bullet u(a',0)\,2da' \\
&\quad = (\mathfrak{G}^\bullet u)(a,0).
\end{aligned}$$

$D(\mathfrak{G}^\bullet)$ can now be described as the class of functions $u \in C(R^2)$ such that

$$\begin{aligned}
u^\bullet &= \Delta u/2 && \text{off } R^1 \times 0 \\
&= \lim_{\varepsilon\downarrow 0}\varepsilon^{-1}[u(a,\varepsilon) - 2u(a,0) + u(a,-\varepsilon)] && \text{on } R^1 \times 0
\end{aligned}$$

is of class $C(R^2)$, Δ being the local BROWNian generator of 7.2.3). $\mathfrak{G}^\bullet u = u^\bullet$ for $u \in D(\mathfrak{G}^\bullet)$.]

---

* $\mathfrak{m}_\infty \equiv (t: x(t-) = \infty)$.

7.20. Right-continuous paths

Consider the general motion of 7.1, but now let the sample path be merely right-continuous. $\mathfrak{G}$ can be introduced as before and DYNKIN's formula applies:

1) $$(\mathfrak{G}\, u)\,(a) = \lim_{b \downarrow a} E_a(\mathfrak{m})^{-1}[E_a[u(x_{\mathfrak{m}})] - u(a)]$$

where $\mathfrak{m}$ is the exit time $\inf(t : x(t) \notin B)$ from a small closed neighborhood B of a. $P_a(x_{\mathfrak{m}} \in db)$ need not be concentrated on ∂B because the path is permitted to *jump* out, and this is reflected in the fact that $\mathfrak{G}$ can be a *global* operator.

Consider, as the simplest case, the 1-sided differential process on $[0, +\infty)$ with LÉVY formula

2) $$E_0[e^{-\alpha x(t)}] = \exp\left(-t\left[m\,\alpha + \int_{0+}^{+\infty} (1 - e^{-\alpha l})\, n(dl)\right]\right)$$

$$m \geqq 0,\ n(dl) \geqq 0,\ \int_{0+}^{+\infty} (1 - e^{-l})\, n(dl) < +\infty$$

and probabilities

3) $$P_a(B) = P_0(x + a \in B) \qquad a \geqq 0$$

(see the note on differential processes placed at the end of 1.8).

Given $f \in C[0, +\infty)$ and $a \geqq 0$, let $f_a(b) = f(a + b)$ and let G_1 be the GREEN operator. Because $(G_1 f)\,(b) - (G_1 f)\,(a) = G_1 (f_b - f_a)\,(0)$, $C^1(\mathfrak{G}) = D(\mathfrak{G}) \cap C^1[0, +\infty)$ includes $G_1\, C^1[0, +\infty)$, *i.e.*, $C^1(\mathfrak{G})$ *is well-populated.*

$\mathfrak{G}$ is to be computed on $C^1(\mathfrak{G})$.

Given $\gamma > 0$, $u = e^{-\gamma l}$ satisfies $G_1 u = \beta\, u$ with $\beta = (G_1\, u)\,(0)$ and so belongs to $D(\mathfrak{G})$, and now a comparison of

4a) $$(\mathfrak{G}\, u)\,(0) = \lim_{\varepsilon \downarrow 0} \int_0^{+\infty} (e^{-\gamma l} - 1)\, n_\varepsilon(dl)$$

$$n_\varepsilon(dl) = E_0(\mathfrak{m}_\varepsilon)^{-1}\, P_0[x(\mathfrak{m}_\varepsilon) \in dl]$$

$$\mathfrak{m}_\varepsilon = \inf(t : x(t) > \varepsilon)$$

and

4b) $$(\mathfrak{G}\, u)\,(0) = \lim_{t \downarrow 0} t^{-1}\, E_0[e^{-\gamma x(t)} - 1]$$

$$= -m\,\gamma + \int_{0+}^{+\infty} (e^{-\gamma l} - 1)\, n(dl)$$

implies that $(1 - e^{-l})\, n_\varepsilon(dl)$ converges as $\varepsilon \downarrow 0$ to the measure

5) $$n^\bullet(dl) = (1 - e^{-l})\, n(dl) \qquad l > 0$$

$$n^\bullet(0) = m.$$

But if $u \in C^1(\mathfrak{G})$, then

6)
$$u^\bullet(l) = \frac{u(l) - u(0)}{1 - e^{-l}} \qquad l > 0$$
$$= u^+(0) \qquad l = 0$$

belongs to $C[0, +\infty)$ so that

7)
$$(\mathfrak{G} u)(0) = \lim_{\varepsilon \downarrow 0} \int_0^{+\infty} [u(l) - u(0)]\, n_\varepsilon(dl)$$
$$= \lim_{\varepsilon \downarrow 0} \int_0^{+\infty} u^\bullet(l)\,(1 - e^{-l})\, n_\varepsilon(dl) = \int_0^{+\infty} u^\bullet\, n^\bullet(dl)$$
$$= m\, u^+(0) + \int_0^{+\infty} [u(l) - u(0)]\, n(dl),$$

and using $(\mathfrak{G} u)(a) = (\mathfrak{G} u_a)(0)$, it is seen that

8)
$$(\mathfrak{G} u)(a) = m\, u^+(a) + \int_0^{+\infty} [u(b + a) - u(a)]\, n(db) \qquad a \geqq 0,$$

completing the identification of $\mathfrak{G}$ restricted to $C^1(\mathfrak{G})$.

Problem 1 after E. B. DYNKIN [4: 58]. Check that for the 1-sided stable process with exponent $0 < \alpha < 1$ and rate $1/2$ ($m = 0$, $\int_{0+}^{+\infty} (1 - e^{-\gamma l})\, dn = \gamma^\alpha/2$),

$$P_0[x(\mathfrak{m}_\varepsilon) < l] = \frac{\sin \pi\alpha}{\pi} \int_{\varepsilon/l}^{1} (1 - t)^{-\alpha}\, t^{\alpha - 1}\, dt \qquad l > \varepsilon$$
$$E_0(\mathfrak{m}_\varepsilon) = \frac{2\varepsilon^\alpha}{\Gamma(\alpha + 1)}.^{*}$$

[Given $\gamma > 0$, $\int_0^{+\infty} E_0[e^{-\gamma x(t)}]\, dt$

$$= \int_0^{+\infty} e^{-\gamma l} \int_0^{+\infty} dt\, P_0(x(t) \in dl) = 2\gamma^{-\alpha}$$
$$= 2\Gamma(\alpha)^{-1} \int_0^{+\infty} e^{-\gamma l}\, l^{\alpha - 1}\, dl,$$

i.e.,

$$\int_0^{+\infty} dt\, P_0(x(t) \in dl) = 2\Gamma(\alpha)^{-1}\, l^{\alpha - 1}\, dl,$$

and it follows that if $f \in C[0, +\infty)$ vanishes near $+\infty$ and to the left

---

* DYNKIN's evaluation of $E_0(\mathfrak{m}_\varepsilon)$ contains an incorrect constant.

of ε, then

$$(G_{0+}f)(0) = E_0\left[\int_0^{+\infty} f(x)\,dt\right] = 2\Gamma(\alpha)^{-1}\int_0^{+\infty} a^{\alpha-1} f(a)\,da$$

$$= E_0\left[\int_0^{+\infty} f(x(t+\mathfrak{m}_\varepsilon))\,dt\right]$$

$$= \int_0^{+\infty} P_0[x(\mathfrak{m}_\varepsilon)\in da]\,(G_{0+}f)(a)$$

$$= \int_0^{+\infty} P_0[x(\mathfrak{m}_\varepsilon)\in da]\,2\Gamma(\alpha)^{-1}\int_a^{+\infty}(b-a)^{\alpha-1} f(b)\,db$$

$$= 2\Gamma(\alpha)^{-1}\int_0^{+\infty} f\,db\int_0^b (b-a)^{\alpha-1} P_0[x(\mathfrak{m}_\varepsilon)\in da],$$

or, what interests us,

$$\int_0^b (b-a)^{\alpha-1} P_0(x(\mathfrak{m}_\varepsilon)\in da) = b^{\alpha-1} \text{ or } 0 \textit{ according as } b \geqq \varepsilon \textit{ or } b < \varepsilon;$$

now take LAPLACE transforms on both sides and solve. Coming to $E_0(\mathfrak{m}_\varepsilon)$, it is clear that the scaling $x(t)\to\varepsilon\, x(t/\varepsilon^\alpha)$ preserves the measure $P_0(B)$. $\mathfrak{m}_\varepsilon$ is therefore identical in law to $\varepsilon^\alpha\,\mathfrak{m}_1$, $E_0(\mathfrak{m}_\varepsilon) = \varepsilon^\alpha E_0(\mathfrak{m}_1)$, and the evaluation $E_0(\mathfrak{m}_1) = 2/\Gamma(\alpha+1)$ follows on putting $u = e^{-l}$ $(l \geqq 0)$ and comparing

$$(\mathfrak{G}\,u)(0) = \lim_{t\downarrow 0} t^{-1} E_0[e^{-x(t)} - 1] = -\tfrac{1}{2}$$

and

$$(\mathfrak{G}\,u)(0) = -\lim_{\varepsilon\downarrow 0} E_0(\mathfrak{m}_\varepsilon)^{-1} E_0\left[e^{-x(\mathfrak{m}_\varepsilon)} - 1\right]$$

$$= \lim_{\varepsilon\downarrow 0} E_0(\mathfrak{m}_1)^{-1}\frac{\sin\pi\alpha}{\alpha}\int_\varepsilon^{+\infty}\frac{e^{-l}-1}{(l-\varepsilon)^\alpha}\frac{dl}{l}$$

$$= \int_0^{+\infty}\frac{e^{-l}-1}{l^{1+\alpha}}\,dl/E_0(\mathfrak{m}_1)\,\Gamma(\alpha)\,\Gamma(1-\alpha)$$

$$= -1/E_0(\mathfrak{m}_1)\,\alpha\,\Gamma(\alpha).]$$

7.21. Riesz potentials

We know that the NEWTONian potentials

1) $$\int |b-a|^{2-d} e(db) \qquad e \geqq 0,\ d \geqq 3$$

are linked to the d-dimensional BROWNian motion via the formula

2) $$\int_0^\infty (2\pi t)^{-d/2} e^{-|b-a|^2/2t}\,dt = \text{constant}\times|b-a|^{2-d}.$$

What about the fractional-dimensional (RIESZ) potentials

$$3)\qquad \int |b-a|^{\gamma-d}\, e(db) \qquad e \geqq 0$$

$$0<\gamma<1 \qquad d=1$$

$$0<\gamma<2 \qquad d\geqq 2?$$

Given $0<\gamma\leqq 2$, the FOURIER transform

$$4)\ p_\gamma(t,a,b) = (2\pi)^{-d}\int_{R^d} e^{i(b-a)\cdot c}\, e^{-\frac{t}{2}|c|^\gamma}\, dc \qquad (b-a)\cdot c = |b-a|\, r\cos\theta$$

$$= (2\pi)^{-d}\int_0^{+\infty} e^{-\frac{t}{2}r^\gamma}\, r^{d-1}\, dr \int_{S^{d-1}} e^{i|b-a|r\cos\theta}\, do$$

$$= (2\pi)^{-d/2}\int_0^{+\infty} e^{-\frac{t}{2}r^\gamma}\, r^{d/2}\, J_{\frac{d}{2}-1}(|b-a|r)\, dr \qquad t>0,\ a,b\in R^d$$

is non-negative. p_2 is the d dimensional GAUSS kernel[1], and imitating the construction of d-dimensional WIENER measure from p_2, it is possible to construct an d-dimensional (strict) MARKOV process with right-continuous paths and transition probabilities $p_\gamma(t,a,b)\,db$; this is the so-called *isotropic stable process* with characteristic exponent γ.

Given $f\in C(R^d)$ such that $\hat f = \int_{R^d} e^{-ic\cdot a} f\, da$ is small near ∞, the GREEN operator $G_\alpha: f\to E_\bullet\left[\int_0^{+\infty} e^{-\alpha t} f(x_t)\, dt\right]$ sends f into $(2\pi)^{-d}\int_{R^d} e^{ia\cdot c}$ $(\alpha+\frac{1}{2}|c|^\gamma)^{-1}\hat f\, dc$, and introducing the operator

$$\mathfrak{G}: f\to -\tfrac{1}{2}(2\pi)^{-d}\int_{R^d} |c|^\gamma \hat f\, dc,$$

it is clear that $(\alpha-\mathfrak{G})\, G_\alpha f = f$; in short, the generator of the isotropic stable process is

$$5)\qquad -\tfrac{1}{2}(-\Delta)^{\gamma/2},$$

apart from the technical definition of its domain. $(-\Delta)^{\gamma/2}$ means the non-negative root of the non-negative operator $-\Delta$ in the sense of spectral resolution (FOURIER transform) over the HILBERT space $L^2(R^d, db)$; see S. BOCHNER [*1*].

$\mathfrak{G}$ is a *global* operator if $\gamma<2$.

---

* $a\cdot b$ is the inner product in R^d.

[1] Here, the nuisance factor 1/2 in the GAUSS kernel is dropped.

Coming to the potential kernel,

$$6)\quad \int_0^{+\infty} p_\gamma(t,a,b)\,dt$$
$$= 2(2\pi)^{-d/2}\int_0^{+\infty} r^{\frac{d}{2}-\gamma} J_{\frac{d}{2}-1}(|b-a|\,r)\,dr$$
$$= 2(2\pi)^{-d/2}\int_0^{+\infty} r^{\frac{d}{2}-\gamma} J_{d/2-1}(r)\,dr \times |b-a|^{\gamma-d}$$
$$= +\infty \qquad d=1,\ \gamma \geqq 1$$
$$< +\infty \qquad d=1,\ \gamma<1 \ \text{ or } \ d\geqq 2,\ \gamma<2,$$

and this is just the kernel for the $(d-\gamma)$-dimensional potentials of M. RIESZ [1]; compare W. FELLER [13].

Given compact $B\subset R^d$ and $\gamma<1$ $(d=1)$ or $\gamma<2$ $(d\geqq 2)$, if $\mathfrak{m}_B$ is the hitting time $\inf(t: t>0,\ x(t)\in B)$, then $P_\bullet(\mathfrak{m}_B<+\infty)$ is the electrostatic potential, *i.e.*, it is the greatest of the potentials

$$7)\qquad p=\int_B \frac{e(db)}{|a-b|^{d-\gamma}},\ e\geqq 0,\ p\leqq 1;$$

the electrostatic distribution

$$8)\qquad e(db) = \lim_{\varepsilon\downarrow 0}\varepsilon^{-1}\,db\times P_b(\mathfrak{m}_B<+\infty,\ \mathfrak{m}_B(w_\varepsilon^+)=+\infty)$$

and the electrostatic $(d-\gamma)$-dimensional RIESZ capacity $C(B)=e(B)$ enjoy most but not all of the properties of the NEWTONian capacity $(\gamma=2)$; for instance, the electrostatic distribution of a solid sphere B spreads out over the whole interior and $C(\partial B)=0$ (see problem 2 for detailed information on this point).

WIENER's test for the singular points of the electrostatic potential takes the form

$$9)\qquad P_a(\mathfrak{m}_B=0)=0 \ \textit{or}\ 1 \ \textit{according as} \sum_{n\geqq 1} 2^{(d-\gamma)n}\,C(B_n) \gtreqless \infty,$$

where $a\in B$ and B_n is the meet of B and the spherical shell $2^{-n-1}\leqq|a-b|<2^{-n}$; the proof is the same as in the BROWNian case.

Problem 1. Given a 1-sided stable process with exponent $0<\alpha<1$ and rate $1/2$, prove that for compact $B\subset R^1$, $P_\bullet(\mathfrak{m}_B<+\infty)$ is the $(1-\alpha)$-dimensional electrostatic potential, *i.e.*, the greatest of the potentials

$$p(a)=\frac{2}{\Gamma(\alpha)}\int_{b\geqq a}\frac{e(db)}{(b-a)^{1-\alpha}}\leqq 1,\ e\geqq 0,\ e(R^1-B)=0.$$

Problem 2. Compute the $(1-\alpha)$-dimensional capacity

$$C_{1-\alpha}(B) = \max e(B): e\geqq 0,\ e(R^1-B)=0,\ \frac{2}{\Gamma(\alpha)}\int_{b\geqq a}\frac{e(db)}{(b-a)^{1-\alpha}}\leqq 1$$

of an interval $B = [a, b]$.

$\Big[e(db) = db/(1-b)^{\alpha}\, 2\Gamma(1-\alpha)$ is the solution of

$$\frac{2}{\Gamma(\alpha)} \int_a^1 \frac{e(db)}{(b-a)^{1-\alpha}} = 1 \quad (0 < \alpha < 1),$$

$C_{1-\alpha}[0, 1]$ is the total charge $e[0, 1] = 1/2\Gamma(2-\alpha)$, and

$$C_{1-\alpha}[a, b] = C_{1-\alpha}[0, l] = l^{1-\alpha} C_{1-\alpha}[0, 1] = \frac{l^{1-\alpha}}{2\Gamma(2-\alpha)} \quad l = b - a.\Big]$$

Problem 3. Given an *isotropic* d ($\geqq 2$)-dimensional stable process with exponent $0 < \alpha < 2$ and $E_0[e^{i\gamma x(t)}] = e^{-t|\gamma|^{\alpha}/2}$ ($\gamma \in R^d$), prove that for compact $B \subset R^d$, $P_\bullet(\mathfrak{m}_B < +\infty)$ is the $(d-\alpha)$-dimensional electrostatic potential, *i.e.*, the greatest potential

$$p(a) = \frac{\Gamma\left(\frac{d-\alpha}{2}\right)}{\pi^{d/2}\, 2^{\alpha-1}\, \Gamma\left(\frac{\alpha}{2}\right)} \int \frac{e(db)}{|b-a|^{d-\alpha}} \leqq 1, \; de \geqq 0, \; e(R^d - B) = 0.$$

Problem 4 after PÓLYA and SZEGÖ [*1*: 39]. Prove that

$$\int_0^1 \int_0^{\pi} \int_0^{2\pi} (1-r^2)^{-\frac{\alpha}{2}} (\xi^2 - 2r\,\xi\cos\psi + r^2)^{\frac{\alpha-3}{2}} r^2 \sin\psi \, dr\, d\psi\, d\theta$$

$$= 2\pi\, \Gamma\left(\frac{\alpha}{2}\right) \Gamma\left(\frac{2-\alpha}{2}\right) \qquad 0 \leqq \xi \leqq 1$$

and conclude that $(3-\alpha)$-dimensional capacity $C_{3-\alpha}(B)$ of the 3-dimensional ball $B: |b| \leqq 1$ is

$$\pi 2^{\alpha-3} / \Gamma\left(\frac{3-\alpha}{2}\right) \Gamma\left(\frac{5-\alpha}{2}\right).$$

Draw a picture of the electrostatic distribution

$$e_{3-\alpha}(db) = \frac{\sqrt{\pi}\, 2^{\alpha-2} (1 - |b|^2)^{-\alpha/2}\, db}{\Gamma\left(\frac{2-\alpha}{2}\right) \Gamma\left(\frac{3-\alpha}{2}\right)} \qquad |b| \leqq 1$$

for α near 0, $\alpha = 1$, and α near 2 and give a probabilistic interpretation of its changing shape.

8. A general view of diffusion in several dimensions

8.1. Similar diffusions

Given a (conservative) diffusion D on a space Q as described in *7.1*, its generator $\mathfrak{G}$ can be expressed in terms of the hitting probabilities and mean exit times

1a) $\quad h(a, db) = h_{\partial D}(a, db) = P_a[x(\mathfrak{m}_{\partial D}) \in db, \mathfrak{m}_{\partial D} < +\infty]$

1b) $\quad e(a) = e_D(a) = E_a(\mathfrak{m}_{\partial D})$

for open $D \subset Q$ via E. B. DYNKIN's formula

2) $\quad (\mathfrak{G} u)(a) = \lim_{D \downarrow a} e(a)^{-1} \left[\int_{\partial D} h(a, db)\, u(b) - u(a) \right] \quad u \in D(\mathfrak{G});$

to borrow a phrase of W. FELLER's, *h is the road map, i.e., it tells what routes the particle is permitted to travel, and e governs the speed.*

A diffusion is said to be *similar* to another if there is a homeomorphism between their state spaces mapping the road map of one onto the road map of the other.

Given a non-singular conservative diffusion on the open line $Q = R^1$, its hitting probabilities can be expressed as

4) $$P_\xi[\mathfrak{m}_a < \mathfrak{m}_b] = \frac{s(b) - s(\xi)}{s(b) - s(a)} \qquad a < \xi < b,$$

and two such diffusions are similar if and only if $s(Q)$ and $s^\bullet(Q)$ are similar under the euclidean group, *i.e.*, if and only if $s(Q)$ and $s^\bullet(Q)$ are both R^1, or both half-lines, or both bounded.

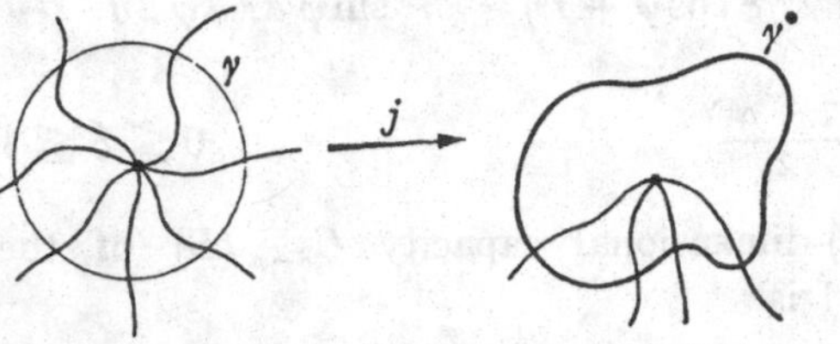

Diagram 1

Consider, as a second example, the standard 2-dimensional BROWNian motion and the space-time BROWNian motion.

Given a standard BROWNian path starting at the center of a circle γ and a homeomorphism j mapping its hitting probabilities into spacetime BROWNian hitting probabilities as in the diagram, each open arc of the image $\gamma^\bullet$ of γ under j would have positive space-time hitting probabilities. But this is absurd, *so the two motions are dissimilar*; as a matter of fact, in the plane, there is an infinite number of dissimilar diffusions.

Using the language of road maps and speeds, the major problems of higher-dimensional diffusion can be stated as follows:

Q 1) to express the basic dissimilar road maps in geometrical terms (see 8.6 for the meaning of *geometrical*),

Q 2) to single out from the diffusions with the same road map some especially simple one (*Brownian motion*) and to show how to construct its sample paths (the meaning of *simple* will depend upon the geometrical picture of the road map),

Q 3) to establish between two diffusions with the same road map a change of speed (time substitution) sending one into the other (see 8.3 for an expression for this speed change).

Q 3) is solved (see 8.3); the others are not, but a number of special cases are tractable; in the pages before us, the reader will find an informal account of the present state of the art and some rough ideas about what should be done.

8.2. $\mathfrak{G}$ as differential operator

Given a diffusion on $Q = R^d$ $(d \geqq 1)$ with standard BROWNian hitting probabilities, the mean exit times $e_D = E_\bullet(\mathfrak{m}_{\partial D})$ are the potentials $\int_D G\,dm$ of a (positive) *speed measure* m, independent of D (G is the classical GREEN function of D). Each $u \in D(\mathfrak{G})$ can be expressed in the small as the potential $\int G(a, b)\, e^u(db)$ of a signed measure $e^u(db)$ plus a harmonic function, and

1) $$\mathfrak{G}\, u = -\frac{e^u(db)}{m(db)}$$

(see 7.19).

T. UENO [*1*] discovered another case in which $\mathfrak{G}$ can be expressed as a differential operator $-e^u(db)/m(db)$.

Consider a persistent diffusion, *i.e.*, suppose that

2a) $$P_\bullet(\mathfrak{m}_D < +\infty) \equiv 1 \qquad D \neq \varnothing \textit{ open},$$

and add the technical conditions

2b) $$\int_{\partial D} h(\cdot, db)\, f(b) \in C(Q - \partial D) \qquad f \in B(\partial D)$$

2c) $$P_\bullet(\mathfrak{m}_{\partial D_1} < \mathfrak{m}_{\partial D_2}) > 0 \quad a \notin \bar{D}_1 \cup \bar{D}_2,\ \bar{D}_1 \cap \bar{D}_2 = \varnothing,\ Q - \bar{D}_1 \cup \bar{D}_2 \textit{ connected}.$$

Given D_1 and D_2 as in 2c) with compact boundaries ∂D_1 and ∂D_2, if

3) $$\begin{aligned} &\mathfrak{m}_1 = \mathfrak{m}_{\partial D_1} \qquad \mathfrak{m}_2 = \min(t: x(t) \in \partial D_2; t > \mathfrak{m}_1) \\ &\mathfrak{m}_3 = \min(t: x(t) \in \partial D_1, t > \mathfrak{m}_2) \\ &\mathfrak{m}_4 = \min(t: x(t) \in \partial D_2, t > \mathfrak{m}_3) \\ &\textit{etc.}, \end{aligned}$$

then

$$P_\bullet[\mathfrak{m}_1 < \mathfrak{m}_2 < \mathfrak{m}_3 < \textit{etc.} \uparrow +\infty] \equiv 1,$$

and the hitting places

$$x(\mathfrak{m}_{2n-1})\ (n \geqq 1) \quad \textit{and} \quad x(\mathfrak{m}_{2n})\ (n \geqq 1)$$

on ∂D_1 and ∂D_2 constitute ergodic MARKOV chains with one-step transition probabilities

4a) $$\int_{\partial D_2} h_{\partial D_2}(a, d\xi)\, h_{\partial D_1}(\xi, db) = p_1(a, db)$$

4b) $$\int_{\partial D_1} h_{\partial D_1}(a, d\xi)\, h_{\partial D_2}(\xi, db) = p_2(a, db).$$

A condition of K. YOSIDA is satisfied, permitting the introduction of stable distributions e_1 and e_2:

5a) $$\int_{\partial D_1} e_1(da)\, p_1(a, db) = e_1(db) \qquad e_1(\partial D_1) = 1$$

5b) $$\int_{\partial D_2} e_2(da)\, p_2(a, db) = e_2(db) \qquad e_2(\partial D_2) = 1,$$

and it turns out that

6) $$m(db) = \int_{\partial D_1} e_1(da)\, E_a[\text{measure}(t : x(t) \in db, t < \mathfrak{m}_{\partial D_2})] + \int_{\partial D_2} e_2(da)\, E_a[\text{measure}(t : x(t) \in db, t < \mathfrak{m}_{\partial D_1})]$$

is positive on opens, finite on compacts, and *stable*:

7) $$\int_Q m(da)\, P_a[x(t) \in db] = m(db) \qquad t > 0;$$

moreover, a second such stable measure is a constant multiple of m (see also G. MARUYAMA and H. TANAKA [2] and HAS'MINSKIĬ [1] for 6) and 7), and E. NELSON [1] for a different approach to stable measures).

Given $D \subset Q$ with mean exit time $E_\bullet(\mathfrak{m}_{\partial D}) < +\infty$, UENO proves that the GREEN measure

8) $$G(a, db) = E_a[\text{measure}(t : x(t) \in db, t < \mathfrak{m}_{\partial D})]$$

factors on $D \times D$ into the product of a *Green function* $G(a, b)$ and the stable measure $m(db)$, and expresses $\mathfrak{G}\, u$ as $-e^u(db)/m(db)$, using

9) $$\begin{aligned}\int G(a, b)\, e^u(db) &= -\int G(a, b)\, \mathfrak{G}\, u\, m(db) \\ &= -E_a\left[\int_0^{\mathfrak{m}_{\partial D}} (\mathfrak{G}\, u)\, (x_t)\, dt\right] \\ &= u(a) - \int_{\partial D} h(a, db)\, u(b).\end{aligned}$$

$G(a, b)$ should depend on the road map alone, but this is not proved in general; it need not be symmetric (see 8.4 for an account of the potentials of such unsymmetrical GREEN functions).

Bearing UENO's result in mind, it is plausible that each diffusion should have a positive (speed) measure $m(db)$ and GREEN functions $G(a, b)$ depending upon its road map alone such that $\mathfrak{G}u = -e^u(db)/m(db)$. $e^u(db)$ will depend upon $u \in D(\mathfrak{G})$ and the road map alone; see 8.5 for a probabilistic formula for it.

8.3. Time substitutions

Given two diffusions with the same road map, common GREEN functions, sample paths x and $x^\bullet$, and generators

1 a) $$\mathfrak{G}u = -\frac{e^u(db)}{m(db)} \qquad u \in D(\mathfrak{G})$$

1 b) $$\mathfrak{G}^\bullet u = -\frac{e^u(db)}{m^\bullet(db)} \qquad u \in D(\mathfrak{G}^\bullet),$$

it is evident that there should be a time substitution $t \to \mathfrak{f}^{-1}$ mapping $x \to x^\bullet$, as conjectured in 8.1; in fact, if $m^\bullet(db) = f(b)\, m(db)$ with $0 < f \in C(Q)$ and if $t \to \mathfrak{f}^{-1}$ is the classical time change with $\mathfrak{f}(t) = \int_0^t f(x_s)\, ds$, then it is a simple matter to show that $x(\mathfrak{f}^{-1})$ is identical in law to $x^\bullet$.

But if $m^\bullet$ is singular to m, the above prescription loses its sense, and the HELLINGER integral

2) $$\mathfrak{f}(t) = \int_Q \frac{\text{measure}(s : x(s) \in db, s \leqq t)\, m^\bullet(db)}{m(db)},$$

which appears hopeful at first glance, turns out to be intractable, though it seems that it should converge.

B. VOLKONSKIĬ [1] in the 1-dimensional case and H. P. MCKEAN, JR. and H. TANAKA [1] in the case of BROWNian hits in $d \geqq 1$ dimensions got over the obstacle by approximating $t \to \mathfrak{f}^{-1}$ by suitable classical time substitutions, and R. BLUMENTHAL, R. GETOOR, and H. P. MCKEAN, JR. [1] used the same method for a wide class of motions including the general diffusions..

MCKEAN and TANAKA noted that, in the case of BROWNian hits, the mean hitting times $e^\bullet$ are excessive in the sense of G. HUNT:

3) $$E_a[e_D^\bullet(x_t), t < \mathfrak{m}_{\partial D}] \uparrow e_D^\bullet(a) \qquad t \downarrow 0,\ a \in D$$

and defined $\mathfrak{f}$ for $t < \mathfrak{m}_{\partial D}$ as the limit of

4) $$\mathfrak{f}_\varepsilon(t) = \int_0^t \varepsilon^{-1}[e_D^\bullet(x(s)) - E_{x(s)}(e_D^\bullet(x_\varepsilon), \varepsilon < \mathfrak{m}_{\partial D})]\, ds$$

for suitable $\varepsilon \downarrow 0$, the idea being that since

5) $$\mathfrak{G}_\varepsilon u = \varepsilon^{-1}[E_\bullet(u(x_\varepsilon), \varepsilon < \mathfrak{m}_{\partial D}) - u] \qquad u \in D(\mathfrak{G})$$

converges on D as $\varepsilon \downarrow 0$ to $\mathfrak{G}\, u$ and since $\mathfrak{G}^\bullet e_D^\bullet = -1$, it should be that

$$6)\qquad \lim_{\varepsilon \downarrow 0} \mathfrak{G}_\varepsilon e_D^\bullet\, dm = \lim_{\varepsilon\downarrow 0} \mathfrak{G}_\varepsilon^\bullet e_D\, dm^\bullet = -dm^\bullet$$

and

$$7)\qquad \mathfrak{f}_\varepsilon(t) = -\int_Q \frac{\text{measure}\,(s : x(s) \in db,\, s \leq t)\, \mathfrak{G}_\varepsilon e_D^\bullet\, m(db)}{m(db)}$$

should converge to the HELLINGER integral 2). M. G. ŠUR [1] has put this idea into an elegant and useful form.

Because the 1-dimensional BROWNian motion has local times, all positive measures are speed measures, but in $d \geqq 2$ dimensions the situation is not so simple; for example, in the case of standard BROWNian hits, a speed measure has to have continuous potentials and its null sets have to be *thin* for the BROWNian motion in the sense that if $Z_1 \supset Z_2 \supset$ *etc.* are open and if $m\Big(\bigcap_{n\geqq 1} Z_n\Big) = 0$, then

$$P_a[\mathfrak{m}_{\partial Z_n} \downarrow 0,\, n \uparrow +\infty] = 1 \qquad a \in \bigcap_{n \geqq 1} Z_n$$

(see H. P. MCKEAN, JR. and H. TANAKA [1]).

8.4. Potentials

A few of the deeper properties of potentials due to G. HUNT [2 (3)] will now be explained; it will be seen that the standard BROWNian motion was too simple an example because the symmetry of its transition densities obscured the significance of the backward motion described below.

Given a *transient* diffusion on a non-compact space Q with a positive stable measure e and continuous transition densities $p(t, a, b)$ such that

$$1\text{a})\qquad \int_Q e(da)\, p(t, a, b) \equiv 1,^*$$

$$1\text{b})\qquad \int_Q p(t-s, a, \xi)\, p(s, \xi, b)\, e(d\xi) \equiv p(t, a, b),$$

$$1\text{c})\qquad G(a, b) \equiv \int_0^{+\infty} p(t, a, b)\, dt < +\infty \text{ off the diagonal of } Q \times Q$$

as in the 3-dimensional BROWNian case, the backward kernel $p^*(t, a, b) \equiv p(t, b, a)$ satisfies

$$2\text{a})\qquad \int_Q p^*(t, a, b)\, e(db) \equiv 1$$

$$2\text{b})\qquad \int_Q p^*(t-s, a, \xi)\, p^*(s, \xi, b)\, e(d\xi) \equiv p^*(t, a, b),$$

---

* 1a) is almost automatic because e is stable.

i.e., the $p^*(t, a, b)$ can be considered as the transition densities of a conservative *backward* motion starting afresh at each instant $t \geq 0$; *backward* because if (positive) infinite probabilities are permitted, then the *stationary backward motion* with probabilities

3a) $$P^*[x(t_1) \in db_1, x(t_2) \in db_2, \ldots, x(t_n) \in db_n] \\ = e(db_1)\, p^*(t_2 - t_1, b_1, b_2)\, e(db_2) \ldots p^*(t_n - t_{n-1}, b_{n-1}, b_n)\, e(db_n) \\ -\infty < t_1 < t_2 < \cdots < t_n$$

is nothing but the *stationary forward motion* with probabilities

3b) $$P[x(t_1) \in db_1, x(t_2) \in db_2, \ldots, x(t_n) \in db_n] \\ = e(db_1)\, p(t_2 - t_1, b_1, b_2)\, e(db_2) \ldots p(t_n - t_{n-1}, b_{n-1}, b_n)\, e(db_n) \\ -\infty < t_1 < t_2 < \cdots < t_n$$

run backward in time.

Assuming that the backward motion is a diffusion, it is easy to see that its generator $\mathfrak{G}^*$ is dual to the generator $\mathfrak{G}$ of the forward motion; in detail, if $u \in D(\mathfrak{G})$ and $u^* \in D(\mathfrak{G}^*)$ vanish far out, then

4) $$\int u^* \mathfrak{G} u\, de \\ = \lim_{\varepsilon \downarrow 0} \int_Q u^*(a)\, e(da)\, \varepsilon^{-1} \left[\int_Q p(t, a, b)\, u(b)\, e(db) - u(a) \right] \\ = \lim_{\varepsilon \downarrow 0} \int_Q u(b)\, e(db)\, \varepsilon^{-1} \left[\int_Q p^*(t, b, a)\, u^*(a)\, e(da) - u^*(b) \right] \\ = \int_Q u\, \mathfrak{G}^* u^*\, de$$

A non-negative measurable function, u, is said to be *excessive for the forward motion* if $\int_Q p(t, a, b)\, u\, e(db) \uparrow u$ as $t \downarrow 0$.

Given such an excessive function u, the measure $e^*(db) = u\, e(db)$ is *excessive* for the backward motion, *i.e.*, $\int_Q e^*(da)\, p^*(t, a, b)\, e(db) \uparrow e^*(db)$ as $t \downarrow 0$. Besides, if $e^* \geq 0$ is excessive for the backward motion, then $u = \lim_{t \downarrow 0} \int_Q e^*(da)\, p^*(t, a, b)$ is an excessive function for the forward motion and $e^*(db) = u\, e(db)$; in brief, there is a perfect correspondence between excessive functions for the forward motion and excessive measures for the backward motion. HUNT exploits this fact to prove that an excessive function u for the forward motion can be expressed as the *forward potential* $\int_Q G(a, b)\, e^u(db)$ of a non-negative BOREL measure e^u plus $u_\infty = \lim_{t \uparrow +\infty} \int_Q p\, u\, de$ (see 8.5 for additional note on e^u).

Given a bounded closed or open subset A of Q, the forward and backward hitting probabilities

5a) $$p_A = P_\bullet[\mathfrak{m}_A < +\infty]$$

5b) $$p_A^* = P_\bullet^*[\mathfrak{m}_A < +\infty]$$

are excessive; thus, p_A is a forward potential

$$\int G(a, b)\, e_A(db)$$

plus $\lim_{t\uparrow+\infty} P_\bullet(\mathfrak{m}_A(w_t^+) < +\infty)$ (= 0 because of transience), and p_A^* is a backward potential

$$\int G^*(a, b)\, e_A^*(db)^*$$

plus $\lim_{t\uparrow+\infty} P_\bullet^*(\mathfrak{m}_A(w_t^+) < +\infty)$ (= 0).

G. Hunt [2 (3): 175] discovered the pretty fact that $e_A(Q) = e_A^*(Q)$ and called this common total charge the *natural capacity of A*.

Here is a simple proof.

Given compact A contained in bounded open B and bounded open $D \supset \overline{B}$,

6) $$\begin{aligned} e_A(Q) &= \int_Q p_B^*(b)\, e_A(db) \\ &= \int_Q \left[\int_Q e_B^*(da)\, G(a, b)\right] e_A(db) \\ &= \int_Q e_B^*(da)\, p_A(a) \\ &\leq \int_Q e_B^*(da)\, p_D(a) \\ &= \int_Q p_B^*(b)\, e_D(db) \\ &\downarrow \int_Q p_A^*(b)\, e_D(db) \qquad B \downarrow A \\ &= \int_Q e_A^*(da)\, p_D(a) \\ &= e_A^*(Q), \end{aligned}$$

where $p_B^* \downarrow p_A^*$ on ∂D and $e_D(Q - \partial D) = 0$ were used in the next to last step. Because $e_A^*(Q) \leq e_A(Q)$ for similar reasons, $e_A(Q) = e_A^*(Q)$, and, letting $A \uparrow B$ and using

7) $$\begin{aligned} e_A(Q) &= \int_Q e_D^*(da)\, p_A(a) \\ &\uparrow \int_Q e_D^*(da)\, p_B(a) \qquad A \uparrow B \\ &= e_B(Q) \end{aligned}$$

---

* $G^*(a, b) = G(b, a)$.

and its counterpart for $e_A^*(Q)$, it is found that $e_B(Q) = e_B^*(Q)$ for open B also.

$e_A(Q) = e_A^*(Q)$ is an alternating function of A in the sense of G. CHOQUET [see 7.8.13)].

The reader will find additional information on potentials in E. B. DYNKIN [7], G. HUNT [2 (1, 2, 3)], BEURLING and DENY [1], F. SPITZER [2], and M. G. ŠUR [1].

8.5. Boundaries

Consider the standard BROWNian motion on the open disc $E^2: |b| < 1$ stopped on ∂E^2, with transition probabilities

1 a) $g^\bullet(t, a, b)\, 2db = P_a[x(t) \in db, t < \mathfrak{m}_{\partial E^2}]$

$$(t, a, b) \in (0, +\infty) \times E^2 \times E^2,$$

GREEN function

1 b) $$G(a, b) = \int_0^{+\infty} g^\bullet(t, a, b)\, dt = \frac{1}{2\pi} \lg \left| \frac{1 - a b}{b - a} \right| \qquad (a, b) \in E^2 \times E^2,$$

and hitting distribution (POISSON kernel)

1 c) $$h(a, e^{i\theta})\, d\theta = P_a[x(\mathfrak{m}_{\partial E^2}) \in d\theta]$$
$$= \lim_{r \uparrow 1} (1 - r)^{-1} G(a, r e^{i\theta})\, d\theta$$
$$= \frac{1}{2\pi} \frac{1 - |a|^2}{|a - e^{i\theta}|^2}\, d\theta \qquad |a| < 1, \quad 0 \leq \theta \leq 2\pi.$$

Given $0 \leq \theta \leq 2\pi$, $u_\theta(a) = h(a, e^{i\theta})$ is a *minimal* harmonic function, *i.e.*,

2a) $$0 \leq u \leq u_\theta$$

2b) $$u \text{ harmonic in } E^2$$

implies

3) $$u \text{ is a constant multiple of } u_\theta.$$

On the other hand, if $0 \leq u$ is a minimal harmonic function, then it can be expressed as $\int_{\partial E^2} u_\theta\, e(d\theta)$ $(e \geq 0)$, and e has to be concentrated at a single point $\theta \in \partial E^2$, *i.e.*, u is a constant multiple of u_θ; in brief, the map $\theta \to u_\theta$ establishes a 1 : 1 correspondence between the points of ∂E^2 and minimal harmonic functions.

∂E^2 is now seen to have two properties:

4a) *it is not too large, i.e., the sample path tends as $t \uparrow \mathfrak{m}_{\partial E^2}$ to a single point of ∂E^2;*

4b) *it is not too small, i.e., if* $0 \leqq u$ *is harmonic on* E^2 *then*

$$u(a) = \int_{\partial E^2} h(a, e^{i\theta})\, e(d\theta) \qquad |a| < 1, e \geqq 0.$$

Given a BROWNian motion on a GREENian domain $Q \subset R^d$ stopped on ∂Q, we cannot expect the *geometrical boundary* ∂Q to be large enough for the POISSON representation 4b) as is clear from the simple example of the slit disc

$$Q: 0 < r < 1, \qquad 0 < \theta < 2\pi,$$

but as R. S. MARTIN [1] discovered, 4a) and 4b) can be salvaged if ∂Q is properly ramified.

Consider for this purpose, the classical GREEN function $G(a, b)$ of Q.

Given $b \in Q$, if $\partial B \subset Q$ is a little spherical surface with center at b excluding a second point $a \in Q$, then for c near ∂Q,

5)
$$\begin{aligned} G(a, c) &\geqq E_a\left[\int_{\mathfrak{m}_{\partial B}}^{+\infty} g^\bullet(t, x(\mathfrak{m}_{\partial B}), c)\, dt\right] \\ &= E_a[\mathfrak{m}_{\partial B} < +\infty, G(x(\mathfrak{m}_{\partial B}), c)] \\ &\geqq \textit{a constant multiple of} \int_{\partial B} G(\bullet, c)\, do \\ &= \textit{a constant multiple of } G(b, c), \end{aligned}$$

i.e.,

6)
$$u_{ab} = \frac{G(b, \bullet)}{G(a, \bullet)} \quad \textit{is bounded near} \quad \partial Q,$$

and using 6), the MARTIN boundary Q° of Q is defined as the space of ideal points added to Q in the smallest compactification $\overline{Q} = Q + Q^\circ$ making u_{ab} continuous on $\overline{Q} - (a \cup b)$ for each choice of a and $b \in Q$.

Given $\theta \in Q^\circ$ and fixing $a \in Q$,

7)
$$h(b, \theta) = \lim_{c \to \theta} \frac{G(b, c)}{G(a, c)}$$

is harmonic, but perhaps not minimal; however, the map $\theta \to h(\bullet, \theta)$ establishes 1 : 1 correspondence between a BOREL subset $Q^\bullet$ of Q° and the minimal harmonic functions u with $u(a) = 1$, and 4b) holds, *i.e.*, if $0 < u$ is harmonic, then $u(b) = \int_{Q^\bullet} h(b, \theta)\, e^u(d\theta)$, and $e^u (\geqq 0)$ is unique.

$Q^\bullet$ is therefore *not too small* and as DOOB [6] proved, *it is not too large, i.e.*, almost all sample paths converge as $t \uparrow \mathfrak{m}_{\partial Q}$ to a *single point* of $Q^\bullet$ (see also G. HUNT [3]).

DOOB also found a pretty expression for e^u:

8)
$$e^u(d\theta) = \frac{u(b)\, P_b^u[x(\mathfrak{m}_{\partial Q} -) \in d\theta]}{h(b, \theta)}$$

where $P^u_\bullet(B)$ is the probability of the event B for the u-motion with transition probabilities

9) $$P^u_a[x(t) \in db] = \frac{P_a[x(t) \in db, t < \mathfrak{m}_{\partial Q}]}{u(a)} u(b);$$

esp., taking $u \equiv 1$,

10) $$h(b, \theta) = \frac{P_b[x(\mathfrak{m}_{\partial Q} -) \in d\theta]}{e^1(d\theta)},$$

which is the analogue of the classical expression of the POISSON kernel as a hitting density, just as 7) is the analogue of its expression as normal derivative of GREEN function [see 1 c)].

DOOB's expression of e^u in terms of the u-motion has applications to potentials as well; in detail, if $0 < u$ is excessive, *i.e.*, if $2\int g^\bullet(t, a, b)\, u(b)\, db \uparrow u$ as $t \downarrow 0$ and if u is not pure harmonic, then the u-motion will sometimes be killed at a time $\mathfrak{m}_\infty < \mathfrak{m}_{\partial Q}$, and the RIESZ measure

11) $$e^u(db) = \lim_{\varepsilon \downarrow 0} \varepsilon^{-1}[u(b) - E_b(u(x_\varepsilon), \varepsilon < \mathfrak{m}_{\partial Q})] \times 2db$$

used in the expression of u as a potential $\int_Q G(a, b)\, e^u(db)$ plus a non-negative harmonic function (see 7.5) will be

12) $$e^u(db) = \frac{u(a)\, P^u_\bullet[x(\mathfrak{m}_\infty -) \in db, \mathfrak{m}_\infty < \mathfrak{m}_{\partial Q}]}{G(a, b)} \quad \textit{inside } Q.$$

$Q^\bullet = Q^\circ$ is a single point if $Q = R^d$ $(d \geq 3)$ (see 8.7 for additional information on 1-point boundaries).

MARTIN boundaries have also been treated in connection with MARKOV chains (see DOOB [*7*], DOOB, SNELL, and WILLIAMSON [*1*], HUNT [*3*], T. WATANABE [*1*], and for a different boundary, W. FELLER [*8*], and there is no doubt that similar boundaries can be attached to any decent MARKOV process (see BISHOP and DE LEEUW [*1*] and D. RAY [*3*]).

Given a conservative diffusion on a space Q with a nice boundary $Q^\bullet$ (neither too large nor too small), run the sample paths with so slow a clock that $x(t)$ tends to $Q^\bullet$ as $t \uparrow \mathfrak{m} < +\infty$, let $\mathfrak{D}$ be the corresponding *local* generator inside Q, and pose the problem:

to find all conservative generators $\mathfrak{G} \subset \mathfrak{D}$, i.e., to find all boundary conditions that can be imposed upon $\mathfrak{D}$.

Consider, as an illustration, the standard BROWNian local generator $\mathfrak{D}$* the open unit disc $Q = E^2$. $Q^\bullet$ is the unit circle ∂E^2, and it is probable that the most general conservative diffusion operator $\mathfrak{G} \subset \mathfrak{D}$ can be expressed in terms of a conservative 1-dimensional differential operator

---

* $\mathfrak{D} u = \Delta u/2$ if $u \in C^2(E^2)$.

$\mathfrak{D}^\bullet$ on ∂E^2 and a BOREL function $0 \leqq p(\theta) \leqq 1$: *to wit*, it is conjectured that $\mathfrak{G}$ coincides with $\mathfrak{D}$ applied to the class $D(\mathfrak{G})$ of functions $u \in C(\overline{E^2})$ such that $\mathfrak{D}\,u$ has a continuous extension to $\overline{E^2}$ and

13) $$p(\theta)\frac{\partial u}{\partial r}(1-,\theta) + [1-p(\theta)](\mathfrak{D}\,u)(1-,\theta) = (\mathfrak{D}^\bullet u^\bullet)(\theta)$$
$$0 \leqq \theta < 2\pi, \quad u^\bullet = u(1,\theta).$$

Given a reflecting BROWNian motion x in the disc $\left[\frac{\partial u}{\partial r}(1-,\bullet) = 0\right]$, if $\mathfrak{t}(t)$ is local time at $r = 1$ for its radial (BESSEL) part and if ψ is an independent circular BROWNian, then the motions corresponding to

14a) $$\frac{\partial u}{\partial r}(1-,\theta) = \frac{1}{2}\frac{\partial^2 u}{\partial \theta^2}(1,\theta)$$

14b) $$\frac{\partial u}{\partial r}(1-,\theta) = \frac{\partial u}{\partial \theta}(1,\theta)$$

will be

15a) $$x\,e^{i\psi(\mathfrak{t})}$$

15b) $$x\,e^{i\mathfrak{t}}$$

[see problem 7.15.4]. 15b) should be considered as a reflecting BROWNian motion because 14b) can be expressed as $\frac{\partial u}{\partial o} = 0$, where $\frac{\partial}{\partial o}$ means differentiation along the oblique line pointing at $-45°$ to the outward-pointing normal.

As to the original problem, similar boundary conditions $p\frac{\partial u}{\partial n} + (1-p)\,\mathfrak{D}\,u = \mathfrak{D}^\bullet u^\bullet$ should occur, and the corresponding motions should be expressible in terms of a reflecting barrier motion $\left(\frac{\partial u}{\partial n} = 0\right)$, a distribution $\mathfrak{t}(t, d\theta)$ of (reflecting barrier) local time on $Q^\bullet$, and a boundary motion associated with $\mathfrak{D}^\bullet$; see A. VENTCEL [*1*], T. UENO [*2*], N. IKEDA [*1*], K. SATO [*1*], and K. SATO and H. TANAKA [*1*] for some information about this. A nice problem is to explain the meaning of reflecting BROWNian motion on the (MARTIN) closure of a GREENian domain $D \subset R^3$ (see J. DOOB [*8*] for information on this point).

8.6. Elliptic operators

Consider a nice differentiable manifold Q and, on it, an elliptic differential operator $\mathfrak{D}$ expressed in local coordinates as

1) $$\mathfrak{D} = \frac{1}{2}\sum_{i,j \leqq d} e_{ij}\frac{\partial^2}{\partial x_i \partial x_j} + \sum_{j \leqq d} f_j \frac{\partial}{\partial x_j}.$$

$\mathfrak{D}$ generates a diffusion if it has HÖLDER continuous coefficients and if $e = [e_{ij}]$ is symmetric and positive definite at each point of Q; to be precise, $\mathfrak{Q}$ agrees on $C^2(Q)$ with the generator of a diffusion which is permitted to run out to the point ∞ before time $+\infty$ if Q is non-compact (see POGORZELSKI [*1*]). K. ITÔ [*3*] proved that the desired motion satisfies the stochastic differential equation

2) $$dx = \sqrt{e(x)}\, db + f(x)\, dt$$

up to the time it leaves the neighborhood on which 1) holds. $\sqrt{e}$ is the symmetric positive definite root of e, $f = (f_1, f_2, \ldots, f_d)$, $b(t)$ is a standard d-dimensional BROWNian motion, and it is supposed that the HÖLDER exponent of the coefficients $= 1$.

Under a change of local coordinates $x \to x^\bullet$, e^{-1} changes according to the rule:

3) $$e^{\bullet -1}_{ij} = \sum_{k,l \le d} e^{-1}_{kl} \frac{\partial x_k}{\partial x_i^\bullet} \frac{\partial x_l}{\partial x_j^\bullet},$$

so

4) $$\sum_{i,j \le d} e^{-1}_{ij}\, dx_i\, dx_j$$

induces a RIEMANNian distance on Q, and $\mathfrak{Q}$ can be expressed as the LAPLACE-BELTRAMI operator

5) $$\frac{1}{2}\Delta = \frac{1}{2} \sum_{i,j \le d} \sqrt{|e|}\, \frac{\partial}{\partial x_i} \frac{e_{ij}}{\sqrt{|e|}} \frac{\partial}{\partial x_j}$$

plus a first order differential operator (*vector field*) f corresponding to a *drift* (see, for example, E. NELSON [*1*]). 2) now simplifies to $dx = db + f(x)\, dt$, b being the (BROWNian) motion associated with $(1/2)\,\Delta$.

Because the differential operator $\mathfrak{G}\, u = -e^u(db)/m(db)$ of 8.2 shares with elliptic operators of degree ≤ 2 the property that $\mathfrak{G}\, u \le 0$ at a local minimum of $u \in D(\mathfrak{G})$, it is to be hoped that some feature of the above geometrical picture can be salvaged in a general setting.

Besides the papers cited above, the reader can consult HAS'MINSKIĬ [*1*], NELSON [*2*], YOSIDA [*3*], and 8.7 for additional information about elliptic operators.

Problem 1. Prove that if a differential operator $\mathfrak{Q}$ with continuous coefficients has the property that $\mathfrak{Q}\, u \ge 0$ at a local minimum of $u \in C^\infty(R^d)$, then it is elliptic of degree ≤ 2 with no part of degree 0.

8.7. Feller's little boundary and tail algebras

Given a diffusion on a space Q such that

1) $$\int h_{\partial D}(a, db)\, f(b) \in C(D) \qquad f \in B(\partial D),$$

let $\mathsf{B}^\bullet$ be the *tail algebra* $\bigcap_{n \ge 1} \mathsf{B}[x(t) : t \ge n]$ modulo the ideal of negligible events.

Given $B \in \mathsf{B}^\bullet$, $p = P_\bullet(B)$ satisfies

2) $$p(a) = P_a(w_{\mathfrak{m}}^+ \in B) = \int h_{\partial D}(a, db)\, p(b) \qquad a \in D,$$

i.e., $p \in C(Q)$, and because of

3) $$p = \int e^{-t} P_\bullet(w_t^+ \in B)\, dt = G_1 p,$$

p is a solution of

4a) $$0 \leqq p \leqq 1$$

4b) $$p \in D(\mathfrak{G})$$

4c) $$\mathfrak{G} p = 0.$$

On the other hand, if p is a solution of 4), then

5) $$\begin{aligned} E_\bullet[p(x_t) \mid \mathsf{B}_s] &= E_{x(s)}[p(x_{t-s})] \\ &= p(x_s) + E_{x(s)}\left[\int_0^{t-s} (\mathfrak{G} p)(x_\theta)\, d\theta\right] \\ &= p(x_s) \qquad t \geqq s, \end{aligned}$$

i.e., $p(x_t)$ $(t \geqq 0)$ is a bounded martingale; as such, it converges to a $\mathsf{B}^\bullet$ measurable limit p_∞ as $t \uparrow +\infty$, and one gets

6a) $$p(x_t) = E_\bullet(p_\infty \mid \mathsf{B}_t) = E_{x(t)}(p_\infty) \qquad t \geqq 0;$$

in particular, if $p = P_\bullet(B)$ $(B \in \mathsf{B}^\bullet)$, then

6b) $$\begin{aligned} \lim_{t \uparrow +\infty} p(x_t) &= \lim_{t \uparrow +\infty} P_\bullet(B \mid \mathsf{B}_t) = P_\bullet(B \mid \mathsf{B}) \\ &= \textit{the indicator of } B. \end{aligned}$$

G. Hunt pointed out to us that 6) expresses a 1 : 1 correspondence between the events $B \in \mathsf{B}^\bullet$ and the extreme points of the (compact convex) class of solutions of 4).

W. Feller [*8*] had used this correspondence to express $\mathsf{B}^\bullet$ as an algebra of subsets of a topological space $Q^\bullet$ attached to Q as its boundary. $Q^\bullet$ is what we are calling Feller's little boundary (see J. Feldman [*1*] for its relation to the Martin boundary).

Given a non-constant solution p of 4), it is possible to choose $a < b$ and open A and $B \subset Q$ such that $p < a$ on A and $p > b$ on B, and if the motion is persistent (*i.e.*, if $P_\bullet[\mathfrak{m}_D < +\infty] \equiv 1$ for each open $D \subset Q$), then the sample path enters both A and B *i.o.* as $t \uparrow +\infty$, and the bounded martingale $p(x_t)$ $(t \geqq 0)$ cannot converge because $p(x_t) < a$ and $p(x_t) > b$ *i.o.*; in brief, p must be constant and the 01 law

7) $$P_\bullet(B) = 0 \textit{ or } 1 \qquad B \in \mathsf{B}^\bullet$$

must hold.

Now let us suppose that $Q = R^d$ and that our generator $\mathfrak{G}$ is an elliptic operator:

8)
$$\mathfrak{D} = \sum_{i,j \leqq d} e_{ij} \frac{\partial^2}{\partial x_i \partial x_j}$$

with continuous coefficients and let us pose the problem: *to find conditions on the quadratic form* $(e\,\xi, \xi) = e_{11}\xi_1^2 +$ *etc. such that the* 01 *law* 7) *holds, i.e., such that all the solutions of* 4) *are constant*; in more detail, let γ_+ and γ_- be the greatest and smallest eigen-values of e at a point and let us ask *what is the maximal rate of growth of* γ_+/γ_- *such that*

9a) *the motion is persistent,*

9b) *all the solutions of* 4) *are constant,*

9c) *all the non-negative solutions of* $\mathfrak{D}\, p = 0$ *are constant?*

It is known that 9c) holds if γ_+/γ_- is bounded. There is also the extraordinary result of S. Bernstein [*1*] that 9b) holds if $d = 2$ and if $\gamma_- > 0$ at each point. Bernstein also gave the simple example

$$\mathfrak{D} = (2 + 4b^2)\frac{\partial^2}{\partial a^2} + 4b\frac{\partial^2}{\partial a\,\partial b} + \frac{\partial^2}{\partial b^2},$$

in which 9c) fails ($p = e^{a - b^2}$) and $\gamma_+/\gamma_- \sim 4b^4$ as $b \uparrow +\infty$; see E. Hopf [*1*] for other examples connected with Bernstein's result, and for more up to date information, J. Moser [*1*].

Bibliography

AM = Ann. of Math.; BAMS = Bull. Amer. Math. Scc.; IJM = Illinois J. Math.; PNAS = Proc. Nat. Acad. Sci. U.S.A.; TAMS = Trans. Amer. Math. Soc.; TV = Teor. Veroyatnost. i ee Primenen.

ALEKSANDROV, P., and H. HOPF: [1] Topologie, 1. Berlin 1935.

BACHELIER, L.: [1] Théorie de la spéculation. Ann. Sci. École Norm. Sup. **17**, 21—86 (1900).

BERGMAN, S., and M. SCHIFFER: [1] Kernel functions and elliptic differential equations in mathematical physics. New York 1953.

BERNSTEIN, S.: [1] Über ein geometrisches Theorem und seine Anwendung auf die partiellen Differentialgleichungen vom elliptischen Typus. Math. Z. **26**, 551—558 (1927).

BESICOVITCH, A. S.: [1] On linear sets of points of fractional dimension. Math. Ann. **101**, 161—198 (1929).

— [2] On existence of subsets of finite measure of sets of infinite measure. Indagationes Math. **14**, 339—344 (1952).

BESICOVITCH, A. S., and S. J. TAYLOR: [1] On the complementary intervals of a linear closed set of zero Lebesgue measure. J. London Math. Soc. **29**, 449—459 (1954).

BEURLING, A., and J. DENY: [1] Dirichlet spaces. PNAS **45**, 208—215 (1959).

BISHOP, E., and K. DE LEEUW: [1] The representations of linear functionals by measures on sets of extreme points. Ann. Inst. Fourier **9**, 305—331 (1959).

BLUMENTHAL, R.: [1] An extended Markov property. TAMS **85**, 52—72 (1957).

BLUMENTHAL, R., and R. GETOOR: [1] Sample functions of stochastic processes with stationary independent increments. J. Math. Mech. **10**, 493—516 (1961).

BLUMENTHAL, R., R. GETOOR, and H. P. MCKEAN, JR.: [1] Markov processes with identical hitting distributions. IJM **6**, 402—420 (1962).

BOCHNER, S.: [1] Diffusion equation and stochastic processes. PNAS **35**, 368—370 (1949).

— [2] Positive zonal functions on spheres. PNAS **40**, 1141—1147 (1954).

BOREL, E.: [1] Les probabilités dénombrables et leurs applications arithmétiques. Rend. Circ. Mat. Palermo **27**, 247—271 (1909).

BOYLAN, E.: [1] Local times for a class of Markov processes. IJM **8**, 19—39 (1964).

CAMERON, R. H.: [1] The generalized heat flow equation and a corresponding Poisson formula. AM **59**, 434—462 (1954).

CAMERON, R. H., and W. T. MARTIN: [1] Evaluations of various Wiener integrals by use of certain Sturm-Liouville differential equations. BAMS **51**, 73—90 (1945).

CARATHÉODORY, C.: [1] Funktionentheorie. Basel 1950.

CHOQUET, G.: [1] Theory of capacities. Ann. Inst. Fourier, Grenoble **5**, 131—295 (1953—54).

CHRYSTAL, G.: [1] Electricity. Encyclopaedia Britannica, 9th ed. **8**, 3—104 (1879).

CHUNG, K. L., P. ERDÖS, and T. SIRAO: [1] On the Lipschitz's condition for Brownian motions. J. Math. Soc. Japan **11**, 263—274 (1959).

CHUNG, K. L., and W. FELLER: [1] On fluctuation in coin-tossing. PNAS **35**, 605—608 (1949).

CIESIELSKI, Z.: [1] Holder condition for realizations of Gaussian processes. TAMS **99**, 403—413 (1961).

COURANT, R., and D. HILBERT: [1] Methoden der mathematischen Physik 1. Berlin 1931.

— [2] Methoden der mathematischen Physik 2. Berlin 1937.

COURANT, R., and A. HURWITZ: [1] Funktionentheorie. Berlin 1929.

DANIELL, P. J.: [1] A general form of integral. AM **19**, 279—294 (1917—18).

DERMAN, C.: [1] Ergodic property of the Brownian motion process. PNAS **40**, 1155—1158 (1954).

DOBLIN, W.: [1] Sur deux problèmes de M. Kolmogoroff concernant les chaînes dénombrables. Bull. Soc. Math. France **66**, 210—220 (1938).

DOETSCH, G.: [1] Theorie und Anwendung der Laplace-Transformation. Berlin 1937.

DONSKER, M. D.: [1] An invariance principle for certain probability limit theorems. Mem. Amer. Math. Soc. No. 6 (1951).

DOOB, J. L.: [1] Stochastic processes. New York 1953.

— [2] Semi-martingales and subharmonic functions. TAMS **77**, 86—121 (1954).

— [3] Martingales and one-dimensional diffusion. TAMS **78**, 168—208 (1955).

— [4] A probabilistic approach to the heat equation. TAMS **80**, 216—280 (1955).

— [5] Brownian motion on a Green space. TV **2**, 3—33 (1957).

— [6] Conditional Brownian motion and the boundary limits of harmonic functions. Bull. Soc. Math. France **85**, 431—458 (1957).

— [7] Discrete potential theory and boundaries. J. Math. Mech. **8**, 433—458 (1959).

— [8] Boundary properties of functions with finite Dirichlet integrals. Ann. Inst. Fourier Grenoble **12**, 573—622 (1962).

DOOB, J. L., J. SNELL, and R. WILLIAMSON: [1] Application of boundary theory to sums of independent random variables. Contributions to Probability and Statistics, 182—197. Stanford University Press 1960.

DVORETSKY, A., and P. ERDÖS: [1] Some problems on random walk in space. Proc. Second Berkeley Symposium on Math. Stat. and Probability, 353—367. University of California Press 1951.

DVORETSKY, A., P. ERDÖS, and S. KAKUTANI: [1] Double points of Brownian motion in n-space. Acta Scientarum Matematicarum (Szeged) **12**, 75—81 (1950).

— [2] Multiple points of paths of Brownian motion in the plane. Bull. Res. Council Israel **3**, 364—371 (1954).

— [3] Nonincreasing everywhere of the Brownian motion process. Proc. Fourth Berkeley Symp. on Math. Stat. and Probability II, 103—116. Univ. of Calif. Press. 1961.

DVORETSKY, A., P. ERDÖS, S. KAKUTANI, and S. J. TAYLOR: [1] Triple points of Brownian paths in 3-space. Proc. Camb. Phil. Soc. **53**, 856—862 (1957).

DYNKIN, E. B.: [1] Continuous one-dimensional Markov processes. Dokl. Akad. Nauk SSSR **105**, 405—408 (1955).

— [2] Infinitesimal operators of Markov random processes. Dokl. Akad. Nauk SSSR **105**, 206—209 (1955).

— [3] Markov processes and semi-groups of operators. TV **1**, 25—37 (1956).

— [4] Infinitesimal operators of Markov processes. TV **1**, 38—60 (1956).

— [5] One-dimensional continuous strong Markov processes. TV **4**, 3—54 (1959).

— [6] Principles of the theory of Markov random processes. Moscow—Leningrad 1959.

DYNKIN, E. B.: [7] Natural topology and excessive functions connected with a Markov process. Dokl. Akad. Nauk SSSR **127**, 17—19 (1959).
— [8] Markov processes. Moscow 1963.
DYNKIN, E. B., and A. YUŠKEVIČ: [1] Strong Markov processes. TV **1**, 149—155 (1956).
EINSTEIN, A., [1] Investigations on the theory of the Brownian movement. New York 1956.
ERDÉLYI, A. (Bateman manuscript project): [1] Tables of integral transforms. 1, 2. New York 1954.
— [2] Higher transcendental functions. 1,2; 3. New York 1953; 1955.
ERDÖS, P.: [1] On the law of the iterated logarithm. AM **43**, 419—436 (1942).
ERDÖS, P., and M. KAC: [1] On the number of positive sums of independent random variables. BAMS **53**, 1011—1021 (1941).
FELDMAN, J.: [1] Feller and Martin boundaries for countable sets. IJM **6**, 356—366 (1962).
FELLER, W.: [1] Zur Theorie der stochastischen Prozesse (Existenz- und Eindeutigkeitssätze). Math. Ann. **133**, 133—160 (1936).
— [2] On the general form of the so-called law of the iterated logarithm. TAMS **54**, 373—402 (1943).
— [3] An introduction to probability theory and its applications. New York 1957 (2nd ed.).
— [4] The parabolic differential equations and the associated semi-groups of transformations. AM **55**, 468—519 (1952).
— [5] The general diffusion operator and positivity preserving semi-groups in one dimension. AM **60**, 417—436 (1954).
— [6] Diffusion processes in one dimension. TAMS **77**, 1—31 (1954).
— [7] On second order differential operators. AM **61**, 90—105 (1955).
— [8] Boundaries induced by non-negative matrices. TAMS **83**, 19—54 (1956).
— [9] Generalized second order differential operators and their lateral conditions. IJM **1**, 459—504 (1957).
— [10] On the intrinsic form for second order differential operators. IJM **2**, 1—18 (1958).
— [11] Differential operators with the positive maximum property. IJM **3**, 182 — 186 (1959).
— [12] The birth and death processes as diffusion processes. J. Math. Pures Appl. **38**, 301—345 (1959).
— [13] On a generalization of Marcel Riesz' potentials and the semi-groups generated by them. Comm. Semi. Math. Univ. Lund Suppl. Vol. 72—81 (1952).
FROSTMAN, O.: [1] Potentiel d'équilibre et capacité des ensembles avec quelques applications à la théorie des fonctions. Medd. Lunds Univ. Mat. Sem. **3**, 1—118 (1935).
GALMARINO, A. R.: [1] Representation of an isotropic diffusion as a skew product. Z. Wahrscheinlichkeitstheorie **1**, 359—378 (1963).
GAUSS, C. F.: [1] Allgemeine Lehrsätze in Beziehung auf die im verkehrten Verhältnisse des Quadrats der Entfernung wirkenden Anziehungs- und Abstoßungskräfte. Ostwalds Klassiker der exakten Wissenschaften, No. 2. Leipzig 1889.
HALMOS, P. R.: [1] Measure theory. New York 1950.
HARRIS, T. E. and H. ROBBINS: [1] Ergodic theory of Markov chains admitting an infinite invariant measure. PNAS **39**, 860—864 (1953).
HAS'MINSKIĬ, R. Z.: [1] Ergodic properties of recurrent diffusion processes and stabilization of the problem to the Cauchy problem for parabolic equations. TV **5**, 196—214 (1960).

HAUSDORFF, F.: [1] Dimension und äußeres Maß. Math. Ann. **79**, 157—179 (1918).

HEINS, M.: [1] On a notion of convexity connected with a method of Carleman. J. d'Anal. Math. **7**, 53—77 (1959).

HILDEBRAND, F. B.: [1] Advanced calculus for engineers. New York 1949.

HILLE, E.: [1] Representation of one-parameter semi-groups of linear transformations. PNAS **28**, 175—178 (1942).

HINČIN, A. YA.: [1] Asymptotische Gesetze der Wahrscheinlichkeitsrechnung. Ergebn. Math. **2**, No. 4, Berlin 1933.

HOPF, E.: [1] Bemerkungen zu einem Satze von S. Bernstein aus der Theorie der elliptischen Differentialgleichungen. Math. Z. **29**, 744—745 (1929).

— [2] Ergodentheorie. Ergebn. Math. **5**, No. 2, Berlin 1937.

HUNT, G.: [1] Some theorems concerning Brownian motion. TAMS **81**, 294—319 (1956).

— [2] Markoff processes and potentials. 1, 2, 3. IJM **1**, 44—93 (1957); **1**, 316—369 (1957); **2**, 151—213 (1958).

— [3] Markoff chains and Martin boundaries. IJM **4**, 313—340 (1960).

IKEDA, N.: [1] On the construction of two-dimensional diffusion processes satisfying Wentzell's boundary conditions and its application to boundary value problem. Mem. Coll. Sci. Univ. Kyoto, A, Math. **33**, 368—427 (1961).

ITÔ, K.: [1] On stochastic processes. 1. Japanese J. Math. **18**, 261—301 (1942).

— [2] On a stochastic integral equation. Proc. Japan Acad. **22**, 32—35 (1946).

— [3] Stochastic differential equations on a differentiable manifold. Nagoya Math. J. **1**, 35—47 (1950).

— [4] Multiple Wiener integral. J. Math. Soc. Japan **3**, 157—169 (1951).

— [5] On stochastic differential equations. Mem. Amer. Math. Soc. No. 4 (1951).

ITÔ, K., and H. P. MCKEAN, JR.: [1] Potentials and the random walk. IJM **4**, 119—132 (1960).

— [2] Brownian motion on a half line. IJM **7**, 181—231 (1963).

KAC, M.: [1] On some connections between probability theory and differential and integral equations. Proc. Second Berkeley Symposium on Math. Stat. and Probability, 189—215. University of California Press 1951.

KAKUTANI, S.: [1] On Brownian motion in n-space Proc. Acad. Japan **20**, 648—652 (1944).

— [2] Two-dimensional Brownian motion and harmonic functions. Proc. Acad. Japan, **20**, 706—714 (1944).

— [3] Random walk and the type problem of Riemann surfaces. Contribution to the theory of Riemann surfaces, 95—101. Princeton University Press 1961.

KALLIANPUR, G., and H. ROBBINS: [1] Ergodic property of the Brownian motion process. PNAS **39**, 525—533 (1953).

KAMETANI, S.: [1] On Hausdorff's measures and generalised capacities with some of their applications to the theory of functions. Jap. J. Math. **19**, 217—257 (1945).

KARLIN, S., and J. MCGREGOR: [1] The differential equations of birth-and-death-processes and the Stieltjes moment problem. TAMS **85**, 489—546 (1957).

— [2] Random walks. IJM **3**, 66—81 (1959).

— [3] Coincidence probabilities. Pacific J. Math. **9**, 1141—1164 (1959).

— [4] A characterization of the birth and death processes. PNAS **45**, 375—379 (1959).

— [5] Classical diffusion processes and total positivity. Technical Report No. 1, Appl. Math. and Stat. Lab., Stanford University, California 1960.

KELLOGG, O. D.: [1] Foundations of potential theory. Grundlehren der Math. Wissenschaften **31**, Berlin 1929.

KNIGHT, F.: [1] On the random walk and Brownian motion. TAMS **103**, 218–228 (1962).
— [2] Random walks and a sojourn density process of Brownian motion. TAMS **107**, 56–86 (1963).
KOLMOGOROV, A. N.: [1] Über die analytischen Methoden in der Wahrscheinlichkeitsrechnung. Math. Ann. **104**, 415–458 (1931).
— [2] Grundbegriffe der Wahrscheinlichkeitsrechnung. Ergebn. Math. **2**, No. 3. Berlin 1933.
KOLMOGOROV, A. N., I. PETROVSKI, and N. PISCOUNOV: [1] Étude de l'équation de la chaleur avec croissance de la quantité de matière et son application à un problème biologique. Bull. U. Etat. Moscou. **6**, 1–25 (1937).
KOMATSU, Y.: [1] Elementary inequalities for Mills' ratio. Rep. Statist. Appl. Res. Un. Jap. Sci. Engrs. **4**, 69–70 (1955).
LAMPERTI, J.: [1] An invariance principle in renewal theory. Ann. Math. Stat. **33**, 685–696 (1962).
— [2] Wiener's test and Markov chains. Jour. Math. Anal. and Appl. **6**, 58–66 (1963).
LEBESGUE, H.: [1] Sur le problème de Dirichlet. Rend. Circ. Mat. Palermo **24**, 371–402 (1907).
— [2] Conditions de régularité, conditions d'irrégularité, conditions d'impossibilité dans le problème de Dirichlet Cont. Rend. Acad. Sci. Paris (1924) 349–354.
LEEUW, K. DE and H. MIRKIL: [1] A priori sup norm estimates for differential operators. IJM **8**, 112—124 (1964).
LÉVY, P.: [1] Théorie de l'addition des variables aléatoires. Paris 2nd ed. 1954.
— [2] Sur certains processus stochastiques homogènes. Compositio Math. **7**, 283—339 (1939).
— [3] Processus stochastiques et mouvement brownien. Paris 1948.
— [4] Le mouvement brownien. Paris 1954.
— [5] Remarques sur le processus de W. Feller et H. P. McKean. C. R. Paris **245**, 1772—1774 (1957).
— [6] Construction du processus du W. Feller and H. P. McKean en partant du mouvement Brownien. Probability and Statistics (the Harald Cramér volume). Stockholm 1959.
MCKEAN, H. P., JR.: [1] Sample functions of stable processes. AM **61**, 564–579 (1955).
— [2] Elementary solutions for certain parabolic partial differential equations. TAMS **82**, 519–548 (1956).
— [3] The Bessel motion and a singular integral equation. Mem. Coll. Sci. Univ. Kyoto. Ser. A, Math. **33**, 317–322 (1960).
— [4] A Hölder condition for Brownian local time. J. Math. Kyoto Univ. **1**, 195 — 201 (1962).
— [5] Excursions of a non-singular diffusion. Z. Wahrscheinlichkeitstheorie **1**, 230–239 (1963).
— [6] A. Skorohod's integral equation for a reflecting barrier diffusion. J. Math. Kyôto Univ. **3**, 85—88 (1963).
MCKEAN, H. P., JR., and D. B. RAY: [1] Spectral distribution of a differential operator. Duke Math. J. **29**, 281–292 (1962).
MCKEAN, H. P., JR. and H. TANAKA: [1] Additive functionals of the Brownian path. Memoirs Coll. Sci. Univ. Kyoto, A, Math. **33**, 479–506 (1961).
MAGNUS, N., and F. OBERHETTINGER: [1] Formulas and theorems for the special functions of mathematical physics. New York 1949.
— [2] Anwendung der elliptischen Funktionen in Physik und Technik. Berlin 1949.

MARTIN, R. S.: [1] Minimal positive harmonic functions. TAMS **49**, 137—172 (1941).
MARUYAMA, G., and H. TANAKA: [1] Some properties of one-dimensional diffusion processes. Mem. Fac. Sci. Kyushu Univ. **11**, 117—141 (1957).
— [2] Ergodic property of n-dimensional Markov processes. Mem. Fac. Sci. Kyushu Univ. **13**, 157—172 (1959).
MOSER, J.: [1] On Harnack's theorem for elliptic partial differential equations. Comm. Pure Appl. Math. **14**, 577—591 (1961).
MOTOO, M.: [1] Proof of the law of iterated logarithm through diffusion equation. Ann. Inst. Statis. Math. **10**, 21—28 (1959).
MOTOO, M., and H. WATANABE: [1] Ergodic property of recurrent diffusion process in one dimension. J. Math. Soc. Japan **10**, 272—286 (1958).
NELSON, E.: [1] The adjoint Markov process. Duke Math. J. **25**, 671—690 (1958).
— [2] Representation of a Markovian semi-group and its infinitesimal generator. J. Math. Mech. **7**, 977—988 (1958).
PALEY, R., and N. WIENER: [1] Fourier transforms in the complex domain. New York 1934.
PALEY, R., N. WIENER, and A. ZYGMUND: [1] Note on random functions. Math. Z. **37**, 647—668 (1933).
PERRON, O.: [1] Über die Behandlung der ersten Randwertaufgabe für $\Delta u = 0$. Math. Z. **18**, 42—54 (1923).
PETROVSKI, I.: [1] Zur ersten Randwertaufgabe der Wärmeleitungsgleichung. Compositio Math. **1**, 383—419 (1935).
— [2] Lectures on partial differential equations. New York 1954.
POGORZELSKI, W.: [1] Étude de la solution fondamentale de l'équation parabolique. Ric. Math. **5**, 25—57 (1956).
PÓLYA, G.: [1] Qualitatives über Wärmeausgleichung. Z. Angew. Math. Mech. **13**, 125—128 (1933).
PÓLYA, G., and G. SZEGÖ: [1] Über den transfiniten Durchmesser (Kapazitäts konstante) von ebenen und räumlichen Punktmengen. J. Reine Angew. Math. **165**, 4—49 (1931).
RAY, D. B.: [1] On spectra of second-order differential operators. TAMS **77**, 299—321 (1954).
— [2] Stationary Markov processes with continuous paths. TAMS **82**, 452—493 (1956).
— [3] Resolvents, transition functions, and strongly Markovian processes. AM **76**, 43—72 (1959).
— [4] Sojourn times of a diffusion process. IJM **7**, 615—630 (1963).
RIESZ, F.: [1] Sur les fonctions subharmoniques et leur rapport à la théorie du potentiel. 1, 2. Acta Math. **48**, 329—343 (1926); **54**, 321—360 (1930).
RIESZ, M.: [1] L'intégrale de Riemann-Liouville et le problème de Cauchy. Acta Math. **81**, 1—223 (1949).
SATO, K.: [1] Time change and killing for multi-dimensional reflecting diffusion. Proc. Japan Acad. **39**, 69—73 (1963).
SATO, K., and H. TANAKA: [1] Local times on the boundary for multi-dimensional reflecting diffusion. Proc. Japan Acad. **38**, 699—702 (1962).
SEIFERT, H., and W. THRELFALL: [1] Lehrbuch der Topologie. New York 1947.
SPITZER, F.: [1] Some theorems concerning 2-dimensional Brownian motion. TAMS **87**, 187—197 (1958).
— [2] Recurrent random walk and logarithmic potential. Proc. Fourth Berkeley Symp. on Math. Stat. and Probability II 515—534, Univ. of Calif. Press. (1961).
— [3] Electrostatic capacity in heat flow and Brownian motion. Z. Wahrscheinlichkeitstheorie, in press.

Šur, M. G.: [1] Continuous additive functionals of Markov processes and excessive functions. Dokl. Akad. Nauk SSSR **137,** 800—803 (1961).

Sz.-Nagy, B. v.: [1] Spektraldarstellung linearer Transformationen des Hilbertschen Raumes. Ergebn. Math. **5** No. 5. Berlin 1942.

Tanaka, H.: [1] Certain limit theorems concerning one-dimensional diffusion processes. Mem. Fac. Sci. Kyushu Univ. **12,** 1—11 (1958).

Taylor, S. J.: [1] The α-dimensional measure of the graph and the set of zeros of a Brownian path. Proc. Cambridge Philos. Soc. **51,** 265—274 (1955).

Thompson, D'Arcy: [1] On growth and form. Cambridge 1959.

Tihonov, A.: [1] Théorèmes d'unicité pour l'équation de la chaleur. Mat. Sb. **42,** 119—216 (1935).

Trotter, H. F.: [1] A property of Brownian motion paths. IJM **2,** 425—433 (1958).

— [2] An elementary proof of the central limit theorem. Arch. Math. **10,** 226—234 (1959).

Ueno, T.: [1] On recurrent Markov processes. Kōdai Math. Sem. Rep. **12,** 109—142 (1960).

— [2] The diffusion satisfying Wentzell's boundary condition and the Markov process on the boundary. 1, Proc. Japan Acad. **36,** 533—538 (1960).

Ventcel' A. D.: [1] On lateral conditions for multi-dimensional diffusion processes. TV **4,** 208—211 (1959).

Volkonskiĭ, V. A.: [1] Random substitution of time in strong Markov processes. TV **3,** 332—350 (1958).

— [2] Construction of inhomogeneous Markov processes by means of a random time substitution. TV **6,** 47—56 (1961).

Watanabe, S.: [1] On stable processes with boundary conditions. J. Math. Soc. Japan **14,** 170—198 (1962).

Watanabe, T.: [1] On the theory of Martin boundaries induced by countable Markov processes. Mem. Coll. Sci. Univ. Kyoto. Ser. A. Math. **33,** 39—108 (1960).

Weyl, H.: [1] Über gewöhnliche Differentialgleichungen mit Singularitäten und die zugehörigen Entwicklungen willkürlicher Funktionen. Math. Ann. **68,** 220—269 (1910).

— [2] Das asymptotische Verteilungsgesetz der Eigenschwingungen eines beliebig gestalteten elastischen Körpers. Rend. Cir. Mat. Palermo **39,** 1—50 (1915).

Whitman, W.: Some strong laws for random walks and Brownian motion. TAMS, to appear.

Wiener, N.: [1] Differential space. J. Math. Phys. **2,** 131—174 (1923).

— [2] The Dirichlet problem. J. Math. Phys. **3,** 127—146 (1924).

— [3] Generalized harmonic analysis. Acta Math. **55,** 117—258 (1930).

Yosida, K.: [1] On the differentiability and the representation of one-parameter semi-group of linear operators. J. Math. Soc. Japan **1,** 15—21 (1948).

— [2] Integration of Fokker-Planck's equation in a compact Riemannian space. Ark. Mat. **1,** 71—75 (1949).

— [3] A characterization of second order differential operators. Proc. Japan Acad. **31,** 406—409 (1955).

— [4] Fractional powers of infinitesimal generators and the analyticity of the semi-groups generated by them. Proc. Japan Acad. **36,** 86—89 (1960).

List of Notations

α	a LAPLACE transform variable.
B	a BOREL algebra.
$\mathsf{B}[x(t,w): a \leqq t < b]$	the smallest BOREL algebra measuring $x(t)$: $a \leqq t < b$.
B_t	the algebra $\mathsf{B}[x_s : s \leqq t]$.
$\mathsf{B}_{\mathfrak{m}+}$	the algebra $(B : B \cap (\mathfrak{m} < t) \in \mathsf{B}_t,\ \ t \geqq 0)$.
$\mathsf{B}(Q)$	the smallest BOREL algebra measuring all open subsets of Q.
$B(Q)$	the space of bounded BOREL measurable functions $f: Q \to R^1$.
$C(Q)$	the space of bounded continuous functions $f: Q \to R^1$.
$C^n(Q)$	the space bounded, n-times continuously differentiable functions $f: Q \to R^1$.
$C_\infty(Q)$	the space of continuous functions vanishing at ∞.
$C(B)$	the NEWTONian capacity of B (see 7.8).
d	dimension.
D	the space of functions $f \in B(Q)$ such that $f(a\pm) = f(a)$ if $P_a[\mathfrak{m}_{a\pm} = 0] = 1$ in chapters 3–6 (see 3.6), and, in chapters 7–8, a domain (= connected open set).
∂D	the boundary of the domain D.
D	a diffusion.
$D(\mathfrak{G})$	the domain of $\mathfrak{G}$.
Δ	the LAPLACE operator.
do	the uniform distribution on the surface of a ball.
E^2	the unit disk.
$E_\bullet(f)$	the expectation of f based on $P_\bullet$.
$E_\bullet(f, A)$	the expectation of f times the indicator function of A.
$e(d\gamma, a, b) = \mathfrak{e}\,\mathfrak{f}\,\mathfrak{e}$	an eigendifferential (see 4.11).
$\mathfrak{f} = \mathfrak{f}(d\gamma)$	a spectral measure (see 4.11).
$f^- = f^-(a)$	the left derivative.
$f^+ = f^+(a)$	the right derivative.
$f(da)$	the BOREL measure based on the interval function $f(a, b] = f(b+) - f(a+)$.
$\mathfrak{f} = \mathfrak{f}(t)$	a local time integral (see 5.2).
$g = g(t, a, b)$	the GAUSS kernel $(2\pi t)^{-1/2} \exp[-(b-a)^2/2t]$.
g_1, g_2	the increasing and decreasing solutions of $\mathfrak{G} g = \alpha g$ (see 4.6).
$G = G(a, b)$	a GREEN function (see 2.6, 4.11, and 7.4).
G_α	a GREEN operator (see 3.6).
$\mathfrak{G}$	a generator (see 3.7).
$h(a, B)$	the classical harmonic measure of $B \subset \partial D$ as seen from $a \in D$ (see 7.12).
K_-	the left shunts (see 3.4).
K_+	the right shunts (see 3.4).
$\varkappa$	the killing rate for $K_- \cap K_+$ (see 4.8).
k	a killing measure (see 4.3).
$k_\pm$	the killing measures for shunts (see 4.8).
$\mathfrak{k} = \mathfrak{k}(t)$	the killing functional (see 5.6).
Λ^h	the HAUSDORFF measure based on h (see note 2.5.2).

Λ^α the HAUSDORFF-BESICOVITCH α-dimensional measure (see note 2.5.2).
$L^2(Q, m)$ the space of m-measurable functions f with $\|f\|_2 = \sqrt{\int f^2\, dm} < +\infty$. modulo null functions.
m a speed measure (see 4.2).
$\mathfrak{m}$ a MARKOV time (see 1.6 and 3.2).
$\mathfrak{m}_a$ the passage time to a.
$\mathfrak{m}_B$ the entrance time into B (see problem 7.1.1).
$\mathfrak{m}_\infty$ the killing time (see 2.3).
n a LÉVY measure (see note 1.7.1).
$P_a(B)$ the probabilities for paths starting at a.
$P_\bullet(B)$ $P_a(B)$ as a function of a.
$p = p(t, a, b)$ a transition probability kernel (see 4.11).
$\mathfrak{p}$ a POISSON measure figuring in the LÉVY decomposition of an increasing differential process (see note 1.7.1) or a projection (see 4.11).
Q the state space of a diffusion.
R^d the d-dimensional EUCLIDean space.
$s = s(a)$ a scale (see 4.2).
$s_\pm$ the shunt scales (see 4.8).
S^d the d-dimensional unit sphere $\subset R^{d+1}$.
t time.
$\mathfrak{t} = \mathfrak{t}(t, a) = \mathfrak{t}(t, a, w)$ a local time (see 2.2 and 5.4).
W a space of sample paths.
$w: t \to x(t)$ a sample path.
w_s^+ the shifted path: $x_t(w_s^+) = x_{t+s}(w)$.
$w_s^\bullet$ the stopped path $x_t(w_s^\bullet) = x_{t\wedge s}(w)$.
$x(t, w) = x(t)$ the value of the sample path w at t.
Z^1 the set of all integers.
$\mathfrak{Z} = (t : x(t) = 0)$ a visiting set.
$\mathfrak{Z}_n\ (n \geqq 1)$ the (open) intervals of $[0, \infty) - \mathfrak{Z}$.

$a \wedge b$ the smaller of a and b.
$a \vee b$ the greater of a and b.
$\#(C)$ the number of objects in the class C.
$\binom{n}{m} = n!/m!\,(n-m)!$ for $m = 0, 1, \ldots, n$.
$\lg_2 t = \lg(\lg t)$, $\lg_n t = \lg(\lg_{n-1} t)$ $(n \geqq 3)$.
$\|f\| = \|f\|_\infty = \sup_Q |f|$.
$\|f\|_2 = \sqrt{\int f^2\, dm}$.
$a \uparrow b$ a increases to b.
$a \downarrow b$ a decreases to b.
$i.o.$ infinitely often.
$f \in \uparrow$ an increasing ($\equiv$ non-decreasing) function.
$f \in \downarrow$ a decreasing ($\equiv$ non-increasing) function.
$(x(t) < h(t), t \downarrow 0)$ means that there exists $t_1 > 0$ such that. $x(t) < h(t)$ for $0 < t < t_1$.
$(x(t) < h(t), t \uparrow +\infty)$ means that there exists $t_2 > 0$ such that $x(t) < h(t)$ for $t > t_2$.
$\emptyset$ the empty set.
∞ an extra state (see 2.3).
$\pm\infty$ the ends of the line.

Index

d = dimension, $\mathfrak{G}$ = generator, t = local time, $\mathfrak{Z} = (t : x = 0)$

DRUCK: STRAUSS OFFSETDRUCK, MÖRLENBACH

BINDEN: TRILTSCH, WÜRZBURG

∞ CIM

M. Aigner Combinatorial Theory ISBN 978-3-540-61787-7
A. L. Besse Einstein Manifolds ISBN 978-3-540-74120-6
N. P. Bhatia, G. P. Szegő Stability Theory of Dynamical Systems ISBN 978-3-540-42748-3
J. W. S. Cassels An Introduction to the Geometry of Numbers ISBN 978-3-540-61788-4
R. Courant, F. John Introduction to Calculus and Analysis I ISBN 978-3-540-65058-4
R. Courant, F. John Introduction to Calculus and Analysis II/1 ISBN 978-3-540-66569-4
R. Courant, F. John Introduction to Calculus and Analysis II/2 ISBN 978-3-540-66570-0
P. Dembowski Finite Geometries ISBN 978-3-540-61786-0
A. Dold Lectures on Algebraic Topology ISBN 978-3-540-58660-9
J. L. Doob Classical Potential Theory and Its Probabilistic Counterpart ISBN 978-3-540-41206-9
R. S. Ellis Entropy, Large Deviations, and Statistical Mechanics ISBN 978-3-540-29059-9
H. Federer Geometric Measure Theory ISBN 978-3-540-60656-7
S. Flügge Practical Quantum Mechanics ISBN 978-3-540-65035-5
L. D. Faddeev, L. A. Takhtajan Hamiltonian Methods in the Theory of Solitons ISBN 978-3-540-69843-2
I. I. Gikhman, A. V. Skorokhod The Theory of Stochastic Processes I ISBN 978-3-540-20284-4
I. I. Gikhman, A. V. Skorokhod The Theory of Stochastic Processes II ISBN 978-3-540-20285-1
I. I. Gikhman, A. V. Skorokhod The Theory of Stochastic Processes III ISBN 978-3-540-49940-4
D. Gilbarg, N. S. Trudinger Elliptic Partial Differential Equations of Second Order ISBN 978-3-540-41160-4
H. Grauert, R. Remmert Theory of Stein Spaces ISBN 978-3-540-00373-1
H. Hasse Number Theory ISBN 978-3-540-42749-0
F. Hirzebruch Topological Methods in Algebraic Geometry ISBN 978-3-540-58663-0
L. Hörmander The Analysis of Linear Partial Differential Operators I – Distribution Theory and Fourier Analysis ISBN 978-3-540-00662-6
L. Hörmander The Analysis of Linear Partial Differential Operators II – Differential Operators with Constant Coefficients ISBN 978-3-540-22516-4
L. Hörmander The Analysis of Linear Partial Differential Operators III – Pseudo-Differential Operators ISBN 978-3-540-49937-4
L. Hörmander The Analysis of Linear Partial Differential Operators IV – Fourier Integral Operators ISBN 978-3-642-00117-8
K. Itô, H. P. McKean, Jr. Diffusion Processes and Their Sample Paths ISBN 978-3-540-60629-1
T. Kato Perturbation Theory for Linear Operators ISBN 978-3-540-58661-6
S. Kobayashi Transformation Groups in Differential Geometry ISBN 978-3-540-58659-3
K. Kodaira Complex Manifolds and Deformation of Complex Structures ISBN 978-3-540-22614-7
Th. M. Liggett Interacting Particle Systems ISBN 978-3-540-22617-8
J. Lindenstrauss, L. Tzafriri Classical Banach Spaces I and II ISBN 978-3-540-60628-4
R. C. Lyndon, P. E Schupp Combinatorial Group Theory ISBN 978-3-540-41158-1
S. Mac Lane Homology ISBN 978-3-540-58662-3
C. B. Morrey Jr. Multiple Integrals in the Calculus of Variations ISBN 978-3-540-69915-6
D. Mumford Algebraic Geometry I – Complex Projective Varieties ISBN 978-3-540-58657-9
O. T. O'Meara Introduction to Quadratic Forms ISBN 978-3-540-66564-9
G. Pólya, G. Szegő Problems and Theorems in Analysis I – Series. Integral Calculus. Theory of Functions ISBN 978-3-540-63640-3
G. Pólya, G. Szegő Problems and Theorems in Analysis II – Theory of Functions. Zeros. Polynomials. Determinants. Number Theory. Geometry ISBN 978-3-540-63686-1
W. Rudin Function Theory in the Unit Ball of $\mathbb{C}^n$ ISBN 978-3-540-68272-1
S. Sakai C*-Algebras and W*-Algebras ISBN 978-3-540-63633-5
C. L. Siegel, J. K. Moser Lectures on Celestial Mechanics ISBN 978-3-540-58656-2
T. A. Springer Jordan Algebras and Algebraic Groups ISBN 978-3-540-63632-8
D. W. Stroock, S. R. S. Varadhan Multidimensional Diffusion Processes ISBN 978-3-540-28998-2
R. R. Switzer Algebraic Topology: Homology and Homotopy ISBN 978-3-540-42750-6
A. Weil Basic Number Theory ISBN 978-3-540-58655-5
A. Weil Elliptic Functions According to Eisenstein and Kronecker ISBN 978-3-540-65036-2
K. Yosida Functional Analysis ISBN 978-3-540-58654-8
O. Zariski Algebraic Surfaces ISBN 978-3-540-58658-6